# Handbook for I&C System Development Process using Systems Engineering Approach

정재천 지음

인터비전

# Preface

One of the goals of KEPCO International Nuclear Graduate School (KINGS) is to nurture global nuclear energy professionals by equipping them with the tools to become leaders in the nuclear industry. In order to achieve this goal, KINGS has adopted a Systems Engineering (SE) approach to Nuclear Power Plant (NPP) engineering. Systems engineering is an interdisciplinary approach to the development of complex systems that encompasses the entire system's lifecycle: from the concept definition to the system disposal, while ensuring that stakeholder needs are met within the project's constraints. This is particularly relevant in NPPs, which are composed of large and complex systems having stringent safety requirements and long lifetimes.

Most NPPs are designed to operate for at least 60 years. Within their lifetime, some of the systems, structures, and components must be replaced due to degradation in functionality, evolving practices in the nuclear industry, or simply because of improvement opportunity afforded by new technologies. According to the IAEA, operating NPPs will soon be undertaking the replacement or upgrade of existing I&C systems [1]. Thus, there is a need for professionals to be capable of addressing both the technical and project management aspects of I&C systems.

Up until now, both the concepts of Systems Engineering and I&C system design have been independently addressed by multiple texts. However, there is no book that focuses on the application of systems engineering processes to the design and development of NPP I&C systems. The primary intent of this book is to address this need which directed at both the engineers responsible for the technical design of the systems and the technical managers responsible for the system development. In addition, this book serves as a reference to I&C students at the graduate level.

## 1. Boundary and Limitations

This book focuses on the application of the systems engineering approach to the design of NPP Man-Machine Interface System (MMIS). While it addresses most of the technical processes defined by INCOSE, it does not deal with long-term concerns such as operation, maintenance, and disposal of the developed system. The Man-Machine Interface System includes both the Human Machine Interfaces and I&C system which constitutes of the Protection, Control, and Monitoring system. This book addresses primarily the protection systems, and briefly introduces the control systems how it will be optimized through processed approach. Further, the design considerations presented here are based on the Korean APR1400 power plant. Thus, all examples and configurations are specific to that plant.

## 2. Content and Structure

This handbook has been developed based on the work that has been carried out by the Instrumentation and Control Lab at KINGS over the past nine years, under the direction of Professor J.C Jung.

The structure of this textbook comprises of seventeen (17) chapters grouped into three (3) parts:

- ***Part 1***: Planning Phase consists of chapters 1- 5. Chapters 2 and 3 provide a system engineering approach, technical process, and system analysis process for designing MMIS system. In Chapters 4 to 5, the general overview of the APR1400 Man Machine Interface, its configuration and the safety critical functions that the MMIS performs are described.
- ***Part 2***: Design, Implementation and Verification Phase, consists of chapters 6 -12. This section highlights the stakeholder requirements for MMIS development. The chapters focused on the Safety Systems requirements analysis, architecture definition process and design criteria. In addition, the critical issues to be considered when implementing FPGA for MMIS, the verification, and acceptance criteria are discussed.
- ***Part 3***: Non-Safety System, is composed of Chapter 13 to 17. Chapter 13, describes Power Control Systems for NSSS. Based on the systems engineering approach, it deals with the architecture of power control system in the

perspective of requirements definition and system design analysis. In Chapters 14 and 15, two non-safety systems, Feedwater Control System and Reactor Regulating System, are analyzed and modified to enhance their performance. The concepts of system analysis and re-engineering are used in a practical example. In Chapter 16, Cyber Security in the Nuclear Power Plant is discussed. Lastly, Chapter 17 gives a brief introduction into the Human Machine Interface and its requirements in NPPs.

At the end of each chapter, homework assignments and further reading references are provided.

## 3. Acknowledgement

The author of this book gratefully acknowledges the contribution from the 2018 KEPCO International Nuclear Graduate School Winter OJP team members for their support in developing the outline and contents this book.

In addition, the author acknowledges the previous students of the I&C Laboratory whose research has been used to develop the content of this book.

Adebena Oluwasegun
Ahmad Salah Ibrahim
Dong-hwi Jeon
Joyce Mayaka
Tran Thanh Cong
Utah Michael
Satrio Pradana
Ibrahim Fathy Awad
Ibrahim Ahmed
Mostafa Mousa
Thiago Nascimento
Hyo-sub Shim
Ganzorig Norovsambuu
In-sok Hwang
Jun-sung Kim
Kim Sang-moon
Mohamed Elakrat
Mwongeera Murungi
Nam-chual Kim
Hae-kyung Kim
Restu Maerani
Shin Jong-wook
Zulhelmi Zulkifly
Erol Bicer
Md. Mehedi Hasan

**Jae Cheon, Jung**

**Spring 2022 in KINGS**

# Contents

# Acronyms

| | |
|---|---|
| AFAS | Auxiliary Feed Water Actuation Signal |
| AHP | Analytical Hierarchy Process |
| AOO | Anticipated Operational Occurrences |
| APC-S | Auxiliary Process Cabinet - Safety |
| ATWS | Anticipated Transients Without SCRAM |
| BOP | Balance of Plant) |
| BTP | Branch Technical Position |
| CAD | Computer Aided Design |
| CCC | Cross Channel Communication |
| CCF | Common Cause Failure |
| CCG | Control Channel Gateway |
| CCS | Component Control System |
| CEA | Control Element Assembly |
| CFR | Code of Federal Regulations |
| CIAS | Containment Isolation Actuation Signal |
| CIM | Component Interface Module |
| CIS | Containment Isolation System |
| CM | Configuration Management |
| CPCS | Core Protection Calculation System |
| CPM | Control Panel Multiplexer |
| CSAS | Containment Spray Actuation Signal |
| CWP | CEA Withdrawal Prohibit |
| DCD | Design Control Document |
| DCS | Distributed Control System |
| DMA | Diverse Manual Actuation |
| DNBR | Departure from Nucleate Boiling |
| DPS | Diverse Protection System |
| DR | Design Requirements |
| DRCS | Digital Rod Control System |
| DS | Design Specification |
| EFFBD | Enhanced Functional Flow Block Diagram |
| ENFMS | Ex-core Neutron Flux Monitoring System |
| ESCM | ESF Soft Control Module |
| ESF | Engineered Safety Features |
| ESFAS | Engineered Safety Features Actuation System |

| | |
|---|---|
| ESF-CCS | Engineered Safety Features Component Control System |
| EURD | European Utility Requirements Document |
| FDP | Flat Display Panel |
| FMEA | Failure Mode Effects Analysis |
| FPGA | Field Programmable Gate Array |
| FSMD | Finite State Machine With Data Path |
| FTA | Fault Tree Analysis |
| FWCS | Feedwater Control System |
| GC | Group Controllers |
| GDC | General Design Criteria |
| HDL | Hardware Definition Language |
| HIS | Human System Interface |
| HMI | Human Machine Interface |
| HPD | HDL Programmable Device |
| HVAC | Heating, Ventilation, and Air Conditioning |
| I&C | Instrumentation and Control |
| IAEA | International Atomic Energy Agency |
| IDE | Integrated Development Environment |
| INCOSE | International Council on Systems Engineering |
| IPS | Information Processing System |
| ITP | Interface and Test Processor |
| KURD | Korean Utility Requirements Document |
| LC | Loop Controllers |
| LCL | Local Coincidence Logic |
| LOCA | Loss of Coolant Accident |
| LPD | Local Powcr Density |
| LPPT | Low Pressurizer Pressure Trip |
| MCR | Main Control Room |
| MI | Minimum Inventory |
| MMIS | Man-Machine Interface System |
| MOE | Measures of Effectiveness |
| MOP | Measures of Performance |
| MSIS | Main Steam Isolation Signal |
| MTP | Maintenance and Test Panel |
| NPP | Nuclear Power Plant |
| NSSS | Nuclear Steam Supply System |
| OM | Operator Module |

| | |
|---|---|
| P-CCS | Process-Component Control System |
| PCS | Power Control System |
| PLC | Programmable Logic Controller |
| PLCS | Pressurizer Level Control System |
| PLD | Programmable Logic Device |
| PnR | Place and Route |
| PPCS | Pressurizer Pressure Control System |
| PPS | Plant Protection System |
| PRDBE | Performance Related Design Basis Event |
| QA | Quality Assurance |
| QIAS-N | Qualified Indication and Alarm System – Non-Safety |
| QIAS-P | Qualified Indication and Alarm System – Post Accident Monitoring Instrumentation |
| RAM | Reliability, Availability, and Maintainability |
| RCS | Reactor Coolant System |
| RG | Regulatory Guide |
| RMS | Radiation Monitoring System |
| RPCS | Reactor Power Cutback System |
| RPS | Reactor Protection System |
| RRS | Reactor Regulating System |
| RSPT | Reed Switch Position Transmitter |
| RSR | Remote Shutdown Room |
| RT | Reactor Trip |
| RTD | Resistance Temperature Detector |
| RTL | Register Transfer Level |
| RTM | Requirement Traceability Matrix |
| RTSS | Reactor Trip Switchgear System |
| SBCS | Steam Bypass Control System |
| SCC | Structures, Systems or Components |
| SDD | Software Design Description |
| SDL | Serial Data Link |
| SDN | Safety System Data Network |
| SDP | Software Development Plan |
| SE | Systems Engineering |
| SEMP | Systems Engineering Management Plan |
| SIAS | Safety Injection Actuation Signal |
| SIS | Safety Injection System |
| SMP | Software Management Plan |

| | |
|---|---|
| SOI | System of Interest |
| SPM | Software Program Manual |
| SRDC | Safety-Related Divisional Cabinet |
| SRGM | Software Reliability and Growth Model |
| SRP | Standard Review Plan |
| SRR | Software Requirements Review |
| SRS | Software Requirements Specification |
| SSP | Software Safety Plan |
| SysML | System Modeling languages |
| T/G | Turbine Generator |
| TCB | Trip Circuit Breaker |
| TPM | Technical Performance Measures |
| UML | Unified Modeling Language |
| URD | Utility Requirements Document |
| USNRC | United States Nuclear Regulatory Commission |
| V&V | Verification and Validation |
| VAC | Alternating Current Voltage |
| VBPSS | Vital Bus Power Supply System |
| VHDL | VHSIC Hardware Description Language |
| VOPT | Variable Overpower Trip |

PART 1

# PLANNING PHASE

# Chapter 1 OVERVIEW OF MAN-MACHINE INTERFACE SYSTEM (MMIS)

The Man-Machine Interface System (MMIS) is the system that protects, monitors and controls all the sub-systems required for safe plant operation, and provides devices from which action can be taken to operate the plant safely under all operating conditions and to maintain the plant in a safe state under accident conditions [2]. It is composed of the Human Machine Interface (HMI) and the Instrumentation and Control (I&C) Systems.

The HMI displays data received from the sensors and plant components, allowing the operator to have a clear idea of the state of the plant. In addition, it is through those interfaces (touchscreens, push buttons, switches etc.) that the operator is able to send commands to the systems and components e.g. open a pump or reduce flow through a valve.

The I&C system receives process information from sensors, feedback information from plant components and command signals from the plant operators. Based on these inputs, the control systems carry out the decision/control logic and generate the signals required to actuate the components. These systems also generate alarm signals when predefined plant parameters are exceeded, faulty components are detected or to signal an abnormal plant condition.

Figure 1-1 indicates the scheme of MMIS Interface as discussed above.

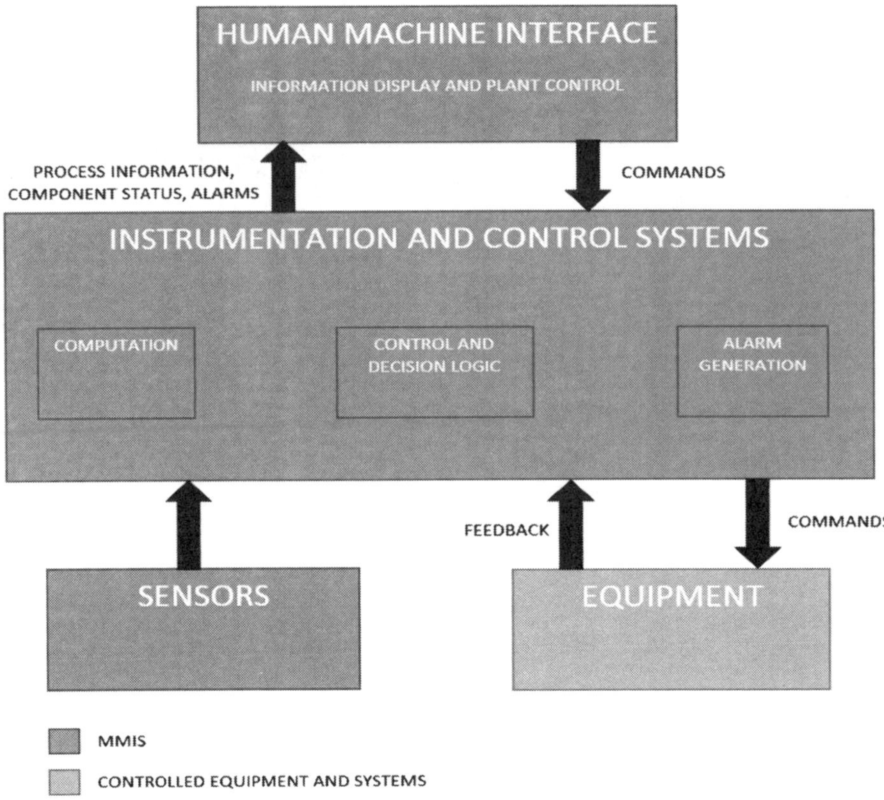


Figure 1-1. Man-Machine Interface System

The MMIS of the APR1400 is illustrated in Figure 1-2. From this figure, we can see that the MMIS can be further classified into safety and non-safety systems. The non-safety systems are designed to maintain the reactor within operational limits for effective power generation. These operational limits are kept sufficiently below plant safety limits to ensure safety during power demand variations or anticipated operational disturbances[3]. The safety I&C systems are those, which are necessary "to automatically initiate the operation of systems and components to assure that fuel design limits are not exceeded as a result of anticipated operational occurrences, and to sense accident conditions and initiate the operation of systems and components important to safety".[4]

In this book, the system engineering principles discussed shall be applied primarily to the safety I&C systems. In Part 1 to Part 2, only the safety systems are addressed. This approach was taken as the safety critical systems have more exacting requirements than the non-safety systems. However, in Part 3 a brief introduction into several of the non-safety systems is given.

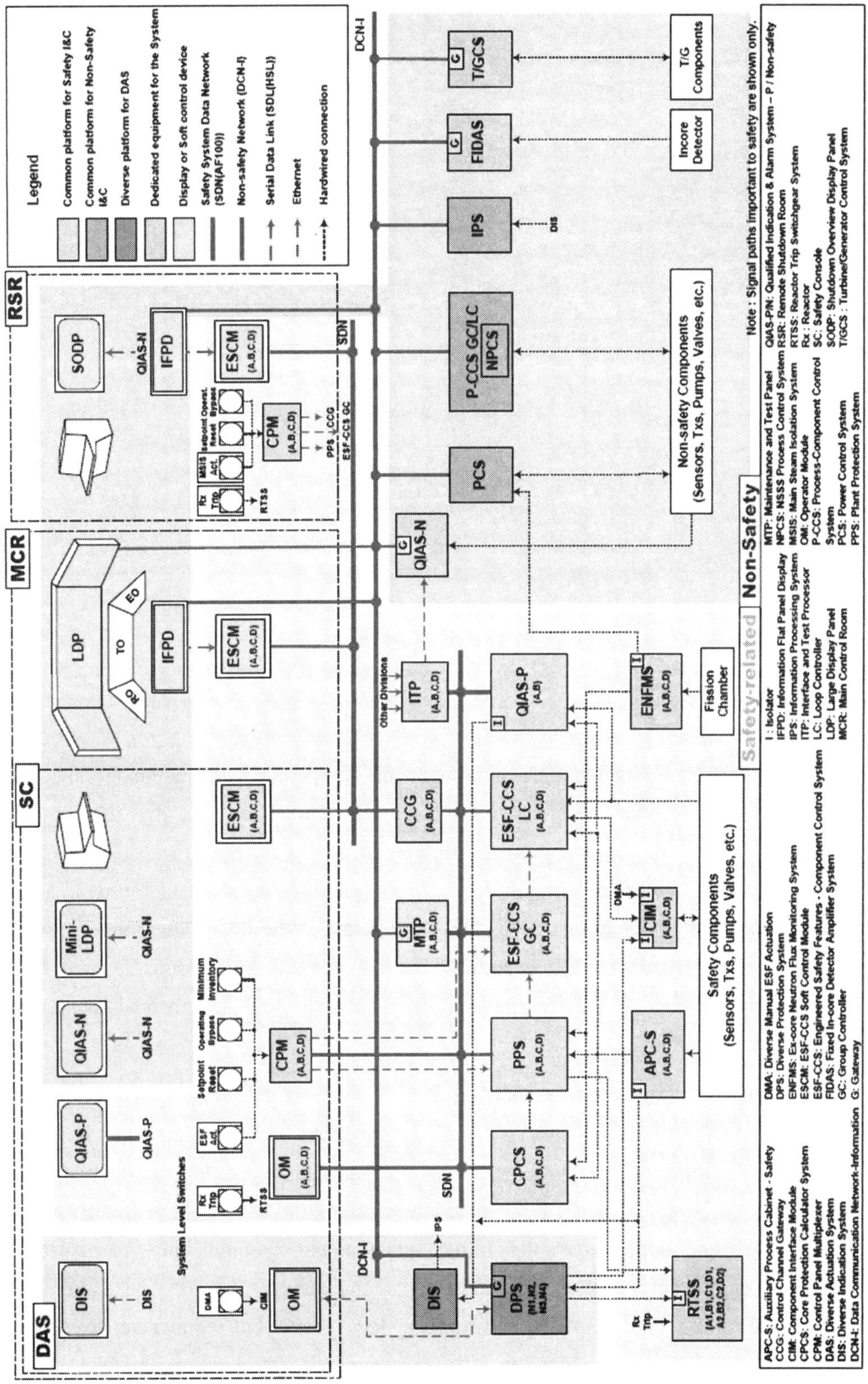


Figure 1-2 APR1400 MMIS [5]

# Chapter 2 SYSTEMS ENGINEERING APPROACH

**SUMMARY**

A System is an aggregation of components cooperating for common objective. The objective cannot be achieved by a sole component but by the synergy emerging from the cooperation of those components. Systems engineering is therefore a discipline and an approach to developing a successful system. Many references say engineering is based on scientific facts or laws, but systems engineering is based on the experience of the developers. Lessons learned from failure and success of the development are abstracted and documented in systems engineering standards.

The ability of systems engineering can be gained by many years of project development experience. However, what is important is that humankind has been supreme creature in the world through the inheritance of the experience from ancestors. Therefore, the new generations of humankind have much more knowledge and ability than the former generations. Through the inheritance of systems engineering knowledge, the systems engineer of the new generation can have the ability and experience within a few years, but equivalent to several years of experience gained by experienced engineers in previous generations.

The systems engineer serves as a conductor in an orchestra when it comes to developing a complex system. An example of a complex system developed by humankind is the nuclear power plant. In developing a nuclear power plant, several engineers from different disciplines (architect, civil, electrical, control, mechanical) should collaborate closely. The systems engineer, therefore, brings together the activities to be performed by all these experts and specialties to develop a successful system.

There are fundamental questions that must be answered when developing a system. The objective of these questions is to unearth relevant details about the end product or service of the system. Therefore, an inability to clearly define what the system is, or what its boundaries are, can impose significant challenges on the system development activities. Some of these fundamental questions include:

- What is a system?
- What are the boundaries and limitations of the system?
- What function is the system expected to perform within the organization?
- What stated missions are the systems expected to perform? and
- What are the expected outcomes from the system?

These fundamental questions are sometimes difficult to answer, but, by clearly and concisely answering these questions, the system definition and analysis process becomes much easier.

In this chapter, we will discuss best practices for defining a system by examining the systems characteristics and attributes. Further, we will take a close look at what systems engineering approach involves and its effectiveness in system development process. Finally, we will discuss the management aspect of systems engineering approach and its application in the development of the MMIS. The structure of the chapter is defined in the mind map below.

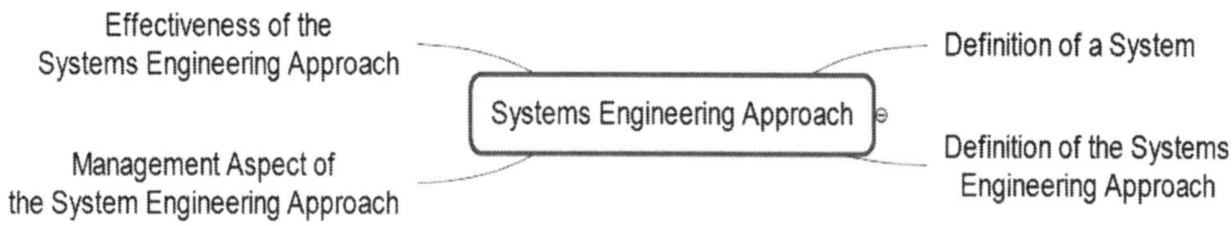


Figure 2-1 Mind Map of Chapter 2

**OBJECTIVES**

- Definition of Systems and related attributes and characteristics
- Introduction of the Systems Engineering approach and a discussion of its effectiveness
- Discussion of the Systems Engineering Management aspects, specifically the Systems Engineering Management Plan, its purpose and structure

## 2.1 Definition of a System

The term "system" originates from the Greek term “systema,” which means to "place together.” Multiple business and engineering domains have varying definitions of a system. This textbook defines a system as:

> “A System is an integrated set of interoperable elements, each with explicitly specified and bounded capabilities, working synergistically to perform value-added processing to enable a User to satisfy mission-oriented operational needs in a prescribed operating environment with a specified outcome and probability of success” [6].

From the definition, we can deduce four key terms that are essential in defining a system. These terms are combination (aggregation, set), interaction, stated (common) purpose, and end product. These terms are explained in detail below.

### (1) Combination (aggregation, set)

A system is composed of two or more elements. Thus, everything that can be divided into multiple elements, can be considered as a system. The universe, earth, molecule, and atom can be considered as system. A molecule is composed of atoms, and then atom is composed of nuclei and electrons. We can find that an element of a system can be a system by itself. For example, elements of the system can be hardware, software, facilities, equipment, material, and procedures. If the NPP (nuclear power plant) system needs trained operator, then the human is also the element of the NPP system.

We are sometimes confused with the word "subsystem", "element", "component", "assembly", and the relationships among them. There are no universal and agreed relationships among those terminologies. Therefore, an organization should define, agree, and use the hierarchy of the system and the name of each layer. What is important is that the whole participants of the project use the same terminology on the structure of the system. The elements of a system could be either hardware, personnel, software, or facilities. Furthermore, a subsystem (element) in one organization can be regarded as a system in other organization that develops the subsystem. In addition, the systems engineering principle are applied to system and subsystem in the same way. After all, the system is a combination of two or more element to create emergent properties, which an element cannot give by itself.

### (2) Interaction

As mentioned above, a system is composed of two or more elements. However, if these elements do not interact with each other, then this set of elements should not be considered as a system from the viewpoint of systems engineering. The interactions between the elements are the core of the emergent property. The interaction can be the interchange of energy, material, or information. For instance, if you carry two stones in your hands, it is not considered as a system unless there is some form of interaction between these stones. To satisfy the common purpose, the elements must be compatible with each other and at least a simple interaction to ensure the desired function is achieved.

### (3) Stated (common) purpose

The purpose of the system cannot be achieved by the sole element of the system. If two elements of the system interact, there can be emergent property or synergy, which is the purpose of the system. Therefore, the elements of a system interact for common objectives. Designing a system is identifying the purposes (requirements), design elements, the functions, their physical characteristics, and interactions between them. In Figure 2-2 the purpose of the system in to process the input to output.

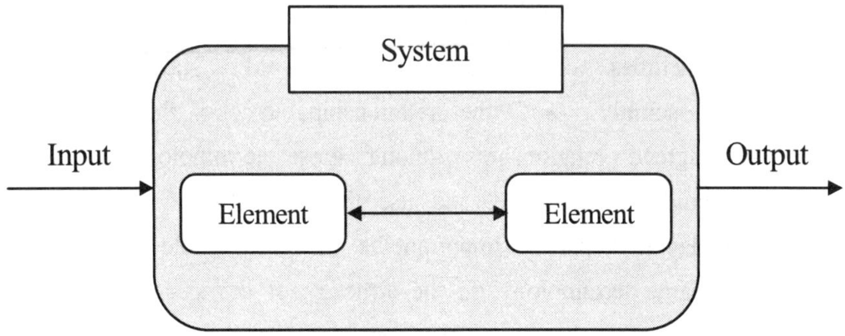


Figure 2-2 System element, interaction, and purpose

### (4) End products and Enabling products

In general, systems are composed of end product and enabling products, which we referred to as elements in Figure 2-2 above. The end products perform operational need or mission and the enabling products perform life cycle sustainment functions. The life cycle sustainment functions are related with development, production, test, deployment, training, operational support, and disposal.

For nuclear power plants, the NSSS (Nuclear Steam Supply System), BOP (Balance of Plant) and T/G (Turbine Generator) perform the operational requirement, which is to generate electricity. From the life cycle perspective, we need enabling products such as CAD (Computer Aided Design), analysis software for development and simulator for training. It is very important that the life cycle sustainment requirements are identified early in the development stage and are considered in planning the project and designing the overall system.

### 2.1.1 Characteristic of system

Considering a system as an aggregation of components, it obviously has some characteristics. Generally, there are four main characteristics of a system and these are shown in Figure 2-3. The first characteristic is the ability to effectively combine resources such as equipment, material, software, facilities, data, memory, human beings, etc. to accomplish desired function. Secondly, a system must be contained within some form of hierarchy. For example, a nuclear power plant may be included in a company that is a part of the electric power system of a country. In such a situation, the system being addressed is highly influenced by the performance of the higher-level system, and its external factors.

Thirdly, a system may be broken down into subsystems and related components; and the extent to which the system can be broken down depends on its complexity and the function being performed. Dividing the system “into smaller units allows for a simpler approach relative to the initial allocation of requirements and the subsequent analysis of the system and its functional interfaces” [7].

Lastly, a system must have a purpose. It must be functional, able to respond to some identified need, and able to achieve its overall objective in a cost-effective manner. It should be noted that, there might be instances where objectives may conflict or be influenced by the higher-lever system in the hierarchy; in such a situation, the system must be characterized in such a way that it is capable of meeting its primary purpose in the best way possible.

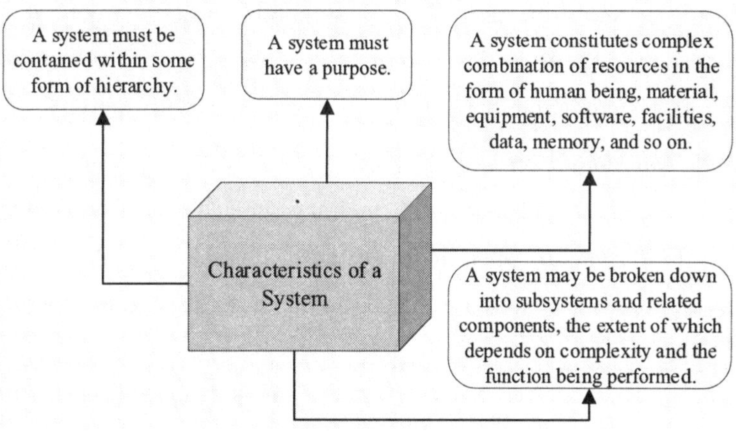


Figure 2-3: Characteristics of a system [7]

### 2.1.2 Categories of Systems

There are different types of systems and based on general characteristics of these systems, it can be categorized further for analysis. The main categories of systems are described in detail in INCOSE handbook [8], but for quick reference, these categories are summarized below:

i. ***Natural and man-made systems*** : natural systems came into existence through natural phenomena. Examples include river system, solar systems, etc. Man-made systems on the other hand are developed by human processes and have wide variety of capabilities. An example of man-made system is the electric power systems, which is the largest man-made system. Man-made systems generally have some form of interfaces with natural systems and these interfaces must be addressed during the system development.

ii. ***Physical and conceptual system*** : physical systems comprise components residing within a predefined space. Unlike physical systems, conceptual systems are mainly formulation of ideas and do not have any physical attribute. It should be noted that physical systems are developed based on conceptual systems.

iii. ***Static and dynamic systems*** : systems that have structure but do not have any activity are called static systems. This includes buildings, roads, etc. Dynamic systems on the other hand combine both structural components and activity together. For instance, a vehicle or conveyer systems is a dynamic system combining both structural part with activity.

iv. ***Closed-loop and open-loop systems*** : closed loop system is a system that is self-contained and does not interact with its environment [7]. Unlike closed loop system, open loop systems interact with the environment by receiving feedback, which is used on input to tuning its operation.

### 2.1.3 Characteristics, Attributes and Properties of MMIS

An important concept in systems engineering is to define the systems characteristics, properties and attributes. In a general term, a system's characteristics are its behavioral and physical qualities that uniquely identify the system. Attributes are physical features of the system, whereas system properties refer to mass properties of the system. For simplicity, these terms are illustrated with examples for the MMIS.

The **characteristics** of the MMIS are those behavioral and physical qualities that uniquely describe the MMIS. Physical qualities of the MMIS are large display panel, safety consoles, and workstations. Some behavioral characteristics of the MMIS workstation is to use multiple devices that allow for simultaneous access to plant data and information. Other behavioral and unique qualities are:

- Ability to perform defense in depth concept;
- Capable of providing diversity. This means equipment for same function, but with diverse mechanism
- Redundancy
- Independency
- Fail in safe
- Capability for multiple barrier protection;

**The properties** of the MMIS are the mass properties of the MMIS and these include the following:

i. ***General features*** - These are related to diverse fault-tolerant distributed control systems and the ability for common control through universal soft-controller for safety and non-safety systems.

ii. ***Process-component control system*** - This feature is responsible for control of plant process systems and ESF-CCs, etc.

iii. ***Information processing system*** - This relates to processing of data for large display panels and optical disks for historical data storage.

iv. ***Plant protection system*** - This feature integrates reactor trip and engineered safety feature actuation; manual initiation of reactor power cutback to prevent SCRAMS etc.

The **attributes** of the MMIS refers to its physical features. Examples include:

i. ***System architecture***: Every system consists of multi-level, logical (functional) and physical structure or architecture that provides the framework for its form, fit, and function [7]. The MMIS architecture is designed in such a way that the operator interface functions and control functions for NSSS or BOP are integrated in common design standard; and implemented in Distributed Control System (DCS) or in programmable logic controllers [9]. The DCS provides multi-loops and logical control function segmentation for cost-effective and reliable implementation of the MMIS architecture.

ii. ***System interfaces***: the system interfaces enable interaction within the system itself and between external operating environments.

iii. ***System operating domain***: the operating domain describes the "sphere of influence" of the system with regards to its operation, interaction with the environment and effectiveness.

## 2.2 Definition of Systems Engineering Approach

The fundamental requirement for systems engineering is to have a good understanding of how to describe a system, service, or product in terms of performance, attributes, and properties. Therefore, to understand clearly the concept and approach adopted in systems engineering, it is essential to understand the underlining principles and the right definition of what the system entails. Thus, the purpose of adopting a system engineering approach is to guide the engineering of complex system due to the intricately diverse interrelation between the various components or elements of the system. Systems engineering differs from traditional discipline, because system engineering focuses on the system as whole and is highly concerned with meeting the customers' needs in the specified operating environment.

Systems engineering bridges the gap between specialty engineering and traditional engineering. It is an integral part of project management but places more emphasis on using systematic and analytic procedures to plan and guide all engineering efforts in system development. Three standards are widely used in systems engineering (ISO/IEC/IEEE 15288, EIA 632, and IEEE 1220) and although these are three different standards, the common objective for each of these standards is to facilitate the effective and efficient management of engineering aspect of the system being developed. The objectives of these standards are to facilitate the following:

- Benchmark system engineering practices;
- Development of standard systems engineering process or policies of an organization;
- Development of standard process of specific industrial sector or area;
- Development of evaluation model of the maturity of a process;
- Communications on the terms and concepts of the systems engineering;
- Development of the training material; and
- Planning for development of products.

## 2.3 Effectiveness of Systems Engineering Approach

One of the most difficult factors in applying systems engineering in an organization is how to demonstrate its effectiveness. This is because systems engineering is not a new technology, in which case its advantages could be easily seen; but rather a process with the sole purpose of reducing unnecessary work or mistake in system development. The systems engineering approach focuses on defining the right problem and developing the right solutions to address these problems. By using the SE approach, the stakeholder and system requirements are defined clearly. These two requirements serve as the basis for subsequent design activities. Defining these requirements clearly, eliminates redesign or rework, which often occurs in a traditional system development approach. This is the main effectiveness of systems engineering approach. The methodologies used in evaluating SE effectiveness are discussed thoroughly in “A survey on systems engineering effectiveness, NDIA, 2007”.

In traditional engineering approach, the system development process depends on one experienced manager. However, in todays’ world, systems are becoming increasingly complex with intricate requirements and constraints to be fulfilled. Therefore, it becomes more difficult for one person to manage all the facets of the system development. The traditional method where only one experienced project manager is depended on results in general problems such as [10]:

- Poorly defined stakeholder requirements;
- Confusion over customer requirements;
- Not achieving technical requirements;
- Point design without consideration of other options;
- Conflict between systems and design engineering;
- Failure to document decisions in engineering memos; and
- Floating baseline, without rigid formal change control.

### 2.3.1 Reduction in Rework and Cost Savings

Studies show that by eliminating design changes or reworks in complex projects results in significant savings in cost and schedule. Generally, defects are mostly generated in the early phase of the system development due to poor definition of stakeholder or systems requirements. Most of these defects are discovered during the integration phase of the system. Meanwhile, studies show that, the cost to repair the system defects increases logarithmically as the development phase proceeds; and thereby having significant impact on the project cost. In addition, the project schedule is affected and the probability to satisfy the customer becomes lower and lower. In severe cases, some defects identified at late phase of the development may not be corrected.

The effectiveness of SE approach is to reduce the number of design change at the early phase of the system development by focusing on clear definition of stakeholder requirements from the whole system life cycle perspective. SE approach facilitates design of logical solution before designing the physical solution; and ensures that traceability requirements, verification requirement, trade-off consideration are clearly established during the initial phases of the system development.

### 2.3.2 Effective System Lifecycle Management

From Figure 2-4 , we can see that almost 70% of the cost of the system life cycle is committed in the basic design phase, while about 15% of the total cost is expended during the basic design stage. As the phases of the system lifecycle proceeds, opportunity for design change decreases, which means that as the opportunity decreases the potential for cost savings, schedule management and improvement in system quality decreases. This eventually affects the overall lifecycle of the system. The Core of systems engineering is development of the system architecture, which is usually referred to as basic design, and performing system analysis to ensure holistic system development (detailed design, production, operation).

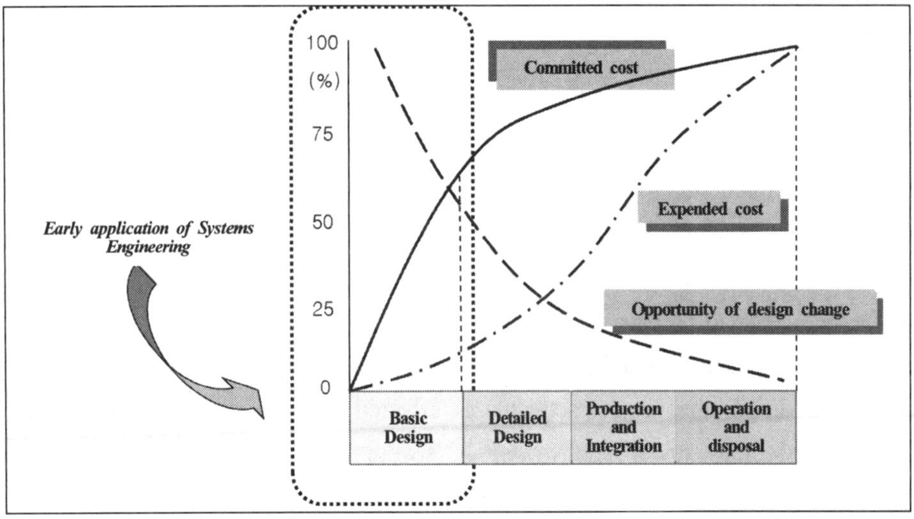


Figure 2-4: Effectiveness of early application of SE [8]

## 2.4 Management Aspect of System Engineering Approach

The Systems Engineering Technical Management Processes are technical planning, configuration management, Interface management, technical data management, requirement management, risk management, technical assessment, and decision analysis [11]. These processes are used to establish plans, to execute the plans, to assess actual achievement and progress against the plans and to control execution through to fulfillment.

Technical management processes include project planning, project assessment and control, decision management, risk management, configuration management, information management, measurement, and quality assurance (QA). These processes are found throughout an organization, as they are essential to generic management practices and apply both inside and outside the project context [8]. Figure 2-5 illustrates Technical management processes.

Figure 2-5: Technical management processes

## 2.4.1 System Engineering Management Plan (SEMP)

The System Engineering Management Plan is an outcome of the project planning process. This document describes the overall plan for the systems engineering management of the project as well as the processes invoked to accomplish the plan. It defines the technical organization, as well as the roles and responsibilities of the project team. In addition, it establishes the high-level guidelines (and root of procedural work instructions) for the technical control of the project. The SEMP ensures that the management and engineering processes include the necessary structure required for the success of the project [12].

It is a key document for the system engineering approach and represents the highest-level technical plan that includes the master plan, which identifies the planning activities, milestones, work breakdown structure, project schedule, and resources requirements necessary to ensure the system accomplishes its intended goals and objectives. It is intended to guide the work of all those involved in the project. To develop SEPM the following activities should be performed in the planning process as defined by INCOSE [8]:

- Analyze the project proposal and related agreements to define the project objectives, scope, and constraints;
- Establish tailoring of organization procedures and practices to carry out planned effort;
- Establish Work Breakdown Structure based on the evolving system architecture;
- Define and maintain a life cycle model that is tailored from the defined life cycle models of the organization;
- Establish roles and responsibilities for project;
- Define top-level work packages for each task and activity identified;
- Develop project schedule based on objectives and work estimates;
- Define the infrastructure and services required;
- Define costs and estimate project budget;
- Plan the acquisition of materials, goods, and enabling system services;
- Generate quality management, configuration management, risk management, information management, and measurement plan to meet the needs of the project; and
- Establish the criteria to be used for major milestones, decision gates, and internal reviews.

In addition, INCOSE [8] suggests that the SEMP should include the following:

- Organization of the project and how SE interfaces with the other parts of the organization;
- Responsibilities and authority of the key positions;
- Clear system boundaries and scope of the project;
- Project assumptions and constraints;
- Key technical objectives;
- Infrastructure support and resource management;
- Approach and methods used for planning and executing;
  - The Technical Processes;
  - Technical Management Processes;
  - The Applicable Specialty Engineering Processes.

SEMP should state that a verification plan will be written to define the lifecycle products to be verified as well as the verification methods. In addition, the verification activities may also receive special attention in SEMP by defining the Technical Performance Measures (TPM). The critical path for the master schedule should be defined in the SEMP.

## PROBLEMS

1. Select a System in the NPP. For this system define the:
   a. Sub-systems
   b. Interaction
   c. Stated Purpose
   d. End Products and Enabling Products
2. Explain the benefits of adopting systems engineering approach to MMIS development as compared to traditional engineering method.
3. Taking the NPP system as an example, explain how systems engineering approach could provide useful solutions and acceptability of the systems product across all stakeholders.
4. Explain how the APR1400 is physically structured using a diagrammatic representation.
5. Identify four (4) physical systems of MMIS and state the characteristics, attributes, and properties for those systems.
6. Discuss the main contents of the SEMP and explain the effectiveness of using a SEMP for system development.

## FURTHER READING


- Charles S. Wasson, System Analysis, Design, and Development Concepts, Principles, and Practices, Wiley-Interscience, 2006
- INCOSE, Systems Engineering Handbook Ver. 4.0
- Guide to the Systems Engineering Body of Knowledge (SEBoK)

# Chapter 3 TECHNICAL PROCESS FOR MMIS DEVELOPMENT

## SUMMARY

As defined in the previous chapter, System Engineering is a structured approach to the development of systems. As such, there exist predefined processes and established best practices in the use of this approach. In this chapter, these processes shall be discussed, with emphasis on the techniques applicable to the development of the MMIS. The chapter is structured as in the mind map in Figure 3-1.

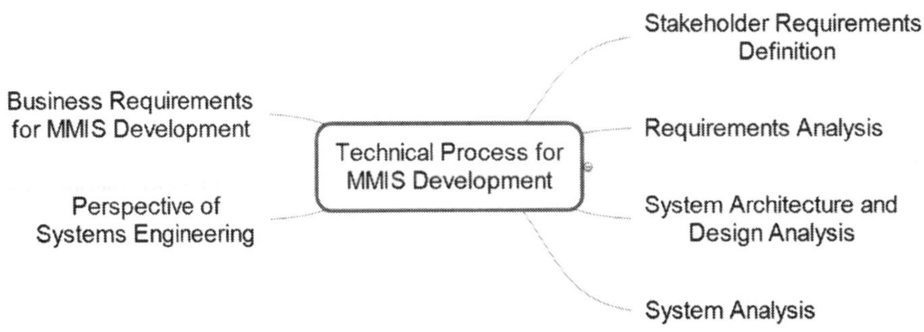


Figure 3-1 Mind Map of Chapter 3

**OBJECTIVES**

- Definition of the technical processes used in the systems engineering approach
- System requirements definition and analysis methods
- Discussion of system acceptance criteria: Measures of Effectiveness, Measures of Performance and Technical Performance Measures

## 3.1 Technical Process Overview

The first priority for effective and efficient development of any system is to evaluate the stakeholder's requirements. The stakeholder is the starting point and the ending point of the development process. This is because the stakeholder defines the problems to be solved (requirements) and performs validation to ensure that the problems identified were adequately solved. The technical process from the system design perspective involves three main processes as discussed below and illustrated in Figure 3-2.

The technical process for system development starts with defining the stakeholder's requirement and success criteria that will be used to evaluate the system solution. The stakeholder's requirements usually emanate from enterprise-level strategies, which sometimes are an approach to address mandated legislation, enacted laws, regulatory requirements, or even public opinion. The enterprise-level strategies are translated into business needs and operational concepts are established for system solutions identified to address the business need.

The second process is the system design requirement. Based on the stakeholder requirement and the various system solutions selected, design requirements are performed for the selected systems. These requirements are decomposed into common requirements, constraints, performance requirements, functional requirements and verification requirements. As part of the decomposition process, functional requirements of the system are developed and serve as the input for the third process.

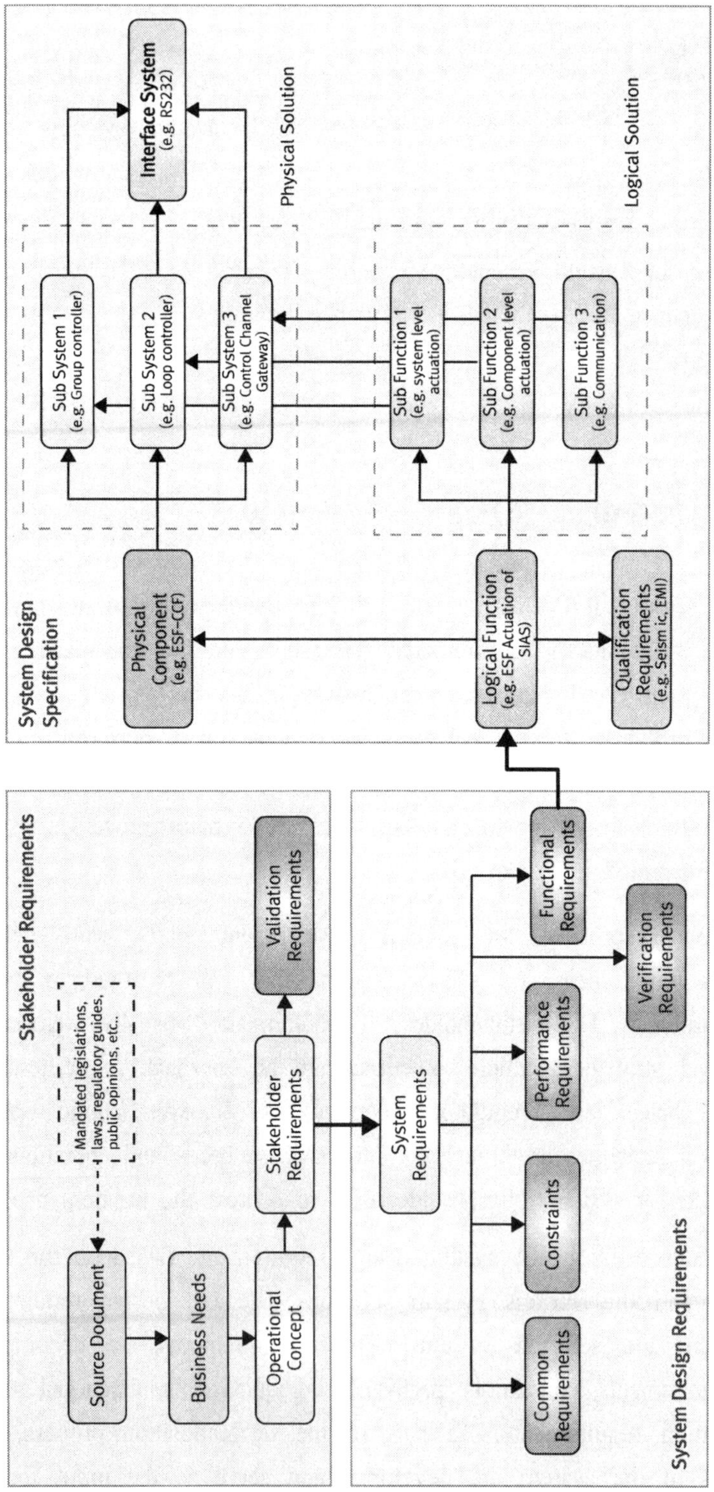


Figure 3-2: Technical Processes for Systems Engineering

The third process is the development of the system design specification. It comprises of two sub-process: logical solution and physical solutions. The logical functions of the system are mapped with corresponding physical components to perform those system functions. For example, in Figure 3-2, we can see that the logical function such as "system level actuation" is performed by the physical component "group controllers" and these physical components requires some form of interface to relate with other subsystems. An example of the interface system is RS-422 high-speed communication links to the adjacent channels.

## 3.2 Business Requirements for MMIS Development

The development of any complex system requires a thorough analysis of the specific business in which the system is to be deployed; more importantly, the role of the system in achieving the overall operational objective defined in the enterprise strategy. Business requirements have a major impact on Return on Investment. Thus, it is an integral part of system development process.

In the nuclear power industry, the fundamental business objective is to generate electricity in a safe, reliable, and economical way. However, various factors can influence how these business objectives are achieved. Therefore, enterprise needs are developed to comply with these factors i.e. mandated legislation, enacted laws, regulatory requirements, public opinions, etc. A concept of operation is developed from the enterprise viewpoint and it describes how the system will operate and who will operate it. This concept of operation is refined during the business needs level and it feeds into how the multiple systems will operate to achieve the desired business solution. Figure 3-3 shows the relationship between enterprise needs, business management levels, concept of operation, and operational concept.

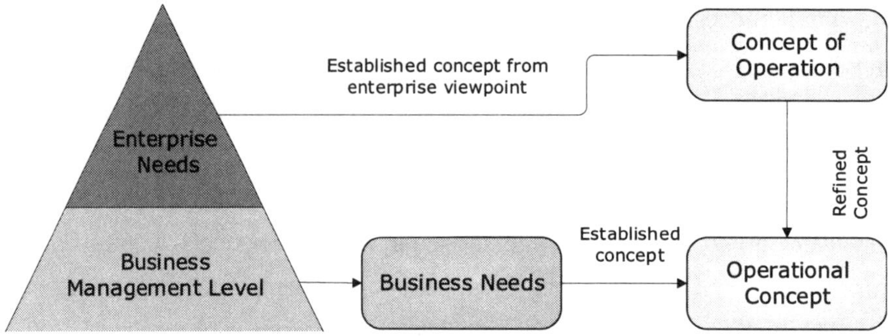


Figure 3-3 Business viewpoint of system development

### 3.2.1 Difference between Concept of Operation and Operational Concept

Most often concept of operation and operational concept are used synonymously; however, there is a slight difference in both terminologies. Operational concept focuses on a broader description of the capability of the system encompassing multiple systems solutions to address the business need. Whilst concept of operation describes how the system will operate and who will use the system.

For example, an MMIS operational concept would include Plant Protection System, Process Component Control System, Power Control System, and ESF Component Control System. Operational concept can even be applied to subsystems control such as Feedwater Control System, reactor regulating system, or even Field Programmable Gate Array (FPGA) which constitutes the functional logic. Operational concept describes how a large collection of multiple systems will operate, therefore in defining operational concept a "system" is not refer to a single system but multiple systems.

Concept of operation on the other hand refers to only a single system and it relates to a specific scenario in connection with a single system. In defining the concept of operation, the following must be considered:

- Complete description of the mission and its success criteria;
- The relationship or interaction of the system (the system referred to here is a single system not multiple systems) with other systems or entities;
- The sources and destination of information for the system; and
- The constraints of the system.

### (1) Considerations for Business Requirements Development

The MMIS business requirement must be integrated to the overall business goal. As an example, a key requirement for introducing MMIS is to take advantage of operator capabilities and at the same time provide intelligent feedback to check operator errors. This requirement must fit into the overall goal of reliable power generation. The following questions are important for the development of business requirement for MMIS:

i. ***What functionality is being provided or replaced as a way of resolving an existing problem;***

This is an important consideration because it defines clearly the exact problem to be solved with the system. For example, if the business requirement is improved reliability, it means the development of the MMIS should be done with highly reliable components and equipment. Moreover, obsolescence is the critical problem which should be considered how reliable equipment could be replaced.

ii. ***What is the allocated budget?***

The business requirement should outline all relevant cost implications for the system development. For instance, in MMIS development an important cost component is continuous education and training of operating staff because these have a great impact on the level of automation, design of control facilities and even arrangement of controls consoles or displays.

iii. ***What is the current hardware infrastructure?***

This question address, the infrastructure and technology that will be used in the system development. The nuclear power industry is very conservative in accepting new technology, however, if there are proven technologies available in reference plants this should be factored in the business requirement. The use of modern and proven technology provides significant benefits to the system design and overall business case.

iv. ***Who will serve as project team?***

This requirement should be stated clearly in the business requirement document. In the KURD for instance, the business requirement for project team for the MMIS design states that, it should be directed by a single organization responsible for the entire MMIS design and should be carried out by a multi-discipline design team. Providing clear business requirement eliminates ambiguity and enhances implementation of effective business solutions.

## 3.3 Perspective of Systems Engineering

Systems and processes are essential for an efficient operation of any business. However, the concept of system engineering is not used effectively in many industries although this field of engineering has matured rapidly in recent times. As people within an organization turn to learn more and more about the benefits of using a systems engineering approach, their views and needs continue to vary. Figure 3-4 shows an example of each viewpoint and the corresponding system that needs to be addressed in developing a system. We can see that the needs and perspectives vary at different levels throughout the organization. These viewpoints from each level must be integrated into the system development process. So, using SE approach for system development looks at the system from a holistic viewpoint and this helps in fulfilling the expectations of stakeholders at each level throughout the organization.

The views throughout the organization and corresponding needs are shaped based on three emergent perspectives. These are system thinking, systems engineering, and engineering systems. The perspective of system thinking is to focus on the respective processes and consider critical issues that might influence or enhance the system development. This concept also evaluates multiple factors and influences that the system might have on its environment and vice versa. In system thinking, it is essential to establish common understanding and foster relationships.

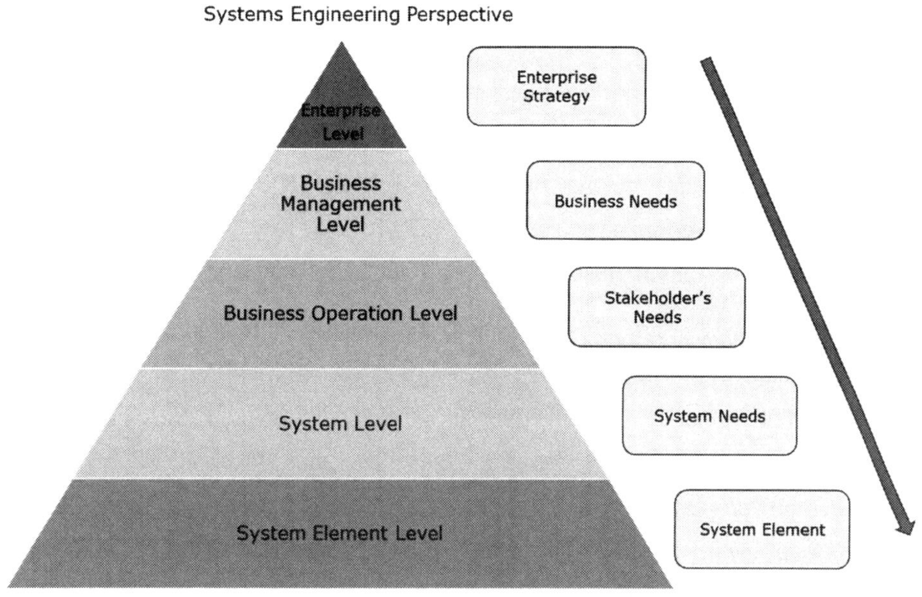


Figure 3-4 System engineering views throughout organizational levels

Unlike system thinking which focuses on the process, systems engineering focuses on the whole product or outcome of the system development. In the systems engineering approach, emphasis is placed on solving complex technical issues and developing tangible system solutions. These activities are carried out with the sole purpose of meeting stakeholder requirements and performance indicators such as Measure of Effectiveness (MOE), Measure of Performance (MOP), and Technical Performance Measure (TPM). These measures (MOE, MOP & TPM) will be discussed in subsequent sections.

The perspective of engineering systems is a combination of both processes and product and emphasis are placed on interdisciplinary technical and management issues. Unlike systems thinking, which evaluates factors and their influences, engineering systems rather influence policy and process and uses systems engineering to develop systems solutions. In addition, by integrating human and technical document approaches.

### 3.3.1 Technical Measures

There are four (4) technical measures commonly used in evaluating a system development to ascertain the appropriateness of the various technical solutions adopted. These measures are measure of effectiveness (MOE), Measure of Performance (MOP), Technical Performance Measures (TPM) and Key Performance Parameters (KPP). The discussion in this textbook will focus on the first three measures.

Figure 3-5 shows an illustration of the relationship among these technical measures. The MOEs and MOPs are expressed with different intentions. For example, MOE refers to the effectiveness of the solution and does not depend on a particular solution. MOE therefore considers the stakeholders intention and states certain properties, which a potential solution must possess to meet the stakeholder's intention [13]. The MOP focuses on the capability of a specific solution and its contribution to achieving the stakeholders stated intention (in this case MOE). It is often used to verify and validate if the system meets the stated requirement.

TPM provide lower level view of certain aspects of the solution evaluated in MOP. It is important to note that the TPM and MOP are of a lower hierarchy than the MOE; and the TPM and MOP should be defined with MOE in mind: to ensure that these measures are integrated and aggregated to provide the necessary acceptability of the systems operation.

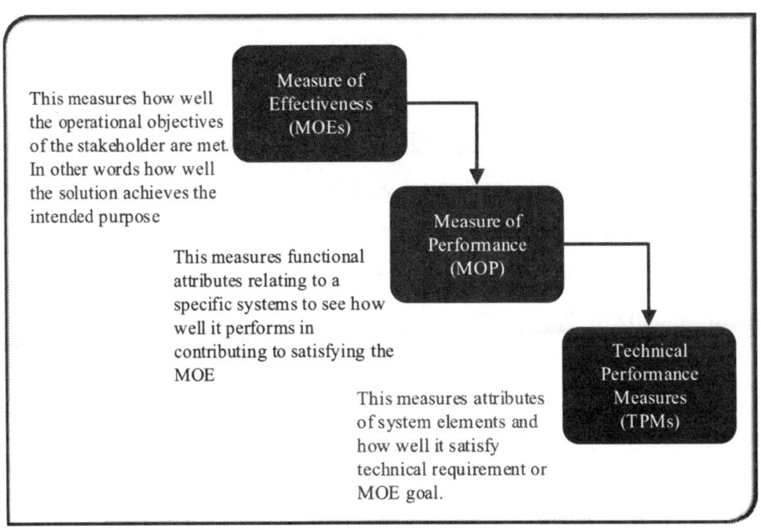


Figure 3-5 Relationship between MOE, MOP, and TPM

## 3.4 Stakeholder Requirement Definition

A stakeholder is any person, organization that is interested in the development of the system. Stakeholders are involved throughout the lifecycle of the system and therefore for any system to be successful it must satisfy not only the end user (customer) but all other stakeholders as well. The MMIS development process begins with identification of the stakeholders and eliciting their requirements because this serves as the corner stone in the development of the system. A typical stakeholder requirement document is the Utility Requirement Document (URD). Several approaches are used to elicit stakeholder requirements. Some of these methods are:

- Survey;
- Focus group;
- Functional analysis; and
- Brainstorming;

The stakeholder requirement definition process starts with identifying the system users and describing the business process, which the system is intended to support and the people involved in these processes. It is essential to identify people who may be affected by the introduction of the system at the stage of stakeholder definition. Further, the operators and maintenance staff responsible for the development and maintenance of the system should be considered. External bodies such as regulators, and certification authorities who may wish to place requirements on the system should also be defined clearly during the stakeholder definition process [14].

### 3.4.1 Identification of Stakeholders

As discussed earlier, stakeholders are people who have interest in the development of the system, and their interests may range from benefit to loss or harm. The customer may be the most important in the development of a system but the customer is also one of the stakeholders. Examples of stakeholders are shown in Figure 3-6. For a successful development of the system, the project should identify all the stakeholders, elicit, and analyze their requirements. It is necessary to keep in mind that the source requirement is not from the customer alone, but from all other stakeholders that have expectation for the system's product or outcome.

The customer can be an individual or organization that commissions the engineering of a system, or a prospective purchaser of the product. A user on the other hand is either an individual or organization that uses the end product of a system [15].

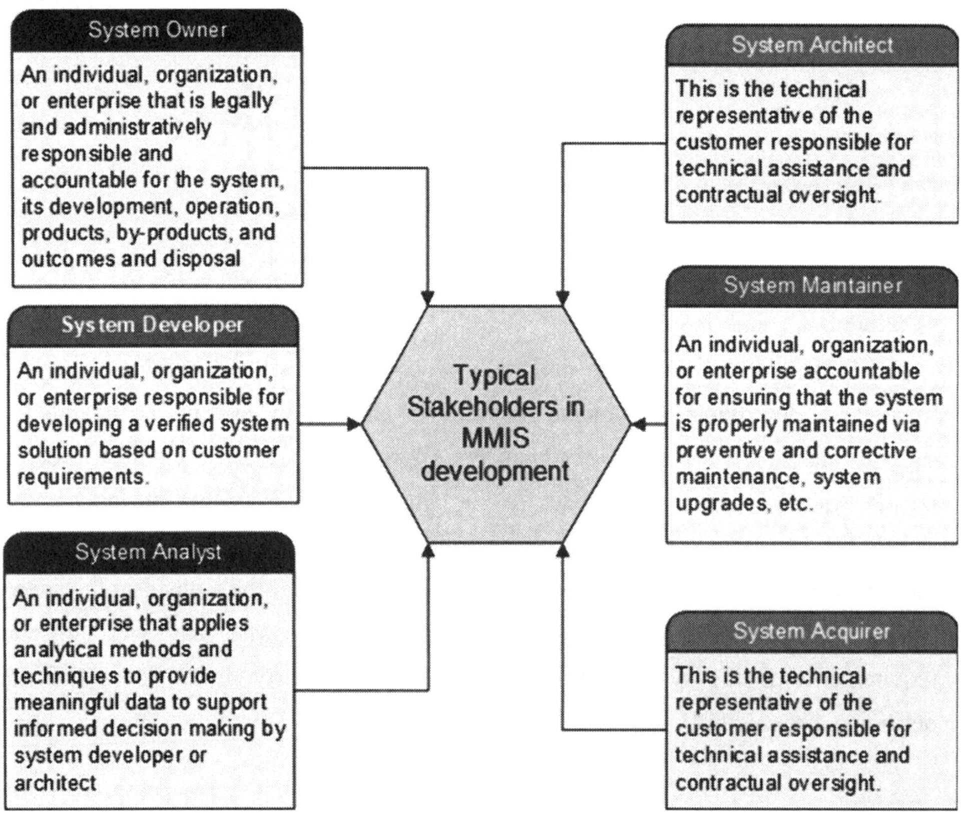


Figure 3-6 Examples of Stakeholders in MMIS Development

PART 1

### 3.4.2 Elicitation of Stakeholder's Requirements

After identifying the stakeholder, the requirements of the stakeholders should be elicited using any of the requirement elicitation techniques listed in Table 3-1. Requirement elicitation is typically the responsibility of the systems engineer because most often stakeholders do not express their requirements well. In pragmatic principles of systems engineering [16] a pragmatic approach to requirements gathering is described in detail.

Base on the elicited requirements the following key documents are developed:

- Operational concept;
- Constraints;
- External interface requirements; and
- Other requirements (Quality factor, training, packing etc.).

Table 3-1: Techniques for stakeholder requirement elicitation

| Technique | Description |
|---|---|
| Survey/Questionnaire | Asking stakeholders to answer the pre-defined question or subjective question. This method can be used to obtain the opinion of many stakeholders with little effort. However, one should be careful when defining the questions. |
| Brainstorming | A group interview for generation of ideas. Can generate the information on what is useful for the stakeholders. This method can gather many requirements in short time, but it is questionable if all requirements are elicited. |
| Card sorting | Classify the characteristics of the system and record the criteria of classification. This method is easy to be automated but exact performance requirement may not be elicited. |
| Focus group | A group interview to find out the interest of the participants. Usually use file, story board, or mock-up to facilitate the interview. This method is very flexible but should not be in a suppressive environment |
| Functional analysis | Using structured diagram identify what the system should do with the stakeholder. This method can provide clear and practical requirements, but constraints or performance requirements may not be elicited. |
| Prototyping | Make prototype to elicit uncertain requirements. This method helps the stakeholder to tell requirements easily but needs effort in building prototype. |

### 3.4.3 Establishment of Lifecycle Concept

Every system-of-interest (SOI) goes through a life cycle; that is from its inception to its retirement. The lifecycle of the system is established in the development stages and aligned with management planning and decision- making.

As discussed earlier, a system is composed of end product and enabling products and these are called elements. The end products perform operational needs or mission and the enabling products perform lifecycle sustainment functions. Enabling systems are systems that support the lifecycle activities of the SOI by providing services that are needed by the SOI during one or more lifecycle stages, such as production systems, logistics support systems, etc. These enable progress of the SOI in one or more of the lifecycle stages. In establishing the lifecycle for an SOI, it is important to take into account the availability of the enabling systems. This is because the unavailability of these systems may lead to significant issues and delays in the progress of the SOI development [8].

### 3.4.4 Establishing of an Operational Concept

Operational concept describes how the user will use the system; therefore, it is essential to understand the operational needs of the user before commencing the system development. By understanding the user's needs, the system engineer can clearly distinguish between the important criteria from the user's perspective for a successful system. Developing the most elegant system designs that do not meet the user's acceptance criteria is worthless. To under the user's operational need the following should be considered:

- How does the user intend to use the system or its product?
- What are the applicable measures adopted for the success of the system? and
- What are the consequences and ramifications for system failure?

Generally, operational concept is usually defined by the customer, but the customer sometimes does not describe this concept clearly. In this case, the systems engineer should clearly define the operational concept with the customer. It is necessary to keep in mind that operational concept is the basis for further analysis of the system function.

### 3.4.5 Measures of Effectiveness

The Measures of Effectiveness (MOEs) are important criteria to evaluate how well the system accomplished its mission objectives. MOEs typically are objective measures that denote the most critical measures that contribute to the success of the system's outcome. It is measured in qualitative or quantitative form and represents the degree to which the entire system performs the specified task or meets the operational requirement under specified conditions [7].

The objective of the MMIS is to provide instrumentation, monitoring control and protection functions for the nuclear power plant under all operating conditions: for safe, secure, and reliable electric power generation. Therefore, in defining the MOE for the MMIS, it must relate to the nuclear power plant as a whole and describe aspects of its performance in achieving the primary objective. The MOE is derived from stakeholder's requirement and for the MMIS this can be found in KURD Chapter 10.

Regardless of how MOEs are stated, it must consist of three parts: metric, unit, and the context under which the metric is applicable. The measurement can be defined in three broad categories as follows:

- Measurement (i.e. either deterministic or random parameters obtained from actual system or mathematical models);
- Likelihood (probability of an even that may occur and affect other MOEs); and
- Binary (logical variable of occurrence of an event, i.e. event occurring or not).

Examples of MMIS measure of effectiveness statements are:

- Mission requirement 1: The MMIS must be capable of providing automatic and manual controls for all safety and non-safety systems under all operating conditions with an average service availability index of above 92%.
- Mission requirement 2: MMIS must provide 100% reactor protection functions in cases of operational anomalies to limit fuel damage to less than 10% in case of severe accident.

### 3.4.6 Validation and Agreement for the Final Acceptance

Usually, stakeholder requirements are defined in non-technical terms such as needs, desires, wants, and expectations, which are not normally adequate for system design [8]. In addition, the stakeholder requirements stated in non-technical terms are often difficult to satisfy and may become unverifiable. This can bring about uncertainties during the final acceptance of the designed product. In some cases, stakeholder requirements are stated in technical terms but are often expressed in higher technical levels and often need to be translated into technical requirements in clear, unambiguous, and measurable technical terms.

Table 3-2 shows an example of how to validate stakeholder requirements to be sure that they meet the system development objective.

Table 3-2: Requirement validation criteria

| Criteria | Description | Proposed Evaluation Method |
|---|---|---|
| Necessity | Is the requirement really needed? | Backward requirement traceability (i.e. previous stages of development) |
| Clarity | Is the requirement understandable? | Document review<br>(e.g. IEEE Std. 830-1998 Clause 4.3.2) |
| Correctness | Is there any wrong content? | Compare requirement with applicable standards, system requirement, project document and specifications |
| Completeness | Is there any more information needed? | Document review<br>(e.g. IEEE Std. 830-1998 Clause 4.3.3) |
| Conflict | Is there any conflict between requirements? | Traceability matrix |

PART 1

### Documenting and maintaining stakeholder requirements

In documenting requirements, you should be careful to describe all the facets of the requirements. Each requirement statement should consider the following elements described in Table 3-3.

Table 3-3: Sample requirement statement elements

| Element | Description | Example |
|---|---|---|
| Actor | This is the subject of the sentence. Normally the thing being specified | The system |
| Condition | This defines the precondition under which the action is required to take place, or the condition during which the characteristics are to be present. | upon receipt of message |
| Action | This is a verb- the action is to be taken by the actor (subject). | shall switch |
| Object of action | This is a noun, which may be the subject. The objective of action is the thing acted upon by the actor. | that message |
| Constraints of action | This qualifies the action. For functional requirements, constraints of action required performance (how well the action is to be performed) | within 10 milliseconds of receipt |
| Refinement of object | This qualify the object. | for messages in ACP format having a valid routing indicator, |
| Source of object | This defines the source | from the message input port |
| Destination of Action | This defines the destination | to a message output port |
| Refinement of action | This further qualifies the action and may be additional to constraints of action. | corresponding to the routing indicator in the message |

## 3.5 Requirements Analysis

Requirements analysis is a process to translate stakeholder requirements into system requirements. The system requirements define what the system should do, and are generally written in technical terms. Stakeholder requirements are not as detailed as system requirements and delete are written in plain language by focusing more on the operational level requirements.

***"System requirements are developed based on stakeholder requirement and to be able to develop system requirement it is important to first perform requirement analysis."***

Through the requirements analysis we develop system function, and technically stated requirements. In the requirement analysis phase, stakeholder requirements are analyzed and system/engineering requirements are ultimately defined. In addition, all external interface requirements are adequately defined in addition to other requirements related with all the life cycle of the system. It is essential at this stage to keep traceability between the stakeholder requirements and system requirements to ensure that all the stakeholder requirements are satisfied by the system requirements. The results of the requirements analysis are specified in the system specification.

### 3.5.1 Definition of Functions

System functions are what the system is expected to do and form the functional architecture, which is sometimes called logical solution. The function of the system is usually defined by analyzing the operational scenarios that be defined logically and hieratically. Generally, modeling is helpful in defining system function because this aids effective functional analysis. The steps involved in the definition of functions are:

- Establish top level function (e.g. mission or operational function);
- Decompose top level functions into system functions; and
- Decompose system functions into component function.

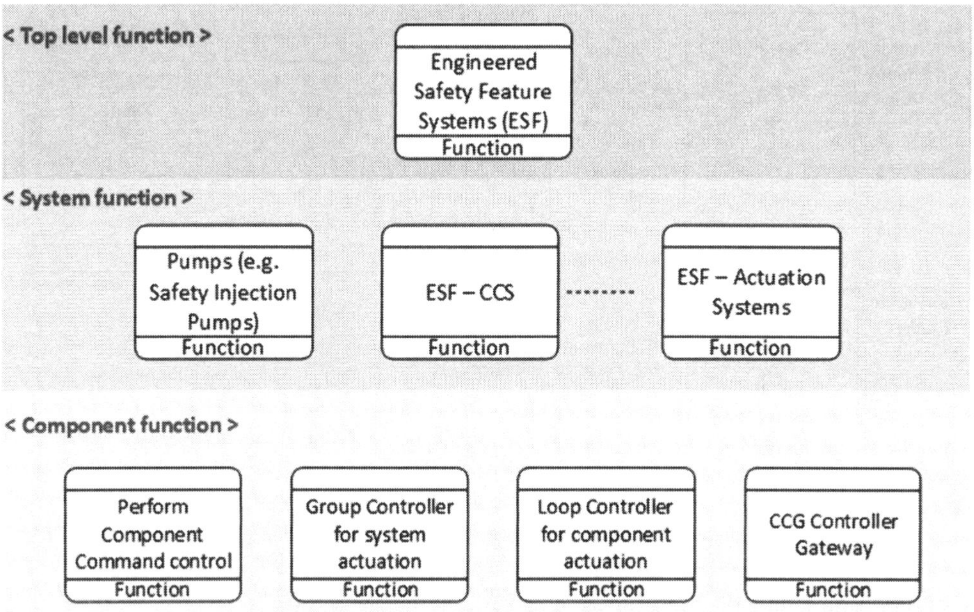


Figure 3-7 Functional decomposition for ESF system function definition

From a systems engineering perspective, a system function defines the transformation of a system input to an output. Therefore, in defining the system function for MMIS system, it is necessary to analyze the input, the processing mechanism, and the expected output. Figure 3-7 shows an ESF-CCS function defining by decomposition.

Figure 3-7 shows the logical representation of I&C function. It should be kept in mind that from systems engineering perspective, a function is transformation of input to output. So, to describe the function of any system we need to define the input, out, and the processing of the input as well as the performance of the function. In Figure 3-8, the inputs to the system could be a command from operators in the control room, these commands are processed by controllers, actuators, etc., and the outputs are in the form of plant performance or other information needed for operator action.

System performance means how well the function should be done. Thus, every function must have performance requirement. By modeling or using diagrams, the systems engineer should discuss the validity of the operational scenario with the system users, designers, and other stakeholders. After the diagram is agreed upon, the systems engineer can derive the function that system must perform.

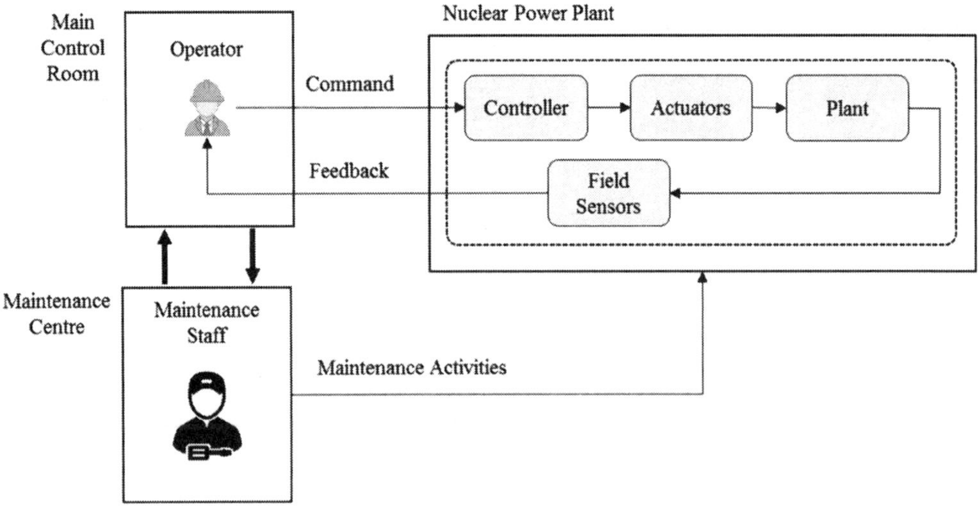


Figure 3-8 Logical Representation of I&C system function

A detailed functional analysis procedure is listed below:

- Identify top level function, then decompose the top-level function define system functions;
- Translate higher-level performance requirements into detailed functional and performance design criteria or constraints. That is, identifying how well the functions have to be performed [17];
- Identify and define all internal and external functional interfaces;
- Identify functional grouping to minimize and control interfaces;
- Define the functional characteristics of existing or directed components in the system and incorporate them in the analyst and allocate;
- Examine all life cycle functions, including the design, develop, operate, maintain, train, disposal as appropriate for the specific project;
- Perform trade study to determine alternative functional approach to meet requirements; and
- Revisit the stakeholder requirements analysis step as necessary to resolve functional issues.

### 3.5.2 Measure of Performance

Measures of Performance (MOP) measures the lowest level of performance of the system and describes distinctively quantifiable performance features of the system under consideration. As discussed earlier, MOEs are of higher hierarchical order than MOPs, thus MOPs are a subset of MOE. The purpose of establishing MOP is to provide detailed performance of the system to ensure that it has adequate capability to attain the expected operational objectives. MOPs are used in collaboration with MOE to ascertain that the system meets the design or performance requirements of the MOE.

In MMIS development, the MOPs are mapped against system level requirements or system specifications and are monitored or used during project execution as inputs to management systems for technical risk management. MOP criteria in MMIS development are used to:

- "Compare alternatives to quantify technical or performance requirements as derived from MOEs; and
- Investigate performance sensitivities to changes in assumptions from the technical viewpoint" [13].

In defining a MOP, it must fit in a specific stakeholders top-level requirement. For example, the lifetime design goal of the availability factor for APR1400 NPP should be above 92%. This means that, when selecting systems to satisfy the overall availability criteria, each systems MOP must meet or contribute to achieve the overall goal of 92%.

### 3.5.3 Definition of System Verification Criteria

According to the definition of ISO 9000, the aim of a system is to satisfy the customer. The qualification means verifying if the specified requirements are met. It is important to perform verification activities early in the development phase of the system. Verification method can enhance the derivation of any requirements to be applied in the design or preparation of the test facility or equipment.

The system is verified based on predefined criteria, MOP and TPM, which are traceable to theme. Generally, verification and qualification methods are classified as:

i. ***Demonstration***: This involves operating the system or function to ensure that it performs its function as intended.

ii. ***Test***: This involves the use of test special instrumentation or equipment. Data is collected and analyzed to verify the system operation.

iii. ***Analysis***: This is the review of data obtained from other verification methods

iv. ***Inspection***: Visual Examination

v. ***Special qualification***: Any special qualification methods

### 3.5.4 Documentation of Design Requirements

Requirement documents should have a common structure, which should be defined as a company standard and should be checked as part of the document quality assurance process. It should encapsulate what the organization thinks is the best way to organize a requirement document. Figure 3-9 shows a typical document structure.

There is no generally agreed template for documenting requirements, but there is some basic information, which should be provided in the requirements document. For example, the document should provide an overview of the system and outline the benefits of developing the system. Also, the service or function requirements for the system must be stated clearly and all non-technical requirements such as reliability and safety must be defined adequately. Constrains on the systems operation and even the system development process should be highlighted in the requirements documentation. The characteristics of a standard requirement document are:

- The standard for requirements document should accommodate the differences / specifics of systems;
- Should make provision for flexibility to omit either parts of the document or to add new sections to the document; and
- Provide introductory page should explain the document and type of variances allowed.

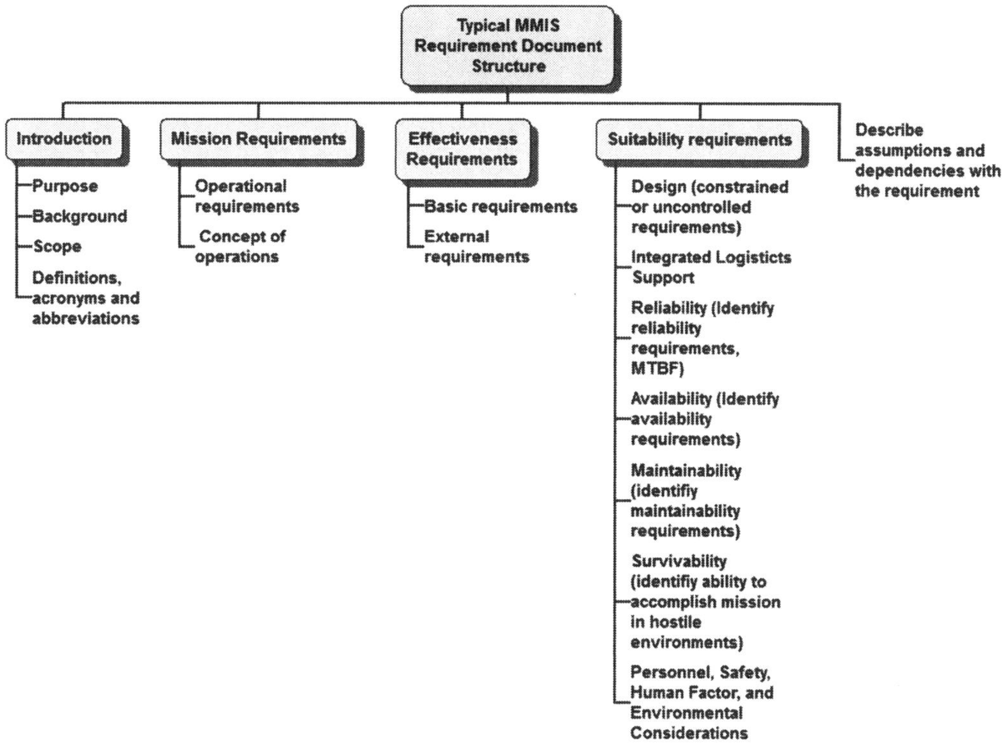


Figure 3-9: Sample Requirement Document Structure for MMIS Development

## 3.5.6 Traceability Analysis

Traceability is one of the important system requirement analysis processes. For effective analysis of the system's requirement, it is necessary to establish and maintain traceability between system elements (for example stakeholder requirements, architecture elements, interface definition, decomposed elements, etc.) [8] and the system requirements. This kind of traceability ensures that all system level requirements and their impacts on the architecture are satisfied in the final design. Table 3-4 shows some of the traceability information that you may want to maintain; and three techniques are often used to maintain traceability information. These are traceability tables, traceability list, and automated traceability links.

Table 3-4 Types of Traceability

| Traceability Type | Description |
|---|---|
| Requirements-sources traceability | This traceability identifies the source of the requirements. The source could be the stakeholder requirements or a system mission statement etc. |
| Requirements-rationale traceability | This traceability identifies why the requirement was included/specified. |
| Requirements-requirements traceability | This traceability involves identifying related requirements and linking them |
| Requirements-architecture traceability | Once the system architecture is defined, this type of traceability links the requirements to the sub-systems in which they will be implemented. |
| Requirements-design traceability | This traceability allocates the requirements to the components in which they are implemented. The components may be either hardware or software |
| Requirements-interface traceability | This traceability links the requirements to the system interfaces. |

### Traceability Tables

"Traceability tables show the relationships between requirements or between requirements and design components. The requirements are listed along the horizontal and vertical axes and relationships between requirements are marked in the table cells. They can be implemented using word processor or spreadsheet tables; a requirements database is not necessary" [14].

Traceability table for showing requirements dependencies should be defined with requirement numbers used to label the rows and columns of the table. In Table 3-5, the Safety System Criteria from chapters 5, 6 and 7 of IEEE 603 are traced to the lifecycle requirements defined in IEC 61513 Chapters 5 and 6. This table shows an example of traceability matrix and where there are some kind of relationship between the cell row and column the symbol (✓) is placed in that cell to signify existences of dependency. By reading down a column, you see all the requirements that have dependency; by reading across a row, you see all requirements that are depended on.

Table 3-5: Example of traceability table

| | | Sample traceability matrix for IEC 61513- general requirements for I&C safety systems | | | | | | | | | | |
|---|---|---|---|---|---|---|---|---|---|---|---|---|
| | | 5. Overall I&C safety life cycle | | | | | | 6. System safety life cycle | | | | |
| | | 5.1 General | 5.2 Deriving the I&C requirements from the plant safety design base | 5.3 Output documentation | 5.4 Design of the overall I&C architecture and assignment of the I&C functions | 5.5 Overall planning | 5.6 Output documentation | 6.1 General | 6.2 Requirements | 6.3 System planning | 6.4 Output documentation | 6.5 System qualification |
| 5 | **Safety system criteria** | | | | | | | | | | | |
| 5.1 | Single-failure criterion | | | | | ✓ | | | ✓ | | | |
| 5.2 | Completion of protective action | | | ✓ | | | | | | | | |
| 5.3 | Quality | | | | ✓ | ✓ | | | ✓ | ✓ | ✓ | ✓ |
| 5.4 | Equipment qualification | | | | | | | | | | | |
| 5.5 | System integrity | | | | | | | | | | | |
| 5.6 | Independence | | ✓ | ✓ | ✓ | | | | ✓ | ✓ | | |
| 5.7 | Capability for test and calibration | | | | | ✓ | | | ✓ | ✓ | ✓ | |
| 5.8 | Information displays | | | | | | | | | | | |
| 5.9 | Control of access | | | | | | | | | | | |
| 5.1 | Repair | | | | | | | | ✓ | | | |
| 5.11 | Identification | | | | | ✓ | | | ✓ | | ✓ | ✓ |
| 5.12 | Auxiliary features | | | | | | | | | | | |
| 5.13 | Multi-unit stations | | | | | | | | | | | |
| 5.14 | Human factors considerations | | ✓ | | ✓ | | | | | | | |
| 5.15 | Reliability | | ✓ | | ✓ | ✓ | ✓ | | ✓ | ✓ | ✓ | ✓ |
| 5.16 | Common cause failure criteria | | | | | | | | ✓ | | | |
| 6 | Sense and command features—functional and design requirements | | ✓ | | | | | | ✓ | ✓ | | |
| 7 | Execute features—functional and design requirements | | ✓ | | | | | | ✓ | ✓ | | |

Possible relationships between requirements, which might exist, are defined by [14] as follows:

i. ***Specifies/is-specified-by***: This relationship indicates that some requirement "B" adds detail to another requirement "A". For example, if "A" is a general security requirement stating that data should be encrypted, "B" might specify the characteristics of the encryption algorithm that should be used.

ii. ***Required/is-required-by***: This relationship indicates that some requirement "B" requires the result provided by some other requirements "A". For example, "A" might specify that the system should maintain a record of the current time and date in some specific format; "B" might specify that each transaction processed by the system should be data stamped.

iii. ***Constrains/is-constrained-by***: This relationship indicates that some requirement "B" is constrained by some other requirement "A". For example, "B" might specify that some real value should be displayed and "A" might state that all real numbers should be implemented as 12-digit fixed-point number.

## 3.6 System Architecture and Design Analysis

The core of systems engineering is developing architecture of the system. This means stakeholders requirements are flown down to system requirements and the system requirements are flown down to subsystem requirements. Figure 3-10 shows the flow down of requirement and the relationship with the requirements and the physical hierarchy of the NPP. The concept of flow down is the basis of requirements traceability.

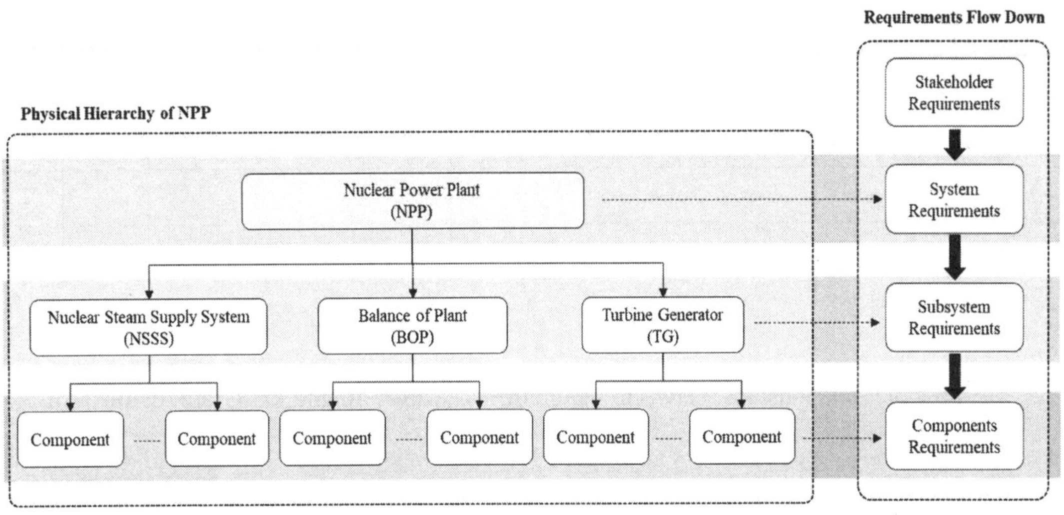


Figure 3-10: Requirements flow down

### 3.6.1 Trade-off Analysis

Generally, trade-off analysis is the process of evaluating various alternatives and solutions and making a decision on what to give up to achieve the intended purpose of the desired system. We can therefore say that trade-off analysis is a decision-making process for selecting the best alternative solution for various requirements. During the system architecture definition process, alternative architectures are created through several models. To evaluate the properties of these alternatives to select the appropriate system elements there must be some form of flexibility in the design. An effective architecture therefore must have this flexibility and support trade-offs for the design definition and analysis. After performing the trade-off analysis, the workable alternate architectures must fulfill all required features of the system.

PART 1

A common trade-off analysis during architecture design is either to have enabling items inside or outside the system. Most often, the decision to have the enabling system either inside or outside the system is decided during the architecture definition process. However, for effective architecture analysis, it is appropriate to make this decision during the definition stage since it depends on design trade-offs and design decisions that are made throughout the system development. After completion of the trade-off analysis, the design decisions and trade-offs must be analyzed to ensure that they do not negatively impact the architecture [8].

### 3.6.2 Technical Performance Measure

Technical performance measures involve two main considerations; these are Measures of Suitability (MOS) and the Operational Effectiveness of the system. MOS assesses the appropriateness of the system to the organization and its intended application. Whereas operational effectiveness answers the questions of whether the system is operationally effective in terms of cost, technical performance, and satisfy its intended mission application and objectives [7].

### 3.6.3 Documentation of Design Specification

One important question, which often arises, is the difference between design requirements and design specifications. Design requirements document defines the system level requirements for a system. The document is typically produced by the system designer during the initial stage of the design process and used throughout the design process. On the other hand, the design specification is used by the component designer for the system and provides more detailed fabrication requirements. Figure 3-11 shows an example of the specifications document structure.

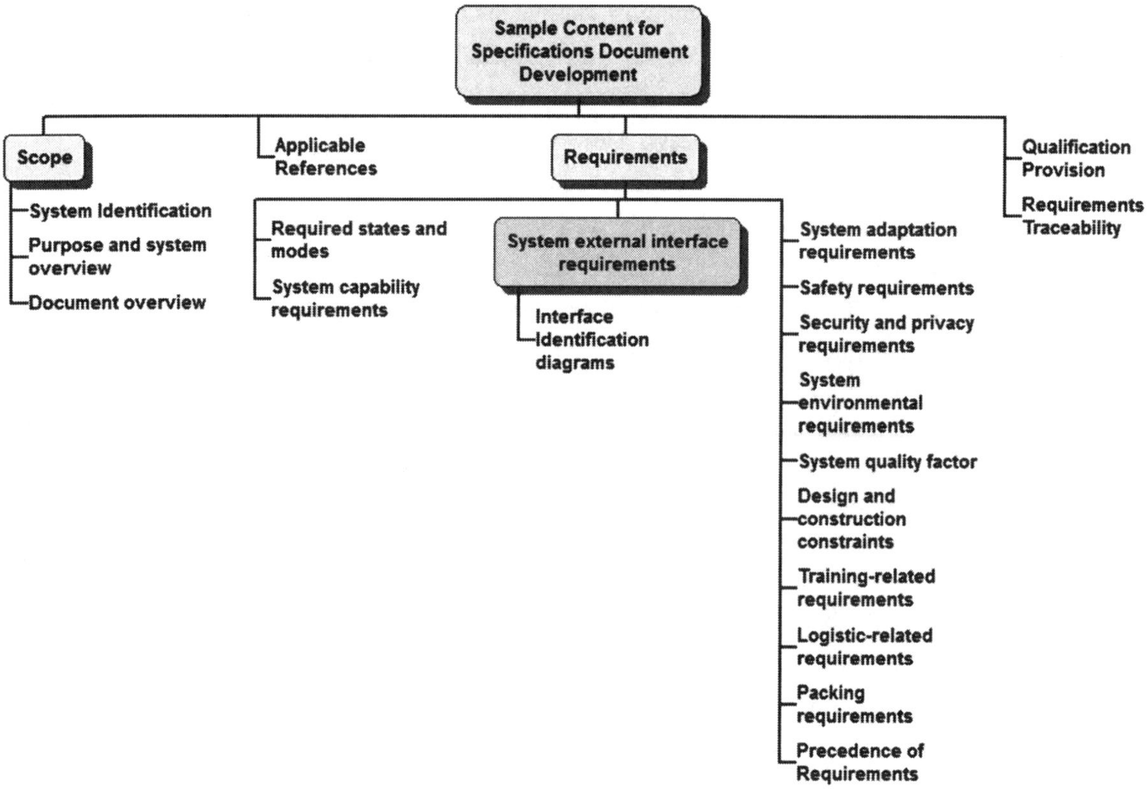


Figure 3-11 Sample Content structure for specifications document

## 3.7 System Analysis

System analysis and decision management are an integral part of the technical process for any system of interest because system analysis provides a detailed understanding of the possible risk associated with the system and this information is useful in technical assessment. Therefore, to make an efficient technical decision on the system development, a system analysis process must be performed. Usually, the analysis process involves mathematical modeling of the system of interest and simulation analysis.

We should keep in mind that modeling is a mathematical representation of the real system and models cannot simulate all the behaviors of the system especially the environmental influences. This is a general constraint in modeling and therefore, when system models are being used for system analysis it is necessary to ensure that parameters and variables used are field data otherwise the results would have more deviations from actual.

In addition, the fundamental purpose of system analysis is to facilitate decision making across the lifecycle of the system of interest. Results obtained from the system analysis process are usually quantitative, representing either technical risk indexes, effectiveness, cost indexes, quality characteristics, etc. These results serve as main inputs to various technical decision processes to ensure that system development satisfies all requirements and possesses the synergy between all the facets of system development. A typical IPO for system analysis from the INCOSE handbook is shown in Figure 3-12.

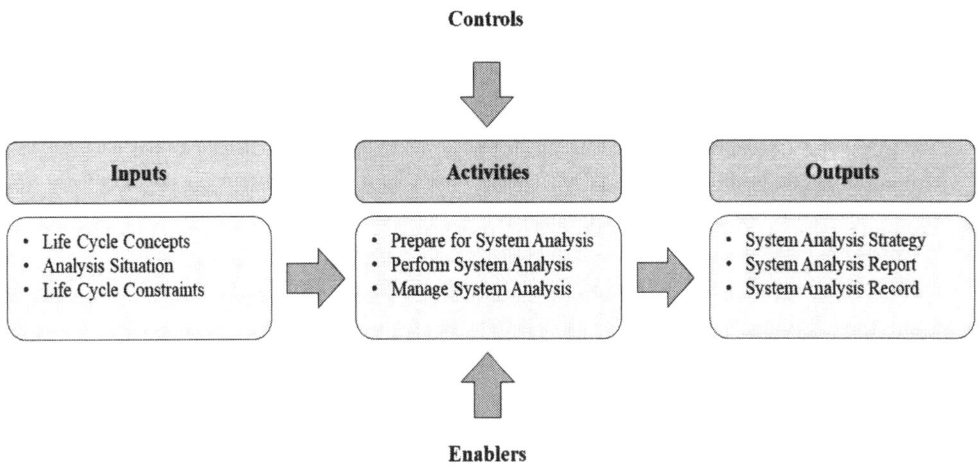


Figure 3-12 System analysis process [8]

## 3.7.1 Specialty Engineering Techniques (RAM)

Reliability, Availability and Maintainability (RAM) of a system is crucial in the system analysis process. The system of interest must perform all required functions under stated operating conditions throughout its life. Thus, reliability analysis must be performed to ascertain the risk of failure of the system during its lifecycle. By performing failure rate analysis, the system availability indexes can be computed and appropriate maintenance requirements or procedures should be provided in the system architecture design analysis. RAM is not a system characteristic but rather nonfunctional requirement and the activities involved in performing RAM should be integrated with other technical process.

For the MMIS development process, Failure Mode Effects Analysis (FMEA) will be discussed for the system of interest to determine the various modes of failure for the system, subsystem, and components; the mechanism that leads to the failure of this systems and how the failure affects the overall system performance. FMEA is a qualitative approach and for quantitative analysis Failure Mode Effects and Criticality Analysis (FMECA) is often performed.

### 3.7.2 System Modeling

System modeling is the "process of developing abstract models of a system, with each model presenting a different view or perspective of that system" [14]. System modeling is typically understood to mean illustrating the system using some kind of graphical notation, such as the Unified Modeling Language (UML). Modeling enables the functionality of the system to be clearly understood by the designers, users and all stakeholders.

System modeling can serve throughout the system lifecycle to define all aspects about the SOI such as:

- Characterizing an existing system
- Formulation and analysis of the system concept and objectives
- System architecture design and requirements allocation;
- Support for systems integration and verification;

Different types of models are needed to support the analysis, specification, design, and verification of systems. System Modeling languages (SysML) are generally intended to be both human interpretable and computer interpretable and are specified in terms of both syntax and semantics. A partial list of representative modeling standards can be found under the modeling standards section of SEBoK (2014) [18].

Figure 3-13 shows the SysML diagram types that can be used to represent systems.

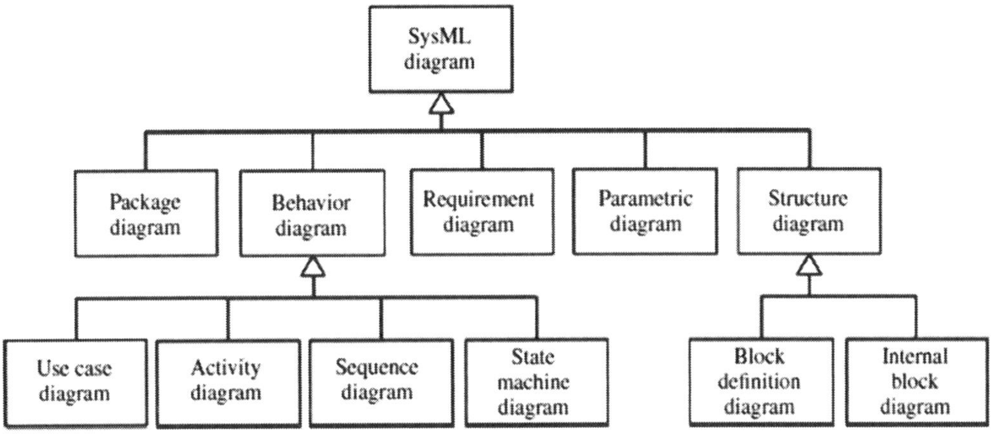


Figure 3-13 SysML diagram types

### 3.7.3 System Simulation

The term "simulation" is the "implementation of a model (or models) in a specific environment that allows the model's execution (or use) over time. In general, simulations provide a means for analysing complex dynamic behaviour of systems, software, hardware, people, and physical phenomena" [18]. Figure 3-14 illustrates the different types of simulation.

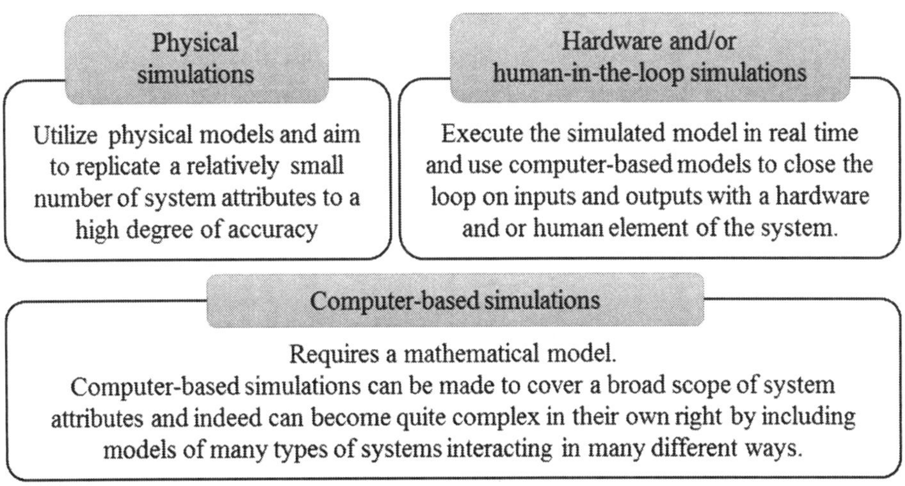


Figure 3-14 Simulation Types

### 3.7.4 System Prototyping

Prototyping is the development of a preliminary version of a system or subsystem with a degree of functionality that is comparable to that of the final system product. The development of prototypes can increase likelihood of the system meeting the user's needs.

There are two types of prototyping defined by INCOSE [8]:

- Rapid prototyping is a particular type of simulation quickly assembled from a menu of existing physical, graphical, or mathematical elements.
- Traditional prototyping is used to verify critical elements of the SOI. A full prototype is a complete representation of the system. It must be complete and accurate in the aspects of concern. Objective and quantitative data on performance times and error rates can be obtained from these higher-fidelity interactive prototypes.

### 3.7.5 Application of System Analysis

As mentioned in section 3.6, system analysis is used to perform the evaluation of technical choices. There are multiple modeling techniques that may be used for system analysis depending on the aspect being analyzed and the stage of the system lifecycle. The different types of models are physical models, structural models, behavioral models, functional models, cost models, probabilistic models, simulations, mathematical models, and prototypes. Multiple models may be used during the lifecycle of a single system. Table 3-6 shows some of models that can be used for different level of application.

Table 3-6 Application Level on System Analysis

| Application level | Modeling Technique |
|---|---|
| Mission and business analysis | Internal Block Diagram, Block Definition Diagram |
| Stakeholder requirements<br>System Requirements | Use case Diagram<br>Enhanced Functional Flow Block Diagram (EFFBD), State Machine Diagram, Activity Diagram, Sequence Diagram |
| Architectural definition process and design definition process | Activity Diagram, Sequence Diagram, Enhanced Functional Flow Block Diagram |
| Integration process, | State Machine Diagram, Enhanced Functional Flow Block Diagram |
| Verification process, and Validation process | Traceability Diagram and Analysis |

## PROBLEMS

1. What is the difference between the Concept of Operations and the Operational Concept?
2. Who are the key stakeholders for the NPP?
3. Why is requirements traceability important?
4. What are the attributes of a good requirements statement?
5. What is the difference between upward and downward requirements traceability?
6. Select a system in the MMIS. For this system define possible
   a. Measures of Effectiveness
   b. Measures of Performance
   c. Technical Performance Measures
7. Which method do you think would be most efficient for the elicitation of stakeholder requirements in NPP systems? Give reasons why.
8. Trade-off analysis can be used in the selection of a system architecture. Give an example using an MMIS system.
9. What are the benefits of system modeling?
10. For a single MMIS system, select a behavioral modeling technique and develop it for the chosen system?

## FURTHER READING


- INCOSE, Systems Engineering Handbook, 4th ed., Wiley, 2015.
- INCOSE, Technical Measurement, INCOSE-TP-2003-020-01
- B. S. Blanchard, System Engineering Management, Wiley.
- C. S. Wasson, System Analysis, Design, and Development.,Wiley, 2006.


PART 1

# Chapter 4 SAFETY CRITICAL MMIS

## SUMMARY

The objective of this book is to illustrate how the systems engineering approach can be applied to the development of the Instrumentation and Control systems. To achieve this end, the operation of the I&C systems must be clearly defined. In addition, the concepts that are integral to the development of these systems such as Defense in Depth and Redundancy should be understood. In this chapter, the safety critical systems are introduced. Their role in the protection of the Nuclear Power Plant is defined, as well as the criteria that are incorporated in their design. In addition, as these systems work together, their interfaces are also discussed. The chapter is structured as illustrated in the mind map below(Figure 4-1).

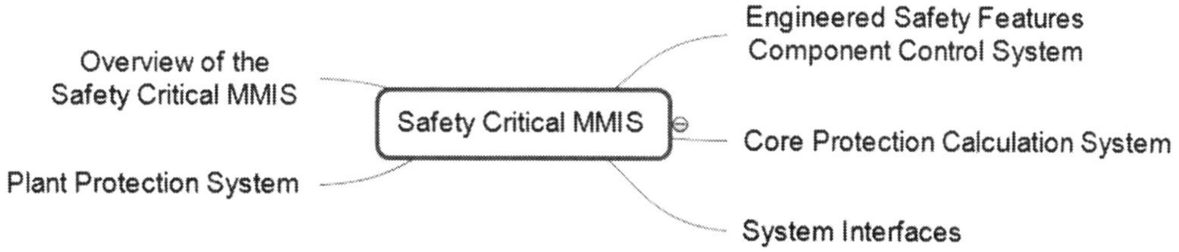


Figure 4-1 Mind Map of Chapter 4

**OBJECTIVES**

- Introduction of the safety critical systems and their functions
- Definition of the system interfaces
- Description of the system flow from operator to actuated equipment

## 4.1 Overview of the Safety Critical Systems

According to the NRC, Defense in Depth is defined as "An approach to designing and operating nuclear facilities that prevents and mitigates accidents that release radiation or hazardous materials." Defense in Depth is achieved by designing a system with multiple independent layers of defense so that the failure (software, mechanical or human) of a single layer does not prevent the execution of a safety function. Defense in depth includes the use of access controls, physical barriers, redundant and diverse key safety functions, and emergency response measures.

Multiple barriers are designed to make sure that in the event of an accident, the release of radioactive material is prevented, mitigated or contained [19]. The Swiss cheese model illustrates in Figure 4-2 illustrates how multiple layers of defense can be utilized. If one defensive barrier fails to eliminate or mitigate a risk/ accident, the next barrier is utilized. For an accident to occur, multiple layers of defense must fail. The defensive barriers are designed such that the probability of multiple failures (all the holes in the Swiss cheese aligning) is extremely low.

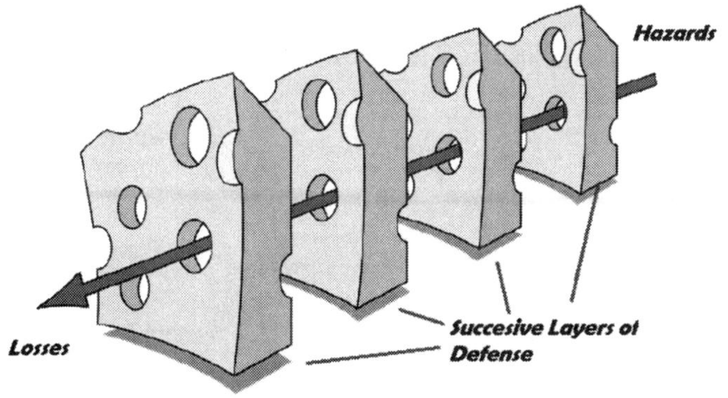


Figure 4-2 Swiss cheese Model [19]

P A R T 1

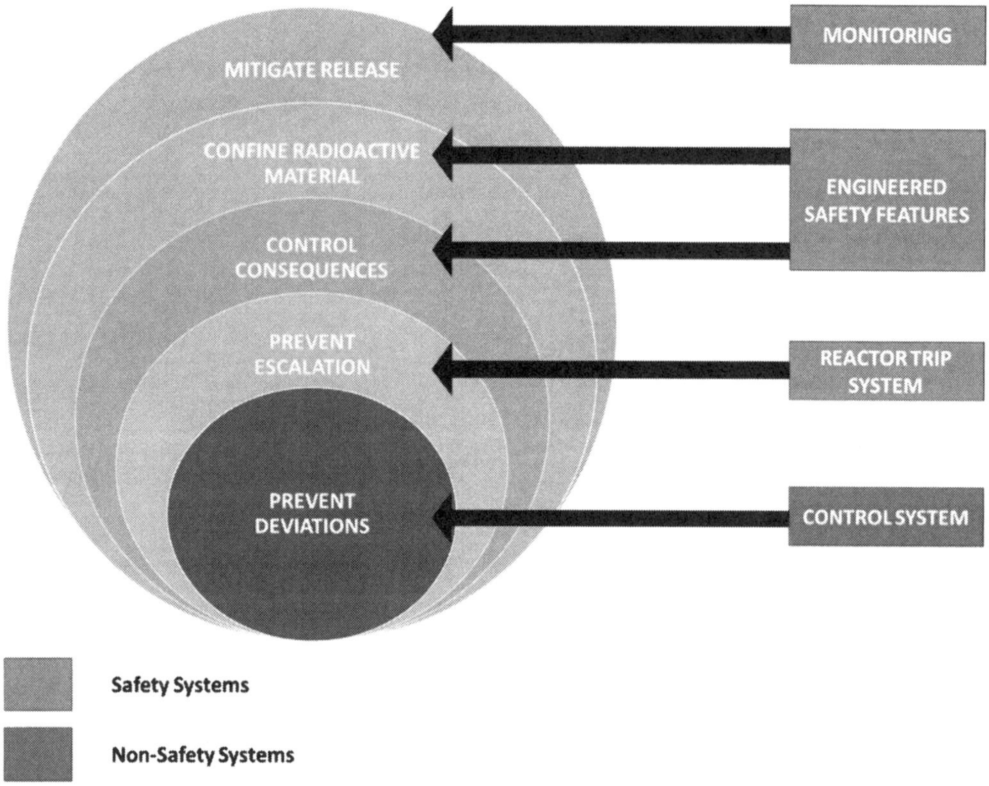


Figure 4-3 I&C systems relationship to plant defense

In the NPP, the safety critical systems complement other safety features within the reactor so that many independent errors must occur before an accident happens. Figure 4-3 illustrates how the different I&C systems operate to provide multiple levels of defense in the NPP, should an accident occur.

The safety systems are of particular importance as they are responsible for shutting down the reactor to prevent further escalation of an accident. When an accident or hazardous state is detected by the safety systems, based on sensor input and predefined set-points, these systems trip the reactor to reduce the impact of the accident. In addition, the safety systems actuate the Engineered Safety Features (ESF) in order to reduce/mitigate the consequences of a postulated accident and to confine all radioactive material within the containment.

Figure 4-4 shows how different limits for the I&C systems are set so as to ensure that corrective action is taken before the plant enters the severe accident space. The following terms are defined as:

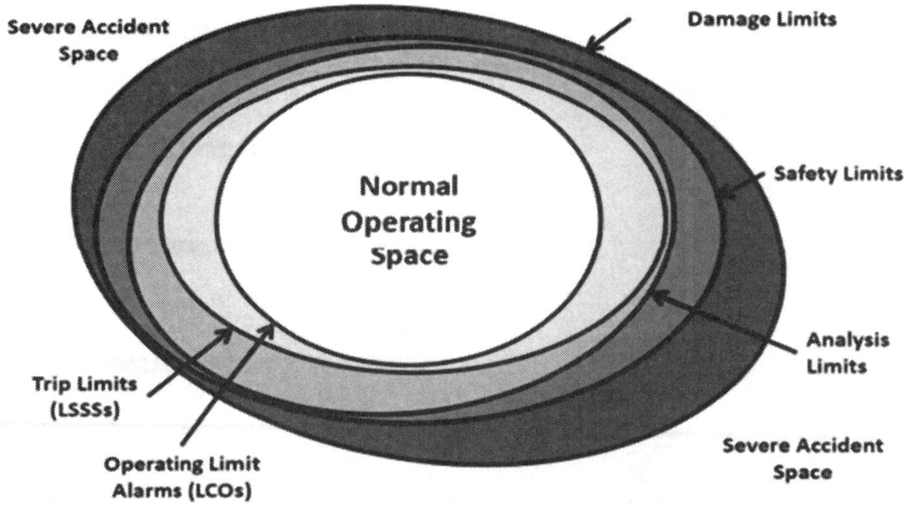


Figure 4-4: Limits for the I&C System

- **Limiting Condition for Operation**: The lowest functional capability or performance level of equipment required for safe operation of the facility [20]
- **Limiting Safety System Setting**: The threshold for protective system action to prevent acceptable limits being exceeded during design basis accidents. The LSSS ensures that automatic protective action will correct the abnormal situation before a safety limit is exceeded [21]
- **Analytical limit**: Limit of a measured or calculated variable established by the safety analysis to ensure that a safety limit is not exceeded.
- **Safety Limit**: A limit upon important process variables to reasonably protect the integrity of certain of the physical barriers that guard against the uncontrolled release of radioactivity [22]

From the diagram, it can be observed that the limit at which plant damage occurs (Damage Limit), is far removed from the trip limits, i.e. the set-points used to shut down the plant. Further, the trip set-points are also set below the safety limits and analysis limits. By so doing, whenever an accident occurs, the reactor is tripped before the plant can enter into an un-safe state.

In the APR1400, the safety systems are as follows:

- Plant Protection System (PPS) - Responsible for tripping the reactor once the Trip Limit is exceeded, thus shutting down the plant

- Core Protection Calculation System (CPCS) – This system monitors the core condition and generates a generates causes the reactor to trip when the trip limits are exceeded
- Engineered Safety Features Component Control System (ESF-CCS) This system actuates the Engineered Safety Features preventing the plant from exceeding the safety limit, and preventing the release of radioactive material
- Qualified Indication and Alarm System – Post Accident Monitoring Instrumentation (QIAS-P) - are necessary for monitoring plant conditions and safety system performance, and deciding preplanned operator actions in response to events [23]

In this chapter, the structure and functions of the PPS, ESF-CCS, and CPCS are discussed. The QIAS-P is beyond the scope of this book.

## 4.2 Plant Protection System

The severity of a reactor accident depends on the extent of fuel damage, the extent of any release of radioactive material from the Reactor Coolant System (RCS) boundary, and the extent of any release of this radioactive material to the environment where it then threatens the health and safety of the public. The severity of this accident can be limited by; preventing or minimizing the extent of fuel damage during an accident, another is to contain the radioactive releases so that it is not released to the environment.

The Plant Protection System (PPS) is designed to sense abnormal occurrences and/or accidents in the reactor plant and to initiate automatic actions to place the plant in a safe condition to maximize the capability of plant systems to maintain the integrity of the three fission product barriers. In addition, its purpose is to automatically generate signals for reactor trip (RT) initiation and/or engineered safety features actuation, whenever monitored process exceeds predefined limits. It does it by monitoring the plant temperatures, pressure levels, flows, and nuclear instrumentation outputs. Figure 4-5 indicates PPS black diagram.

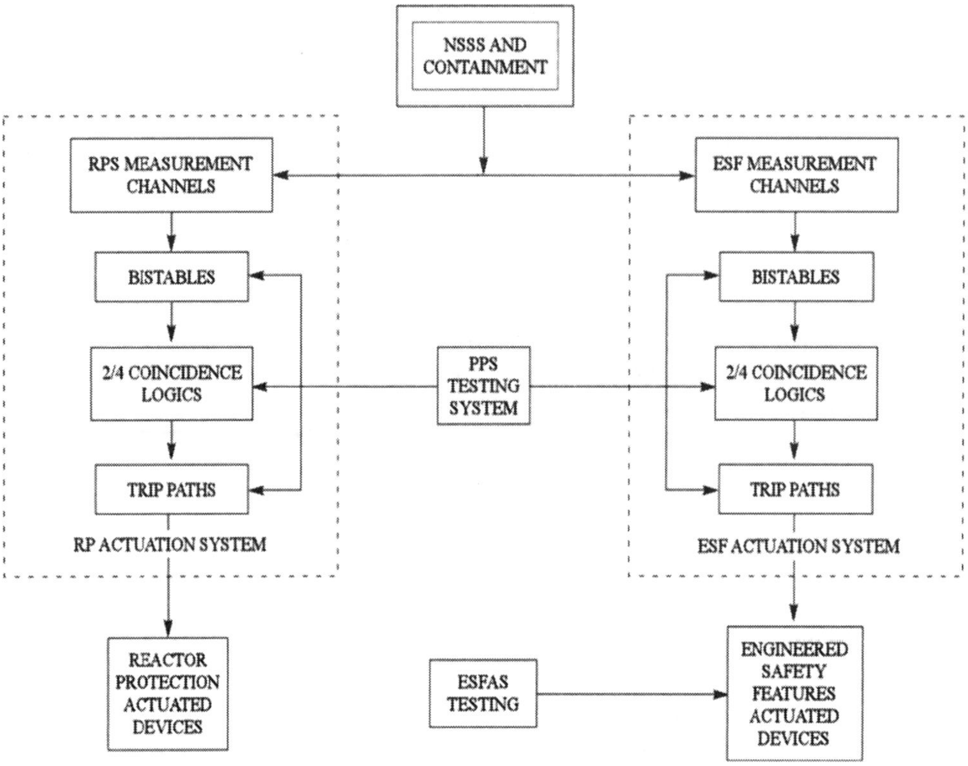


Figure 4-5 Plant Protection System Block Diagram

The PPS has the following main functions:

- Reactor Protection System (RPS) Function
- Engineered Safety Features Actuation System (ESFAS) Function
- Control Function
- Alarm Function
- Bypass Function
- Test Function
- Interlock Function
- Operator Module Function
- MTP/ITP Function

### (1) Reactor Protection System (RPS)

The RPS functions "protect the core fuel design limits and the RCS pressure boundary following Anticipated Operational Occurrences (AOOs) and provide assistance in mitigating the consequences of postulated accidents" [24][25]. Postulated accidents in the sense that the offsite radiation dose is within the limits provided by 10CFR100.

The RPS variables are the measured variables from RPS sensors, which are measured with four independent and redundant process instruments. These variables are continuously monitored before being transmitted to the RPS logic.

- ***Process measurements***: Monitored process parameters such as pressures, levels, and temperatures are input to the Bistable Processors. Figure 4-6 and Table 4-1 show the process protective parameters and the RPS sensors that carry out these measurements.
- ***Control Element Assembly (CEA) Position measurements***: The CEA are used to control the reactivity of the core, based on their level of insertion into the core. The position of these elements is input to the CPCS.
- ***Ex-core neutron flux measurements***: They provide neutron flux information from near start up neutron flux levels (2 x $10^{-8}$ percent) to 200 percent of rated power. It consists of four redundant safety channels.
- ***Reactor coolant flow measurements***: The Reactor Coolant Pump(RCP) Flow is computed based on the speed of each pump.
- ***Core protection calculators***: There are four CPC, one per channel. The outputs from the CPC are the high Local Power Density (LPD) and low Departure from Nucleate Boiling (DNBR) signals.

The figure 4-6 shows the separation of signals between the CPCS, RPS, and ESFAS.

The RPS carries out two types of logic: Bistable logic and Local Coincidence Logic (LCL). The Bistable Logic receives the process variables and compares them to predefined set-points. If these variables exceed the set-points, the bistable logic generates trip signals which are transmitted to the Local Coincidence Logic. The LCL receives signals from all channels (4 redundant channels A, B, C and D). It performs selective two-out-of-four (2oo4) logic in order to generate trip signals. The 2oo4 logic is performed to ensure that the reactor is not tripped by spurious signals. Therefore, the Bistable Logic in at least two channels must

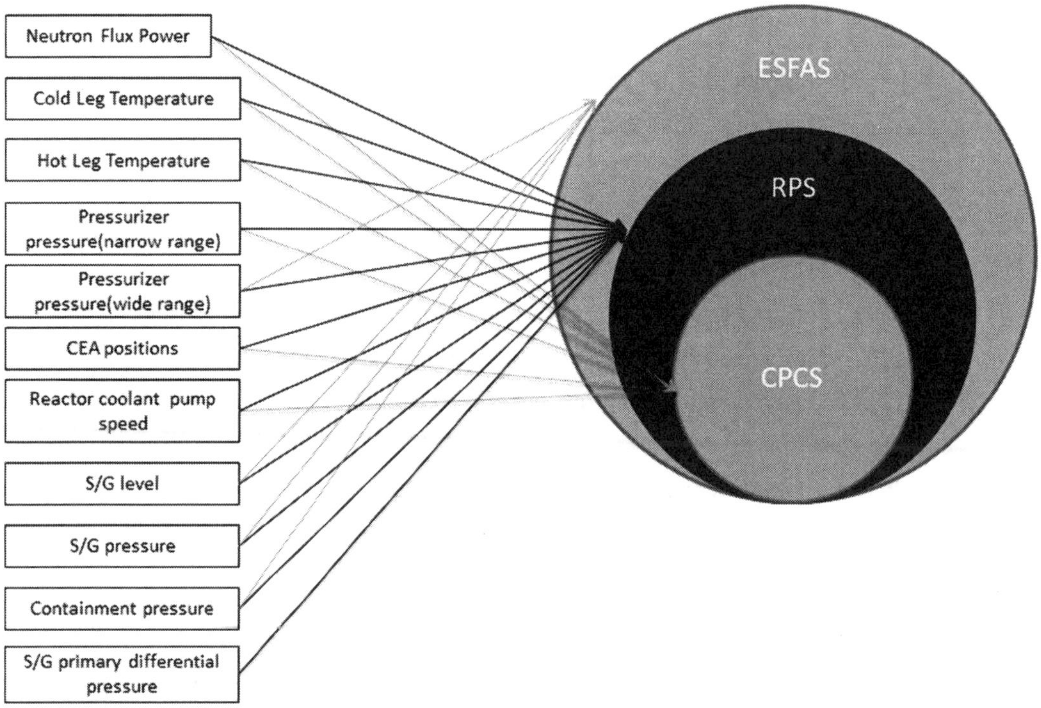


Figure 4-6 Separation of signals for RPS, ESFAS and CPCS

generate a signal for the reactor to be tripped.

Once trip signals are generated by the LCL, they are transmitted to the initiation circuit. The initiation circuit is composed of initiation relays, interposing relays, contacts from the manual initiation switches and wirings. This circuit, on receiving a trip signal, disconnects the power supply to the control rods, causing them to drop, thus tripping the reactor.

Detailed trip description and set-point information is provided in the APR1400 DCD Chapter 7 [27].

Figure 4-7 shows the functional block of the RPS trip logic. The RPS monitors plant parameters such as reactor power, pressurizer pressure, steam generator water levels, and Reactor Coolant System (RCS) temperatures. In addition, the PCS receives input signals from the CPCS (DNBR and LPD trip and pretrip signals). The Bistable comparator then compares the input signals to predefined set-points. If the plant variables exceed the set-point (for increasing process variable) or if the variable falls below the set-point (for a decreasing process variable) a trip signal is generated.

The trip signal is then transmitted to the Local Coincidence Logic (LCL) module. The LCL receives signals from all four channels, and performs 2 out of 4 coincidence logic. If a valid trip signal is received for a given trip parameter from at least two channels, a reactor trip signal is generated. The 12 trip signals are combined (logical OR) and input to the Reactor Trip Circuit.

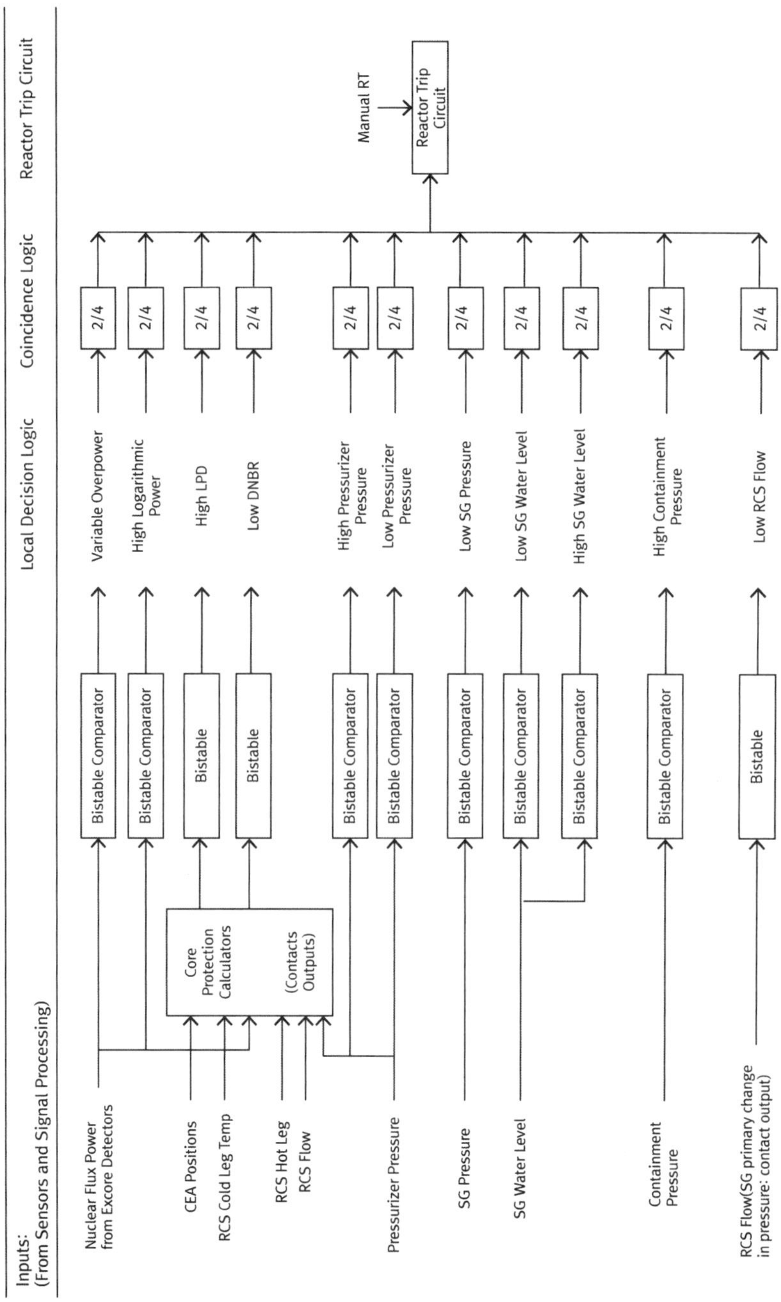


Figure 4-7: Reactor Trip Logic

### (2) Engineered Safety Features (ESF) Actuation System

The Engineered Safety Features Actuation System (ESFAS) and associated Engineered Safety Features (ESF) systems are designed to ensure that accident consequences are kept within acceptable limits. The Figure 4-8 below shows the constituents of the ESF system.

Figure 4-8 ESF System

The ESFAS signals activate various ESF components such as pumps, valves, and fans, unlike the RPS that sends its signals to the RTSG.

In the advanced power reactor APR1400, the ESF are: [26]

i. Safety Injection System (SIS) for the injection of borated water into the reactor vessel for core cooling and diverse reactivity control.
ii. Containment isolation system (CIS) which provides automatic isolation of the containment which prevent uncontrolled fission product releases.
iii. Containment spray system (CSS) for reducing containment pressure and temperature
iv. Safety depressurization and vent system, which provides depressurization of the RCS and post-accident venting of non-condensable gases from the RCS.
v. Auxiliary Feedwater system (AFWS).
vi. Combustible gas control system for hydrogen control in the containment building.
vii. Heating, Ventilation, and Air Conditioning (HVAC) system.

PART 1

The ESFAS signals are generated by the bistable processor logic and the LCL in a similar manner to the reactor trip signals. As seen in Figure 4-11, the RPS and ESFAS functions utilize the same sensors and bistable processors. However, once a pretrip/trip signal is generated by the bistable processor, the LCLs which process the ESFAS signals are different from those of the RPS. On receiving a trip signal from at least two channels, the ESFAS LCLs generate their respective ESFAS signals (e.g. SIAS or MSIAS) which are then transmitted to the ESF-CCS.

### (3) Other PPS Functions

In addition to tripping the reactor (RPS) and generating ESF actuation signals (ESFAS), the PPS performs the following functions:

i. Control Function
ii. Alarm and Indication Function
iii. Bypass Function
iv. Test Function
v. Interlock Function
vi. MTP/ITP Function
vii. Operator Module Function

#### ① Control Function

The control function deals with the generation of CEA Withdrawal Prohibit (CWP) signals, which are transmitted to the DRCS to prevent the withdrawal of the CEAS.

The PPS performs 2oo4 coincidence logic (for CPC CWP or High Pressurizer Pressure Pretrip) and transmits the resultant signal to the DRCS to prevent withdrawal of the control rods.

#### ② Alarm and Indication Function

Reg. Guide 1.47 and IEEE 603 require that alarms be provided to indicate system abnormalities. In order to meet these requirements, the PPS generates the following alarms/ indications:

- Pretrip/Trip display
- Alarms to the QIAS-N and IPS
- Sequence of Events (SOE)

**③ Bypass Function**

Bypass circuits and signals are used to disable system trips for the purposes of reactor start-up, shutdown, testing, and maintenance. IEEE 603 defines two types of bypasses:

i. Maintenance bypass: Removal of the capability of a channel, component, or piece of equipment to perform a protective action due to a requirement for replacement, repair, test, or calibration

ii. Operating bypass: Inhibition of the capability to accomplish a safety function that could otherwise occur in response to a particular set of generating conditions. Different modes of plant operation may necessitate an automatic or manual bypass of a safety function.

In the APR1400, bypasses are categorized as either trip channel bypasses or operating bypasses. The trip channel bypass is equivalent to the maintenance bypass defined by IEEE 603.

***Trip Channel Bypass***

A trip channel bypass allows manual bypass of any combination of system parameters on one (and only one) channel at any given time. Trip channel bypasses are provided to remove a trip channel from service for purposes of maintenance or testing. These bypasses are manually initiated and manually removed at the PPS cabinet [20].

***Operating Bypass***

Operating bypasses are provided to permit orderly startup and shutdown of the plant and to allow low power testing [5].

④ **Test Function**

The PPS is capable of performing both manual and automatic testing of the PPS during normal operation or when the reactor is shut down. These tests cover the trip actions from sensor input through the reactor trip function [5]. To achieve this, the channel under test is bypassed and interlocks generated to ensure only one channel is tested at a given time.

⑤ **Interlock Function**

The PPS has the following interlock capability:

i. Trip Channel Bypass Interlock: Interlocks ensure that the operator can only bypass a given trip parameter in only one channel. E.g High Pressurizer pressure can be bypassed in Channel A but not Channels B, C, or D at the same time. However, different parameters, High Pressurizer Pressure and High Containment Pressuure can both be bypassed in Channel A.
ii. Manual Test Interlock: During system testing, an interlock allows only a single test function to be operated at a given time or cause that test function to assume a trip condition.
iii. CPCS Test Interlock: The low DNBR and high LPD channel trips are interlocked such that they must be bypassed in order to test the CPCS
iv. CEA withdrawal prohibit interlock: The PPS generates CWP signals for the DRCS when the CPC CWP or High Pressurizer Pressure Pretrip meet 2-out-of-4 coincidence logic. This signal prevents withdrawal of the CEA.

⑥ **MTP/ITP Function**

The MTP is the primary local PPS HMI used to monitor PPS status, perform PPS control functions such as the, insertion of bistable trip channel bypasses, modification of set-points and resetting of ESFAS initiation signals. By using the MTP, the operator is able to perform surveillance, corrective maintenance, initiate tests, and display detailed system diagnostic messages [24].

**⑦ Operator Module (OM) Function**

The OM provides an interface between the operator and the PPS, allowing the transmission of commands and the receipt of feedback/ information from the MCR. Through the OM, the operator is able to perform functions such as setting bypasses and the re-setting of trip set-points. In addition, information such as trip, pretrip, bypass status etc. is displayed on the OM for use by the operator.

## 4.3 Engineering Safety Features Component Control System(ESF-CCS)

In an NPP, dedicated structures, systems, and components (SSC) which provide protection during an accidental release of radioactive materials from the RCS boundary, as a result of design basis accidents, are referred to as Engineered Safety Features (ESF). These features are provided to mitigate the consequences of events such as LOCA or Main Steam Line Break by ensuring long-term core cooling, maintaining the containment integrity, and by controlling the release of radioactive substances. The ESF consists of core cooling systems, containment systems, fission product removal and control systems, and habitability systems [26].

The Component Control System (CCS) of the ESF (ESF-CCS) provides actuation and control of plant process components such as valves and pumps for performing ESF functions. ESF-CCS therefore serves as a control interface between ESF actuation signal (ESFAS) that emanates from the PPS and the associated field actuation devices connected through the CIM. The ESF-CCS receives ESF initiation signals from a variety of control sources including soft controls and manual switches from the MCR, RSR, PPS, and the RMS. Based on these signals, it performs coincidence logic (selective 2-out-of-4 coincidence logic for NSSS and 1-out-of-2 for BOP) to automatically actuate plant ESF systems and perform the emergency diesel load sequencing.

The ESF-CCS is a channelized safety system, which is made up of four electrically, and physically independent channels. Figure 4-9 illustrates the main components and interconnections in the ESF-CCS.

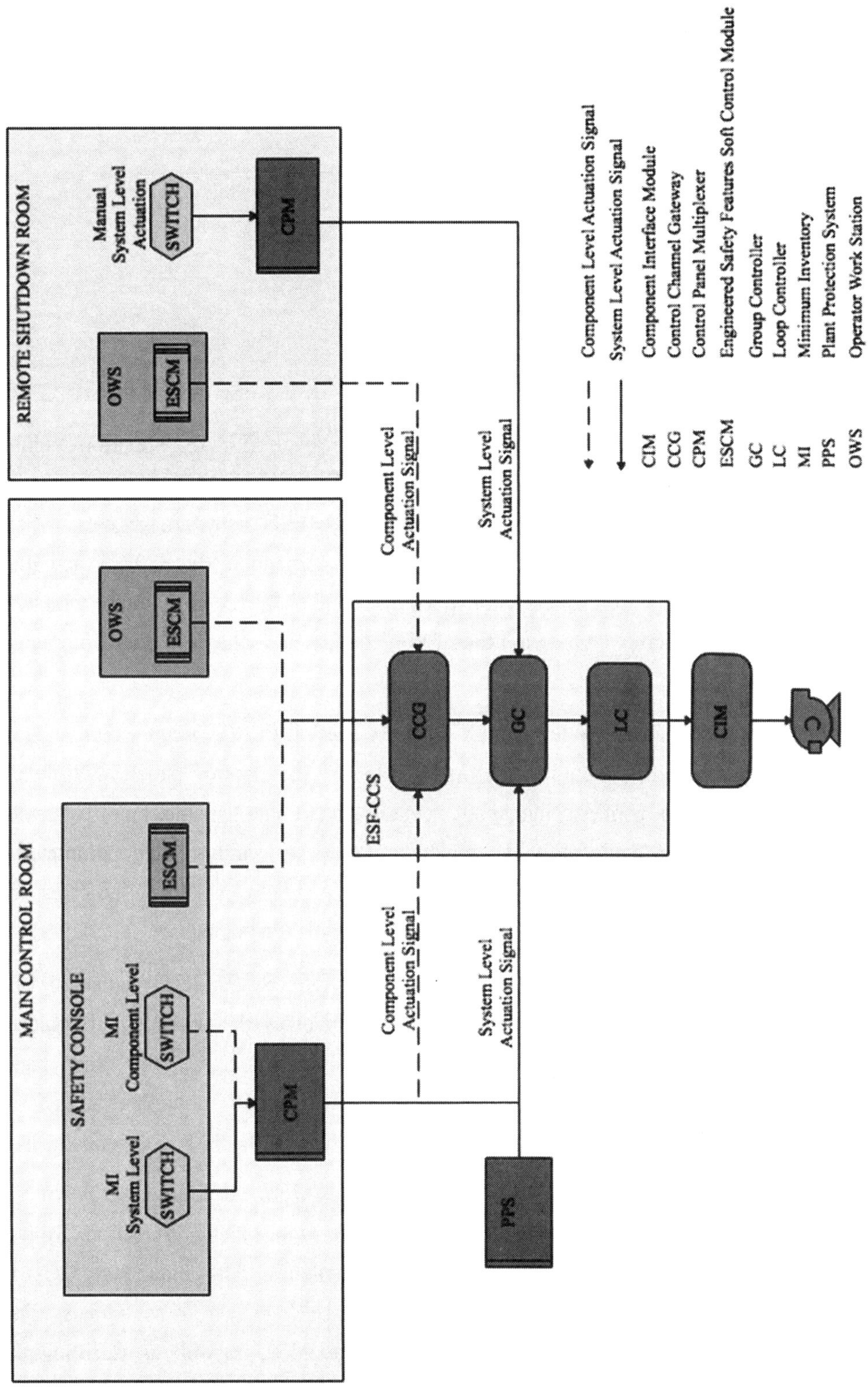


Figure 4-9: ESF-CCS Schematic

The ESF-CCS comprises the:

- Group Controllers (GC)
- Loop Controllers (LC)
- Control Channel Gateway (CCG)
- Control Panel Multiplexers (CPM)
- Maintenance and Test panels (MTP)
- Interface and Test Processors (ITP)

In each channel, the ITP and MTP are shared between the ESF-CCS, PPS, and CPCS.

The Group Controllers (GCs) receive system level ESFAS initiation signals from four (4) channel PPS and performs a selective coincidence voting logic. Upon logic processing, it sends component level ESF actuation signals to the Loop Controllers. The Loop Controllers receive component level ESF actuation signals from Group Controllers and perform component control logic which is used to send control signal to the ESF components through the component interface module (CIM).

The Control Channel Gateway (CCG) interfaces with the main control room (MCR) and the remote shutdown room (RSR) via the control panel multiplexers (CPMs) to acquire confirmation input switch status data. It interfaces with the ESF Soft Control Module (ESCM), which are used for safety-related components. The general system design features are shown in Figure 4-10. It indicates the channelization of the ESF-CCS and the level of initiation signals.

The ESF-CCS has two main modes of actuation: System-level actuation and Component-level actuation. System-level actuation involves the operation of multiple components in order to perform a particular ESF function. On the other hand, Component-level actuation involves the discrete or continuous control of individual ESF components. Table 4-2 lists the methods of actuating the ESF components [27].

Each ESF channel receives NSSS ESFAS signals from all four channels of the PPS. The GCs process System-level demands from the Plant Protection System (PPS) and the manual System-level actuation switches. The redundant GCs in each channel, then execute 2 out of 4 coincidence logic on these signals. This ensures that spurious actuation of the system is avoided, as well as fulfilling the single failure criterion outlined in IEEE Std. 603-1991 and IEEE Std. 379.

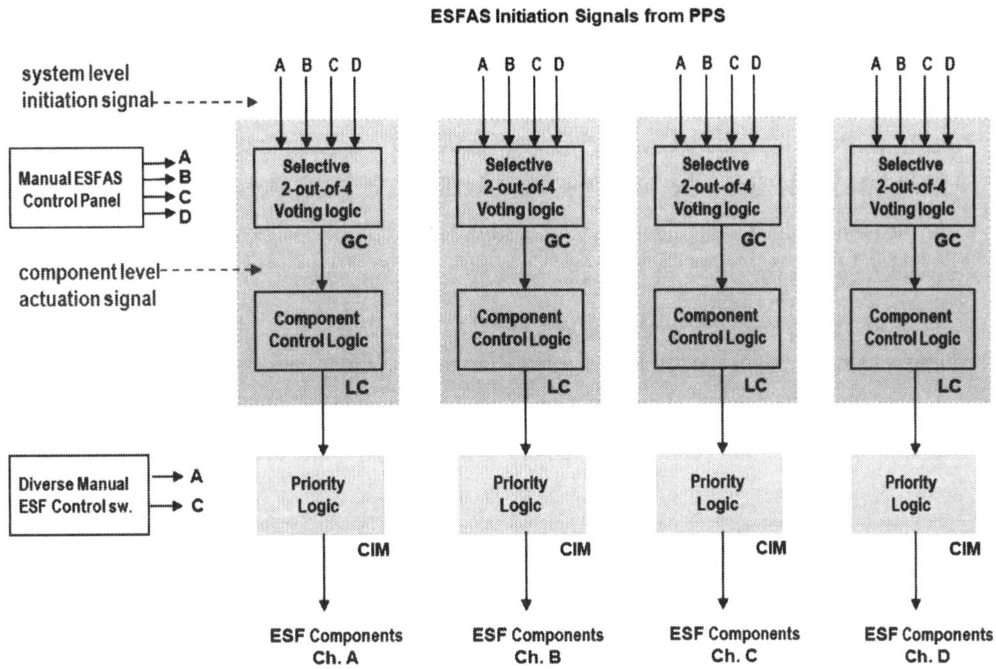


Figure 4-10: ESF-CCS Channelization

Table 4-2: ESF Actuation Control Provisions [27]

| Method | Description | Activated Via |
|---|---|---|
| Automatic | System-level | PPS |
| ESCM | Component-level | Software controllers in the Main Control Room (MCR) and Remote Shutdown Room (RSR) |
| Manual System Actuation Switches | System-level | Minimum Inventory (MI) hardware switches located at the MCR and RSR |
| Manual Component Control Switches | Component-level | Control Switches<br>Component-level Minimum Inventory hardware switches located on the MCR safety console |
| Diverse Manual Actuation Switches | System-level | Hardware switches located at the MCR safety console |
| Local Manual Actuation Switches | System-level | Switches are located in the Maintenance and Test Cabinet |

Figure 4-11 shows the basic ESF-CCS actuation function block for the control of ESF components. To ensure the availability of the ESF-CCS, it is designed to receive component actuation signals from multiple sources. Actuation signals can emanate from the DMA(Diverse Manual Actuation), MI(Minimum Inventory) and soft control switches from the MCR as well as from the PPS. In Figure, the priority of the actuation signals is illustrated. The logic is such that when an actuation signal is received from the PPS, it blocks the any actuation signal from the MI and Soft control switches. In addition, an actuation signal from the DMA blocks the other upper signals from the PPS, MI and the soft control switches. Thus, the highest priority level is given to the DMA switches, then the PPS, and lastly, the soft control (ESCM) and Minimum Inventory switches have the least priority.

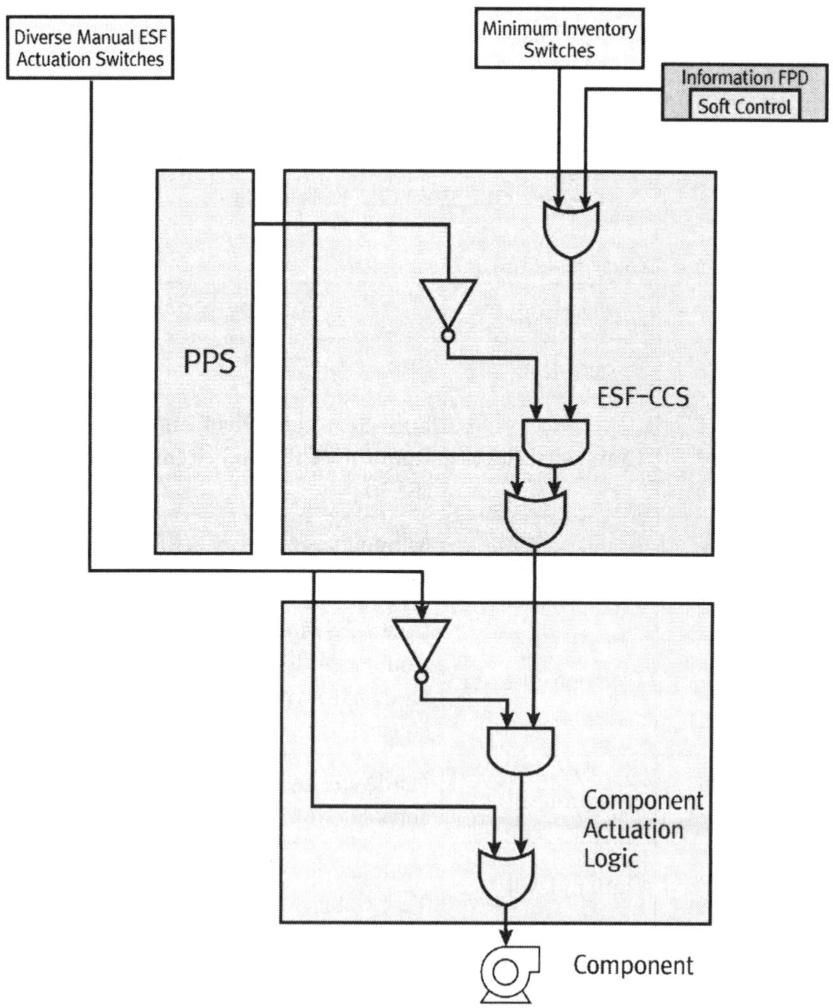


Figure 4-11: ESF-CCS Actuation Signal Prioritization

Based on the output of the coincidence logic, the GCs provide actuation signals to the Loop Controllers. LCs are responsible for the actuation and control of individual components. Therefore, in order to actuate an ESF system, a System-level command is transmitted to multiple LCs. The LCs then pass on the actuation signals to the CIMs. The CIM arbitrates between three possible actuation signals from the PPS, DMA, and Diverse Protection System (DPS), based on the defined priority logic, as illustrated in Figure 2-2. The DMA switches are located on the safety console in the MCR. These switches are not dependent on software, and thereby provide a diverse method of ESF actuation should the ESF-CCS suffer from software Common Cause Failure (CCF).

## 4.4 Core Protection Calculator System

The purpose of the Core Protection Calculator System (CPCS) is to protect the reactor from exceeding the safety limit by monitoring pertinent Reactor Core conditions and to generate trip signals of Departure from Nucleate Boiling Ratio (DNBR) and Local Power Density (LPD) during plant operation. The CPCS computes DNBR and LPD values based on monitored process parameters in each of the four Core Protection Calculators (CPCs). It receives input signals from the primary cycle as illustrated in Figure 4-12 as follow:

- CEA position is provided by two Reed Switch Position Transmitter (RSPT) inputs on each CEA.
- Cold Leg Temperature (two, from diagonally opposite cold legs);
- Hot Leg Temperature, analog voltage (two, one from each hot leg);
- Pressurizer Pressure, analog voltage (one, narrow range);
- Ex-core Flux, analog voltage (three, Upper, Middle, and Lower detectors);
- Reactor Coolant Pump Speed.

The CPCs provide DNBR and LPD Pre-trip and Trip signals when either the calculate DNBR or LPD approaches or exceeds its respective set-points, or when certain auxiliary condition are met. The CPCs channel Pre-trip and Trip outputs are

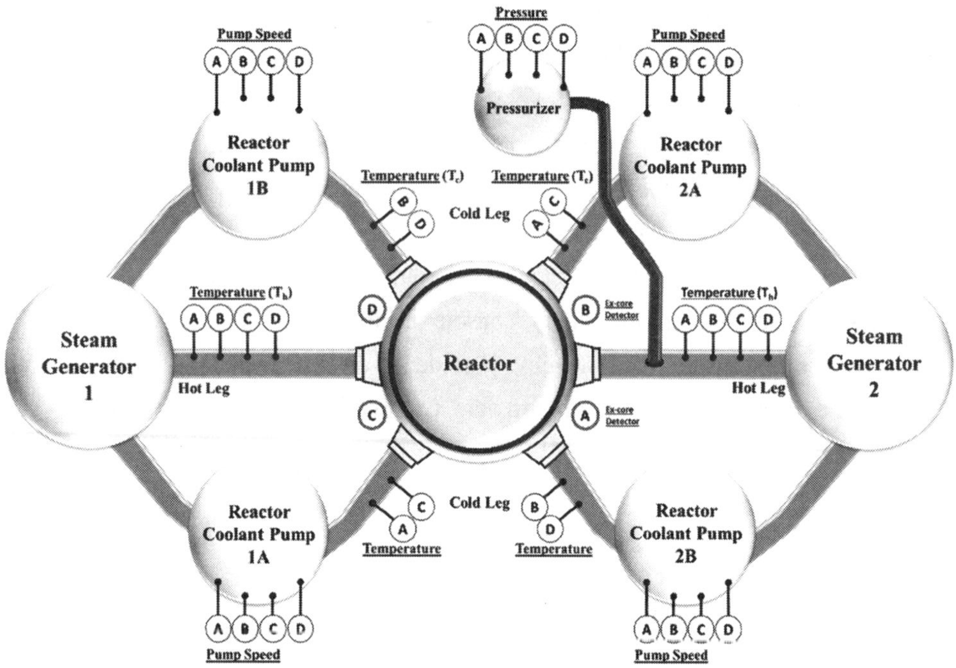


Figure 4-12: CPCS Input Signals

used by the plant protection system (PPS) logic where two-our-of-four possible signals are needed to generate reactor trip signals [25] as shown in the Figure 4-13.

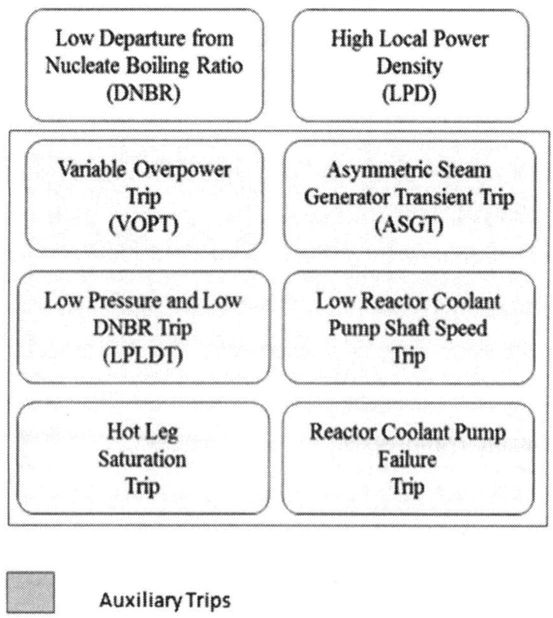


Figure 4-13: CPCS Reactor Trip Signals

The DNBR is the ratio between the critical heat flux (CHF) and the maximum local heat flux. The DNBR is the point at which heat is no longer efficiently removed from the reactor fuel and when fuel damage can occur. If the DNBR drops below than the safety limit, it would cause the formation of vapour layer, causing reduction in heat transfer capability between reactor coolant and fuel rods.

The high LPD trip is provided to trip the reactor when the calculated core peak LPD reaches a pre-set value. The pre-set value is less than the value, which would cause fuel centre-line melting. The calculation of LPD is based on the core average power and the core power distribution and includes a compensation to account for the thermal capacity of the fuel. The calculated trip assures a core peak LPD below the safety limit for peak linear heat rate [24].

In addition, high LPD and low DNBR trip signals are generated when the Reactor Coolant Pump (RCP) Trip, Hot Leg Saturation Trip, Variable Overpower Trip (VOPT), Asymmetric Steam Generator Transient Trip (ASGT), Low Pressure and Low DNBR Trip (LPLDT), out of range trip, or internal processor failure trip. These trips are called Auxiliary Trips.

Furthermore, the CPCS generates a CEA Withdrawal Prohibit (CWP) signals to prevent any abnormal condition in the core due to CEA withdrawal.

## 4.5 System Interfaces

An interface is a shared boundary across which two or more systems or components exchange information. The safety systems in the NPP exchange information with each other as well as transmit information to the non-safety systems. In this section, the systems or components with which the safety systems interact are defined.

### 4.5.1 Data Communication Networks

Figure 4-14 illustrates the interfaces between the different sub-systems of the MMIS. From this image, it is observed that communication between sensors, systems, and channels within a system, is performed using different types of communication links and protocols. In the APR1400, the following data communication networks are used [24]:

i. Safety system data network (SDN) for safety systems
ii. Serial data link (SDL) for safety systems
iii. Data communication network – information (DCN-I) for non-safety systems.

### (1) Safety System Data Network (SDN)

The SDN provides a communication link between safety systems within one channel, with the SDN in a given channel being separated and isolated from that of the other redundant channels. In the APR1400, the SDN is used as follows:

- In the PPS, bistable processors and LCL processors send status data such as bypass and trip status to the MTP, OM an ITP. In addition, they receive testing data from the MTP via the SDN
- In the CPCS, all process variables are received through the SDN. Calculated values are also transmitted to the including inputs and MTP, OM, and ITP through the SDN.
- The ESF-CCS LCs and GCs transmits status information to MTP and ITP through the SDN. Further, the CCG transfers component control signals to the LCs and receives feedback from them through the SDN.

### (2) Serial Data Link (SDL)

The SDL is used to transmit predefined data between each processor within a channel. In addition, this network transmits broadcast data to other channels and non-safety systems.

In the MMIS, it is used in the following cases:

- Data transmission between the bistable processors and LCLs in the PPS
- ESFAS initiation signals from the LCLs in the PPS to the GCs in the ESF-CCS
- In the CPCS, CEA position signals and penalty factors are transmitted to the CPP and CEAC processor through CPPs in the same channel and other channels using the SDLs.

- In the ESF-CCS, the component control signals are transmitted from the GCs to the LCs using this data network. In addition, component control signals from the MI switches are transmitted to the GCs using the SDL

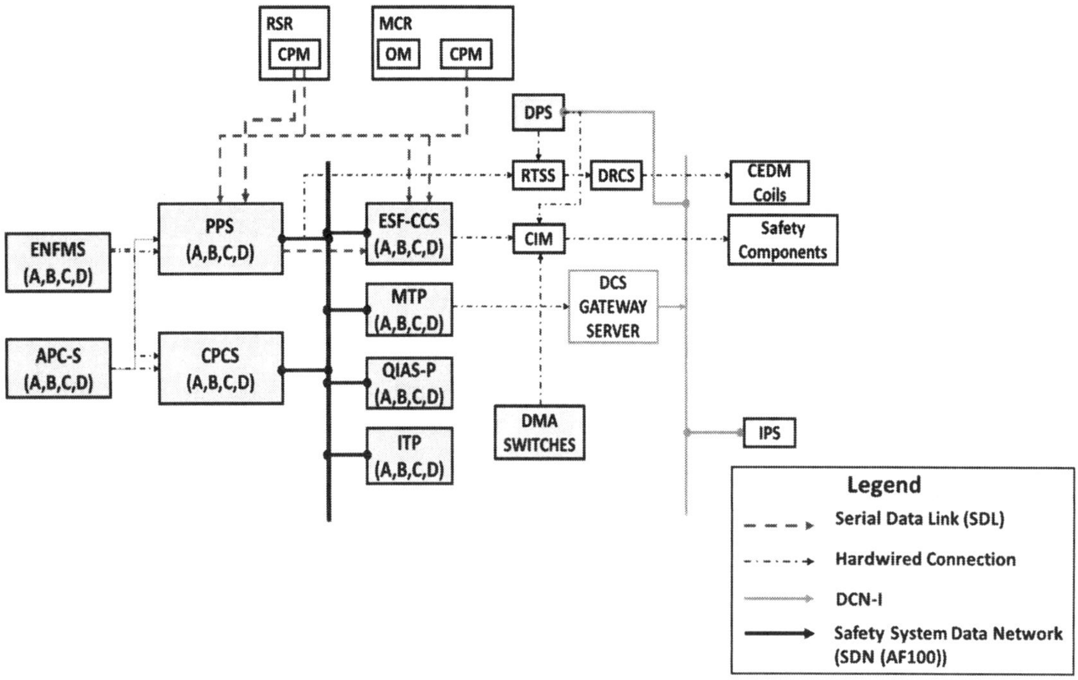


Figure 4-14: MMIS Interface

### (3) Data Communication Network-Information (DCN-I)

The DCN-I is used for non-safety data communication specifically the transmission of data from safety and non-safety systems. To ensure the electric isolation between the safety and non-safety systems, all data paths between the DCN-I and composed of are fiber optic cables.

## 4.5.2 PPS Interfaces

The PPS generates signals to actuate reactor trip and ESF automatically whenever monitored process variables reach predefined limits. It is also equipped with means for manual initiation of each protective action. The interfaces of PPS cabinet are listed in Table 4-3 with the following equipment:

Table 4-3: PPS Interfaces [25]

| System / Component | Description |
|---|---|
| Auxiliary Process Cabinet – Safety (APCS) | The APC-S provides four channels, physically and electrically separate sensor signals for each safety-related plant parameter to the PPS cabinet via hardwired cables. |
| Core Protection Calculator System (CPCS) | The CPCS transmits DNBR and LPD states are to the PPS cabinet via hardwired cables. |
| Ex-core Neutron Flux Monitoring System (ENFMS) | The ENFMS sends log power, calibrated linear power, log power operating bypass permissive, and ex-core trouble annunciation to the PPS. |
| Reactor Trip Switchgear System (RTSS) | The RTSS interfaces are in the form of relay contacts, which are wired to the reactor trip switchgear. |
| Engineered Safety Features - Component Control System (ESF-CCS) | The PPS sends the ESFAS initiation signals to the ESF-CCS GCs in the same channel and the other three redundant ESF-CCS channels. |
| Vital Bus Power Supply System (VBPSS) | Each channel of the PPS is powered from a VBPSS inverter. Each channel of the VBPSS provides a non-interruptible battery backed 120 VAC, single phase, ungrounded power source for essential instrumentation and plant control. |
| Control Panel Multiplexer (CPM) | The conventional switch signals in the MCR and RSR are sent to the PPS by using CPMs and SDL. |
| Operator Module (OM) | The OMs are located on the MCR SC and provide the PPS status (trip/pre-trip/bypass), initiation circuit status, TCB phase status, and operating bypass information to the operator. |

In addition, the PPS interfaces with the DRCS remote I/O cabinet, Information Processing System, QIAS-P and QIAS-N.

The four channels of the PPS separation and mechanical isolation in order to meet the requirements of IEEE Std. 384 and RG 1.75. To ensure this, the communication between the channels and connected systems (e.g ESF-CCS) is implemented using fiber optic cables. The fiber optic cables satisfy the isolation and independence requirements.

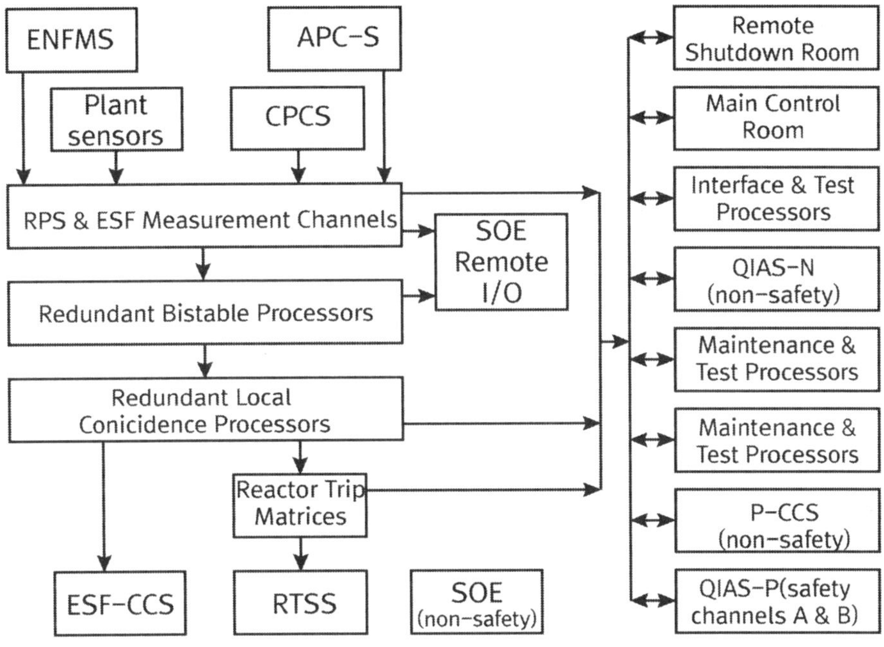


Figure 4-15: Block Diagram of PPS & Interfaces

In Figure 4-15, the signal is flowing down from top to bottom and left side to right side, except signals that are being generated from the main control room, interface test and processors, maintenance tests and processors and remote shutdown room. The signals that flow from top to bottom are meant for protection purposes while the signals that flow from left to right are mostly for monitoring purposes. So we can say that the systems that perform bi-directional functions such as RSR, MCR, ITP, MTP, are having bidirectional capability. Therefore, maximum caution should be kept against cyber-security threats.

### 4.5.3 CPCS Interfaces

The CPCS interface with other system is shown in Figure 4-16 and defined in the table 4-4.

Table 4-4: CPCS Interfaces [25]

| System / Component | Description |
|---|---|
| APC-S | CPC processor receives the Pressurizer pressure signal from APC-S used for DNBR and LPD |
| ENFMS | CPC processor receives the linear sub-channel power signal from ENFMS. These are used to calculate the reactor power calculation and power distribution calculation. |
| Reactor Coolant Pump Shaft Speed Sensing System | CPC processor receives RCP Speed signal from Reactor Coolant Pump Shaft Speed |
| Reed Switch Position Transmitter (RSPT) | CEA position is provided by two RSPT inputs on each CEA. All RSPT inputs are converted to a digital value in the CEA Position Processor (CPP) and are input to all four CPC/CEAC channels over fiber-optically isolated data links. |
| PPS | The CPCS provides the following signals to PPS. Low DNBR Trip/Pre-trip; High LPD Trip/Pre-trip; and CEA Withdrawal Prohibit |
| Field sensors | The CPCS receives the following field sensor signals: Hot leg temperature loop 1; Hot leg temperature loop 2; cold leg temperature loop 1, and cold leg temperature loop 2. |

In addition, the CPCS interfaces with the QIAS-P, through safety system data network. where MTP and ITP are the medium to deliver the safety data to the non-safety system such as IPS and QIAS-N.

VBPSS(Vital bus power supply system) supplies essential dc backup power to CPCS.

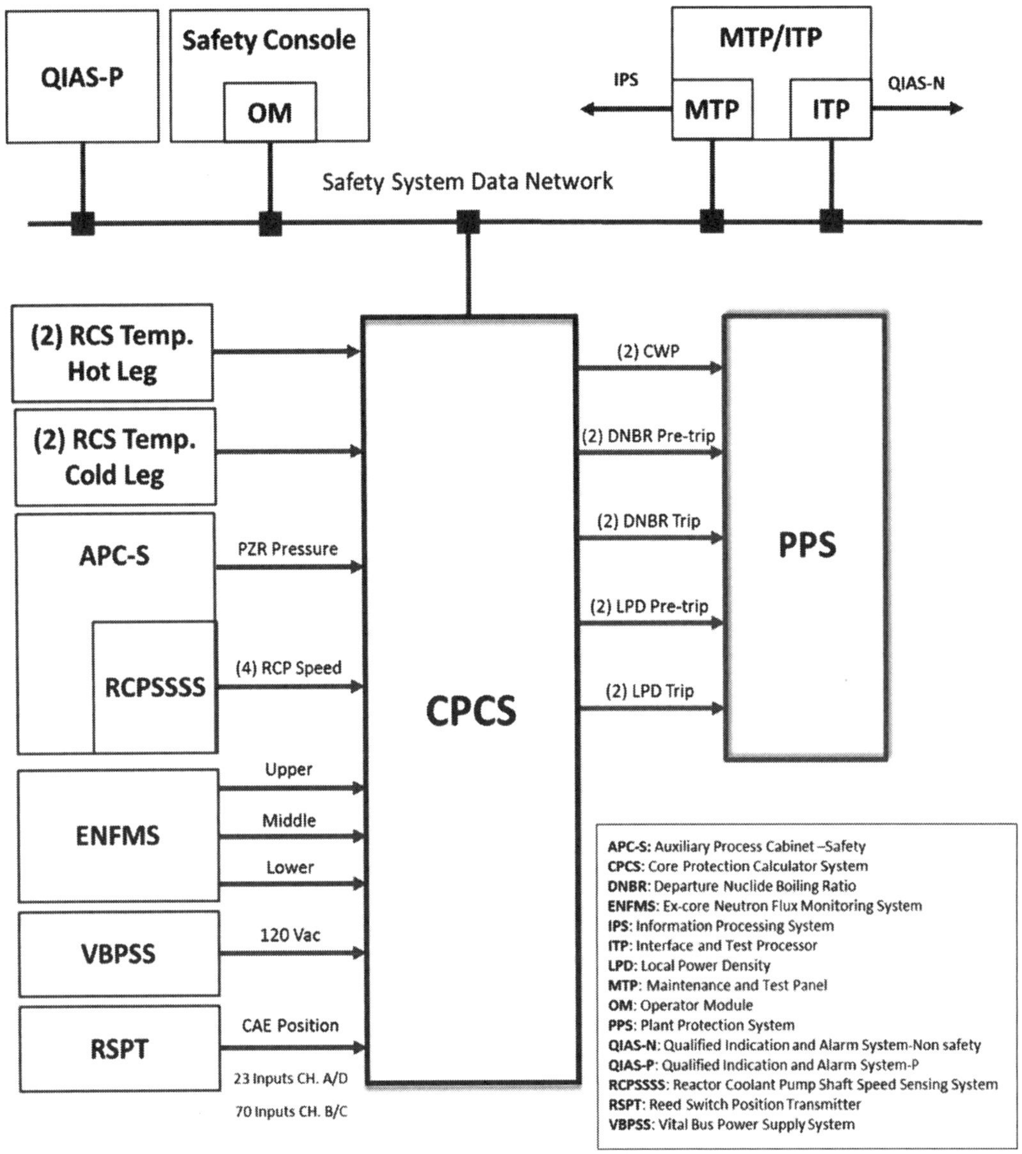


Figure 4-16: CPCS system interfaces

### 4.5.4 ESF-CCS Interfaces

The ESF-CCS interfaces with the following systems: (depicts in Taable 4-5)

As explained in sub-chapter 4.3, ESF-CCS communicates with ESCM and MI which deliver the operator's control command to actuate ESF components such as pumps and valves.

The control command from MI switches on safety console are sent to CPM for transmitting multiple signal by time division multiplexing techique.

Table 4-5: ESF-CCS Interfaces [26]

| System / Component | Description |
|---|---|
| PPS | The PPS ESFAS initiation signals go to the ESF-CCS and form the 2-out-of-4 logic. Fiber optic cables run in conduit provide the isolation and separation between the redundant PPS initiation signals and ESF-CCS channel A, B, C, and D equipment. |
| ENFMS | CPC processor receives the linear sub-channel power signal from ENFMS. These are used to calculate the reactor power calculation and power distribution calculation. |
| CPM | The ESF-CCS receives manual actuation signals from the Safety Console and RSR via the CPM |
| VBPSS | Each channel of the ESF-CCS is powered from Class 1E inverters in each channel of the VBPSS, which provides a reliable, continuous 120 VAC, single phase, ungrounded power for essential instrumentation and plant control |
| CIM | The ESF-CCS provides discrete ESF actuation signals to the associated CIMs. Each ESF-CCS channel is located in a different fire zone, which contains a minimum of combustible materials. The ESF-CCS LC cabinets are located near controlled plant components.. |

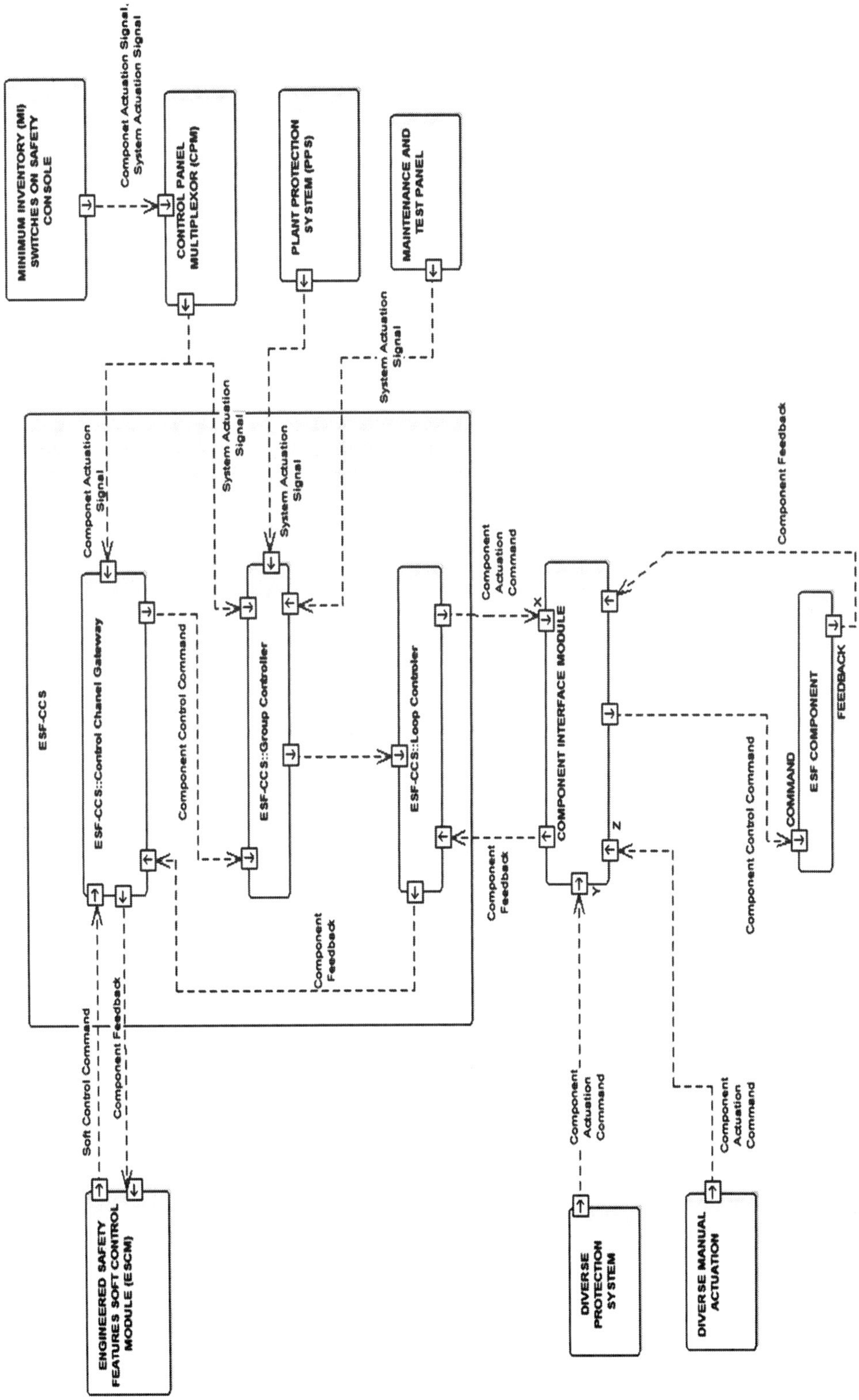


Figure 4-17: ESF-CCS Interfaces

### 4.5.5 Component Interface Module (CIM)

The Component Interface Module (CIM) operates as the interface between the control system and the actuated components (such as motors, valves, circuit breakers etc.). In this capacity, it performs several functions:

- Arbitration between multiple actuation commands from various sources based on a pre-defined priority
- Actuation of system components
- Generation of component feedback and interlock between feedback signals and actuation commands
- Voltage and current level transition between the computer logic and the actuated equipment circuit.
- Test and Diagnostic features

The CIM is a standardized, object-oriented interface used to control components. This means that a single type of CIM is used for all types of components. However, the functions required by specific components (e.g. Valve, Pump, Damper) vary. Therefore, in order to customize the logic for a given component, configuration jumpers are used [24].

In order to fulfill its function, the CIM is composed of several logic blocks such as priority, universal, and test and diagnostic logic, illustrated in Figure 4-18 and described in the following section. The priority logic is focused herein.

#### (1) Priority Logic

For the purpose of improving system reliability, the control system of the Nuclear Power Plant (NPP) is designed to allow for the actuation components using commands from multiple sources, as illustrated in Figure 4-22. Multiple input sources ensure that even if one source fails e.g. the ESF-CCS, the operator will still be able to actuate the plant component using a diverse system e.g. the Diverse Protection System (DPS).

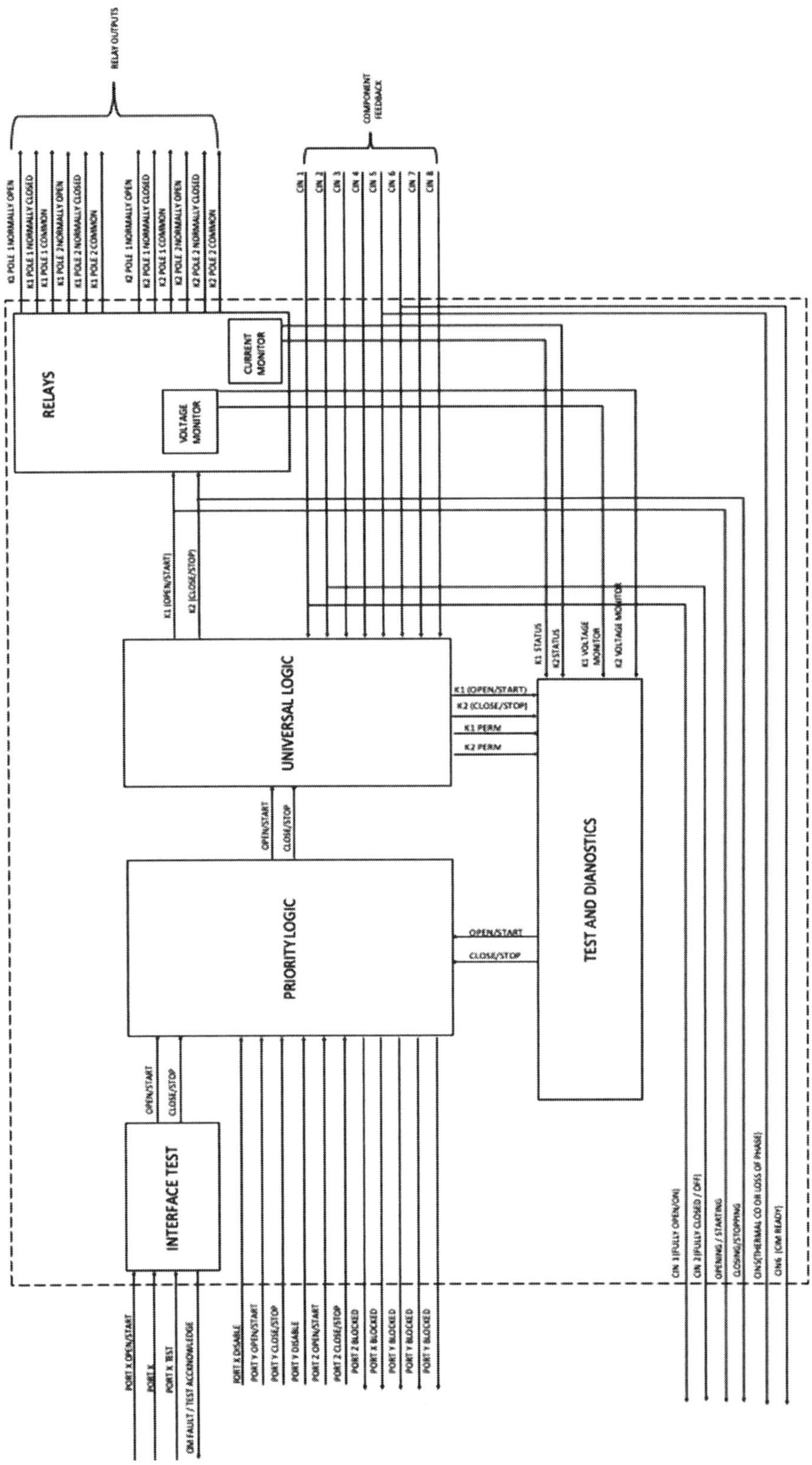


Figure 4-18: CIM Logic Blocks

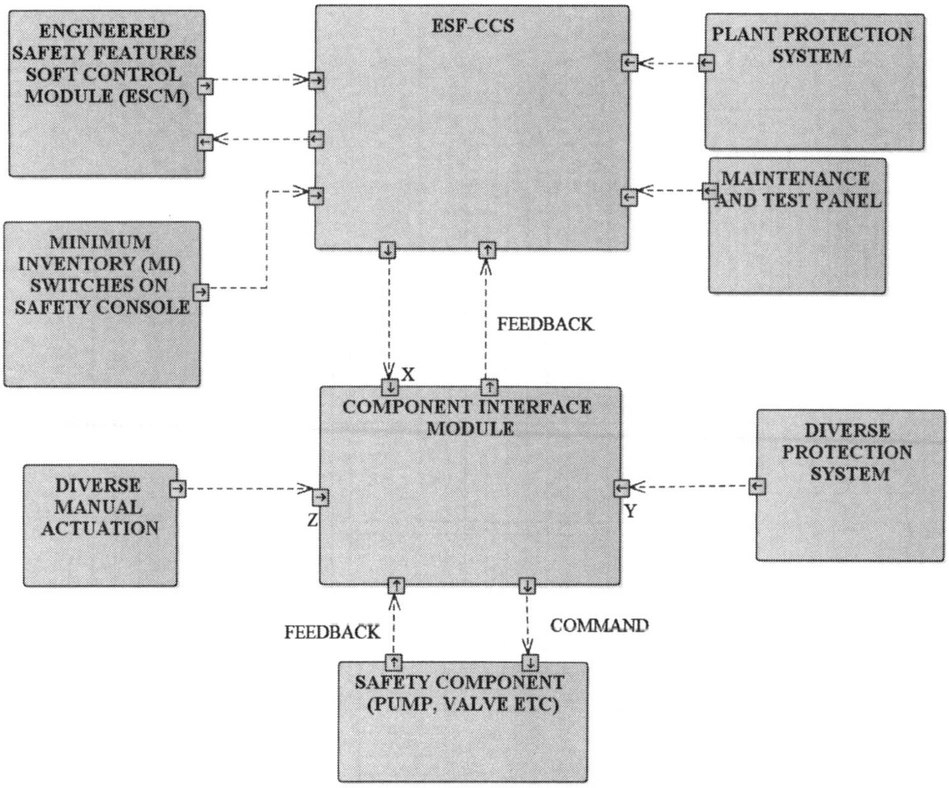


Figure 4-19: Component Actuation Commands and Feedback

The CIM is designed to accommodate inputs from four sources; two host systems and two dedicated hardwired manual controls. The inputs from the two host systems are designated as Port X and Port Y. One manual input, Port Z, corresponds to a remote manual control while the second manual control is local to the CIM. For a typical NPP safety application, Port X corresponds to the PPS, Port Y to the DPS and Port Z to a remote manual control, which may be in the MCR, RSR or some local control panel in the plant. However, a given application may not require all the input sources. Any unused control ports are therefore left unconnected.

When the CIM receives actuation commands from different sources, the decision as to which commands to execute is made based on the configured priority logic. The priority logic is responsible for arbitrating between the command signals received from multiple sources and processing the command with the highest priority. Thus, when a command is received from multiple sources, e.g. from both the PPS and DPS, the priority logic block will process only the command with higher priority.

The priority is defined either based on the system or function. As mentioned, the CIM receives input commands from multiple sources. Thus, in system priority, one of the systems is given priority over the others. With function priority, one command is given priority over another e.g. the open command, regardless from which system it comes, may be set to have priority over the close command. [27].

The output of the priority logic block (start/open or stop/close) is transmitted to the universal logic for further processing.

## PROBLEMS

1. What is the role of the Safety Critical Systems?
2. Discuss how the different MMIS systems operate as multiple levels of defense to prevent the release of radioactive material in the event of an accident.
3. What are the main functions of the PPS?
4. What are the twelve trip signals that can be generated by the PPS? Describe the role of each type of trip.
5. Why does the system utilize selective two out of four logic?
6. What is the difference between an operating bypass and a maintenance bypass?
7. What are the main functions of the ESF-CCS?
8. What are the five main ESF actuation Signals? Describe the role of each.
9. What are the main functions of the CPCS?
10. What is the role of the LPD and DNBR trips?
11. Describe the signal path from Operator Module to actuated ESF component.
12. What is the Component Interface Module? What are its main functions?
13. Draw a block diagram of the three safety critical systems (PPS, ESF-CCS and CPCS) showing the interfaces between these systems?

## FURTHER READING


- KEPCO and KHNP, "APR1400 Design Control Document, Chapter 7 Instrumentationand Controls," [Online]. Available: https://www.nrc.gov/docs/ML1822/ML18228A654.pdf. [Accessed 4 1 2019].
- KHNP, "Engineering Safety Features," in APR1400 Design Description, Korea, 2014.
- USNRC, "Engineered Safety Features," 1 December 2012. [Online]. Available: https://www.nrc.gov/docs/ML0913/ML091380480.pdf. [Accessed 24 December 2018].
- WEC LLC, "Component Interface Module". United States of America Patent US6842669B2, 15 August 2002.

# Chapter 5 SAFETY SYSTEM CRITERIA


## SUMMARY

The safety system criteria are design principles, which ensure that the developed safety critical systems maintain plant parameters within acceptable limits with precision and reliability [25]. These criteria are defined by the stakeholders - regulatory bodies and utility requirements - based on operating experience and accepted 'good practice' from the industry. These criteria therefore form part of the stakeholder requirements to be incorporated in the MMIS design. These criteria are described in this chapter. Mind map of chapter 5 is shown in Figure 5-1.


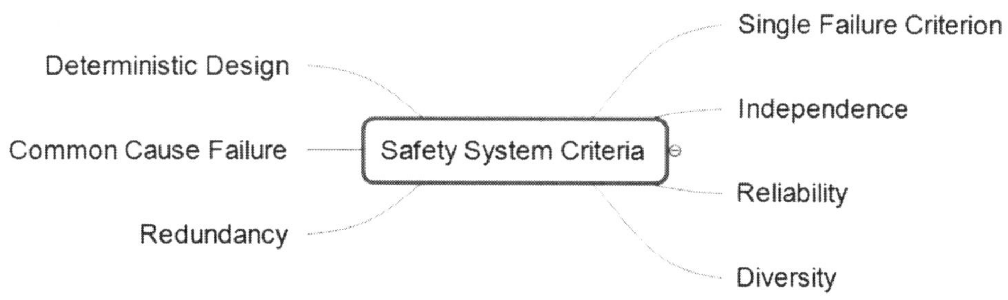


Figure 5-1: Mind map of Chapter 5

Among those criteria which were recommended and reviewed from URD, SRP and IEEE 603, the following limited criteria were selected and are shown in -Figure 5-1. These criteria will be discussed in details in this chapter. The discussion describes how the safety system criteria are evaluated for ensuring system safety by analytical methods.

**OBJECTIVES**

- Discussion of the main Safety System Criteria
- Common Cause Failure definition, sources and design methods for CCF mitigation
- Reliability definition and the introduction of two analysis methods

## 5.1 Overview of Safety System Criteria

One of the most important references for the development of Safety Critical Systems in the NPP is IEEE 603: IEEE Standard Criteria for Safety Systems for Nuclear Power Generating Stations [30]. This standard establishes "the minimum functional and design criteria for the power, instrumentation, and control portions of nuclear power generating station safety systems". The mandated safety system criteria are defined in section 5 of the IEEE 603 as illustrated in Figure 5-2. The subsequent sections of this chapter, looks closer at some of these criteria.

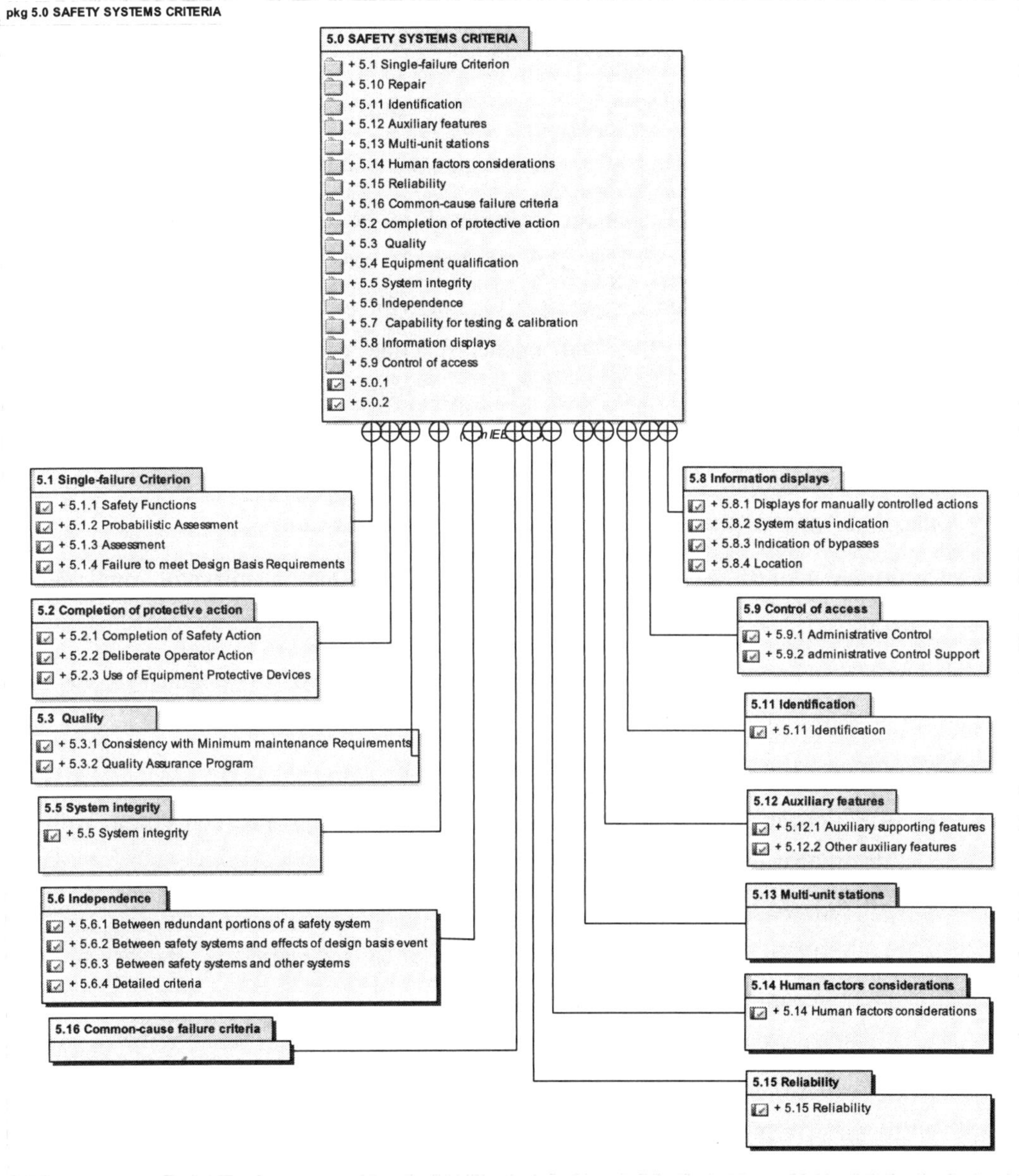


Figure 5-2: Safety system criteria [30]

## 5.2 Single Failure Criterion

The single failure criterion requires that a safety system must remain functional in spite of the failure of any single component within the system.

Fluid and electric systems are considered to be designed against an assumed single failure if neither (1) a single failure of any active component (assuming passive components function properly) nor (2) a single failure of a passive component (assuming active components function properly), results in a loss of the capability of the system to perform its safety functions. All failures must be considered whether it is an active component or a passive component. Active and passive component failures can occur independent of an initiating event and therefore, implies that single failures can occur at any time.

Section 4 of IEEE Std. 379 [31] gives the following statement on single-failure criterion:

"The safety systems shall perform safety functions for a design basis event in the presence of the following:

- Any single detectable failure within the safety systems concurrent with all identifiable but non-detectible failures
- All failures caused by the single failure
- All failures and spurious actions that cause or are caused by the design basis event requiring the safety function"

The single failure could occur prior to, or at any time during, the design basis event for which the safety system is required to function.

Single failure criterion related to deterministic design, which specifies a simple design approach to obtaining a certain minimal redundancy of a system as well as has relation to ensure adequate reliability of safety function. It is based on the general experience that components and equipment that are made to high quality standard may sometimes fail to function, in a way and at a time that is random and unpredictable. For reliability explanation, will be provided in the next section of this chapter.

### (1) Random failures

Failures that occurs statistically independent from failures of other same type devices called random failure. The manufacturing process, statistical variations in material, testing procedures, the operating conditions or the maintenance may affect a device to behave differently from other same type of devices. The cause of the random failure is not known prior to its occurrence.

### (2) Single Failure Analysis

As describe in Safety Series 50-P-1 [32], a single failure analysis (SFA) must be performed to assess compliance with the single failure criterion and associated requirements. SFA must be performed as part of the total safety analysis of the plant and its design. A total plant Probabilistic Safety Assessment (PSA) is an independent methodology for assessing plant safety. For compliance with the regulation and code, a PSA cannot be used in place of an SFA. A PSA may be used to complement the SFA, particularly to cope with postulated initiating events linked mainly to common cause failures.

In the other hand, there is an analysis that assumed a single failure to occur and performed sequentially, considering that failure of the component can affect the performance of large part of the system and it may cause the other consequential failures generally known as a failure modes and effect analysis. Any such failure, including the consequential failures, must be considered as a single failure for the purpose of analysis, which has to continue until all assemblies and all credible failures have been considered.

Certain failures could have severe effects on nuclear safety and result in more severe consequences than active failures. Examples are:

- Rupture of a pipeline or tank (Active Failure)
- Blockage of the containment sump by heat insulation material from the primary circuit. (Active Failure)
- Break of the pipeline from the containment sump to the ECCS pump. The entire water inventory for cooling the core might be lost during the recirculation phase of a LOCA in a PWR (Passive failure)

Due to severe consequences of failures, a cautious attitude to their analysis is required. Thus, failures with severe consequences at least as a minimum must occurs very unlikely by taking an appropriate quality conditions. If feasible, additional redundancy must be taken into account.

For example of single failure criterion, the RPS meets the single failure criterion through four redundant and independent channels. The 2 out of 4 voting logic prevents system level spurious actuation due to single failure. For the RPS to satisfy the requirements of single failure, the design of the RPS shall be such that no single failure of a component will interfere with the proper operation of an independent redundant component or system.

The ESFAS is designed so that any single failure within the system does not prevent proper protective action at the system level.

Due to the 2 out of 4 logic, which is being implemented in the group controller, a single failure of the initiation circuit if it occurs would not interrupt the protective action of the NSSS ESFAS. The worst case would be that one channel or division would fail.

In the case of the BOP ESFAS, a spurious action in one of the channels, would not cause an adverse effect on the plant safety or cause a reactor trip. This is because the BOP ESFAS will automatically actuate valves and dampers of HVAC systems for ventilation of the fuel handling area, operation of the containment purge system, and operation of the MCR HVAC system.

## 5.3 Independence

Independence is the state in which there is no mechanism by which any single design basis event, such as a flood, can cause redundant equipment to be inoperable. According to IEEE Std. 384, independence is achieved by physical separation and electrical isolation. Physical separation can be achieved by the use of safety class structures, separation distance, or barriers or any combination thereof while electrical isolation can be achieved by the use of separation distance, isolation devices, shielding and wiring techniques or combination thereof.

There are two main ways of achieving independence: physical separation, and electrical isolation. Physical separation refers to the systematic use of barriers or distance to separate components or systems performing similar functions. This prevents a failure affecting one component or system, from affecting another. It provides protection against common mode or cross-linked effects, such as fires and missiles, and prevents a fault from migrating from one system to another [33].

Electrical isolation implies that systems have separate electrical power supplies and communication networks. Electrical connections and data connections between redundant divisions within a safety system are designed so that no failure in one redundant division prevents the other from operating. In addition, electrical connections and data connections between safety systems and systems of a lower safety classification are isolated so that no failure in the system of lower safety classification would prevent the safety systems from operating. Isolation is therefore used to control or prevent adverse interactions between equipment and components caused by factors such as electromagnetic interference, electrostatic pickup, short circuits, open circuits, and grounding.

Thus, implementing independence in a system prevents the propagation of failures from system to system, between redundant elements within systems, and caused by common internal plant hazards.

### (1) System Separation and Isolation

In APR1400, separation and isolation are implemented in the following ways:

- Between safety systems and non-safety systems. They are separated both physically and electrically
- Between the different safety systems, that is the PPS, ESF – CCS, and the QIAS - P. They are separated both physically and electrically (Figure 5-3)
- Between the channels of the protection system. The protection system is composed of 4 redundant channels. The four independent channels have a separate connection from a channel sensor to actuation or display. In addition, there are separate power divisions for each channel. (Figure 5-4)

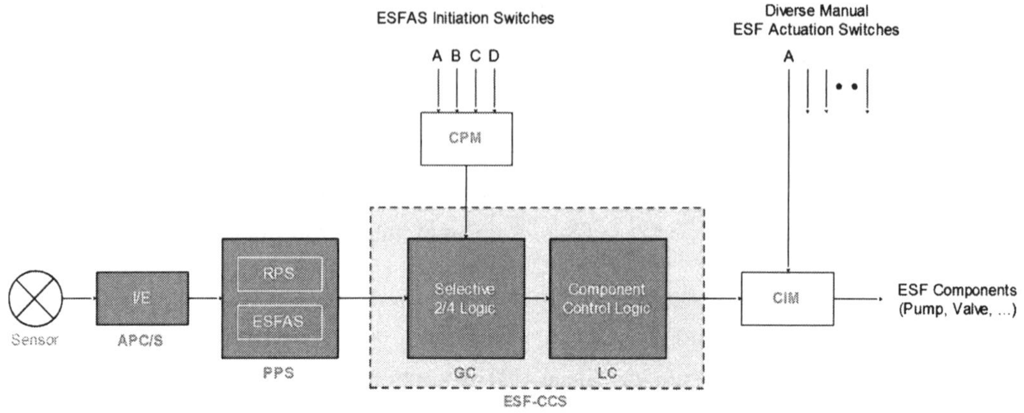


Figure 5-3 Separation in safety system

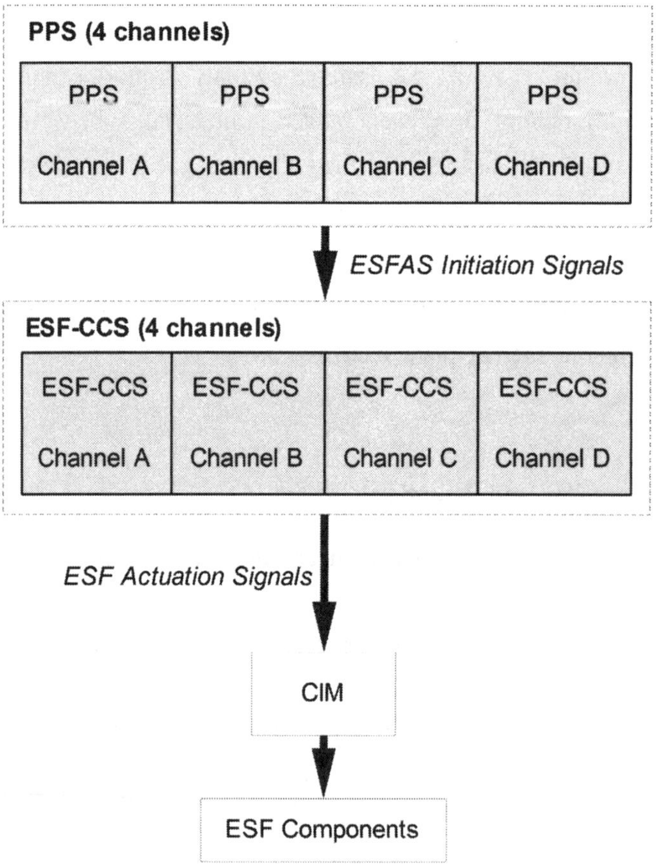


Figure 5-4 Separation between channels in safety system

### (2) Communication Independence

The system is designed to maintain data independence between safety and non-safety systems, thereby guaranteeing that a transmission error in one channel or division will not cause the failure of another channel or division. A buffering circuit is used to implement unidirectional transfer of information from the safety network to the non-safety network. This separation ensures that the non-safety network cannot prevent the safety network from performing its intended function.

Description of how independence is implemented in the APR 1400 MMIS is given with reference to IEEE Std. 603.

Section 5.6 of IEEE Std. 603-1991 describes independence with the following:

i. ***Between Redundant Portions of a Safety System***: Redundant portions of the safety system provided for a safety function shall be independent of and physically separated from each other to the degree necessary to retain the capability to accomplish the safety function during and following any design basis event requiring that safety function.

   The PPS channel receives AC power from vital bus supply system. The PPS does not share the power between channels

ii. ***Between Safety Systems and Effects of Design Basis Event***: safety system equipment required to mitigate the consequences of a specific design basis event shall be independent of , and physically separated from, the effects of the design basis event to the degree necessary to retain the capability to meet the requirements of IEEE Std. 603-1991.

   Independence between the components of the RPS and ESF and the effects of design basis events (IEEE 603) is provided by qualifying the equipment in accordance to ASME NQA-1. All system components are qualified for environmental and seismic conditions in accordance with IEEE Std. 323-1983 and IEEE Std. 344-1987. The system is capable of performing its intended function under the most degraded conditions of the energy supply.

iii. ***Between Safety Systems and Other Systems***: The safety systems shall be such that credible failures in consequential actions by other systems shall not prevent the safety system from meeting the requirements of IEEE Std. 603-1991.

Section 4.8 IEEE 603-1991 describes the actions by other systems as the conditions having the potential for functional degradation of safety system performance and for which provisions shall be incorporated to retain the capability for performing the safety functions (for example, missiles, pipe breaks, fires, loss of ventilation, spurious operation of fire suppression systems, operator error, failure in non-safety-related systems).

The PPS and non-safety systems are isolated using a qualified isolator or fiber optic cable so that any failure in a non-safety system does not cause loss of the safety system function. The PPS signals transmitted to the IPS/QIAS-N are isolated using fiber-optic cable. Data flow is unidirectional from Class 1E systems to non-Class 1E systems.

## 5.4 Common Cause Failure

A common cause failure (CCF) is the concurrent failure of multiple Structures, Systems, or Components (SSCs) that are otherwise considered independent of each other due to a common cause. Common causes include a defective manufacturing process, a common adverse environment, or an inadequate test or maintenance procedure. CCF nullifies the functional objectives of redundancy that are employed to improve the availability of plant functions important to safety. CCF can reduce the effectiveness of system separation and independence of trains of safety systems if they are implemented using common platform such as same vendor or technology type. CCF can generally be avoided by implementing system diversity, but cannot be completely eliminated.

The most common type of common cause failure is software. Although software does not wear out, digital instrumentation and control (I&C) systems are potentially vulnerable to CCF caused by software faults [34]

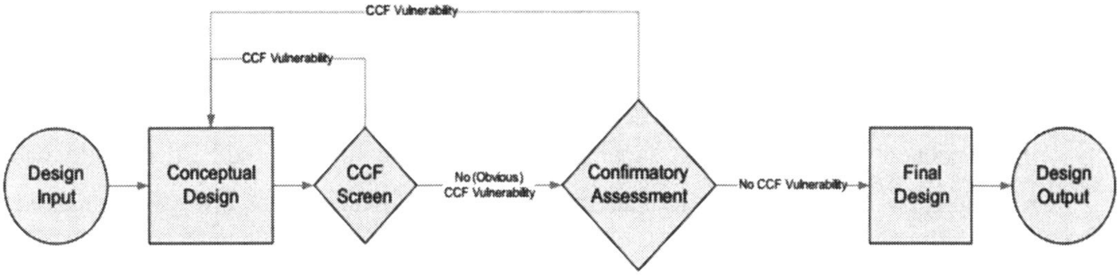


Figure 5-5: CCF Evaluation Process [34]

The evaluation of CCF vulnerability at the early stage of the development life cycle of a project is desirable. This is because serious risk would be faced if the system were determined to be susceptible to CCF late in project development. As such, CCF must be identified and addressed during the systems design process. Figure 5-5 shows the systems engineering approach to the evaluation of CCF vulnerability on system design.

Systems design vulnerability to CCF is assessed during the conceptual design. Sufficient level of detailed evaluation is considered to screen out vulnerabilities until no obvious vulnerability is determined. A confirmatory assessment is performed to ensure that before the final design stage, CCF is resolved.

### (1) Importance of CCF Consideration to MMIS

Defects in software and hardware designs are usually unnoticeable during the design development and implementation phases. However, when these faults are triggered by some mechanisms during operations, functional failures can emanate into CCF. Generally, there are two possible failure states:

- Demand Failure - Outputs that fail as-is
- Spurious state change - Outputs that change states or values

For a safety function such as a reactor trip, a fail as-is condition means that the reactor trip function will not be available and the PPS will not perform its safety function. This could lead to the PPS failure to respond to postulated accident. Failure associated with state changes would normally give indication of an abnormal system behavior.

MMIS plays important role in the safety of NPP by maintaining the integrity of defense-in-depth barriers. As such, the MMIS architecture includes many layers of defenses and safety design characteristics such as diversity, redundancy, independence, and physical separation. Thus, the MMIS are designed with consideration of CCF to assure that the integrities of barriers are always maintained.

### (2) Systems engineering approach to CCF identification

In the CCF evaluation process described in Figure 5-6, faults identification begins at the conceptual design phase. However, at each phase of the life cycle development of an I&C system faults may be introduced. Hence, potential sources of latent faults should be identified at each development phase.

### (3) Sources of CCF

There are many sources of common cause failure. Figure 5-6 shows the general sources of CCF.

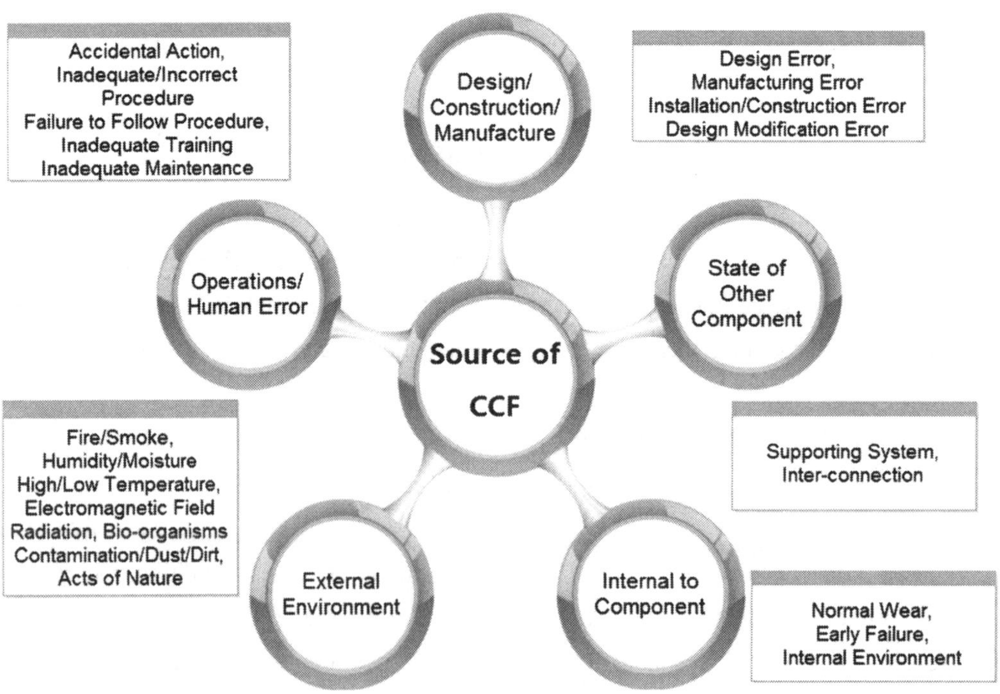


Figure 5-6: Sources of CCF

### (4) Coupling factors of CCF

The coupling factors are the conditions through which failures of multiple units are coupled, and include similarity in design, operation, maintenance, and test [5]. These coupling factors are made up of five major classes. Figure 5-7 shows the coupling factors.

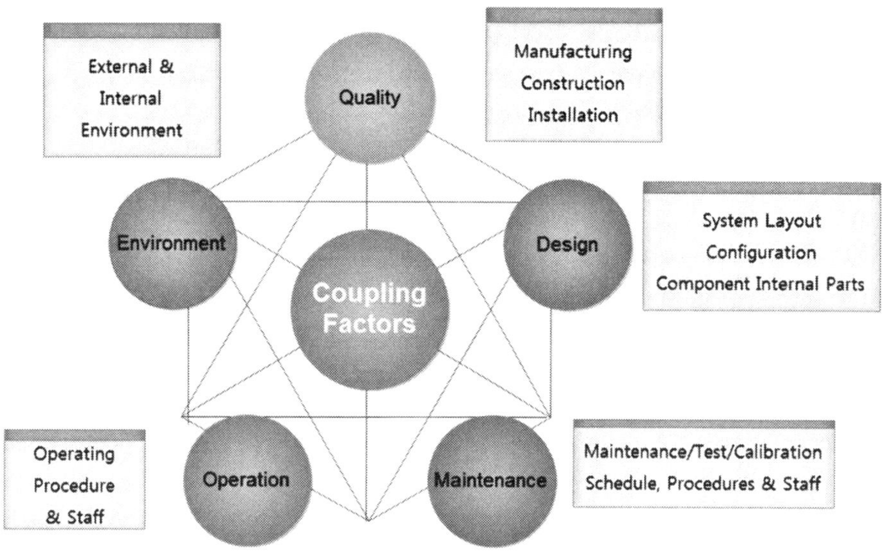


Figure 5-7: Coupling factors of CCF

### (5) Evaluation of MMIS vulnerability to CCF

The evaluation of MMIS susceptibility to CCF is based on the analysis of the MMIS architecture, the CCF sources, and mechanisms, and the inter-dependencies of sub-systems and components. The assessment starts with the decomposition of the MMIS architecture.

The decomposition process includes:

- Top-down decomposition which focuses on the consequences of CCF occurrences with respect to the architecture
- Bottom-up decomposition which focuses on the mechanisms of failure
- Functional decomposition
- Physical decomposition

Based on the decomposed architecture, each block element is analyzed to identify the potential fault, and trigger mechanism that could lead to credible CCF. Analyses that are commonly performed include failure modes and effects analysis (FMEA), fault tree analysis (FTA), hazard analysis, and risk analysis. Once all the fault types and trigger conditions have been identified for the decomposed blocks, a table matrix that captures the fault-trigger combinations, which can lead to CCF, is developed. Blocks in the table matrix that have the same combination of fault and trigger mechanism are vulnerable to CCF.

### (6) MMIS Design Mitigation strategy against CCF

The MMIS design philosophy is based on the fundamental safety principle that safe conditions must be retained under all operational conditions. Prevention methodology entails a defense in depth approach in which human, functional and design diversities play important roles. Mitigation

Strategies include:

- Fault minimization during the system design and development phases to minimize faults. This involves fault avoidance by avoiding unnecessary functional complexity and fault removal by testing, formal inspection, and formal design proofs.
- Avoidance of common faults by employing functional diversity, design diversity, and human diversity.
- Protection against failure propagation on communication channels
- Fault tolerance techniques such as the use of watchdog timers.

There are four models used in computing CCF, namely:

- Basic Parameter Model,
- Beta-Factor Model,
- Multiple Greek Letter Model, and
- Alpha-Factor Model

## 5.5 Diversity

Two common solutions to common cause failures is the use of diversity and defense-in-depth strategies. Although, there is currently no definitive guidance specifying how much diversity is sufficient to mitigate CCF vulnerabilities, many digital I&C systems are designed with this underlying concept. Diversity according to NUREG 6303 [35] is a principle in instrumentation system of sensing different parameters, using different technologies, different logic, or algorithms, or using different actuation means to provide several ways of detecting and responding to a significant even. Figure 5-8 illustrates this concept of a four division of system using different technology to achieve a specified system function.

Diversity is the use of more than one physically or functionally different means of performing a single function. It protects against Common Cause Failure (CCF). A CCF is an event in which a single failure or condition affects the operation of multiple devices that would otherwise be considered independent. CCF causes multiple failures of components from shared root causes. By utilizing systems with different principles of operation, the probability of the same failure occurring in both systems is eliminated, thereby increasing the system reliability

The use of software and equipment from different designs or origin to perform the same function can be considered as diversity. Thus, if the failures occurs due to errors in design of the system will not affect the other performances.

Furthermore, diversity can be achieved by using different physical parameters to initiate an action or by using different physical mechanisms to carry it out.

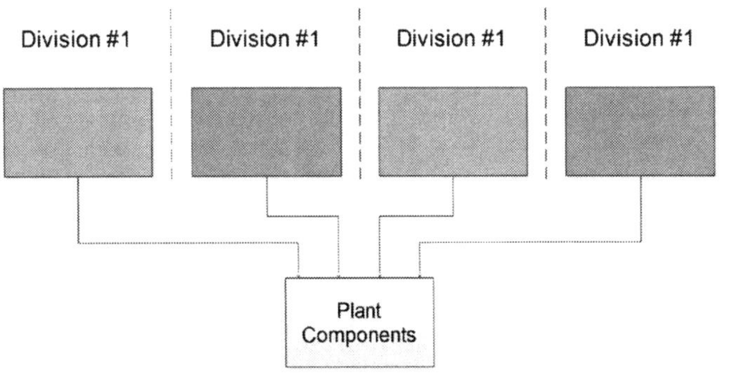


Figure 5-8: Equipment Diversity

The safety system criteria discussed above would address vulnerabilities to single shared failure of equipment. 10 CFR 50 Appendix A (III) gives the criteria for "Protection and Reactivity control systems", several of which deal with issues that are relevant to mitigation of potential CCF vulnerabilities. In particular, GDC 22 requires that "…functional diversity or diversity in component design and principles or operation, shall be used to the extent practical to prevent loss of protection function".

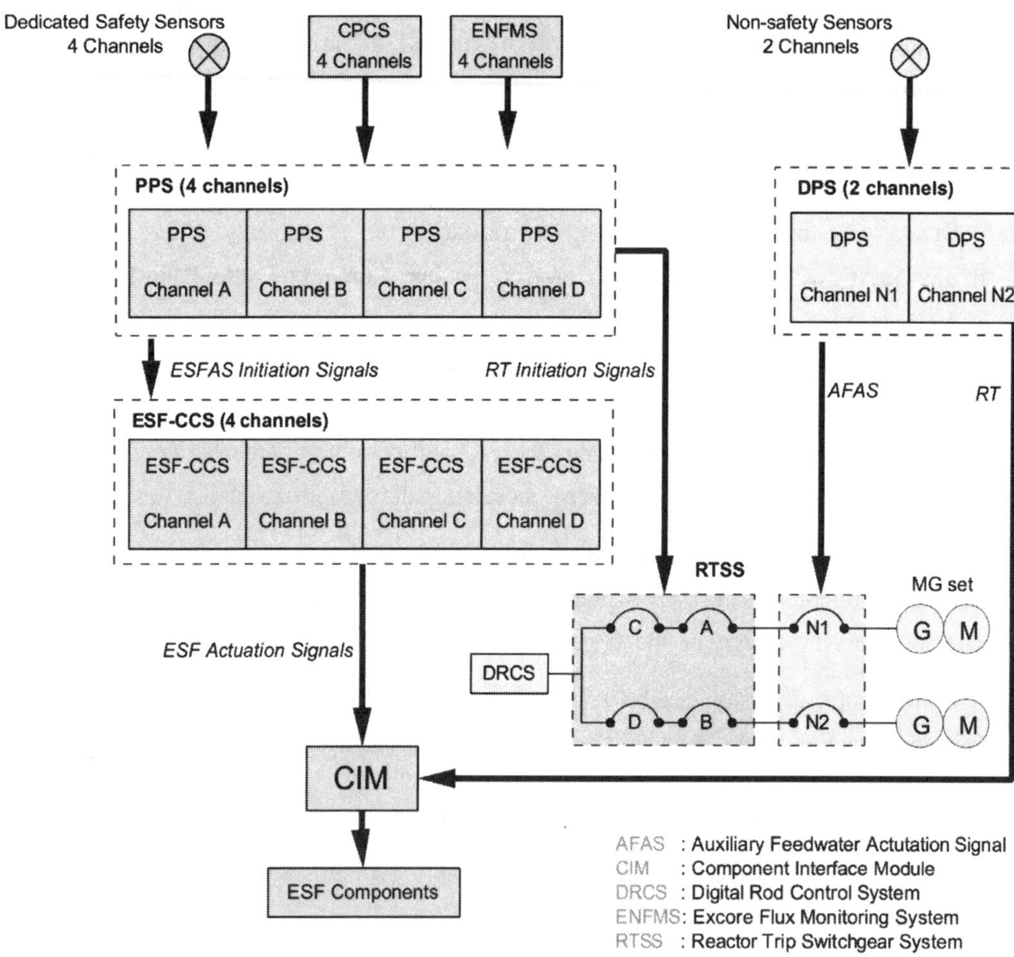


Figure 5-9 Functional Diversity in Safety System

The protection system uses different process variables to detect the design basis accident and initiate the reactor trip. There are 12 possible Reactor Protection System (RPS) trip signals. These signals are generated from different process variables such as containment pressure and pressurizer pressure, to name but a

PART 1

few. As a result, failure of a single process variable would not prevent the reactor from tripping when needed. In addition, both the Plant Protection System (PPS) and the Diverse Protection System (DPS) are capable of tripping the reactor. The PPS trips the reactor by opening the RTSS circuit breakers while the DPS opens the Motor - Generator set output breakers.

In APR1400, two platforms are used, Programmable Logic Control (PLC) and a Distributed Control System (DCS). The Safety systems use a PLC platform, Common Q, while the non-safety systems are implemented on a DCS platform, Ovation. The use of two different hardware and software systems prevents CCF of the digital safety systems. In addition, there are some systems such as the ENFMS and DIS, which are implemented on their own independent platforms. Although the QIAS – N is a non-safety system, it is implemented on the PLC platform due to the fact that it displays important plant parameters and provides diversity from the IPS.

There are generally six important types of diversity: human diversity, design diversity, software diversity, functional diversity, signal diversity, and equipment diversity. These types of diversity are being implemented in the PPS in the following ways:

i. The software execution order in the redundant BPs in a channel is different. In each channel, one BP executes a trip function in sequence 1 through N while the other BP executes trip functions in reverse sequence (N through 1). The reverse trip function execution order between redundant BPs provides software trajectory diversity for the PPS.
ii. Each RTSS circuit breaker has diverse methods of being automatically opened via the shunt trip and under voltage trip devices. The PPS interfaces with the UV trip device and the DPS interfaces with the shunt trip device.
iii. The RTSS consist of one set of four RTSGs (RTSS 1) and another set of four RTSGs (RTSS 2) with diverse design features.
iv. The PPS and the DPS are designed using different hardware and software to address postulated software common cause failures.
v. For the ESF, under LOOP condition, the EDG is started to supply into the safety buses. However, if a disabled condition is initiated by software CCF of PPS/ESF-CCS, necessary power buses are supplied by the AAC DG through manual action.

## 5.6 Redundancy

The fact that single failure criterion is only able to deal with random failures need to be considered as another aspect. Then redundancy, which can be the ultimate outcome for dealing with single failure criterion, are the solution for mitigate the impact of the event.

IEEE Std. 603 describes "redundant equipment or system as a piece of equipment or a system that duplicates the essential function of another piece of equipment or system to the extent that may either perform the required function, regardless of the state of operation or failure of the other. Redundancy can be accomplished by the use of identical equipment, equipment diversity, or functional diversity". Redundancy allows a channel functional test during power operation while still meeting the single failure criteria.

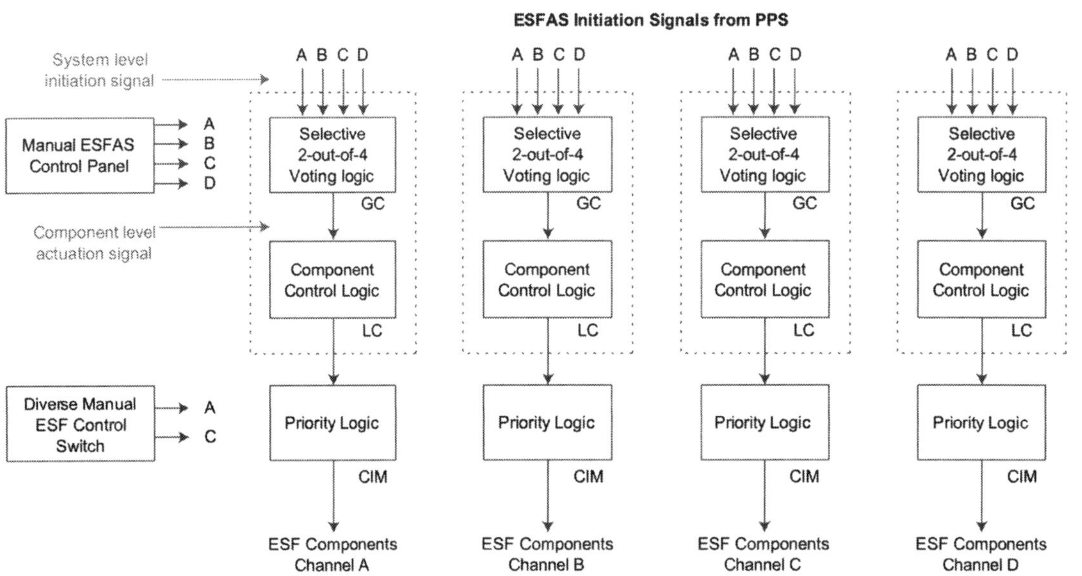


Figure 5-10 Redundancy of ESF-CCS

Redundant features of the PPS include the following:

- Four redundant channels of process sensors
- Four redundant channels of bistable logics
- Redundant BPs on each channel

- Four redundant channels of coincidence logics
- Redundant coincidence logics on each channel
- Four redundant channels of initiation circuits
- Three pairs of manual trip push-buttons with one pair being sufficient to cause a reactor trip
- Four redundant AC power supplies from vital instrument buses
- Four redundant DC power supplies for the RTSG control circuit
- Two independent channels for BOP ESFAS
- The voting logic is changed to 2 out of 3 when a channel is removed for testing without affecting the availability of the system.

## 5.7 Deterministic Design

Deterministic design is a criterion approach to achieve a degree of system safety. Within the deterministic process, several criterion are explained in previous section, including reliability criteria.

A system is deterministic if given a particular input the system will always return the same output by passing through the same sequence of states. In addition, a system has deterministic timing if the time delay between the reception of the input signal and the generation of an output signal has a guaranteed maximum and minimum. A "deterministic" communication system delivers messages within a finite, predictable time delay that is a function of system communication load Deterministic principles are necessary in order to develop “fault free” safety critical systems that can perform their functions with a high level of confidence. This is of particular importance in safety systems where the exact behavior of the system must be known and the system must operate within a predefined time to avoid the escalation of an Anticipated Operational Occurrence (AOO) or Design Basis Accident (DBA).

The way in which determinism is implemented in a system is dependent on the technology used. As an example in PLCs, design practices such as the use of interrupts, multitasking, dynamic scheduling, and event driven design may lead to non-deterministic designs and should be avoided. In FPGAs, asynchronous designs as well as improper definition of states in Finite State Machines may lead to non-determinism in the system behavior.

Thus, BTP 7-14 [31] requires that the algorithms, communication and response time be shown to be deterministic. Guidance on deterministic design and communication is given in NUREG/CR-6083 [37].

## 5.8 Reliability

Reliability is generally defined as the probability that a system or component will perform its function under specified conditions for a specified period without failure. It related to statistical area. Why "Statistics" are used in this reliability method? Because many phenomena that encountered are only partially predictable and it is "Random". Reliability prediction therefore requires a detailed and precise description of the system's function, the environment of operation, the time scale, and the definition of what constitute a failure.

Reliability is based on failure rates. Failure rate, $\lambda$ (t), is the anticipated number of times failure occurs in a specified time interval [0, t], given that it was operating at time t (0). Failure rates are generally categorized as Mean Time Between Failures (MTBF), Mean Time to Failure (MTTF), and Mean Time to Repair (MTTR).

### (1) Mean Time Between Failures (MTBF)

This is the basic measure of reliability for repairable system or component. MTBF is the elapsed time before failure occurs in relation to constant failure rate. It is the expected period between two consecutive failures. For a constant failure rate system, it is expressed as:

### (2) Mean Time to Failure (MTTF)

This is the basic measure of reliability for non-repairable system or component. It is the statistical mean time until the first failure occurs. For a constant failure rate system, it is expressed as:

### (3) Mean Time to Repair (MTTR)

This is the basic measure of reliability for repairable system or component. It is the mean time from failure state to repair state. MTTR is the total corrective/repair time divided by the total repairs/corrective actions.

Both analog and digital I&C systems, which are designed for plant safety and protection functions, must meet reliability criteria. For many years, analog technology has performed these functions. However, due to obsolescence and the benefits that digital systems offer, NPP I&C systems are now transitioning toward digital technology. Unlike analog systems which are characterized by wear & tear, digital technology offers the potential for enhancing the reliability of plant safety and protection systems [39].

The top-level reliability criterion for protection system is articulated as Criterion 21 of the General Design Criteria (GDC) for Nuclear Power Plants in 10 CFR 50 Appendix A. Criterion 21 which is "Protection System Reliability and Testability," requires that the design of protection systems be made for high functional reliability in a manner proportionate to the safety functions to be performed [33].

The industrial standards IEEE Std. 603 and IEEE Std. 7-4.3.2 specify the requirement and guidance for meeting the qualitative and quantitative reliability goals. IEEE Std. 352 and IEEE Std. 577 provide guidance for performing reliability analysis. Both IEEE Std. 603 and IEEE Std. 7-4.3.2 are endorsed by the USNRC regulatory guides RG 1.152 and RG 1.153. The basic framework for satisfying the reliability requirement is presented in Figure 5-11 below.

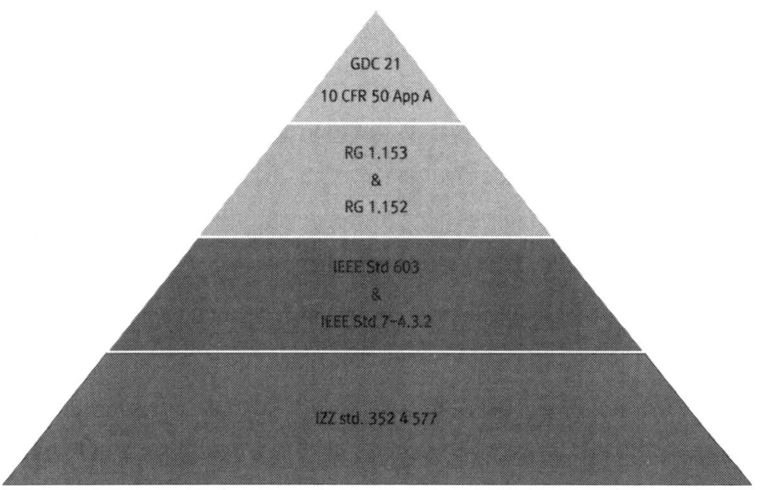


Figure 5-11: Hierarchy framework for digital I&C system reliability portion of requirements

### (4) Probabilistic Safety Assessment

As mentioned in section 5.2, single failure criterion, Probabilistic Safety Assessment can be used for measuring a reliability of system, component, or item. Uncertainty analysis is performed by assigning the distribution for each basic event. In nuclear PSA or lognormal distribution is assumed. PSA is a conceptual and mathematical tool for deriving numerical estimates of risk for nuclear plants and industrial installations (Quantitative Risk Analysis).

The objectives of PSA are to assess the level of safety of the plant and to identify the most effective areas for improvement; assess the level of safety and compare to with explicit or implicit standards; assess the level of safety to assist plant operation. The first comprehensive PSA is WASH-1400 (USNRC, Reactor Safety Study: An Assessment of Accident Risks in US Commercial NPP, NUREG-75/-14)

There are three levels in PSA

- Level 1 : The assessment of plant failures leading to the determination of core damage frequency (CDF)
- Level 2 : The assessment of containment response leading, together with Level 1 results, to the determination of containment release frequencies
- Level 3 : The assessment of off-site consequences leading, together with the result of Level 2 analysis, to estimates of public risks (fatality, cancer, ...)

Major quantitative components of PSA are described as below.

#### ① Fault Tree Analysis

Fault tree analysis is known as a top down analysis. A deductive failure analysis in which an undesired state of a system is analyzed using Boolean logic to combine a series of lower-level events. The “System Failure Condition" is used for the "undesired state" or top event of the fault tree.

A deductive process by means of which an undesirable event (top event), is postulated, and the possible ways for this event to occur are systematically deduced. A fault tree does not necessarily contain all possible failure modes of the components (or units) of the system. Only those failure modes, which contribute to the occurrence of the top event, are modelled.

The example of this is illustrated in Figure 5-12 and more examples are provided in the appendix.

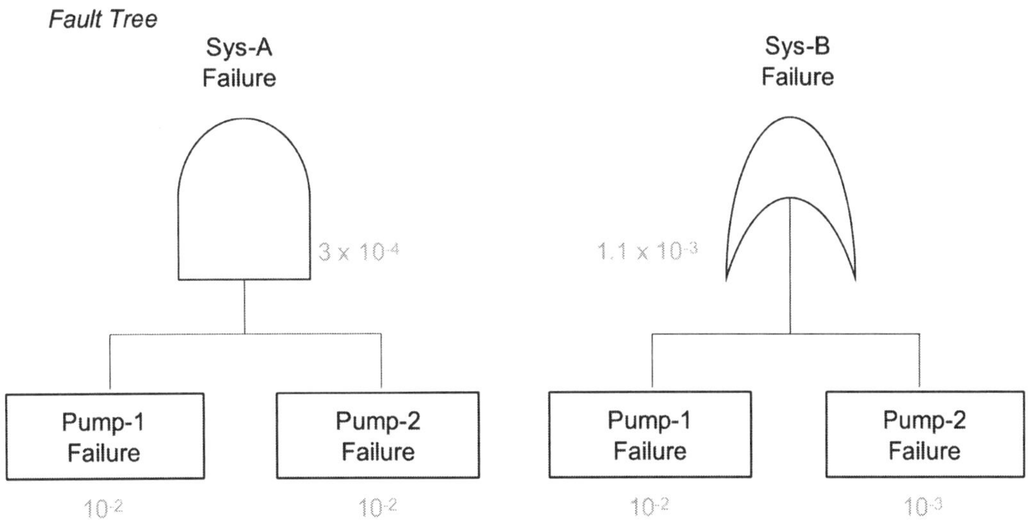


Figure 5-12: Example of Fault Tree Analysis

② **Event Tree Analysis**

Event Tree Analysis is aimed at estimating the probability/frequency of an accident. Estimation of probability/frequency for each accident sequence requires FTA.

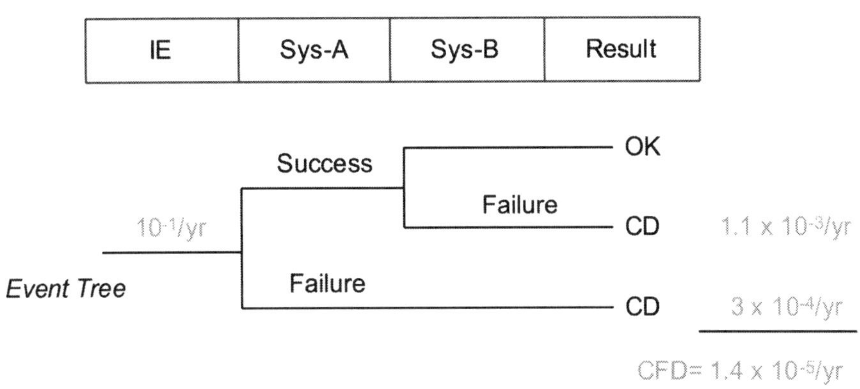


Figure 5-13: Example of Event Tree Analysis

## PROBLEMS

1. What is the importance of the Safety System Criteria?
2. Describe the single failure criterion.
3. What is the importance of Independence in Safety Systems? Give an example of Independence in the MMIS systems
4. What are the main sources of CCF in the MMIS?
5. How can CCF be mitigated?
6. What is the difference between redundancy and diversity?
7. Discuss how the safety system criteria lead to improved system reliability.
8. What is the difference between event tree and fault tree analysis?
9. Select a simple sub-system of the MMIS and develop its Fault Tree
10. Select a simple sub-system of the MMIS and develop its Event Tree

## FURTHER READING


- U.S Nuclear Regulatory Commission, "Instrumentation and Controls," in Standard Review Plan for the Review of Safety Analysis Reports for Nuclear Power Plants: LWR Edition, 2016.
- "IEEE 603 Standard Criteria for Safety Systems for Nuclear Power Generating Stations," 2009.
- IAEA, Protecting against Common Cause Failures in Digital I&C Systems of Nuclear Power Plants, IAEA.
- USNRC, BTP 7-14, Guidance on Software Reviews for Digital Computer-Based Instrumentation and Control Systems, 1997.
- USNRC, "Current State of Reliability Modeling Methodologies for Digital Systems and Their Acceptance Criteria for Nuclear Power Plant Assessments," USNRC, Washington, DC 20555-0001, 2006.
- USNRC, BTP 7-19, "Guidance for Evaluation of Diversity and Defense-In-Depth in Digital Computer-Based Instrumentation and Control Systems Review Responsibilities"

# REFERENCES


[1] IAEA, "Application of Field Programmable Gate Arrays in Instrumentation and Control Systems of Nuclear Power Plants", Vienna, 2016.

[2] USNRC, "PART 50—Domestic Licensing of Production and Utilization Facilities", [Online] [Accessed 3 June 2019].

[3] IAEA, "Modern Instrumentation and Control for Nuclear Power Plants: A Guidebook", [Online] [Accessed 3 June 2019].

[4] USNRC, "Instrumentation and Controls" in *Standard Review Plan for the Review of Safety Analysis Reports for Nuclear Power Plants: LWR Edition*, 2016.

[5] KEPCO and KHNP, "APR1400 Design Control Document, Chapter 7 Instrumentation and Controls," [Online]. [Accessed 4 1 2019].

[6] B. S. Blanchard, System Engineering Management, Wiley.

[7] C. S. Wasson, "System Analysis, Design, and Development.," Wiley, 2006.

[8] INCOSE, Systems Engineering Handbook, 4th ed., Wiley, 2015.

[9] D.W. Kim. H. Y. Chung, "Design of Advanced Power Reactor (APR1400) I&C System," in *IFAC Proceedings Volumes*, Seoul, 2003.

[10] [Online]. http://tti.tamu.edu/product/catalog/reports/ITSRCE-00_01.pdf?cv=1.

[11] W. Parker, Defense Acquisition University Program managers Tool Kit, 2011.

[12] Ocean Observatories Initiative, "Systems Engineering Management Plan," [Online]. [Accessed 3 June 2019].

[13] C. J. Garry J. Roedler, "Technical Measurement," INCOSE, 2005.

[14] Ian Sommerville, Pete Sawyer, "Requirements Engineering: A Good Practice Guide," Wiley, 1997.

[15] DOD, "Guide for Achieving Reliability, Availability and Maintainability", DOD, 2005.

[16] INCOSE, Pragmatic Principles of Systems Engineering, 1993.

[17] Defense Aquisition University, "Systems Engineering Fundamentals"

[18] INCOSE, "Systems Engineering Book of Knowledge 1.9.1," [Online]. [Accessed 5 June 2018].

[19] KHNP, "Engineering Safety Features," in *APR1400 Design Description*, 2014

[20] KEPCO and KHNP, "Safety I&C System," [Online]. [Accessed 5 June 2019].

[21] USNRC, "Engineered Safety Features," 1 December 2012. [Online]. [Accessed 24 December 2018].

[22] J. C. Jung. J. Mayaka, "Complexity reduction of the Engineered Safety Features Component Control System," *Nuclear Engineering and Design,* vol. 331, pp. 194-203, 2018.

[23] P. H. Seong, S. Han, "A methodology for benefit assessment of using in-core neutron detector signals in core protection calculator system (CPCS) for Korea standard nuclear power plants (KSNPP)," *Annals of Nuclear Energy,* vol. 26, no. 6, pp. 471-488, 1999.

[24] WEC, "Component Interface Module". Patent: US6842669B2, August 2002.

[25] IEEE-603, “Standard Criteria for Safety Systems for Nuclear Power Generating Stations", 2009.

[26] IEEE-379, "IEEE Standard for Application of the Single-Failure Criterion to Nuclear Power Generating Station Safety Systems," 2014.

[27] IAEA, "Safety Series No.50-P-1, Application of the Single Failure Criterion," Vienna, 1990.

[28] IAEA, "Modern Instrumentation and Control for Nuclear Power Plants: A Guidebook," TECHNICAL REPORTS SERIES No. 387, Vienna, 1999.

[29] IAEA, “Protecting against Common Cause Failures in Digital I&C Systems of Nuclear Power Plants”,

[30] NUREG/CR-6303, "Method for Performing Diversity and Defense-in-Depth Analyses of Reactor Protection Systems," 1994.

[31] USNRC, "BTP 7-14, Guidance on Software Reviews for Digital Computer-Based Instrumentation and Control Systems," 1997.

[32] USNRC, "Current State of Reliability Modeling Methodologies for Digital Systems and Their Acceptance Criteria for Nuclear Power Plant Assessments," 2006.

[33] USNRC, "Appendix A to Part 50—General Design Criteria for Nuclear Power Plants," USNRC, 29 August 2007. [Online]. [Accessed 4 January 2019].


PART 1

# PART 2

# DESIGN & DEVELOPMENT

# Chapter 6 STAKEHOLDER NEEDS AND REQUIREMENTS DEFINITION


SUMMARY

Stakeholder requirements definition is an important process in the application of systems engineering to the development of any system. The success of the system development is hinged on meeting the needs and requirements of stakeholders throughout the system life cycle. Typically, stakeholder requirements govern the development process of the system and are an important factor in clarifying the scope of the development process.

The concept of stakeholder requirements has been introduced in a previous chapter. Stakeholders are typically system users/operators, regulatory bodies, policy/decision makers, system maintainers, vendors/manufacturers, system designers, organization responsible for standards, technology support organizations, and a large variant of other interested parties who may affect the system life-cycle development. Since the stakeholder-base is quite large for any typical project, it is common in the development process to identify major stakeholders.

This chapter will discuss the source documents from which stakeholder requirements are defined for the MMIS development. The chapter will also provide useful techniques for analyzing the source documents to ensure consistency, correctness, and completeness in order to easily derive the system requirements. Traceability analysis method and document decomposing procedure will be used as validation technique to fulfil stakeholder acceptance.

Mind map of Chapter 6 is illustrated in Figure 6-1.

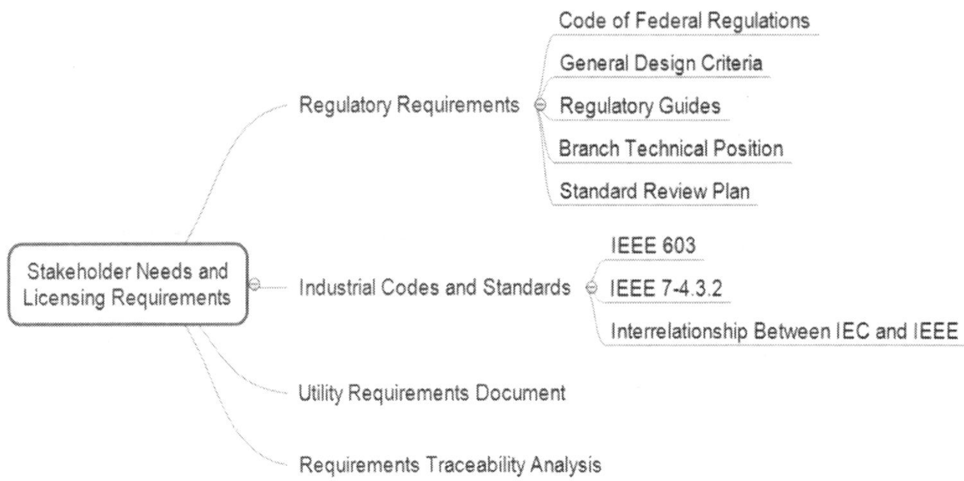


Figure 6-1 Mind Map of Chapter 6

**OBJECTIVES**

- Introduction of Regulatory Requirements
- Discussion of industrial codes and standards
- Traceability analysis of stakeholders requirements

## 6.1 Stakeholder Requirements

From the viewpoint of systems engineering, the MMIS development process begins with stakeholder identification and requirement elicitation. Figure 6-2 shows an example of the different types of stakeholders and their expectations or requirements for a plant protection system. The major stakeholders are typically government, regulatory authorities, industrial organizations, utilities companies, and technology vendors. These stakeholders form the required minimum basis for stakeholder requirements definition. . The expectation of the government is to have

PART 2

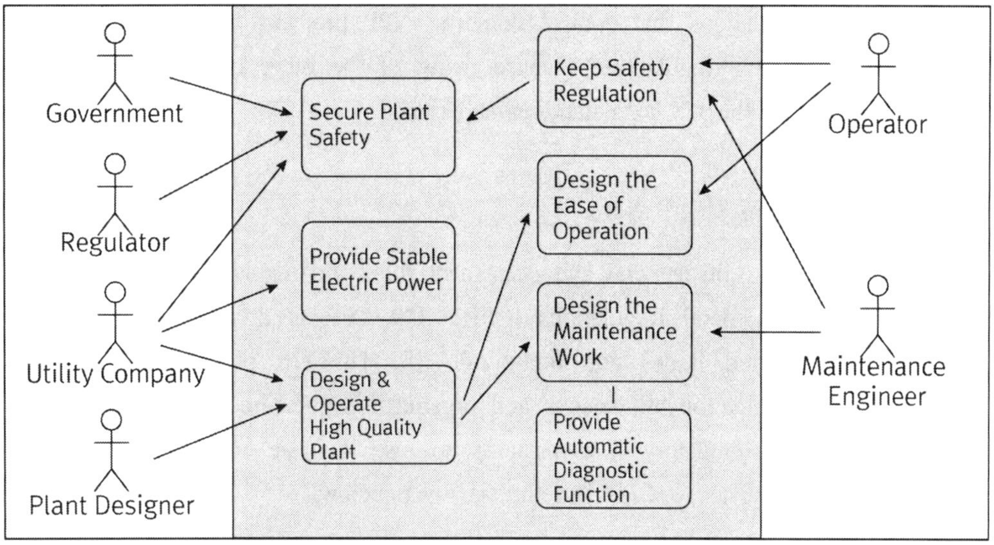


Figure 6-2: Stakeholders and Requirements

plants operated safely and in a manner that will not portend any danger to people and the environment. To achieve this, regulators set the minimum plant design criteria, which the operators must meet. The utilities and plant owners set requirements for plant safety and economics by adopting requisite industrial requirements in the design of reliable and high quality systems, which afford easy maintainability over the operational life of the plant.

The stakeholder requirements documents include the utility requirement documents such as KURD and EURD; industrial codes and standards such as IEEE and IEC; and applicable regulations such as the USNRC regulatory guides.

### 6.1.1 Regulatory Requirements

The USNRC provides comprehensive documentation of applicable laws, regulations, guides, and safety reports that govern the design, construction, commissioning, operation, and decommissioning of a domestic licensed nuclear power plant. Although, the regulatory documents cover a wide range of issues related to NPP project, our focus here is on the issues relating to MMIS life-cycle development, and in particular, our interest is in the safety-critical MMIS regulatory documents. It should be noted that other countries such as

Korea have regulations that control domestic NPP projects, but we have chosen to examine the US regulations because many of the other countries' regulations are derived from the US long-standing experience.

### (1) Code of Federal Regulations

The USNRC has its general and permanent rules and regulations in a document called Code of Federal Regulations (CFR). The CFR is divided into fifty titles and nuclear related issues are under title 10. Title 10 of the CFR contains requirements binding on all persons and organizations who receive a license from NRC (Nuclear Regulatory Commission) to use nuclear materials or operate nuclear facilities. The 10 CFR structure is given below.

Table 6-1: 10 CFR Structure [1]

<table>
<tr><th>Volume</th><th>Chapter</th><th>Parts</th><th>Regulatory Entity</th></tr>
<tr><td>1</td><td rowspan="2">I</td><td>0-50</td><td rowspan="2">Nuclear Regulatory Commission</td></tr>
<tr><td>2</td><td>51-199</td></tr>
<tr><td>3</td><td rowspan="2">II</td><td>200-499</td><td rowspan="2">Department of Energy</td></tr>
<tr><td rowspan="7">4</td><td>500-699</td></tr>
<tr><td>III</td><td>700-999</td><td>Department of Energy</td></tr>
<tr><td>X</td><td>1000-1099</td><td>Department of Energy (General Provisions)</td></tr>
<tr><td>XIII</td><td>1300-1399</td><td>Nuclear Waste Technical Review Board</td></tr>
<tr><td>XVII</td><td>1700-1799</td><td>Defense Nuclear Facilities Safety Board</td></tr>
<tr><td>XVIII</td><td>1800-1899</td><td>Northeast Interstate Low-Level Radioactive Waste Commission</td></tr>
</table>

The following part numbers are tightly related to the NPP

- 10 CFR 20: Standards for Protection Against Radiation
- 10 CFR 50: Domestic Licensing of Production and Utilization Facilities
- 10 CFR 100: Reactor Site Criteria

The parts of 10 CFR Part 50 that are forming the top-tier requirements to I&C safety systems are defined in Figure 6-3.

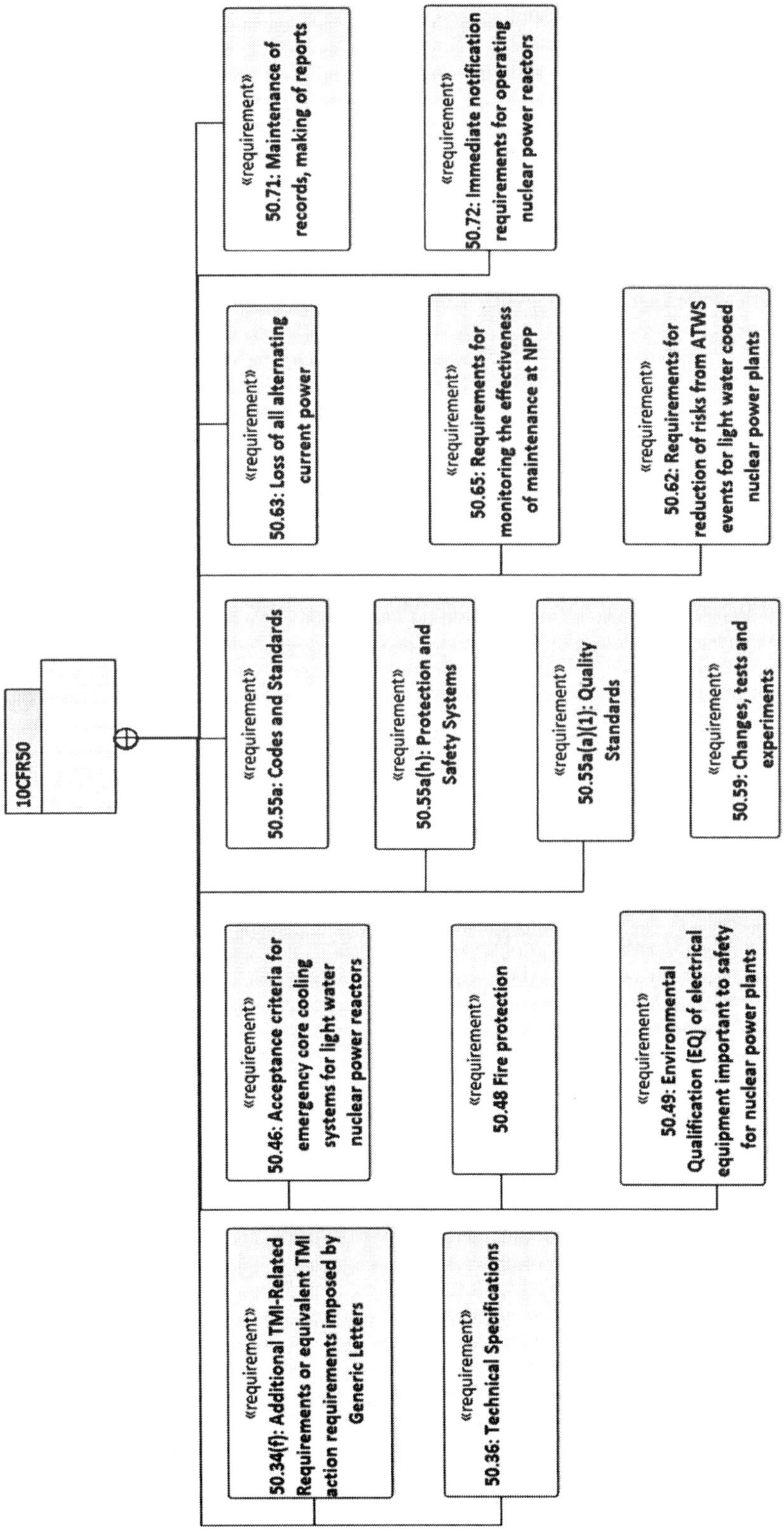


Figure 6-3: 10 CFR 50 Sections relevant to I&C Systems

### (2) General Design Criteria (GDC)

10 CFR 50 Appendix A (General Design Criteria for Nuclear Power Plants) establish minimum requirements for the principal design criteria for water cooled nuclear power plants similar in design location to plants for which construction permits have been issued by the commission. The aim is to provide reasonable assurance that the facility can be operated without undue risk to the health and safety of the public [2]. The GDCs that apply to the safety critical I&C systems are defined in Figure 6-4.

### (3) Regulatory Guides (RGs)

The regulatory guide "provides guidance to licensee and applicants on implementing specific parts of the NRC's regulations, techniques used by the NR staff in evaluating specific problems or postulated accidents, and data needed by the staff in its review of applications for permits or licenses" [3]. This document is not legally binding; equivalent alternative should be proven if not followed. The regulatory guides for protection systems are illustrated in Figure 6-5.

### (4) Branch Technical Paper (BTP)

Just like the RGs, BTPs also outline acceptable approaches to technical issues or questions of interpretation of NRC's regulations that may have arisen in the past. The BTPs that apply to the safety critical I&C systems are defined in Figure 6-6.

### (5) Standard Review Plan

The SRP includes the NRC laws, General Design Criteria (GDC), Regulatory Guides (RGs), and endorsed industry standards related to licensing issues for LWR reactors (e.g., APR1400). The purpose of the SRP is to aid NRC Reviewing Staff in evaluating the applicant's/ Licensee's application, to ensure that it meets the NRC regulations.

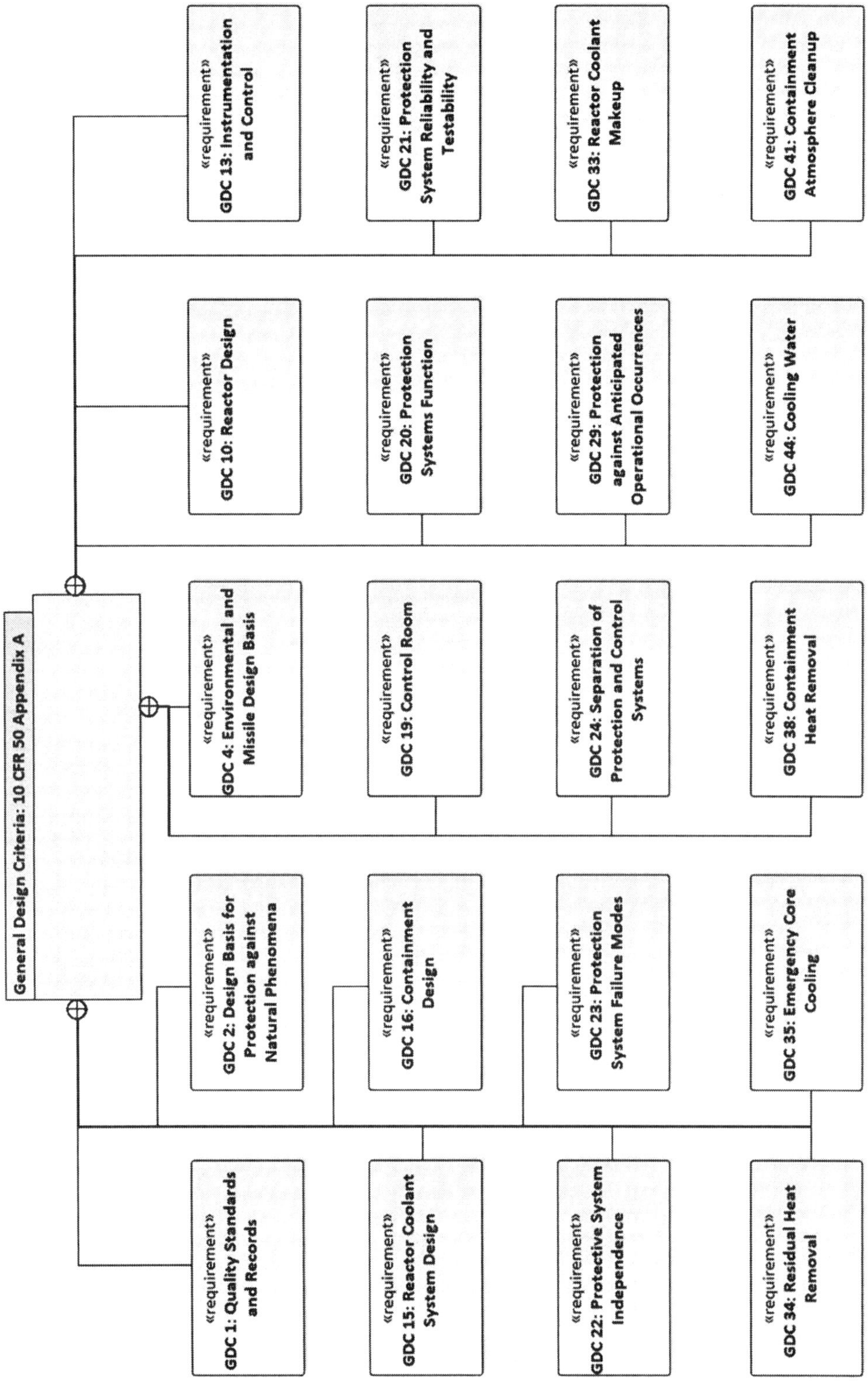


Figure 6-4: GDCs applicable to I&C Systems

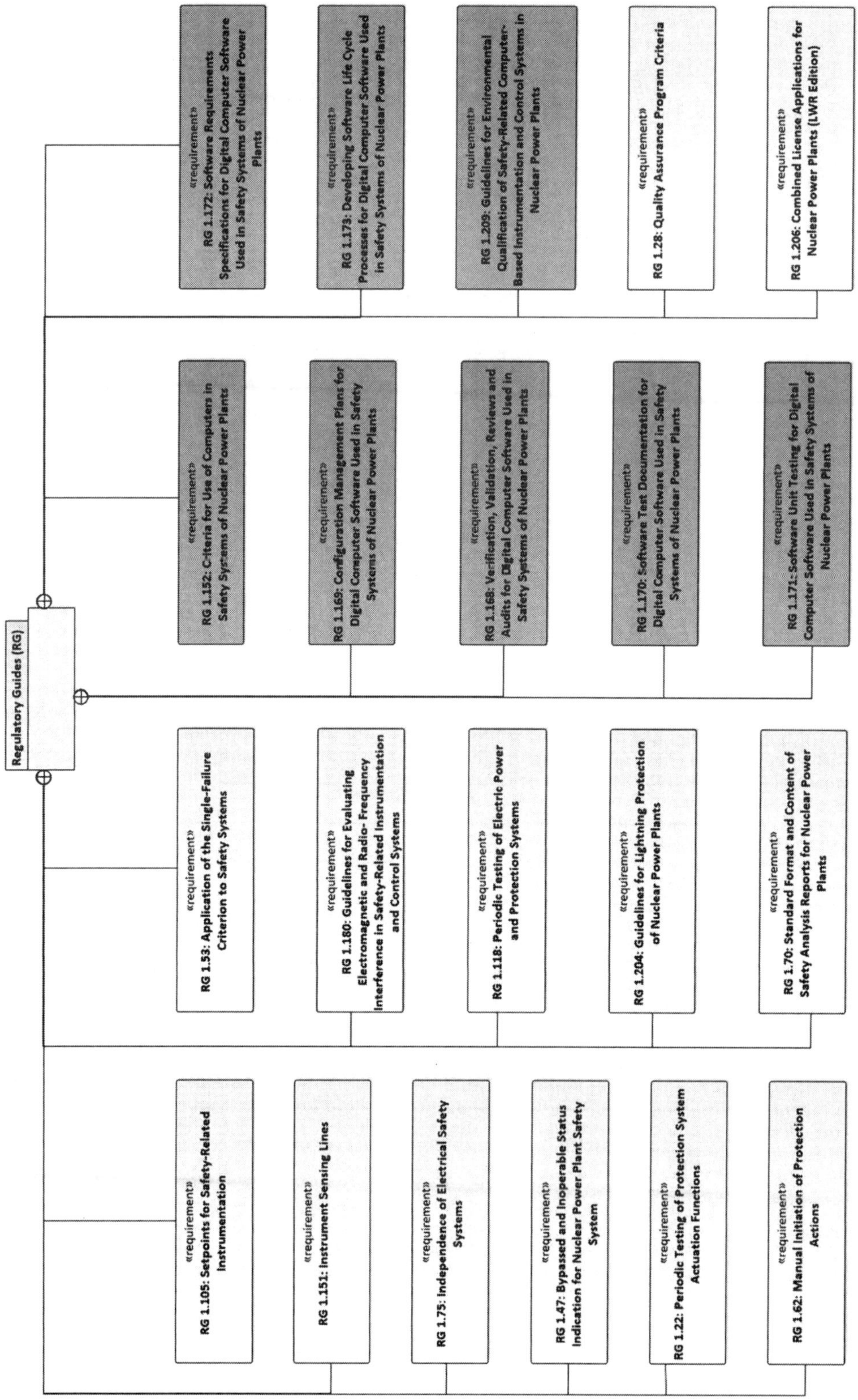


Figure 6-5: Regulatory Guides Related to I&C Systems

PART 2

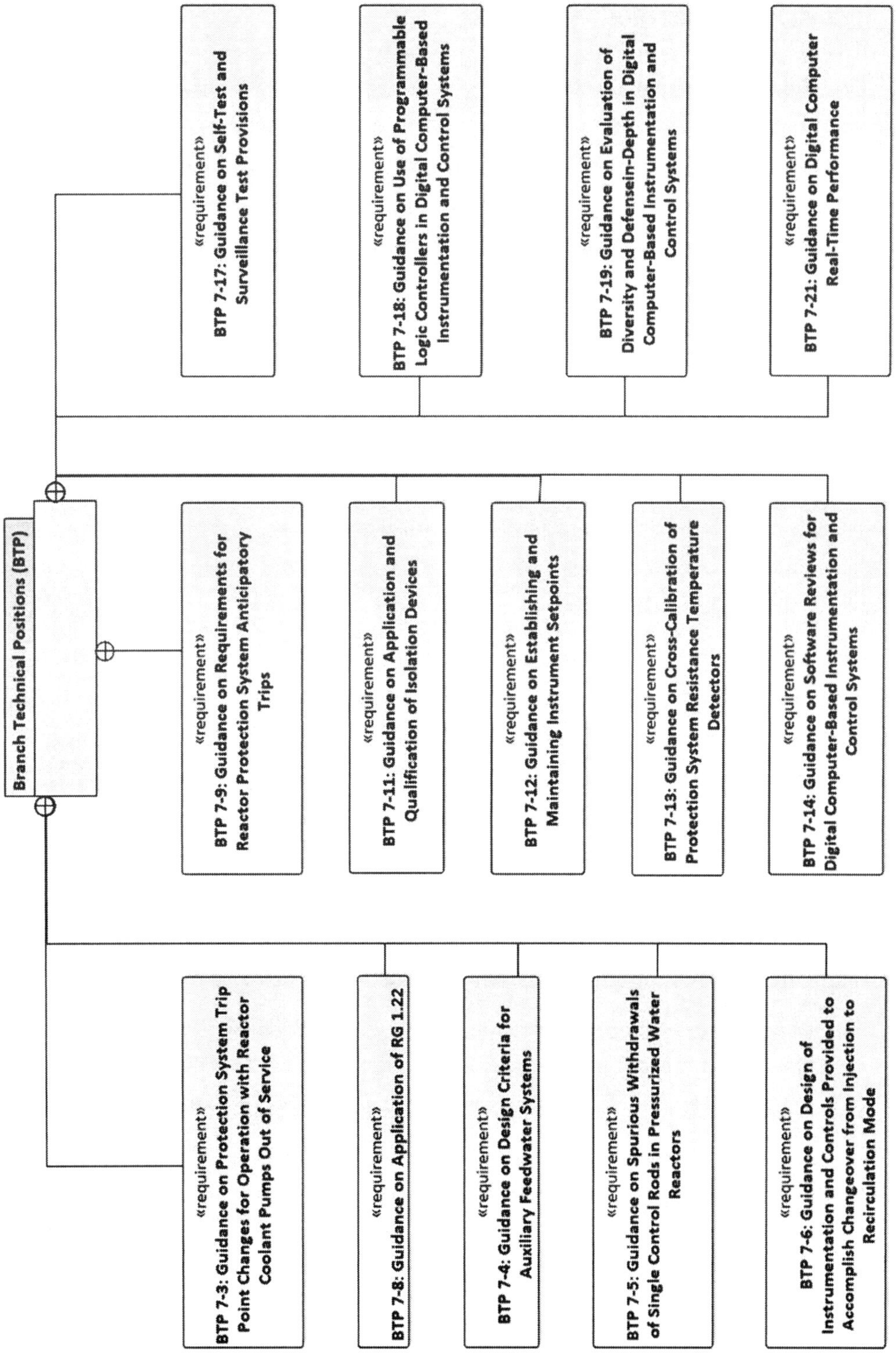


Figure 6-6: BTPs related to I&C Systems

### 6.1.2 Industrial Codes and Standards

The Regulatory requirements listed in the previous section (CFR, RG, BTP) contain very general requirements. Where there are established industrial standards on a given aspect e.g. Single Failure Criteria, the regulatory guide may endorse the relevant standard i.e. IEEE 308, Standard Application of the Single-Failure Criterion to Nuclear Power Generating Station Safety Systems. When a regulatory guide explicitly endorses a given standard, the plant designer must adhere to the standard in order to fulfill the regulatory requirements. NRC NUREG-0800 (SRP) lists standards, RGs and BTPs as documents that contain specific SRP acceptance criteria to meet relevant requirements of NRC's regulations. Figure 6-7 details the standards endorsed by the NRC for safety critical I&C systems. It should however be noted, that while the regulatory guides may endorse a given standard, sometimes clarifications or exceptions are made. In such a case, the designer should carefully review the RG to check if it states any exceptions where the standard should not be followed.

In addition both the regulatory guides and standards are updated periodically by the regulatory body (NRC) and standard issuing body (e.g. IEEE and IEC) respectively. Care must therefore be taken that the correct standard version that is endorsed by the regulatory guide is in use. The standard that is endorsed is not always the most recent version. For example, RG RG 1.168, Revision 2, (Verification, Validation, Reviews, and Audits for Digital Computer Software Used in Safety Systems of Nuclear Power Plants) endorses IEEE 1012 - 2004. The most recent version of IEEE 1012 is IEEE 1012 – 2016. However, in order to fulfill the requirements of RG 1.168, the older standard must be used.

Furthermore, the designer must specify the cut-off dates, that is, the date the beyond which changes to codes and standards and new information will not be considered. As mentioned, the RGs are periodically revised. By defining a cut-off date, the version of the RG that will be used is 'locked', allowing the designer to work with a fixed set of requirements.

### (1) Review of IEEE Std. 603 - 2009

The regulations in 10 CFR 50.55a(h) require that reactor protection and safety systems satisfy the criteria in Institute of Electrical and Electronics Engineers (IEEE) Standard (Std.) 603-1991, "IEEE Standard Criteria for Safety Systems for Nuclear Power Generating Stations" The criteria contained in IEEE 603 establish minimum functional and design requirements for the power, instrumentation, and control portions of safety systems for nuclear power generating stations.

The criteria are to be applied to those systems required to protect the public health and safety by functioning to mitigate the consequences of design basis events. The intent is to promote appropriate practices for design and evaluation of safety system performance and reliability. Although the standard is limited to safety systems, many of the principles may have applicability to equipment provided for safe shutdown, post-accident monitoring display instrumentation, preventive interlock features, or any other systems, structures, or equipment related to safety [4].

- **IEEE 603 Decomposition and Traceability**

The structure and content of this stander is illustrated in Figure 6-8. The requirements to be met are stated in chapters 4 through to 8. Chapter 4 is concerned with requirement for the Design Basis Document, while Chapters 5 to 8 contain the safety system criteria and the requirements for the sense and command, execute, and power source sections of the safety systems.

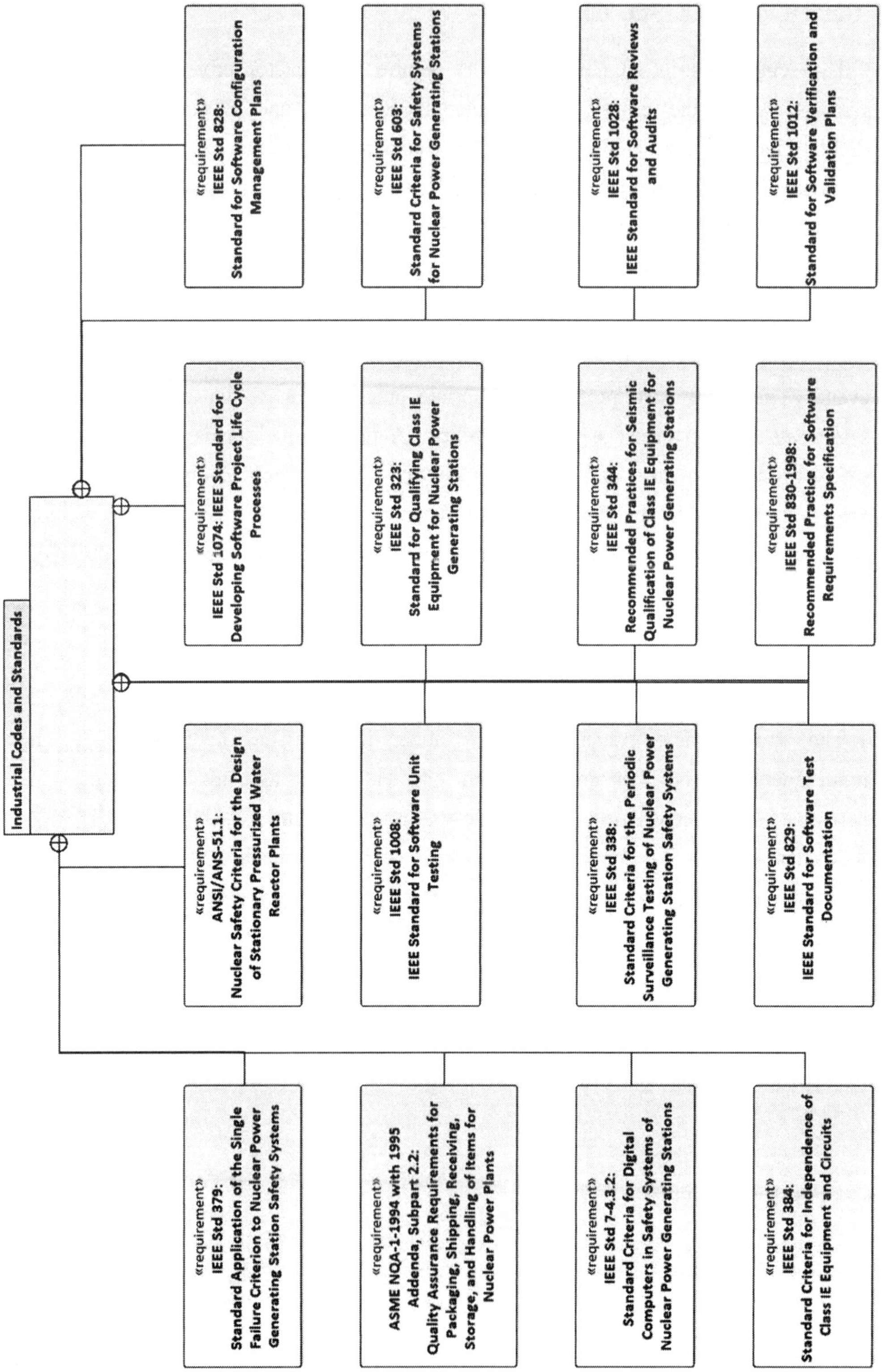


Figure 6-7: Industrial and Standards applicable to I&C Systems

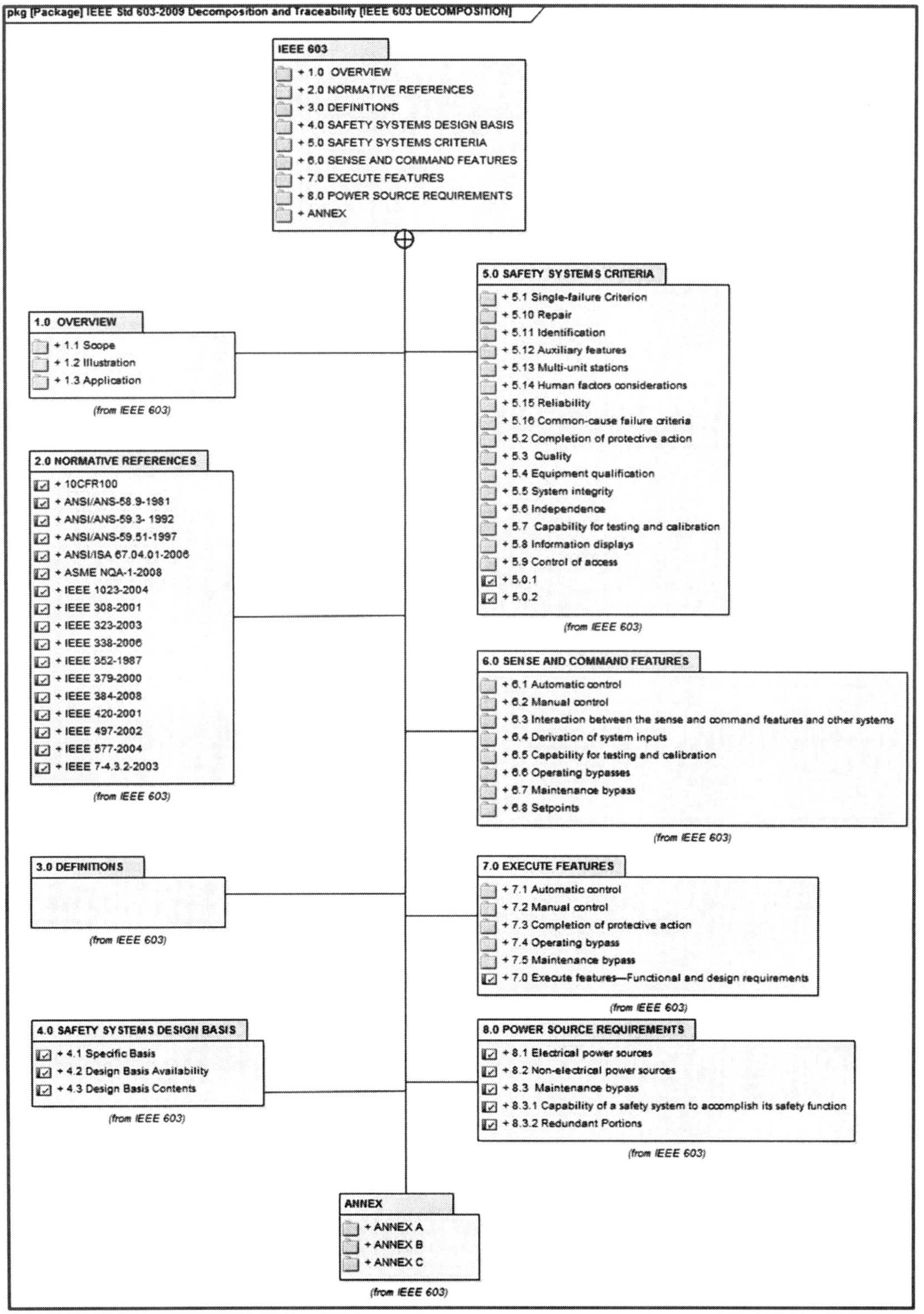


Figure 6-8: Decomposition of IEEE 603

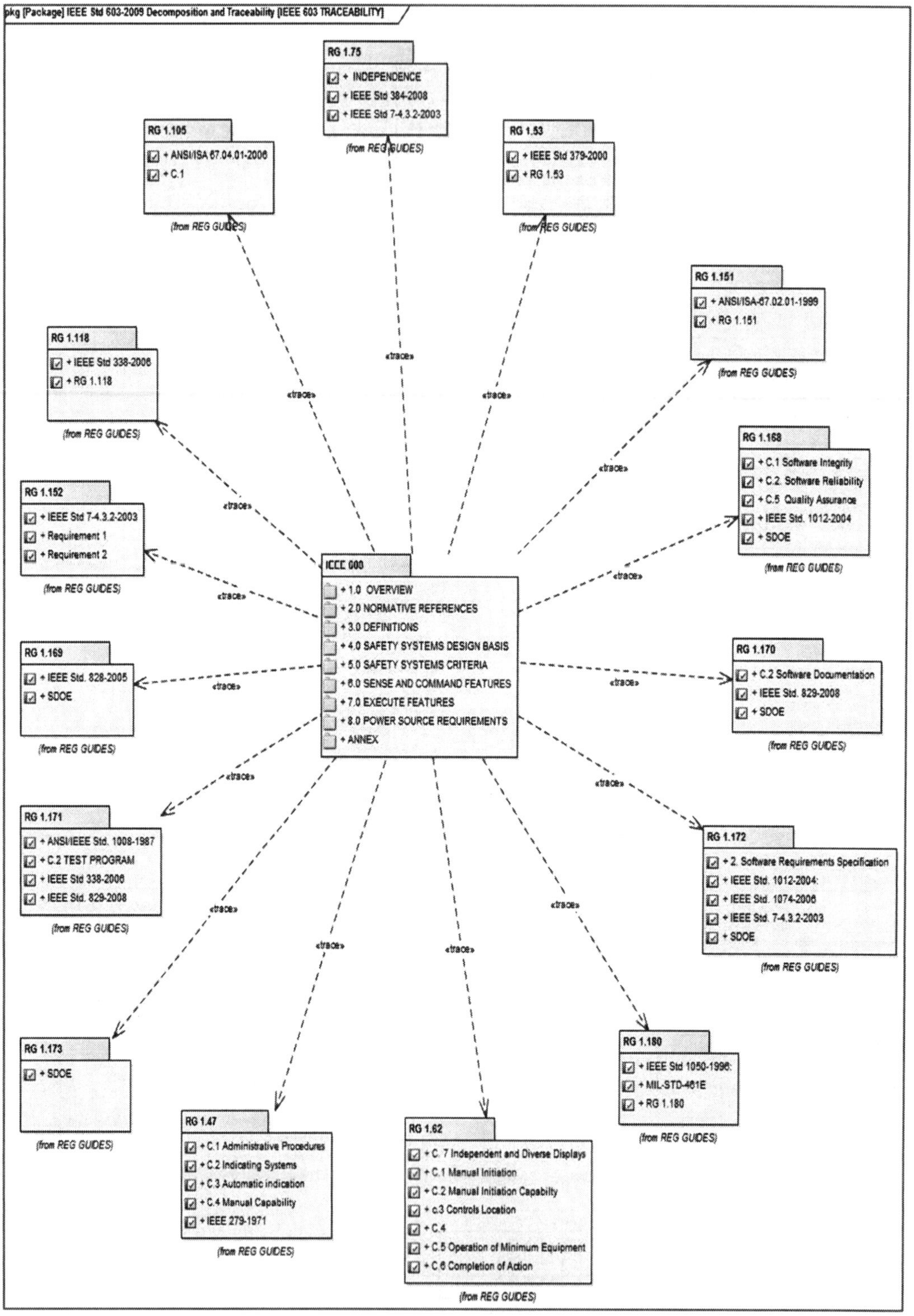


Figure 6-9: Traceability of IEEE 603 to relevant RGs

IEEE 603 can be traced to multiple standards as illustrated in Figure 6-9. In addition, IEEE 603 references other standards. Thus, in order to fulfill the requirements of IEEE 603, the referenced standards must also be adhered to. Table 6-2 details the traceability between the regulatory guides and sections of the IEEE 603, as well as related standards.

Table 6-2: Traceability of IEEE 603 to RGs and other Standards

| NRC RG | RG TITLE | IEEE 603 SECTION | RELATED STANDARD |
|---|---|---|---|
| RG 1.47 | Bypassed and Inoperable Status Indication for Nuclear Power Plant Safety Systems | 5.8.3 Indication of bypasses | IEEE 279-1971: Criteria for Protection Systems for Nuclear Power Generating Stations |
| RG 1.53 | Application of the Single-Failure Criterion to Safety Systems | 5.1 Single Failure Criterion | IEEE Std. 379-2000: Application of the Single-Failure Criterion to Nuclear Power Generating Station Safety Systems. |
| RG 1.62 | Manual Initiation of Protective Actions | 6.2 Manual control - Sense and Command Features | |
| | | 7.2 Manual Control - Execute Features | |
| | | 5.8.1 Displays for manually controlled actions | IEEE Std. 497-2002: Criteria for Accident Monitoring Instrumentation for Nuclear Power Generating Stations. |
| | | 5.6 Independence | IEEE Std. 384-2008: Provides detailed criteria for the independence of Class 1E equipment and circuits.<br><br>IEEE Std. 7-4.3.2-2003: Provides guidance on the application of this criteria for the separation and isolation of the data processing functions of interconnected computers |
| | | 5.1: Single Failure Criterion | IEEE Std. 379-2000: Application of the Single-Failure Criterion to Nuclear Power Generating Station Safety Systems. |

| NRC RG | RG TITLE | IEEE 603 SECTION | RELATED STANDARD |
|---|---|---|---|
| RG 1.75 | Criteria for Independence of Electrical Safety Systems | 5.6 Independence | IEEE Std. 384-2008: Provides detailed criteria for the independence of Class 1E equipment and circuits.<br>IEEE Std. 7-4.3.2-2003: Provides guidance on the application of this criteria for the separation and isolation<br>of the data processing functions of interconnected computers |
| RG 1.105 | Set-points for Safety-Related Instrumentation | 6.8 Set-points: | ANSI/ISA 67.04.01-2006, Set-points for Nuclear Safety-Related Instrumentation |
| RG 1.118 | Periodic Testing of Electric Power and Protection Systems | 5.7 Capability for testing and calibration | IEEE Std. 338™-2006:<br>IEEE Standard Criteria for the Periodic Surveillance Testing of Nuclear Power<br>Generating Station Safety Systems. |
| RG 1.151 | Instrument Sensing Lines | 5.5 System Integrity | ANSI/ISA-67.02.01-1999:<br>Nuclear Safety-Related Instrument-Sensing Line Piping and Tubing Standard for Use in Nuclear Power Plants |
| RG 1.152 | Criteria For The Use of Computers in Safety Systems of Nuclear Power Plants | 5.6.3.1 Interconnected Equipment | IEEE Std. 7-4.3.2-2003:<br>Criteria for Digital Computers in Safety Systems of Nuclear Power Generating Stations |
| | | 5.9 Control of access | |
| RG 1.168 | Verification, Validation, Reviews and Audits for Digital Computer Software Used in Safety Systems of Nuclear Power<br>Plants | 5.3 Quality<br><br>5.4 Equipment qualification | IEEE Std. 1028-2008:<br>Standard for Software Reviews and Audits<br><br>1012-2004:<br>Standard for Software Verification and Validation<br><br>IEEE Std. 7-4.3.2-2003:<br>Criteria for Digital Computers in Safety Systems of Nuclear Power Generating Stations. |
| | | 5.5 System Integrity | |
| | | 5.15 Reliability | |
| | | 5.9 Control of access | IEEE Std. 1012-2004:<br>Standard for Software Verification and Validation |

| NRC RG | RG TITLE | IEEE 603 SECTION | RELATED STANDARD |
|---|---|---|---|
| RG 1.169 | Configuration Management Plans for Digital Computer Software Used in Safety Systems of Nuclear Power Plants | 5.9 Control of access | IEEE Std. 828-2005: Standard for Software Configuration Management Plans |
| RG 1.170 | Test Documentation for Digital Computer Software Used In Safety Systems Of Nuclear Power Plants | 5.9 Control of access<br>5.11 Identification | IEEE Std. 829-2008:Standard for Software and System Test Documentation |
| RG 1.171 | Software Unit Testing for Digital Computer Software Used in Safety Systems of Nuclear Power Plants | 5.7 Capability for testing and calibration<br><br>6.5 Capability for testing and calibration | ANSI/IEEE Std. 1008-1987<br><br>IEEE Std. 338™-2006, IEEE Standard Criteria for the Periodic Surveillance Testing of Nuclear Power<br>Generating Station Safety Systems.<br><br>IEEE Std. 829-2008:<br>IEEE Standard for Software and System Test Documentation |
| RG 1.172 | Initial Test Programs For Water-Cooled Nuclear Power Plants | 4.0 Safety system design basis<br><br>5.9 Control of access | IEEE Std. 7-4.3.2-2003:<br>Standard Criteria for Digital Computers in Safety Systems of Nuclear Power Generating Stations<br><br>IEEE Std. 1012-2004:<br>Standard for Software Verification and Validation<br><br>IEEE Std. 1074-2006:Standard for Developing a Software Life Cycle Process |
| RG 1.173 | Developing Software Life-Cycle Processes for Digital Computer Software Used in Safety Systems of Nuclear Power Plants | 5.9 Control of access | |
| RG 1.180 | Guidelines for Evaluating Electromagnetic and Radio-Frequency Interference in Safety Related Instrumentation and Control Systems | 4.3_g<br><br>Annex B: Electromagnetic compatibility | IEEE Std. 1050-1996:<br>Guide for Instrumentation and Control Equipment Grounding in Generating Stations<br><br>MIL-STD-461E:<br>Requirements for the Control of Electromagnetic Interference Characteristics of Subsystems and Equipment<br><br>IEC 61000 |

### (2) Review of IEEE Std. 7-4.3.2 - 2016

This standard "serves to amplify criteria in IEEE Std. 603, to address the use of programmable digital devices as part of safety systems in nuclear power generating stations. The criteria contained herein, in conjunction with criteria in IEEE Std. 603-2009, establish minimum functional and design requirements for programmable digital devices used as components of a safety system" [5].

- **IEEE 7-4.3.2 Decomposition and Traceability**

The figure bellow illustrates the structure of the IEEE standard.

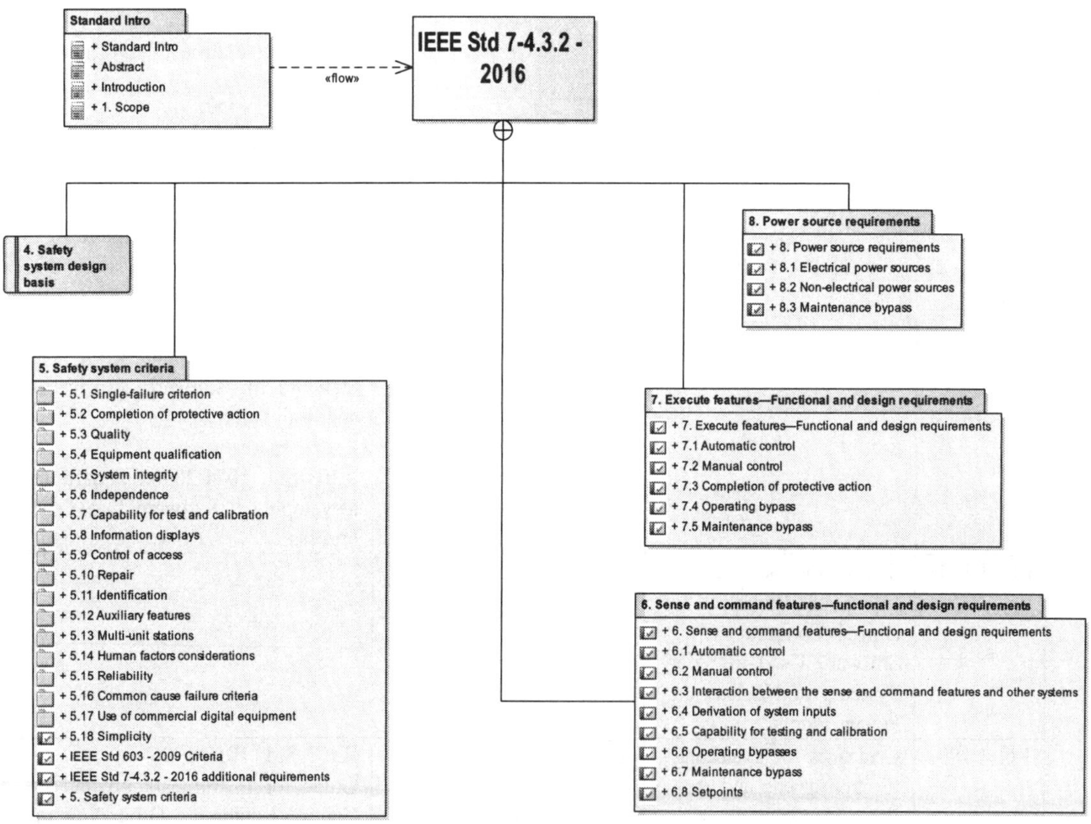


Figure 6-10: IEEE 7-4.3.2 Decomposition

By comparing Figure 6-8 and Figure 6-10, it is obvious that all the chapters are of the same name. This is due to the fact that IEEE 7-4.3.2 is derivative of IEEE 603. Thus, it only provides further detail when extra requirements are needed for digital computers, otherwise it states that "No requirements beyond IEEE Std 603-2009 are necessary". For Example, in Chapter 5 of IEEE 7-4.3.2 which is illustrated in Figure 6-11, all the safety criteria defined by IEEE 603 are included. The standard only concentrates on the aspects that are introduced by the use of computers e.g. Software Configuration Management

Figure 6-11 to Figure 6-15 illustrate the structure/ decomposition of IEEE 1012, while Figure 6-16 illustrates its traceability to the Regulatory Guides 1.47, 1.62, 1.152, respectively.

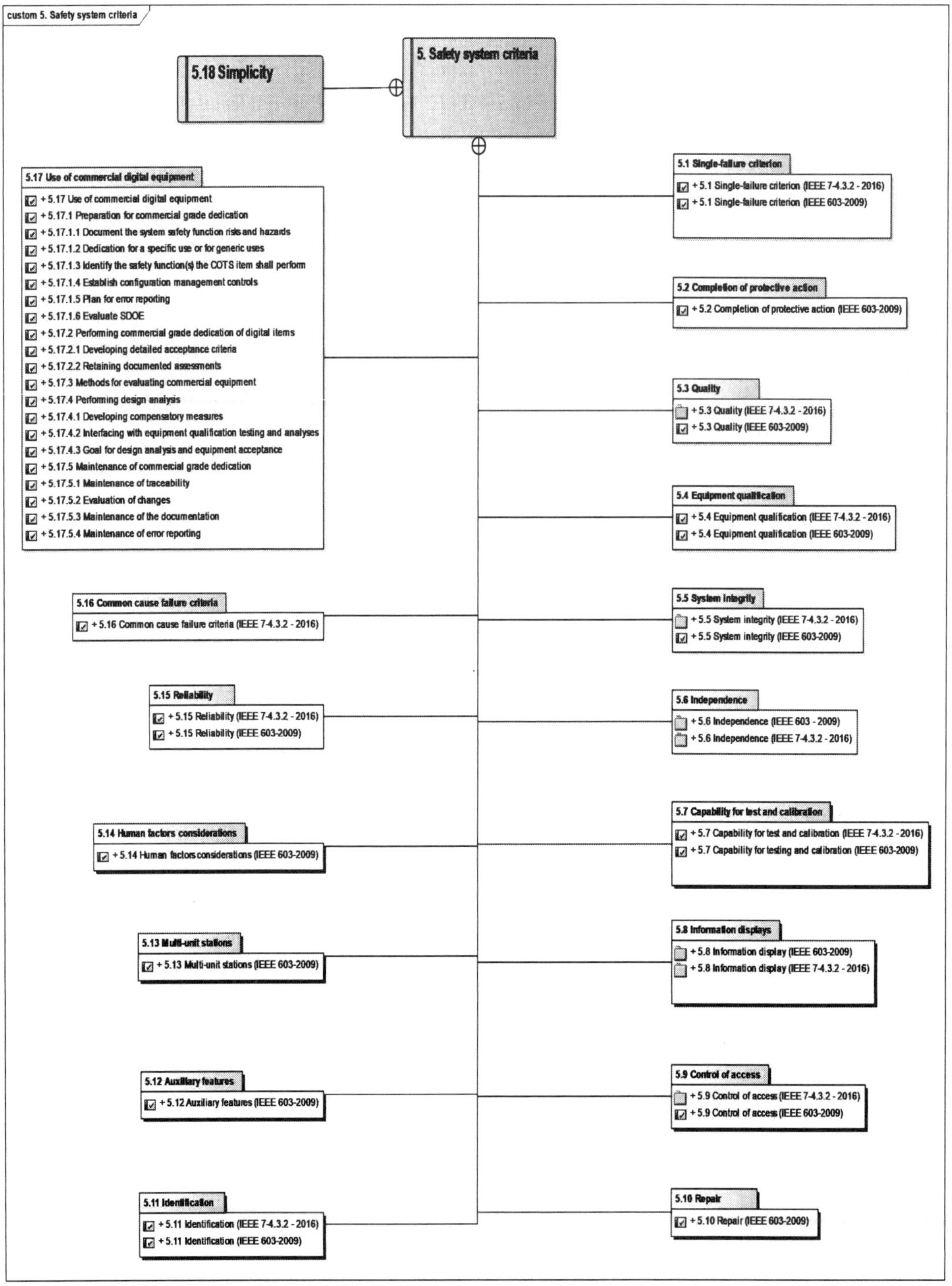


Figure 6-11: Decomposition of IEEE 7-4.3.2 Section 5

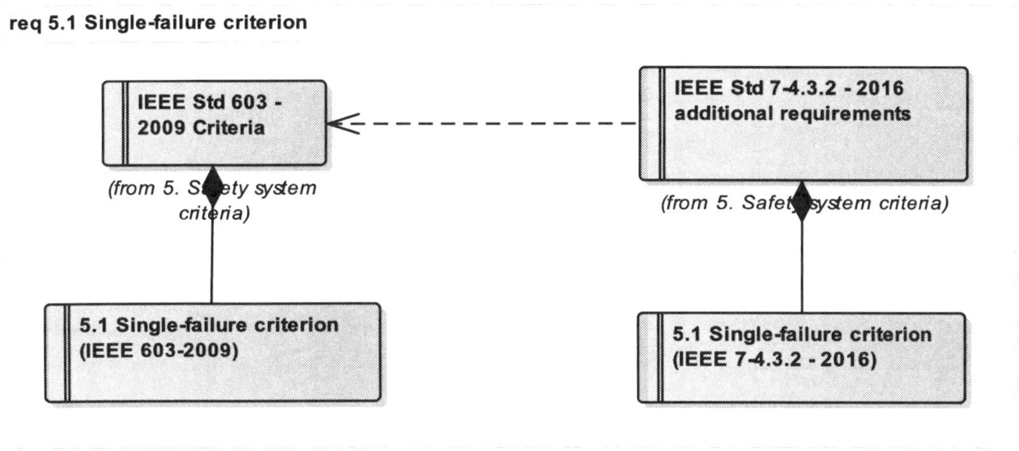


Figure 6-12: Decomposition of IEEE 7-4.3.2 Section 5.1

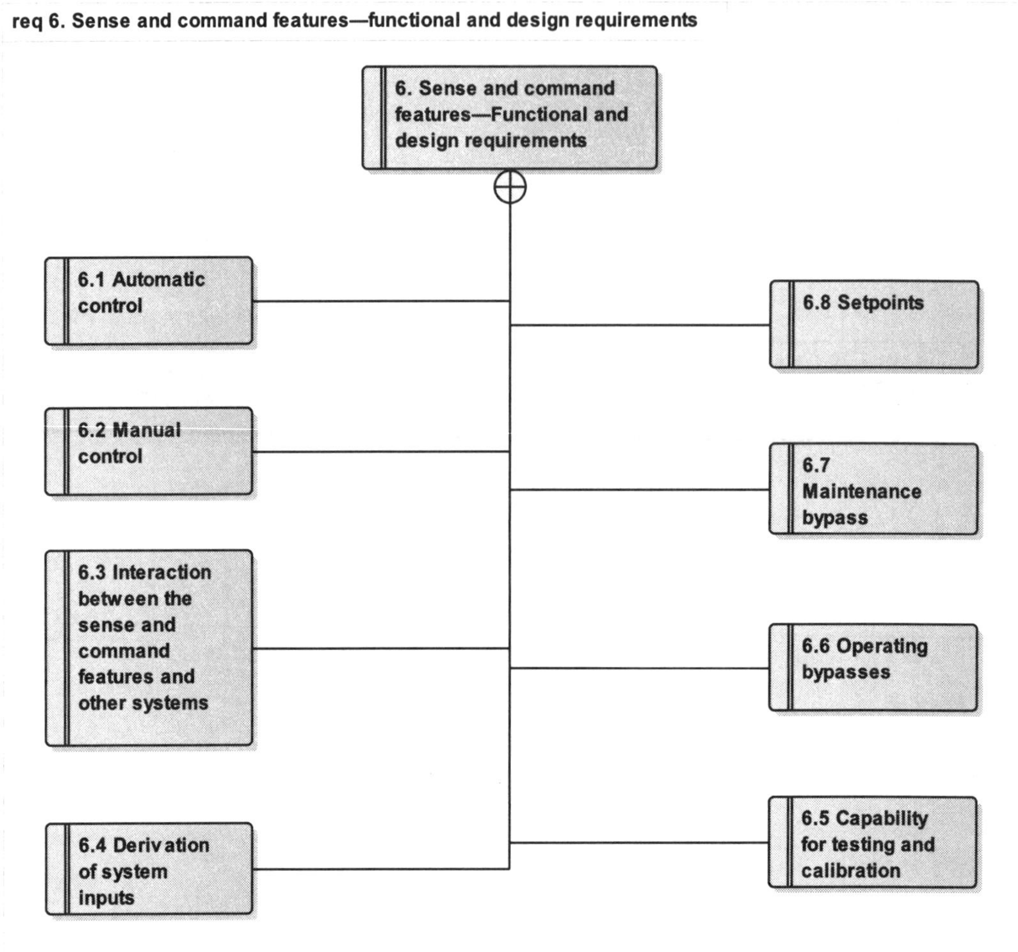


Figure 6-13: Decomposition of IEEE 7-4.3.2 Section 6

P A R T 2

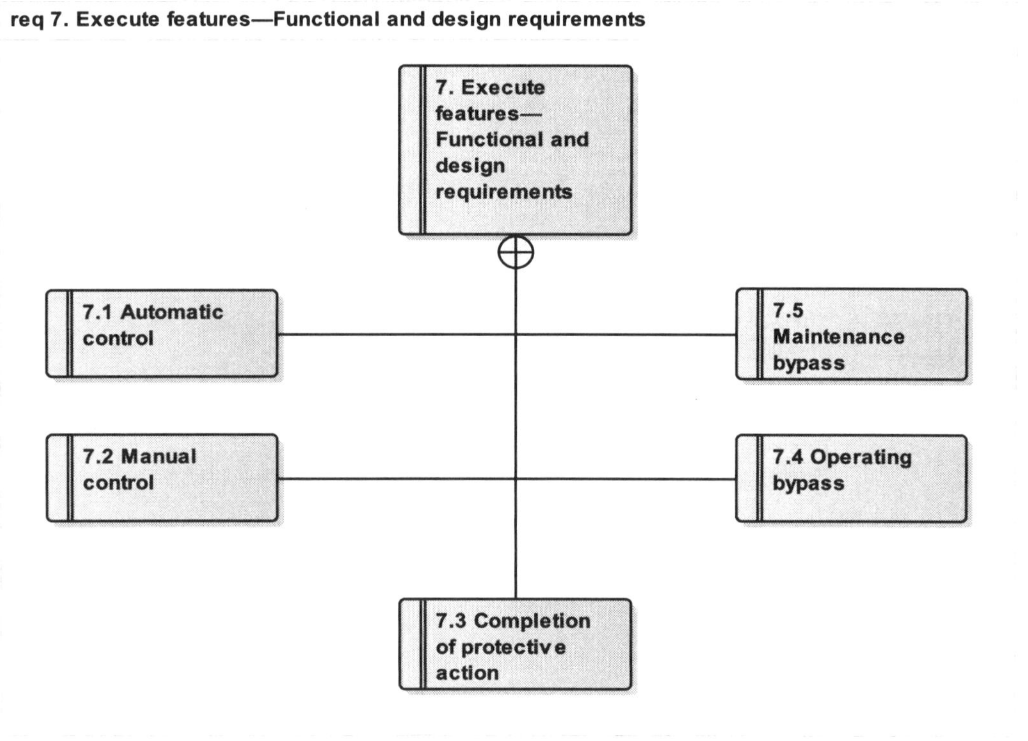


Figure 6-14: Decomposition of IEEE 7-4.3.2 Section 7

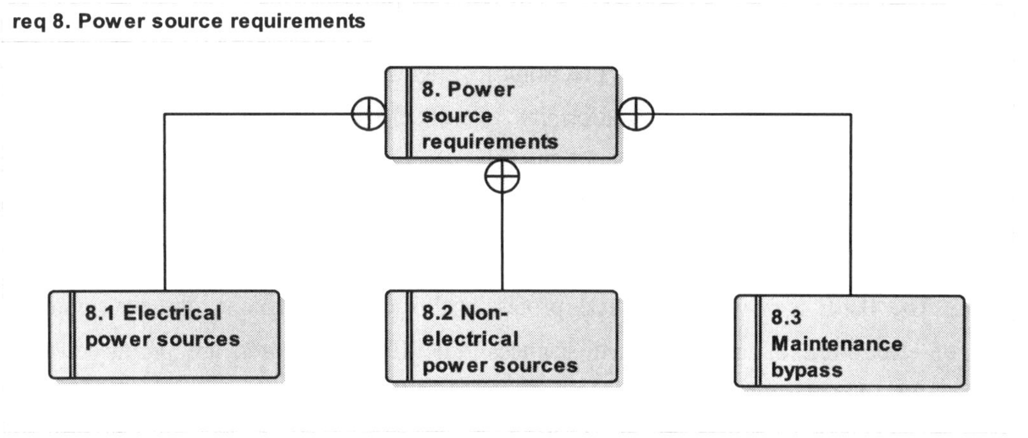


Figure 6-15: Decomposition of IEEE 7-4.3.2 Section 8

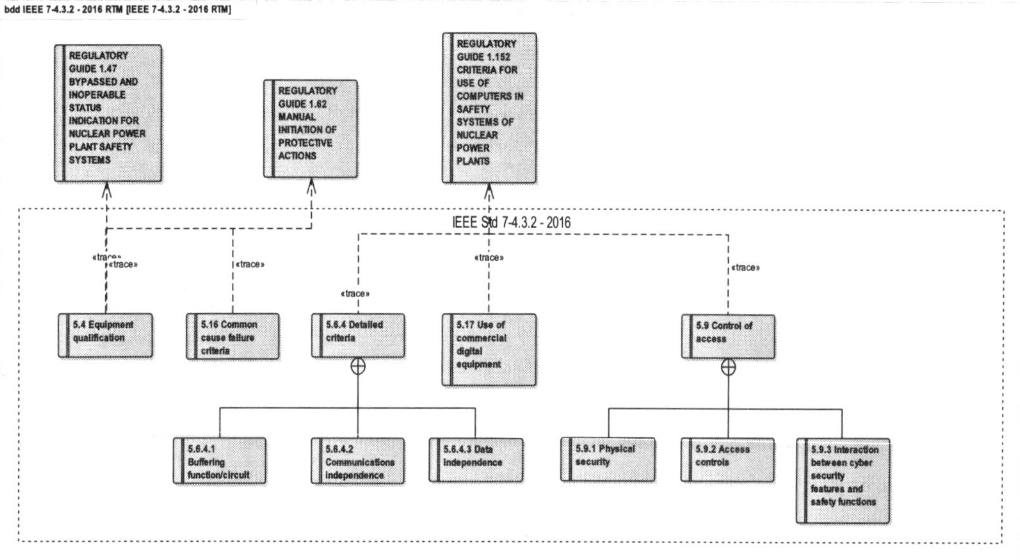


Figure 6-16: IEEE 7-4.3.2 Traceability

## (3) Interrelationship between IEC and IEEE

Worldwide Organizations develop standards used for Instrumentation and Control (I&C) systems in Nuclear Power Plants. Two of the primary standards organizations are the International Electrotechnical Commission (IEC) and the Institute of Electrical and Electronics Engineers (IEEE). The IEC is the international standards and conformity assessment body for all field of electrical technology including NPP technology. The IEC standards cover Europe and other countries around the world and most of NPPs in European countries have been designed based on the IEC standards.

The IEEE is a world technical professional society to promote the development of electro-technology and allied sciences. The IEEE standards are produced with ANSI (American National Standard Institute) and endorsed by USNRC Regulatory Guides. NPPs in USA and South Korea have been designed based on the IEEE standards. The IEEE and IEC standards have some overlap, but in some cases cover different topics. In order to understand the similarities and differences between IEC and IEEE standards for NPP, we analyze interrelationship between IEC and IEEE focus on I&C important to safety systems as stated above.

### ① Interrelationship between IEC 61513 and IEEE standards

The IEC 61513, "I&C for systems important to safety - General requirements for systems", provides requirements and recommendations for the overall I&C systems important to safety. IEEE standards related to IEC 61513 are as follows.

i. ***IEEE 603, "IEEE Standard Criteria for Safety Systems for Nuclear Power Generating Stations"***
ii. ***IEEE 308, "IEEE Standard Criteria for Class 1E Power Systems for Nuclear Power Generating Stations"***
iii. ***IEEE 7-4.3.2, "IEEE Standard criteria for digital computers in safety systems of nuclear power generating stations"***
iv. ***IEEE 379, "IEEE Standard for Application of the Single-Failure Criterion to Nuclear Power Generating Station Safety Systems"***

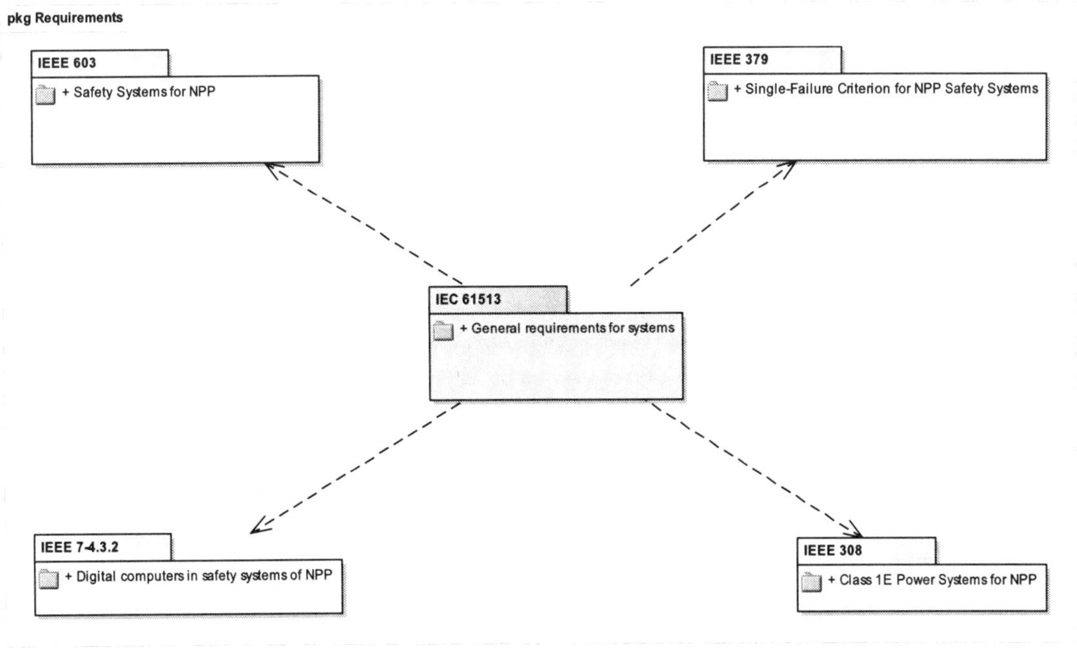


Figure 6-17: Interrelationship between IEC 61513 and IEEE standards

### ② Interrelationship between IEC 61226 and IEEE standards

The IEC 61226 Nuclear power plants - Instrumentation and control important to safety - Classification of instrumentation and control functions "establishes a method of classification of the information and command functions for nuclear power plants, and I&C systems and equipment that provide those functions, into categories that designate the importance to safety of the function" [6]. IEEE standards related to IEC 61226 are as follows.

The interrelation is shown in Figure 6-18.

i. ***IEEE 603, "IEEE Standard Criteria for Safety Systems for Nuclear Power Generating Stations"***

ii. ***IEEE 308, "IEEE Standard Criteria for Class 1E Power Systems for Nuclear Power Generating Stations"***

iii. ***IEEE 7-4.3.2, "IEEE Standard criteria for digital computers in safety systems of nuclear power generating stations"***

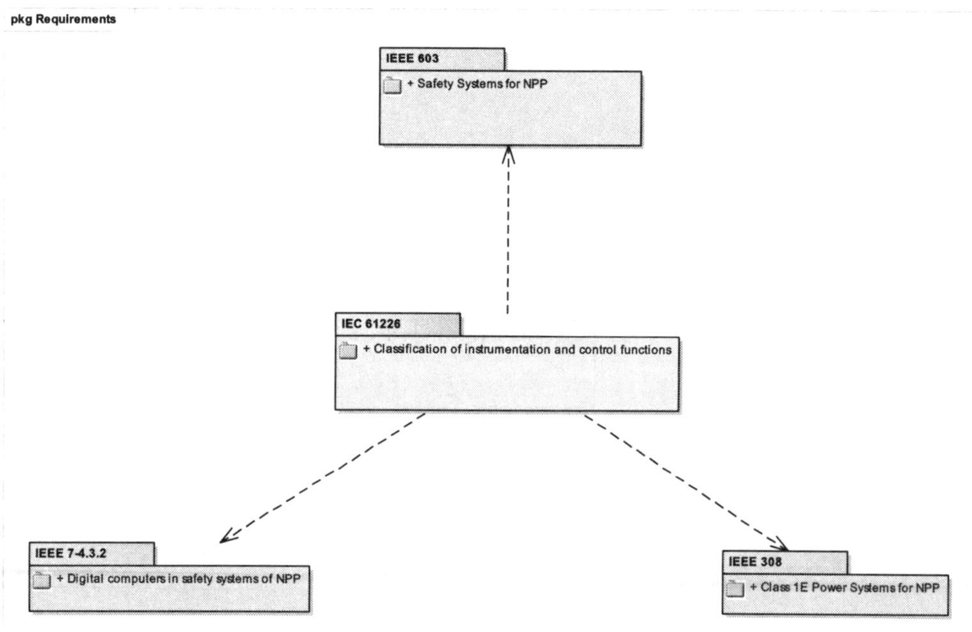


Figure 6-18: Interrelationship between IEC 61226 and IEEE standards

### ③ Interrelationship between IEC 60880 and IEEE standards

The IEC 60880, "Nuclear power plants– Instrumentation and control systems important to safety–Software aspects for computer-based systems performing category A functions", provides requirements for the software of computer-based I&C systems of nuclear power plants performing functions of safety category A as defined by IEC 61226" [7]. IEEE standards related to IEC 60880 are as follows.

Figure 6-19 shows in this relationship.

i. ***IEEE 7-4.3.2, "IEEE Standard criteria for digital computers in safety systems of nuclear power generating stations"***
ii. ***IEEE 1074, "IEEE Standard for Developing a Software Project Life Cycle Process"***
iii. ***IEEE 1228, "IEEE Standard for Software Safety Plans"***
iv. ***IEEE 828, "IEEE Standard for Configuration Management in Systems and Software Engineering"***
v. ***IEEE 1042, "IEEE Guide to Software Configuration Management"***
vi. ***IEEE 1012, "IEEE Standard for System and Software Verification and Validation"***
vii. ***IEEE 1028, "IEEE Standard for Software Reviews and Audits"***
viii. ***IEEE 829, "IEEE Standard for Software and System Test Documentation"***
ix. ***IEEE 1008, "IEEE Standard for Software Unit Testing"***
x. ***IEEE 830, "IEEE Recommended Practice for Software Requirements Specifications"***

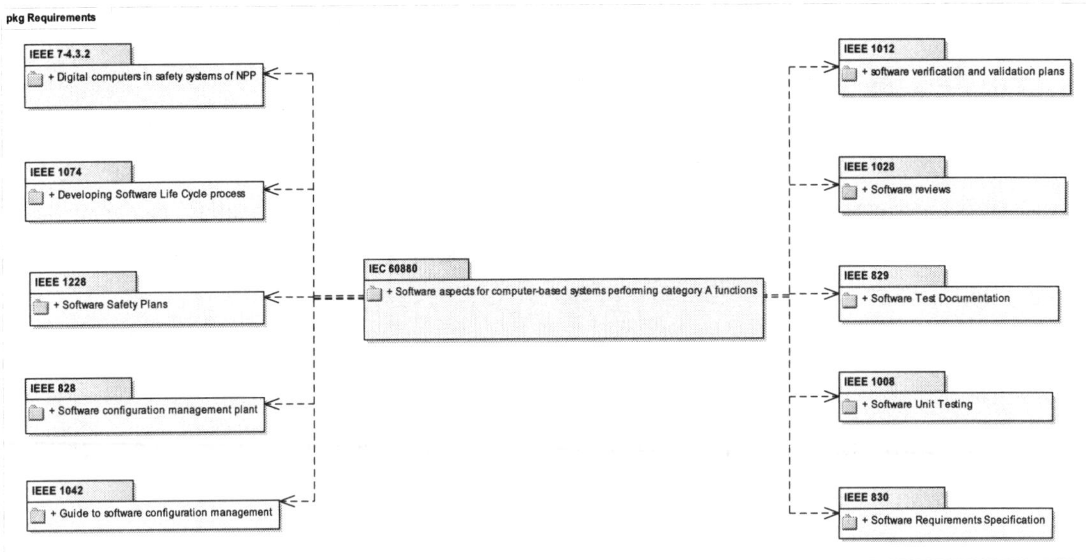


Figure 6-19: Interrelationship between IEC 60880 and IEEE standards

### ④ Interrelationship between IEC 60987 and IEEE standards.

The IEC 60987 Nuclear power plants – Instrumentation and control important to safety – Hardware design requirements for computer-based systems "is applicable to NPP computer-system hardware for systems of Class 1 and 2. Computer hardware facilities used for software loading and checking are not considered to form an intrinsic part of a system important to safety" [8]. IEEE standards related to IEC 60987 are as follows as shown in Figure 6-20.

i. ***IEEE 603, "IEEE Standard Criteria for Safety Systems for Nuclear Power Generating Stations"***
ii. ***IEEE 308, "IEEE Standard Criteria for Class 1E Power Systems for Nuclear Power Generating Stations"***
iii. ***IEEE 7-4.3.2, "IEEE Standard criteria for digital computers in safety systems of nuclear power generating stations"***

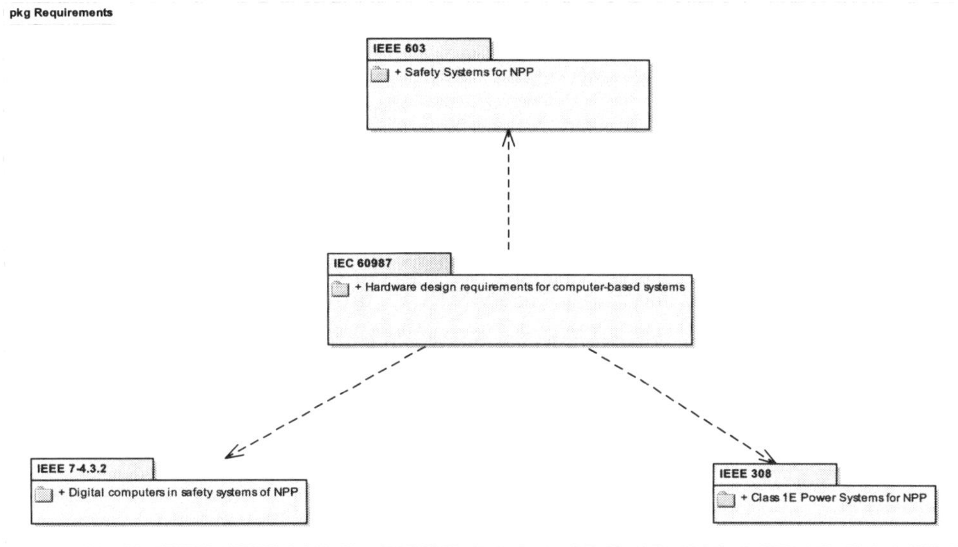


Figure 6-20: Interrelationship between IEC 60987 and IEEE standards

### 6.1.3 Utility Requirement Document (URD)

The URD is a tool that can be used by organizations involved in deploying new nuclear plants to help facilitate and enable a standardized deployment strategy for new nuclear power plants. By providing plant designers a comprehensive set of requirements from a utility perspective, designers are able to better understand utility expectations and strive towards providing compliant designs. Utilities today look to the URD to assist them in making technology assessment decisions by providing a standard that designs can be benchmarked against. (EPRI). The URD contains the expectations of the owner of the plant to the designer. There are different types of URDs such as Electric Power Research Institute (EPRI) URD, European URD (EURD), and Korean URD (KURD) among others.

The KURD represents the requirements of the utility, based on years of operating experience. As a result, these must be implemented in the plant design. There is therefore upward traceability from the NPP design requirements to the KURD Chapter 10 of the KURD deals with the I&C systems of the NPP. Specifically, Section 8 deals with the Reactor Protection Systems and the Engineered Safety features.

### (1) KURD Chapter 10: Man-Machine Interface Systems

This section is concerned about the systems that perform monitoring, control, and protection functions associated with all modes of plant operation. The plant designer designs the MMIS according to the requirements of this chapter. Chapter 10 of KURD has 10 sections which represent the different requirements of the Owner. Section 8: Reactor Protection and Safety Systems which provides requirements for the MMIS to provide reactor protection would be referred to in the discussion of Elicitation of MOE in the next section. The structure of Chapter 10, and Chapter 10 Section 8 is illustrated in Figure 6-21 and Figure 6-22.

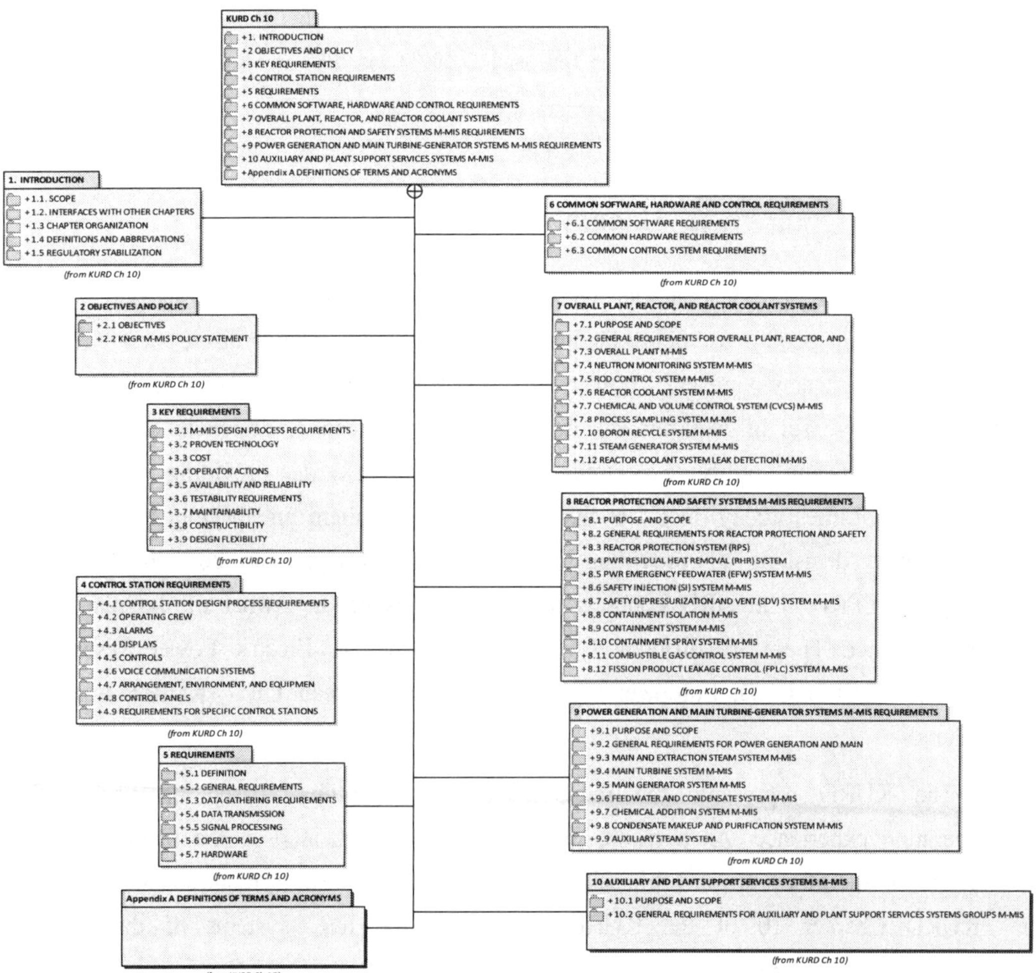

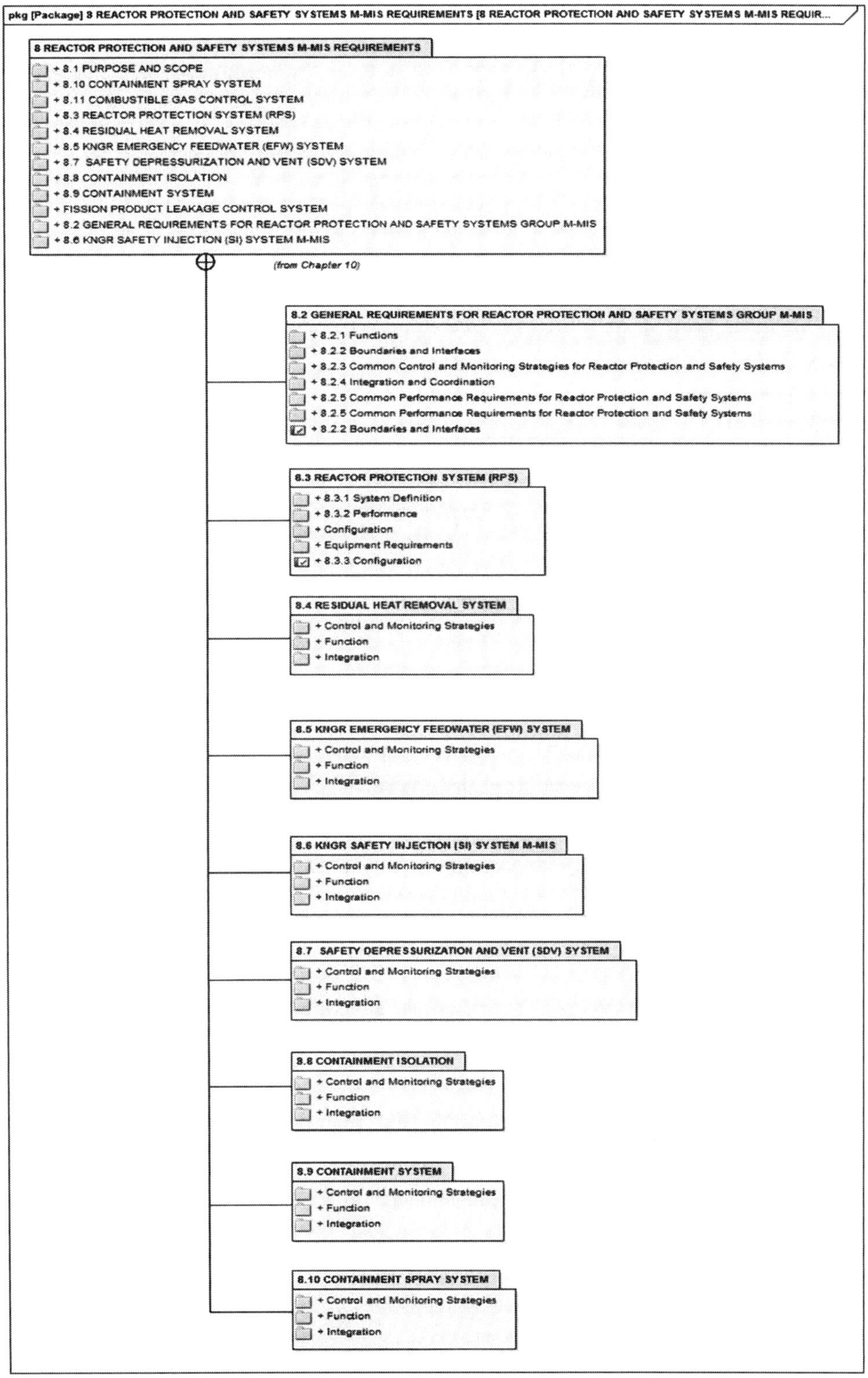


Figure 6-22: KURD Chapter 10, Section 8

### (2) Elicitation of Measure of Effectiveness

Chapter 3.15 has defined what MOE are and how they are measured. Measures can be qualitative or quantitative. The MMIS functions for the reactor protection and safety systems are divided into two major categories as depicted in Figure 6-23.

- ***Functions related to preventing core damage***
- ***Functions related to mitigating the effects if core damage were to occur***

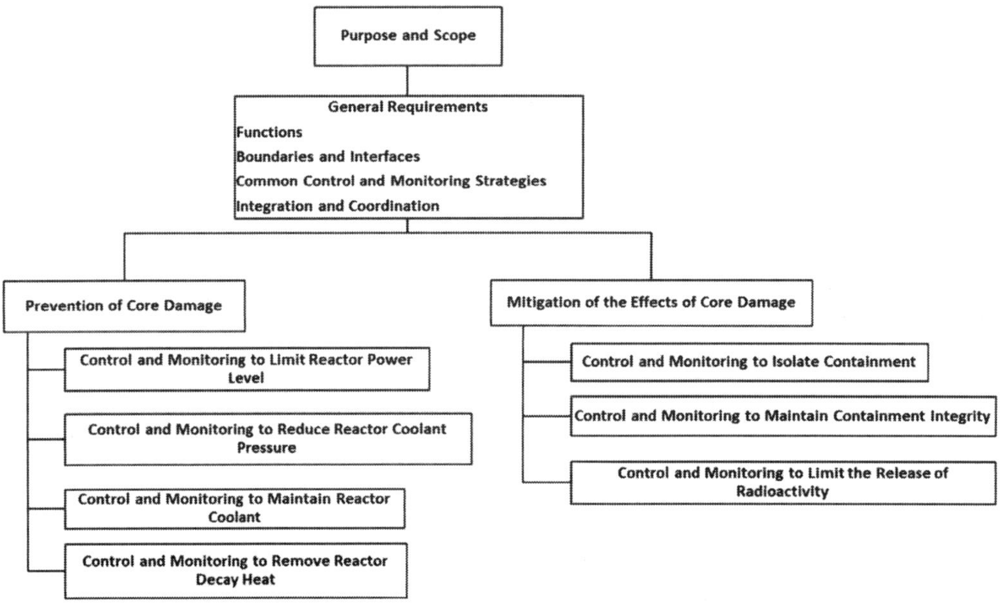


Figure 6-23: MMIS for Reactor Protection and Safety Systems

The systems responsible for performing the two major functions listed above are the PPS and the ESF-CCS. The table 6-3 shows some of the utility requirements and the actual design implemented in the APR1400.

Table 6-3: Utility's Requirement and System Design

| NO. | Utility's Requirements | Implemented Design |
|---|---|---|
| 1 | **Shutdown or Reconfiguration Operation**<br>The MMIS shall provide the minimum interlocks necessary to assure that the manual shutdown or reconfiguration of protection or safety systems takes place only when essential to prevent increasing the seriousness of an event. | Interlock functions in both RPS and ESF satisfy this requirement |
| 2 | **Testing Operation**<br>The MMIS for the protection and safety systems shall normally provide for testing to be initiated at the direction of the operator, but the testing itself to be largely automatic. | This requirement is satisfied by system testing and inoperable surveillance. PPS testing consist of: sensor check, bistable logic test, CPCS test, LCL test, Initiation logic and circuit test, manual trip test, Actuation logic test, selective group test, EDG loading sequencer test and response time test. |
| 3 | **Initiation Signals**<br>The MMIS Designer shall select a set of plant process variables for automatic actuation of reactor trip and other safety functions that provides a balance between the need to provide diversity of actuation signals to reduce the potential for common mode failures and the need to prevent spurious actuations, which can result from having numerous input signals. | The initiation signals for the RPS and ESF are provided through four independent channels that tend to reduce the potential for common mode failure. The channels are independent of each other. The BOP ESFAS is provided via two independent channels. |
| 4 | **Coincidence of Signals**<br>The logic for RPS and safety systems shall be such that it requires coincidence of two or more channels of the same variable or the same combination of variables to initiate system action. | Local Coincidence Logic. This gives initiation signals as output when there is a 2 out of 4 logic. |

## 6.2 Requirements Traceability Analysis

When validating the system, it must be demonstrated to have met all the stakeholder requirements. Thus, an important aspect of requirements definition and analysis is requirements traceability. Requirements traceability can be defined as "the ability to describe and follow the life of a requirement, in both a forward and backward direction". Important benefits from traceability can be realized in the following areas: project management, process visibility, verification and validation (V&V), and maintenance. The most significant benefits provided by traceability can be realized during the V&V stages of a software project. Traceability offers the ability to assess system functionality on a per-requirement basis, from the origin through the testing of each requirement.

Traceability can be used to demonstrate that a system complies with its requirements and that they have been implemented correctly. If a requirement can be traced forward to a design artifact, such as the code or component, it validates that the requirement has been designed into the system. Similarly, if a requirement can be traced to a test case, it demonstrates that the requirement has been verified through testing. Without traceability, it is impossible to demonstrate that a system has been fully verified and validated [9]. The figure 6-24 shows the two approaches to traceability discussed above.

Figure 6-24: Requirements traceability

The hierarchical relationship between the different stakeholder requirements as well as the System Requirements and Specification is illustrated in Figure 6-25 below.

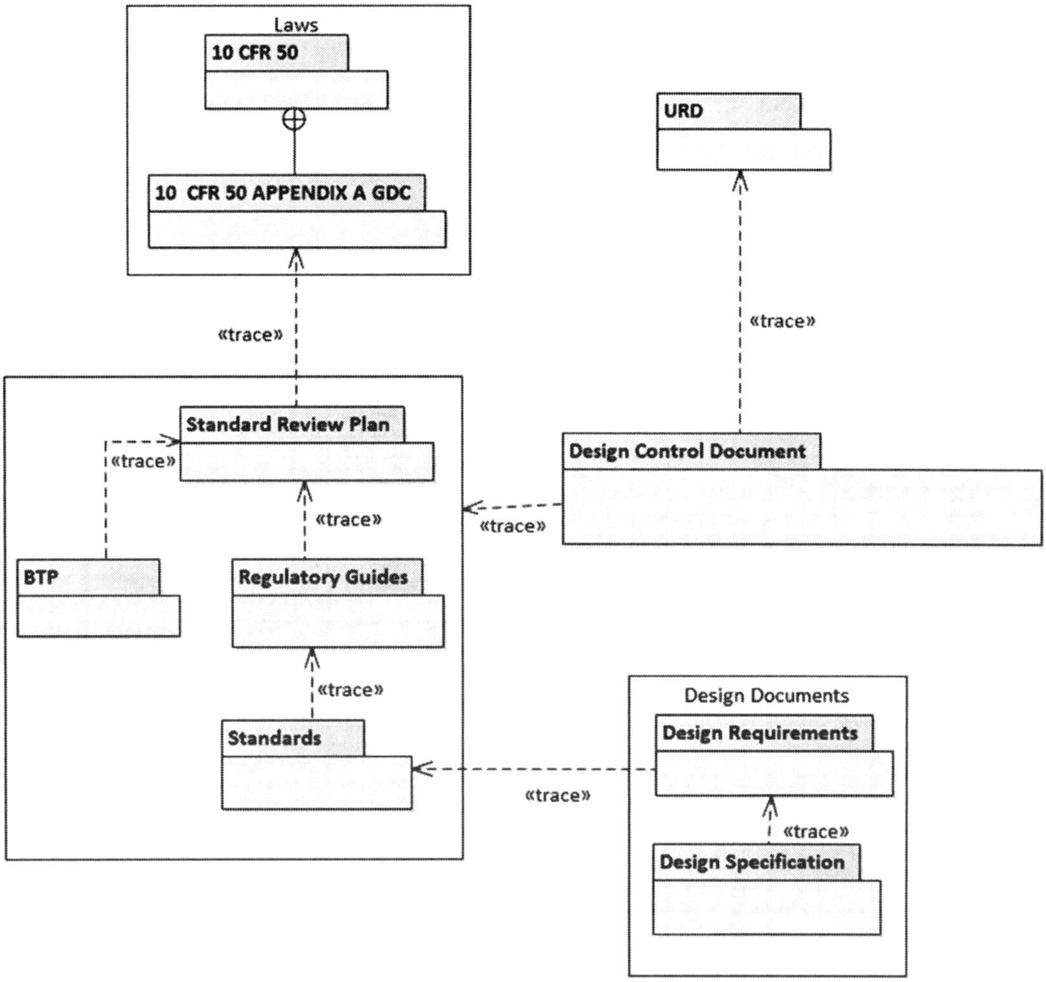


Figure 6-25: Requirements Hierarchy and Inter-relationship

Figure 6-26 illustrates the traceability of the Regulatory Guides and BTPs to the General Design Criteria using a Requirements Traceability Matrix (RTM). Each tick represents a relationship (trace) from an RG to a GDC. A single RG or BTP can be traced to multiple GDC. All the lower level requirements (RG and BTP) should be traceable to at least one GDC. In addition, each GDC should be traced to at least one RG. If there is a lower level requirement not linked to an originating requirement (a column without a tick), it would indicate a lack of backward traceability. If there is an originating requirement not linked to a lower level requirement (a row without a tick), it would indicate a lack of forward traceability. Thus an RTM enables a quick assessment as to whether all the requirements have been included, and no new requirements introduced. The RTM is initiated in the Stakeholder Requirements Phase and maintained and updated throughout the project development.

| GDC/SECY-RG-BTP | | | | | | | | | | | | | | | | | | | | | | | | | | | | |
|---|---|---|---|---|---|---|---|---|---|---|---|---|---|---|---|---|---|---|---|---|---|---|---|---|---|---|---|---|
| GDC \ SECY RG BTP | RG 1.22 | RG 1.47 | RG 1.53 | RG 1.62 | RG 1.75 | RG 1.105 | RG 1.118 | RG 1.151 | RG 1.152 | RG 1.168 | RG 1.169 | RG 1.170 | RG 1.171 | RG 1.172 | RG 1.173 | RG 1.180 | RG 1.204 | RG 1.209 C.17.2 | SRM to SECY 93-087 | BTP 7-8 | BTP 7-11 | BTP 7-12 | BTP 7-13 | BTP 7-14 | BTP 7-17 | BTP 7-18 | BTP 7-19 | BTP 7-21 |
| GDC 1 | | ✓ | | ✓ | | | | ✓ | | ✓ | ✓ | ✓ | ✓ | ✓ | ✓ | ✓ | | ✓ | | | ✓ | | | ✓ | | ✓ | | |
| GDC 2 | | | | | | | | | | | | | | | | ✓ | ✓ | ✓ | | | | | | | | | | |
| GDC 4 | | | | | | | | | | | | | | | | ✓ | | ✓ | | | | | | | | | | |
| GDC 10 | | | | | | | | | | | | | | | | | | ✓ | | | | | | | | | | ✓ |
| GDC 13 | | ✓ | | ✓ | | ✓ | | ✓ | | | | | | ✓ | | ✓ | | ✓ | | | | ✓ | ✓ | | | | | ✓ |
| GDC 15 | | | | | | | | | | | | | | | | | | ✓ | | | | | | | | | | |
| GDC 19 | | ✓ | | | | | | | | | | | | ✓ | | | | ✓ | | | | | | | | | | ✓ |
| GDC 20 | ✓ | | | | | ✓ | | | | | | | | ✓ | | | | ✓ | | ✓ | | ✓ | ✓ | | | | | ✓ |
| GDC 21 | ✓ | ✓ | | ✓ | ✓ | | ✓ | ✓ | ✓ | ✓ | | ✓ | ✓ | | ✓ | ✓ | | ✓ | ✓ | | ✓ | | ✓ | ✓ | ✓ | ✓ | ✓ | ✓ |
| GDC 22 | | ✓ | | ✓ | ✓ | | | ✓ | | | | | | ✓ | | ✓ | | ✓ | ✓ | | ✓ | | | | ✓ | | ✓ | |
| GDC 23 | | | | | | | | | | | | | | ✓ | | ✓ | | ✓ | | | | | | | | | | ✓ |
| GDC 24 | | ✓ | | | | | | ✓ | | | | | | ✓ | | | | ✓ | ✓ | | | | ✓ | | | | ✓ | |
| GDC 25 | | | | | | | | | | | | | | ✓ | | | | ✓ | | ✓ | | | | | | | | ✓ |
| GDC 29 | | | | | | | | | | | | | | | | | | ✓ | ✓ | | | | ✓ | | | | ✓ | ✓ |

Figure 6-26: GDC to RG traceability for PPS

P A R T 2

## PROBLEMS

1. What the most important element in the project success
   A. Stakeholder satisfaction
   B. How the product meets the requirements
   C. Overall product quality
   D. Overall project quality
2. What is the top tier regulation for European Countries? United States?
3. What will happen if there is an originating document not linked to lower lever requirement?
4. Define the hierarchy between the requirements documentation.
5. What is the SRP? How can power plant designers utilize this document?
6. What is the DCD?
7. What is a requirements traceability matrix?
8. How can the system requirements be validated? What is the difference between forward and backward traceability?
9. What are the MOE, MOP, and TPM?

## FURTHER READING

- NUREG-0800, "Standard Review Plan for the Review of Safety Analysis Reports for Nuclear Power Plants: LWR Edition,", Chapter 7.1
- KURD, "Man Machine Interface Systems," Chapter 10
- IEEE-603, "IEEE Standard Criteria for Safety Systems for Nuclear Power Generating Stations"
- IEEE-7-4.3.2, "IEEE Standard Criteria for Programmable Digital Devices in Safety Systems of Nuclear Power Generating Stations"
- Regulatory Guide 1.172, "Software Requirement Specifications for Digital Computer Software and Complex Electronics Used In Safety Systems of Nuclear Power Plants"

Chapter 7

# SYSTEM ANALYSIS FOR ARCHITECTURE AND DESIGN


## SUMMARY

The purpose of the I&C system analysis process for architecture and design is to analyze the system based on rigorous data and assess the quantitative technique in analysis method. The assessment and estimation of the analysis can be based on effectiveness analysis, technical risk analysis, cost analysis, and other critical quality characteristics. The results of analyses and estimations, as data, information, and arguments, are provided to the decision management process for selecting the most efficient alternative or candidate. In some cases, the results may be provided to the project assessment and control process, if the information is needed to monitor the progress of the system against its system objectives, performance thresholds, or growth targets, such as the projected or modeled reliability early in the development of the system as compared to its reliability growth curve. Figure 7-1 illustrates the overall context of system analysis process using IPO diagram.

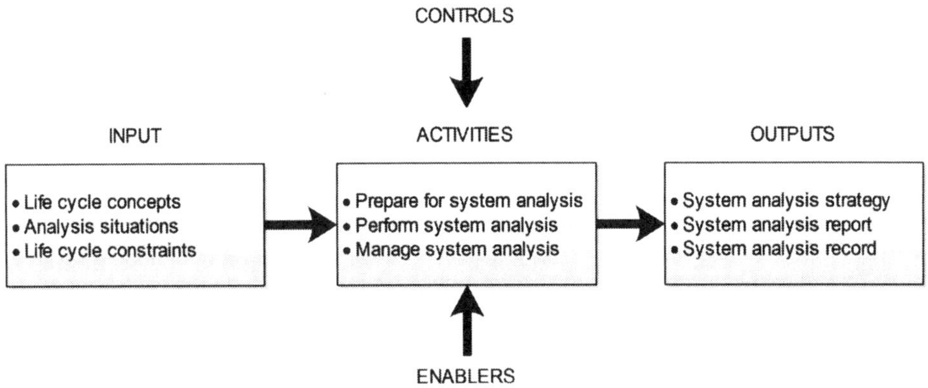


Figure 7-1 Inputs-Process-Outputs diagram of System Analysis Process

These analyses mainly use quantitative modeling techniques, analytical models, and associated simulations, which are applied at varying levels of rigor and complexity depending on the level of fidelity needed. Some cases may necessary to employ a variety of analytic functions or experimentation to obtain the necessary insight. The results serve as inputs into various technical decisions, providing confidence in the adequacy and integrity of the system definition toward achieving the appropriate system balance. Mind-map of chapter 7 is shown in Figure 7-2.

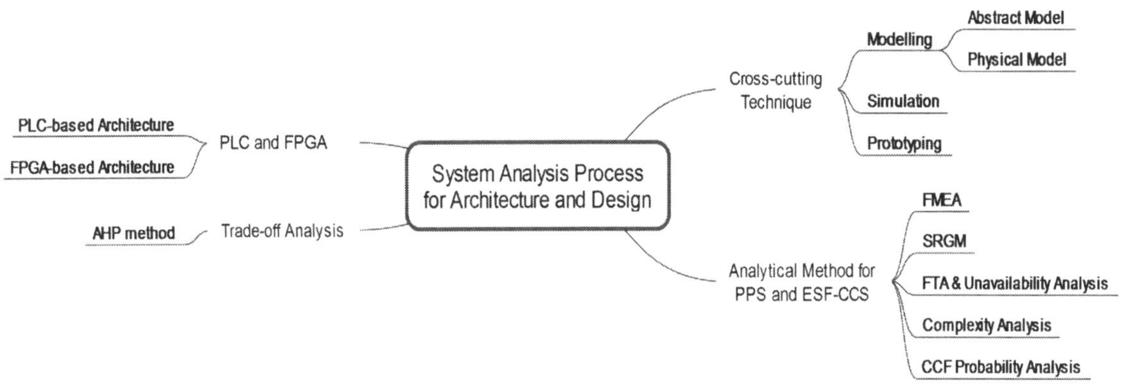


Figure 7-2 Mind-map of Chapter 7

## OBJECTIVES

- Definition of the system analysis
- Technique and methods in system analysis
- Trade-off analysis of PLC and FPGA

## 7.1 System Analysis Overview

Architectural and design system analysis process is performed to analyze and estimate architectural and design characteristics of candidate architectures, the system elements, providing arguments in order to be able to select the most efficient ones. It can be costs, technical risks, effectiveness (e.g., performances, dependability, human factors), and other stakeholder concerns for instance critical quality characteristics, affordability, maintenance, etc.

This process involves transforming the system's functional architecture to physical architecture, defining alternative system concept and system elements, selecting preferred system architecture solution, and defining the physical system's internals and its interfaces. Figure 7-3 shows a simplified approach to I&C system architecture and design process.

From systems engineering viewpoint, requirements analysis is preceded by architectural analysis. Design architecting is often an iterative loop in which system and complexity analyses are applied in other to assess architecture candidates and ultimately select a preferred architecture. Within the loop, functional analysis and allocation to architectural entity is performed.

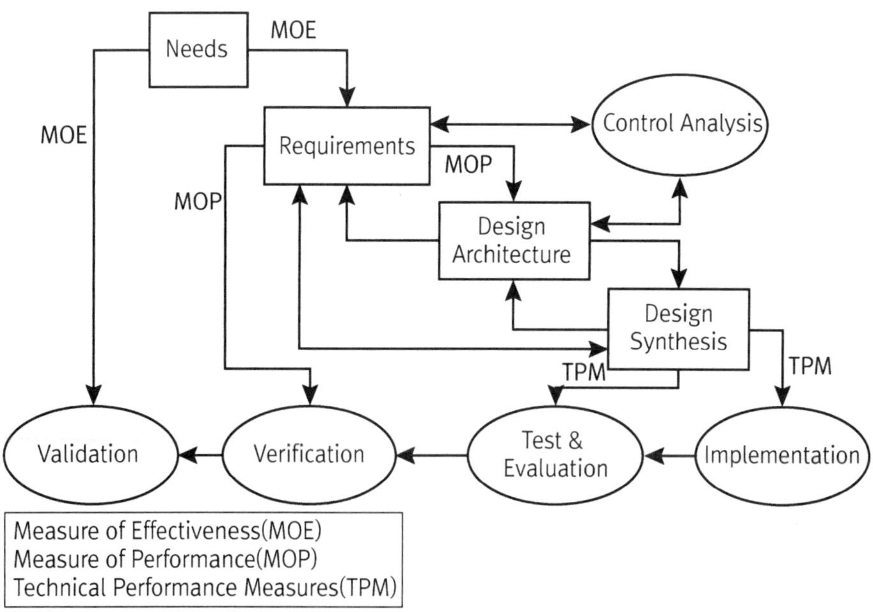


Figure 7-3: Systems Engineering based Architecture Design Development Process

At the end of the loop, the selected architecture must optimally match the complete set of stakeholders' requirements. The design synthesis focuses on the physical architecture of the design on which the system is to be implemented.

In regards of quantitative technique, cross cutting technique shall be used to obtain the result for technical decision and technical processes. This technique provides insight into method that cut across the technical processes reflecting various aspects of the iterative and recursive of system engineering.

Figure 7-4 shows that the process flow of system architecture synthesis might include modeling & simulation as well as trade studies to select the appropriate system architecture and design.

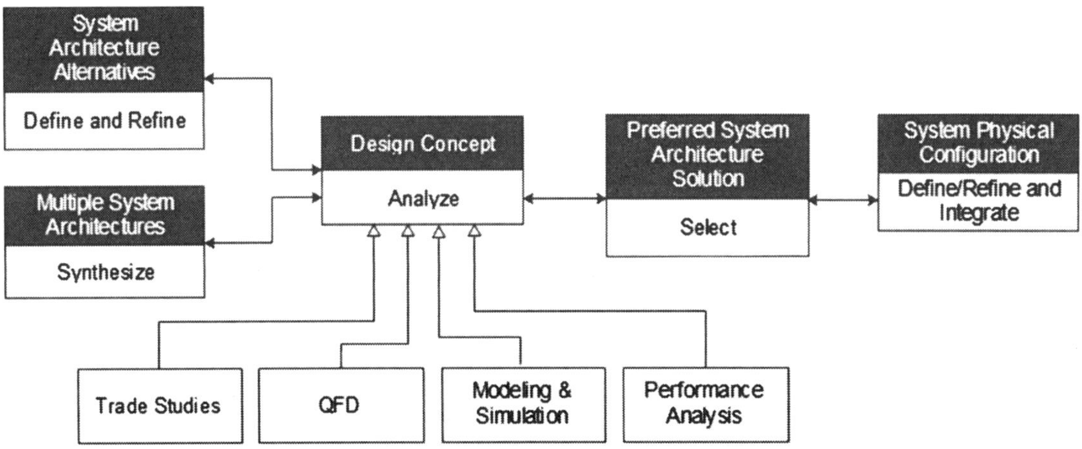


Figure 7-4 System architecture synthesis process flow

## 7.2 Functional Analysis

A function is a characteristic task, action, or activity that must be performed to achieve a desired outcome. It may be accomplished by one or more system elements comprised of equipment (hardware), software, firmware, facilities, personnel, and procedural data.

Functional Analysis/Allocation is an examination of a defined function to identify all the sub-functions necessary to accomplish that function; all usage modes must be included in the analysis. This activity is conducted to the level of depth needed to support required synthesis efforts. Identified functional requirements are analyzed to determine the lower-level functions required to accomplish the parent requirement.

### 7.2.1 Objective of Functional Analysis/Allocation

- Create a functional architecture for which system products and processes can be designed
- Provide the foundation for defining the system architecture through the allocation of functions and sub-functions to hardware/software, databases, facilities and operations (i.e., personnel).

The functional analysis does not describe either the hardware architecture or software architecture of the system; those architectures are developed during System Synthesis in the Architectural Design portion of the SE process. It describe what the system will do, not how it will do it. Ideally, this process begins only after all of the system requirements have been fully identified.

Four Pillars of Modeling shown in Figure 7-5 are based on the integrated dictionary. It used for structured analysis in types of model.

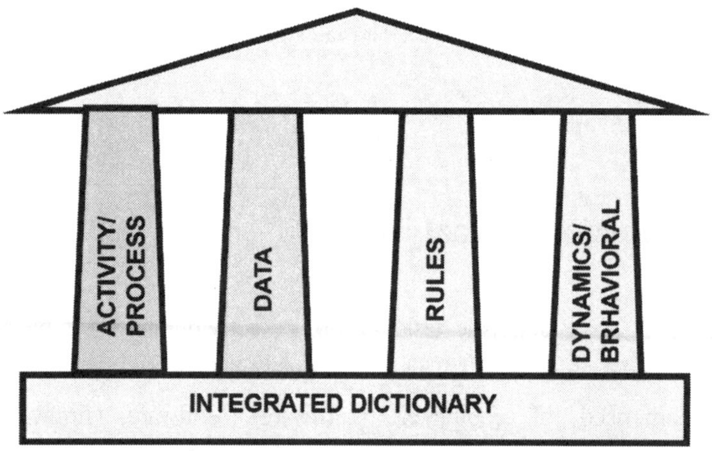


Figure 7-5 Four Pillars of Modeling

PART 2

The different types of those models in structured analysis are as follows:

- Process or Activity model (IDEF0 or DFD or N2)
- Data model (IDEF1x, ERD)
- Rule model (Structured English, Decision Tables and Trees, Logic))
- Dynamics model (STD, EFFBD, State Charts, Behavior Diagrams)

The process activities of functional analysis must be iterative. Even within a single stage in the system life-cycle. The functional architecture begins at the top level as a set of functions that are defined in the applicable requirements document or specification, each with functional, performance, and limiting requirements allocated to it. Figure 7-6 portrayed a process activities of functional analysis. Several tools perform for Functional Analysis/Allocation include Analysis tools, Modeling and Simulation tools, Prototyping tools, and Requirements traceability tools.

Decomposition of a system can be viewed as a top-down approach to problem solving. Functional identification and decomposition can be performed with respect to logical groupings, time ordering, data flow, control flow, state transitions, or some other criterion.

The objective of the functional decomposition activity is to develop a hierarchy of FFBDs that meet all the functional requirements of the system.

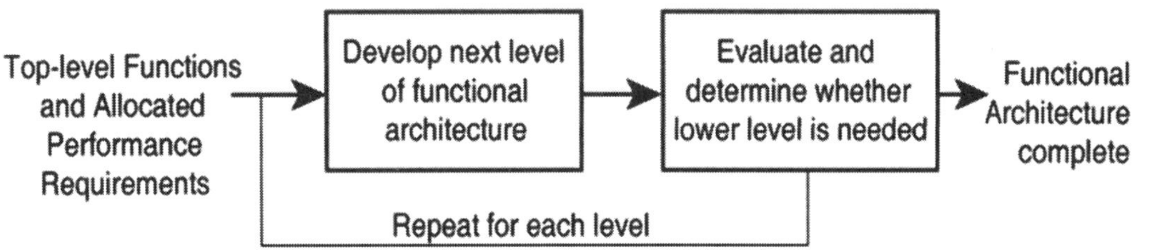


Figure 7-6 Process activities of functional analysis [10]

A description of each function in the hierarchy should be developed to include the following:

- Its place in a network (e.g., FFBD or IDEF0/1 diagrams) characterizing
- The set of functional requirements that have been allocated to it and define what it does
- Its inputs and outputs, both internal and external.

This activity may use various graphical methods to capture the results of the analysis, including structured analysis, such as Data Flow Diagrams, IDEF0/1 diagrams, Control Flow Diagrams, or other modern techniques. These are all forms of the Functional Descriptions.

## 7.3 Trade-off Analysis between PLC & FPGA based PPS and ESF-CCS

From now on, the system analysis will be executed using FPGA and PLC for the purpose of enhancing comprehension for the reader.

For safety-critical RPS and ESF functions, the choice of digital platform for the system architecture and design analysis is often made between Programmable Logic Controller (PLC) and the Field Programmable Gate Array (FPGA) based systems. Although the protection systems of operating reactors are based on PLC design, new NPP digital MMIS designs are tilting towards FPGA-based system. New considerations are being made to transit from PLC-based to FPGA-based system.

In 2010, EPRI carried out an informal survey [11] among member utilities to gather information on current and planned applications of FPGAs in both operating and new nuclear power plants. Utilities were asked to identify the I&C applications of most interest for applying FPGA-based solutions, and what types of functions they expect would be performed by FPGAs. The functions considered most suitable for FPGAs by most respondents included process-related discrete functions such as protective logic functions, bistables, comparators, coincidence logic, and priority logic. A few of the recent implementations of FPGAs in NPPs are [12]:

- Bulgaria: In Units 5 and 6 of the Kozloduy nuclear power plant, RPC Radiy installed six FPGA-based Category A, ESFAS
- United States of America: An FPGA-based control system was installed in 2009 to replace obsolete equipment in the main steam and feed water isolation system at the Wolf Creek Generating Station.
- France: Electricité de France (EDF) in 2009 initiated replacement of obsolete 1970s vintage electronic modules comprising the rod control system (RCS) and the reactor in-core measurement system.
- Japan: Toshiba installed the following FPGA systems in NPPs: startup range and power range neutron monitoring systems for boiling water reactors (BWR), and radiation monitoring systems for both BWRs and pressurized water reactors. In addition, Toshiba has developed reactor trip and isolation systems for ABWR type plants.
- Sweden: In Unit 2 of the Ringhals NPP, FPGAs have been used to implement the CIM.
- Korea: In Shinkori Units 3 & 4, FPGAs have been used in the ESF-CCS to implement the Component Interface Modules. (CIM)

### (1) PLC-Based Architecture

PLC-based protection system is a safety grade industrial digital computer that relies on preloaded software for the execution of its dedicated function. PLCs are widely used for the automation of industrial processes and are built around microprocessor architecture. Therefore, PLC runs on a periodic cycle of execution of sequence of instructions in order to perform its function. Typically, input values are read at the begin of each cycle of instruction and the required variables are generated/stored in memory as the system loops round each cycle of execution. Embedded software in a PLC is typically programmed using specified programming languages such as Function Block Diagram (FBD) and Ladder Diagram (LD).

In comparison with a general-purpose computer, the PLC architecture is similar in many ways. However, the distinction lies in the fact that PLC is designed with high reliability to perform specific function in an industrial environment. The basic PLC architecture, Figure 7-7 consists of the central processing unit (CPU), memory, and the I/O modules. The I/O module types include discrete (on/off), analog (continuous variable), and special modules like motion control or high-speed counters.

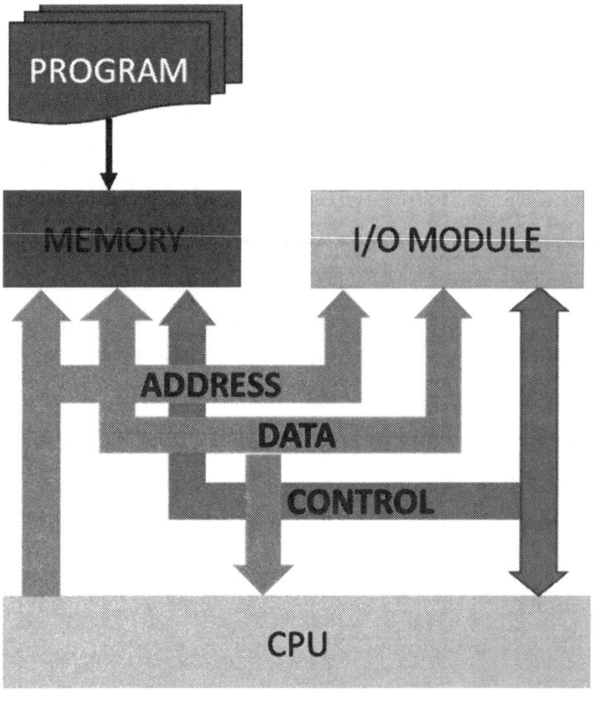


Figure 7-7: Basic PLC Architecture

### (2) FPGA-Based Architecture

A Field Programmable Gate Array (FPGA) is an integrated circuit based on a configurable matrix of logic blocks, designed to be configured in the field to any desired application functionality. The configuration of FPGA is done using a hardware description language (HDL) which describes the functions/logic to be performed and the interconnections similar in principle to how components are hardwired. FPGA architecture is based on static random-access memory (SRAM). This means that the configuration is lost when power loss occurs. Therefore, configuration data is restored using external EEPROM.

A typical FPGA architecture design flow, Figure 7-8, comprises design entry, design synthesis, design implementation, device programming and design verification [13].

PART 2

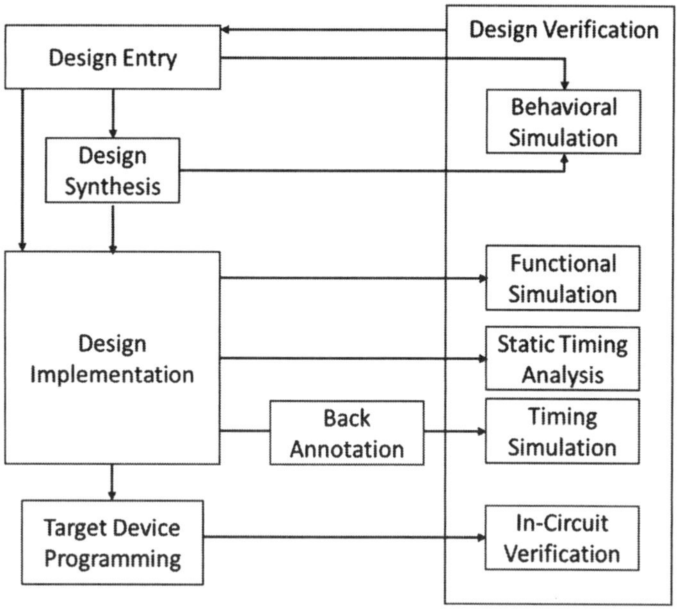


Figure 7-8: Typical FPGA Architecture Design Flow

The design entry can be based on schematics or on algorithm depending on the designers. The design synthesis translates VHDL code into a device netlist format containing the complete circuit logical elements. The design implementation consists primarily of mapping and place & route.

## 7.3.1 Digital Architecture Trade-off Analysis

FPGAs are gaining increased attention worldwide for application in nuclear power plant (NPP) instrumentation and control (I&C) systems, particularly for safety applications [11]. Traditionally, I&C systems have been implemented using relay-based components, analog components, or discrete digital logic components. This is enabled by the fact that most critical I&C functions involve relatively simple processing. As a result, these systems readily lend themselves to implementation on an FPGA platform that is based on discrete logic gates.

Current digital I&C systems are implemented using Programmable Logic Controllers (PLCs). As such, a discussion of the advantages that can be achieved by the implementation of a system on FPGAs, are necessarily discussed in relation to the shortcomings of the PLC-based systems. In the first place, FPGAs exhibit greater portability. This is due to the fact that only the final stages of the development are vendor specific, thereby reducing their vulnerability to obsolescence. In addition, FPGA-based systems have faster, more deterministic response times due to their capability for parallel processing. Due to the capability of native hardware parallelism, FPGAs are able to process more data, provide faster input and output response times and execute more instructions per clock cycle than digital signal processors.

FPGAs exhibit greater resistance to cyber-attacks. This is due to the fact that as they do not make use of an operating system, they do not contain any general-purpose functionality that can be hijacked and exploited. Further, FPGA technologies that require physical access for system modification, such as antifuse technology, can significantly reduce the threat of a cyber-attack. The absence of an operating system also eliminates the risk of software Common Cause failure (CCF). PLCs are susceptible to software CCF due to the fact that software is not 100% testable, and is likely to contain latent faults even when developed under a stringent life cycle process [14]. This threat is not present in FPGA-based systems as the final implementation is devoid of software. FPGA based systems have a relatively easy qualification process due to the simplicity and transparency of system architecture and its design process [15]. This eases the demonstration of the designed system meeting qualification requirements, such as independence, separation, redundancy, and diversity.

However, the FPGAs also have several shortcomings, the foremost being the lack of operating experience in the nuclear industry. Although FPGAs have been used in numerous industries, it is a relatively new technology to the nuclear power industry. As a result, there is a lack of a meaningful database for gaining lessons learned and good practices or to support the basis for acceptance or reliability numbers [12]. It should be noted that the described disadvantages are not intrinsic to the technology but area result of the lack of experience with this hardware, and will be resolved by increased use of FPGAs.

Several criteria should be considered when choosing between design options. For instance, Chapter 10 Section 3.1.4.4.3 of the KURD defines a set of criteria for the determination of suitable architecture for the MMIS system. These criteria are:

- Simplicity
- Standardization
- Reliability and availability
- Protection against common mode failures
- Compatibility with the environment including radiation
- Testability, including the potential that testing could affect plant operation
- Maintainability
- Protection against obsolescence
- Flexibility and Expandability

Table 7-1 shows a comparative analysis between PLC-based and FPGA-based technologies based on these criteria. The USNRC regulations and the IEEE standards provide the requisite applicable regulations and guides for design analysis as follow:

- Regulations:
    - 10 CFR 50 Appendix A, "General Design Criteria for Nuclear Power Plant"
    - 10 CFR 50.55a(h), "Protection and Safety Systems"
    - Regulatory Guide 1.22, "Periodic Testing of Protection System Actuation Functions"
    - Regulatory Guide 1.118, "Periodic Testing of Electric Power and Protection Systems", Rev. 3 endorses IEEE Std. 338-1987
    - Regulatory Guide 1.152, "Criteria for Digital Computers in Safety Systems of Nuclear Power Plants", Rev. 3 - endorses IEEE Std. 7-4.3.2-2003

- Standards:
    - IEEE STD 338-1987, "IEEE Standard Criteria for the Periodic Testing of Nuclear Power Generating Station Safety Systems"
    - IEEE 7-4.3.2-2010, "IEEE Standard Criteria for Digital Computer In Safety Systems of Nuclear Power Generating Stations"
    - IEEE STD 603-1998, "IEEE Standard Criteria for Safety Systems for Nuclear Power Generating Stations"

Table 7-1: Comparison of FPGA and PLC technologies [16]

| CRITERIA | FPGA | PLC |
|---|---|---|
| Simplicity | FPGAs lack software and are therefore simpler to validate. | Microprocessors include a significant amount of relatively elaborate software such as operating systems and application programs resulting in a complex design, V&V, and licensing processes. |
| Protection against CCF | While still susceptible to hardware CCF, they are immune to software CCF due to its absence | Susceptible to both hardware and software CCF |
| Protection against obsolescence | FPGA technology allows a significant degree of design portability, i.e. only the final steps (synthesis and place and route) are dependent on the particular FPGA circuit chosen. This mitigates their vulnerability to obsolescence. | Rapid obsolescence of electronic components results in comprehensive refurbishments of I&C systems during the lifetime of the nuclear power plant and vendors can seldom guarantee full upward compatibility during the long periods between refurbishments |
| Cyber security | FPGA-based implementations increase the level of difficulty that would be faced by a potential attacker for the following reasons:<br>▪ They can be designed not to include high-level, general-purpose components that could be easily tampered with.<br>▪ Where cyber security is a potential concern, the FPGA designs can be based on anti-fuse (one time programmable) type chips. | PLCs are susceptible to cyber security attacks as attackers normally make use of functionalities that are part of the operating system |
| Testability, including the potential that testing could affect plant operation | FPGAs exhibit decreased access to internal signals for monitoring, testing, and troubleshooting. Extra effort on the part of the designer as well as solid communications with those who will be responsible for maintaining the system during plant operation. | PLC systems are equipped with the capability of implementing system as well as self-tests during all modes of operation |
| Flexibility and Expandability | Flexible to accommodate design change and expansion | Modular design allows for ease of expansion |
| Compatibility with the environment | Some FPGAs are susceptible to single event upsets due to radiation | Compatible with the environment |

PART 2

| CRITERIA | FPGA | PLC |
|---|---|---|
| Licensing | Relatively few current applications in the nuclear power industry. There is not a wealth of operating experience on which to build a reliable, meaningful database for gaining lessons learned and good practices or to support the basis for acceptance or reliability numbers. | There are numerous applications well-defined licensing requirements. |
| Program Execution | Parallel Operation. Multiple, independent logical operations can be carried out at the same time | Sequential Operation: The execution of the logical operations is carried out, one after the other. |

Another method of trade-off analysis is the use of Analytical Hierarchy Process (AHP). In this method, a decision matrix using weighted criteria is used and alternatives are evaluated based on the weighted criteria. Table 7-2 shows a sample AHP decision matrix for a trade-off analysis between PLC and FPGA. The AHP analysis process is illustrated in Figure 7-9. The details of this AHP can be found in I&C project “SU 2: Development of FPGA Based Safety Injection Actuation System.” in KINGS.

Table 7-2: Sample AHP Decision Matrix

| Criteria | Weight | PLC | | FPGA | |
|---|---|---|---|---|---|
| | | score | Weighted score | score | Weighted score |
| Maintainability | 0.0845 | 3 | 0.2535 | 5 | 0.4226 |
| Protection against CCF | 0.1549 | 3 | 0.4648 | 4 | 0.6197 |
| Testability | 0.0986 | 3 | 0.2958 | 4 | 0.3944 |
| Etc….. | | | | | |

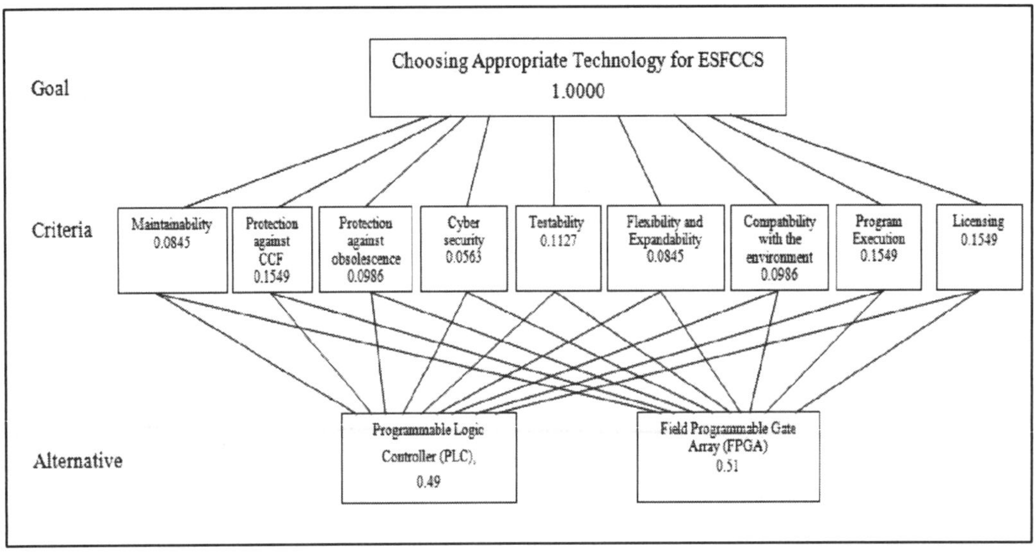


Figure 7-9: Analytical Hierarchical Process

## 7.4 Analytical Method for PPS and ESF-CCS

In general, system analysis involves the activities and processes used in collecting and interpreting system data in order to decompose the system into its components and identify potential failure. System analysis is a problem-solving technique that improves the system and ensures that the elements of the system work efficiently to accomplish their purpose. Many of the system analyses that are often used for the purpose of system reliability and availability include Failure Modes and Effects Analysis (FMEA), Fault Tree Analysis (FTA), and Common Cause Failure (CCF) Analysis. System analysis demonstrates how the safety-critical I&C systems satisfy the requirements of relevant GDC and the IEEE standards.

### (1) Failure Modes and Effect Analysis (FMEA)

FMEA is a systematic and qualitative method for evaluating a system/process in order to identify conditions for failure and to assess the associated effect of the failures. FMEA includes the evaluation of failure modes, causes of failure, symptoms and effects, and method of failure detection. FMEA is often used in evaluating new system design or a new change in existing system design before implementation. IEEE Std. 352 provides guidance on FMEA.

For the RPS, FMEA demonstrates that no single failure result in spurious reactor trip, a single failure cannot prevent a system-level RPS reactor trip due to the implementation of four(4) divisions of redundancy, and a single failure is detectable by diagnostics and periodic testing [17, 18]. Figure 7-12 shows an extract of the FMEA of PPS for the bistable processor module. A complete table of the FMEA can be found in the APR1400 DCD document for reference.

| No. | Name | Failure Mode | Cause | Symptoms and Local Effects Including Dependent Failures | Method of Detection | Inherent Compensating Provision | Effect on PPS | Remarks and Other Effects |
|---|---|---|---|---|---|---|---|---|
| 5 | Bistable processor module (PPS BP rack) | a) Processing section fails to execute program instructions. | Component failure | Affected BP halts.<br>No periodic updates transmitted to SDL and SDN from affected BP | • Lack of BP processor updates detected by MTP/ITP via SDN<br>• Lack of BP processor updates detected by LCL via SDL<br>• Trouble alarm is actuated. | Partial trip/actuation selection logic in all LCLs uses quality data from the redundant BP in affected channel. | • No loss of safety function<br>• Coincidence remains as 2-out-of-3 logic. | N/A |
| | | b) Application program memory failure | Component failure | Affected BP halts.<br>No periodic updates transmitted to SDL and SDN from affected PM | • CRC checks performed on memory.<br>• Lack of BP processor updates detected by MTP/ITP via SDN<br>• Lack of BP processor updates detected by LCL via SDL<br>• Trouble alarm is actuated. | Partial trip/actuation selection logic in all LCLs uses quality data from the redundant BP in affected channel. | • No loss of safety function<br>• Coincidence remains as 2-out-of-3 logic. | N/A |
| | | c) Processing section fails to scan input/output modules. | Component failure | • BP inputs (process values, CPC trips, ENFMS permissives) not updated periodically.<br>• BP outputs (SOE points) are not updated periodically.<br>• Calculated bistable outputs are set to Bad quality. | • Corrupted I/O bus cycles are detected.<br>• Trouble alarm is actuated. | • Partial trip/actuation selection logic in all LCLs uses quality data from the redundant BP in affected channel.<br>• SOE points are provided by redundant BP in affected channel. | • No loss of safety function<br>• Coincidence remains as 2-out-of-3 logic. | N/A |

Figure 7-10: Extract of the FMEA for PPS

### (2) Software Reliability and Growth Model (SRGM)

SRGM is one of the Quantitative Software Reliability Methods (QSRMs) methods that are used to quantify software failure rates and demand failure probabilities of digital system. SRGMs are Time-based methods that use test data to estimate software failure rates that, in turn, are employed to ascertain whether a particular software can be released, by demonstrating that its failure rate meets the desired level [19].

A large number of software reliability growth models (SRGMs) have been developed over the years, and each model has its strengths and limitations. There are different assumptions in different SRGMs on how the failure rate decreases with time; that is, the models specify different empirical formulas, and use test data to estimate their parameters. The empirical formulas and test data are employed to decide if the failure rate objective has been reached.

IEEE Std. 1633 grouped SRGMs into three high-level categories as shown in Figure 7-11 as: exponential Non-Homogeneous Poisson Process (NHPP), non-exponential NHPP, and Bayesian.

The SRGMs categories for Exponential NHPP assume that the failure rate decreases exponentially with time, which is the consequence of the widely used assumption that the failure rate is proportional to the number of current faults in the software.

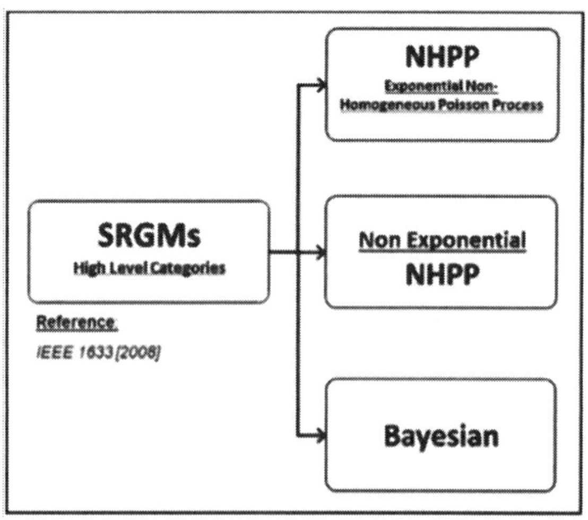


Figure 7-11: SRGM Categories

For Non-Exponential NHPP models, the reliability growth models assume that the way failure rate changes/decreases with time does not follow an exponential function. For example, it may follow the shape of the probability density function of a gamma or Weibull distribution.

Bayesian model developed by Littlewood and Verrall assumes that the failure rate decreases probabilistically/stochastically with time and uses Bayes' theorem in its derivations [20]. The Bayesian model essentially is an exponential NHPP model that explicitly includes the uncertainty of the failure rate in the model.

### (3) Fault Tree Analysis (FTA) and Unavailability Analysis

Unavailability is the probability that an item or system will not be operational at a future instance in time. It may be as a result of the item being repaired (repair unavailability) or it may occur as a result of malfunctions.

Unavailability for I&C protection system, is the frequency of reactor trip failure when the trip parameter deviates from the pre-defined set-point values. For both the PPS and ESF-CCS, the scope of analysis is mostly as shown below in Table 7-3.

Table 7-3: General scope of unavailability analysis

| Module | Major Function | Remark |
|---|---|---|
| Transmitters of Sensor Signals | Transmitters of Process Control System | Include |
| Input Cards | Reception of analog inputs transmitted to two analog input cards of each channel | Include |
| Bistable Processor | -2-out-of-4 logic<br>-Generation of Reactor Trip Signal | Include |
| Coincidence Processor | -2-out-of-4 logic<br>-Initiation of Reactor Trip signal<br>-Generation of the engineered safety feature actuation signal | Include |
| Interface and Test Panel (ITP) | - BP/CP Operation Test and Monitoring<br>- Transmission of PPS status to OM | Not Include |
| Operator Module (OM) | -COM, MCR, RSR<br>-Display of RPS status | Include |
| Trip Circuit Breaker | -Remove the power to Control Element Drive Mechanism (CEDM) for Reactor Trip | Include, but not including CEDM |
| Manual Trip | System level manual trip | Include |

In order to perform FTA and unavailability analysis, component failure probability and CCF probability are calculated.

### (4) Complexity Analysis

IEEE Std. 610.12 defines complexity as the "degree to which a process is difficult to analyze, understand or explain" [21]. When evaluating a system design from a safety perspective, the simpler design options are those that accomplish the function with the most confidence and the least ambiguity [22]. Complexity is a common source of errors in system design logic. At a conceptual level, complexity beyond a certain point challenges the capability of the human mind to perform accurate symbolic manipulations, and errors result [23]. The same factors that limit a person's capability to mental manipulations of more than "7 +/- 2" objects simultaneously [24] apply to software or logic. Limiting complexity results in systems with greater reliability, maintainability, and understandability.

The detection and correction of complexity problems can be achieved if the system architecture is evaluated before going on to the next phase of development. By evaluating the complexity of a system, design choices can be justified or altered if the system complexity is found to be too great. Empirical studies referred to by McCabe [25] show that there is a positive correlation between measured complexity and the number of errors found in the implemented system. Furthermore, the prediction and subsequent reduction of complexity have been found to result in large savings in maintenance costs and efforts [26] [27]. Therefore, the simpler a system is, the easier it is to implement, verify, maintain, and the less likely it is to contain latent errors.

Although there are no regulations, standards, or guidance to address the aspect of simplicity for digital I&C systems in NPPs, complex I&C systems challenge the demonstration of conformance with safety-system design criteria such as independence, reliability, and maintainability. The U.S. Nuclear Regulatory Commission (NRC) [22] considers simplicity "as supporting all fundamental design principles for developing safety systems with high reliability". The system complexity must therefore be taken into account when developing safety critical I&C systems.

Complexity measures the effort necessary to attain a given level of reliability. Therefore, given a fixed level of effort, complexity measures the reliability itself [23]. There are several characterizations of system complexity:

- Task Complexity: The number of tasks in a process
- Control Flow Complexity: The arrangement of tasks in a process and the associated decision points
- Data Flow Complexity: The degree of dependency of the information objects and their mapping to tasks and resources
- Resource Complexity: The level of coupling between resources and their allocation to tasks.

Selecting a metric is therefore dependent on the facet of complexity that is of interest. For real-time safety systems in nuclear plants, NUREG/CR-6083 [28] suggests using either Halstead's, McCabe's or Henry's metrics as they call for fewer subjective judgments. Additionally, different metrics are suited for use at different stages of the lifecycle process. The earlier the metric is used, the easier and more cost-effective it is to identify complex system modules and correct them. Smidts and Li [29] ranked different metrics based on expert opinion for their applicability in different lifecycle phases of the development of safety-critical digital systems. Their results are detailed in Table 7-4. In the table, the Semantic Classification is a grouping of metrics that are used to measure the same quality but using a different means or approach for the evaluation.

Table 7-4: Top 3 Quality Metrics per phase of the development cycle [29]

| Phase | Top 3 Measures | Semantic Classification |
|---|---|---|
| Requirements Phase | Fault Density | Fault Detected Per Unit of Size |
| | Requirements Specification Change Request | Requirements Specification Change Request |
| | Error Distribution | Error Distribution |
| Design Phase | Design Defect Density | Fault Detected Per Unit of Size |
| | Fault Density | Fault Detected Per Unit of Size |
| | Cyclomatic Complexity | Module Structural Complexity |
| Implementation Phase | Code Defect Density | Fault Detected Per Unit of Size |
| | Design Defect Density | Fault Detected Per Unit of Size |
| | Cyclomatic Complexity | Module Structural Complexity |
| Testing Phase | Failure Rate | Fault Detected Per Unit of Size |
| | Code Defect Density | Fault Detected Per Unit of Size |
| | Coverage Factor | Fault-tolerant Coverage Factor |

### ① Fault Density

This measure indicates the fault density of a specific program for the given severity levels that can be used to predict the remaining faults by comparing with the expected fault density, determine if sufficient testing has been completed based on predetermined goals for severity class and establish standard fault densities for comparison and prediction. This is performed by establishing the failure types and the fault types, then observing and logging each failure. Next is to determine the program fault that caused the failure then determine the total lines of executable and non-executable data declaration source code (KLSOC). Finally, the fault density is then calculated for the given severity levels as Fd = F/KLSOC.

### ② Requirements Specification Change Request (RSCR)

This measure indicates the stability of the functional requirement which is very significant in the development of a project as the quality of the system being developed, is dependent on the rate of requirement change. This is the total number of requested changes made to the requirements specification from the beginning of the project to the operations phase of the project.

### ③ Error Distribution

This measure is designed to rank the failure modes according to each classification. This is performed by plotting the errors according to the criteria for classification.

### ④ Cyclomatic Complexity

This measure determines the structure of a coded module which tends to limit the complexity of the module and promote understandability of the module.

### ⑤ Code Defect Density

This is similar to the fault density but the faults counted in this measure are observed directly through the code inspection and walkthrough process which makes it more reliable than the fault density in the sense that the fault locations are identified.

⑥ **Coverage Factor**

This measure reflects the ability of the system to automatically recover from the occurrence of a failure during normal system operation.

Coverage = Probability [system recovers | failure occurs]

### (5) CCF Probability Analysis

In chapter 5, CCF Probability analysis is mentioned and CCF probability are calculated. The CCF probability analysis for PLC based and FPGA based system is provided in Appendix A

## PROBLEMS

1. What is a modeling technique that is mainly used in the analysis process?
2. What is modeling?
3. What is simulation?
4. What is prototyping?
5. What are the benefits from modeling and simulation?
6. What is the objective of modeling and simulation?
7. What system analyses are often used for reliability and availability?
8. Describe FMEA.

## FURTHER READING


- R. Fink, C. Killian and T. Nguyen, "Recommended Approaches and Design Criteria for Application of Field Programmable Gate Arrays in Nuclear Power Plant Instrumentation and Control Systems," EPRI, 2011.
- RPC Radiy, "FPGA-Paved Road to Business Success," September 2017
- IAEA, "Application of Field Programmable Gate Arrays in Instrumentation and Control Systems of Nuclear Power Plants," January 2016.

- Thomas J. McCabe and Charles W. Butler, "Design complexity measurement and testing," Communications of the ACM, vol. 32, no. 12, pp. 1415-1425, December 1989.
- G. Stark, R. Durst and C. Vowell, "Using Metrics in Management Decision Making," IEEE Computer, vol. 27, pp. 42 - 48, September 1994.
- NUREG/CR-6083, UCRL-ID-114565, "Reviewing Real-Time Performance of Nuclear Reactor Safety Systems," 1993.
- C. S. Smidts and M. Li, "Software Engineering Measures for Predicting Software Reliability in Safety Critical Digital Systems," Washington, DC, 2000.

# Chapter 8 PPS REQUIREMENTS AND ARCHITECTURE DEFINITION


## SUMMARY

An overview of the PPS safety functions is described in Chapter 4. The PPS is designed to protect the reactor from abnormal and unsafe conditions. It achieves this by monitoring selected safety-related plant parameters such as pressure, temperature, water level, and flow rate. Depending on predefined parameter set-points, it initiates reactor trip and/or actuates the engineered safety features once a non-permissible plant condition is detected. Each process parameter monitored by the PPS is measured by four redundant instrument channels, such that if any two or more measurements of a given process parameter exceed the set-point, the appropriate initiation signal is generated.

In order to design the PPS, its functionality and design requirements in terms of performance, functions, safety bases, constraints, etc., must be well understood. The description and design bases of PPS can be found in several documents such as the Design Control Document (DCD), and should be read to have a full grasp of the PPS design.

This chapter consolidates on the key concepts presented in the introductory Chapter 1 and Chapter 3 by presenting some level of details around the PPS design using systems engineering approach. Typically, the starting point of systems engineering involves stakeholder definition and requirement analysis. Although this starting point has been somewhat addressed at a relatively higher level in Chapter 6, more depth is given to it here as it relates specifically to the PPS design. Thus, more detailed requirements decomposition and traceability will be demonstrated in this chapter.

In order to demonstrate the concepts of system design, modeling, and analysis, two of the PPS trip functions, Variable Overpower Trip (VOPT) and Low Pressurizer Pressure Trip (LPPT) are redesigned for an FPGA based system.

Mind-map of chapter 8 is shown in Figure 8-1.

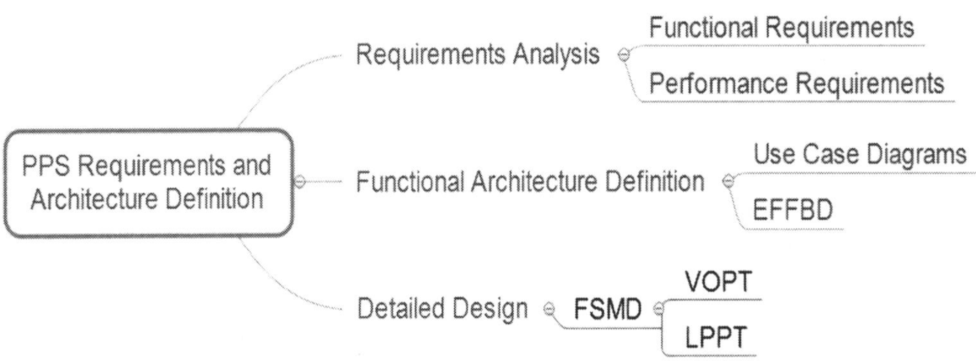


Figure 8-1: Mind-map of Chapter 8

## OBJECTIVES

- Requirement Analysis of PPS including Functional and Performance requirements
- Definition of functional architecture of the PPS
- Discussion of Detailed Design of the PPS

## 8.1 PPS Requirement Analysis

For system modification and redesign, requirements analysis involves the use of reverse engineering and re-engineering concepts in the decomposition and traceability of relevant stakeholder requirements that may have impact on the system design.

From the system level description of PPS trip functions in previous chapters, there are twelve(12) RPS trip functions around the PPS. The requirements for each of the reactor trip function can be analyzed by performing decomposition and traceability of relevant stakeholder requirement documents. Although each trip function operates on a different parameter, they are similar in the method of their requirement decomposition and traceability. The intention here is not to analyze all the PPS trip functions but to present some examples that reflect the common method for requirement analysis. In this respect, the Variable overpower trip (VOPT) and Low Pressurizer Pressure Trip (LPPT) functions will be the focus of our analysis.

Figure 8-2 illustrates the general view of the processes involved in the requirement analysis of the VOPT function with respect to the systems engineering approach that has been discussed thus far. Key elements of Figure 8-2 include the elicitation of stakeholders' requirements, definition of the design requirements, application of relevant regulatory requirements, and the application of requisite SE tools in performing reverse engineering and re-engineering in order to come up with a new design concept.

For determining the VOPT function, a lot of source documents as lifecycle concept, KURD Chapter 10, Concept of Operation, and even the design spec from the previous projects are used for inputs. On the top, there are controls as SEMP, SOW, and Regulation Requirements which should be mandated in the project. Those mandatory requirements will be effective until a single function, VOPT herein, is elicited. As shown in the figure, the final output from this context diagram is the elicited requirements for PPS VOPT function.

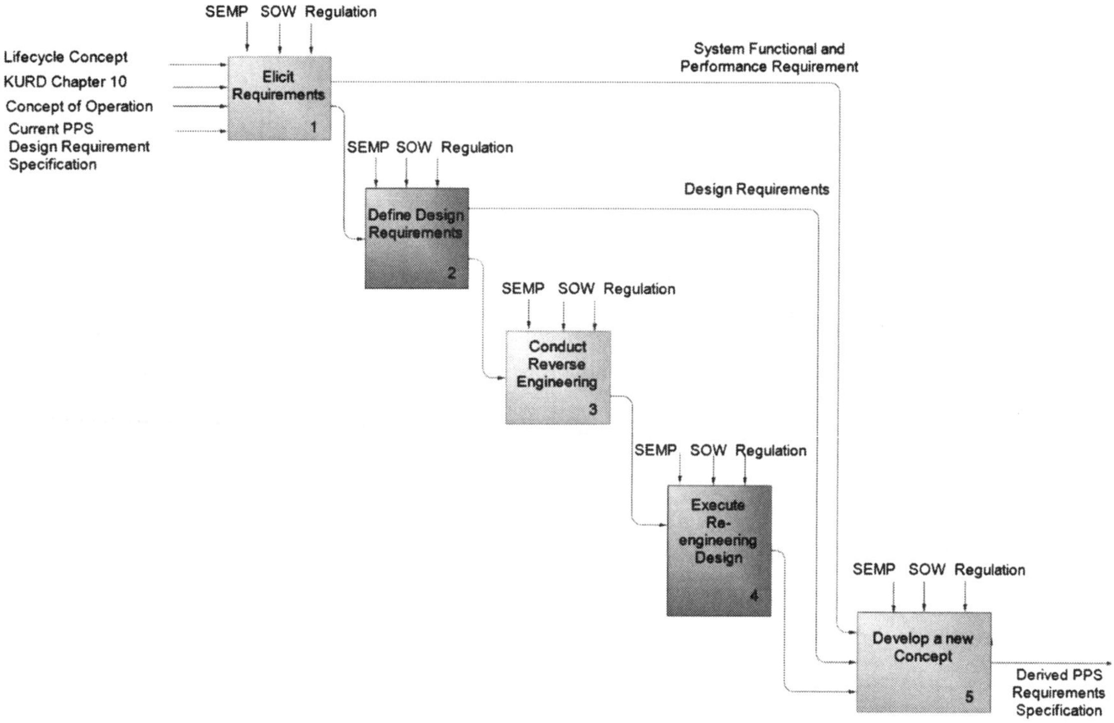


Figure 8-2: General view of requirement analysis for VOPT function

The PPS is a complex system composed of interconnected subsystems, which perform several objectives. The objectives are not limited to trip functions alone but include other functions such as the generation of alarms, testing, and the setting or removal of bypasses. In order to simplify decomposition and traceability, the requirement analyses are often broadly classified into two, namely: the functional requirement analysis, which involves the evaluation of the functions and objectives to be achieved by the system; and the performance requirement analysis which involves the evaluation of how well the functions or objectives must be performed.

### 8.1.1 Functional Requirements

One approach to analyzing the functional requirement of the PPS is by reverse engineering, in which a bottom-up decomposition of the requirement documents is carried out. A structured way of doing this is to first decompose the PPS design requirement (DR) and design specification (DS) documents. The functional requirements can then be analyzed and traced back to the referenced regulatory requirements and the relevant industrial standards. As the decomposition is executed from bottom to top, the applicable regulatory documents and industrial standards are also decomposed so that traceability of all requirements can be done.

Figure 8-3 shows the decomposition of the PPS design requirement document which reveals the system functions, the performance requirements, operational requirements and testing requirements. These requirements can be grouped as functional and non-functional requirements as illustrated in Table 8-1. As depicted in Figure 8-3, there are 10 functional requirements artifacts each of which can be decomposed further. For instance, the RPS functions can be decomposed further into sub-functional requirements as depicted in Figure 8-4. Each of the sub-functional requirements can be further decomposed as illustrated with the case of VOPT function shown in Figure 8-5.

Table 8-1: PPS functional and non-functional requirements

| Functional Requirements | Non-functional Requirements |
|---|---|
| RPS function<br>ESFAS function<br>Control function<br>Alarm function<br>Test function<br>Bypass function<br>Interlock function<br>ITP function<br>MTP function<br>Operator Module function | Performance<br>Expandability<br>Security<br>Maintainability<br>Environmental<br>Design bases<br>Safety criteria<br>Etc. |

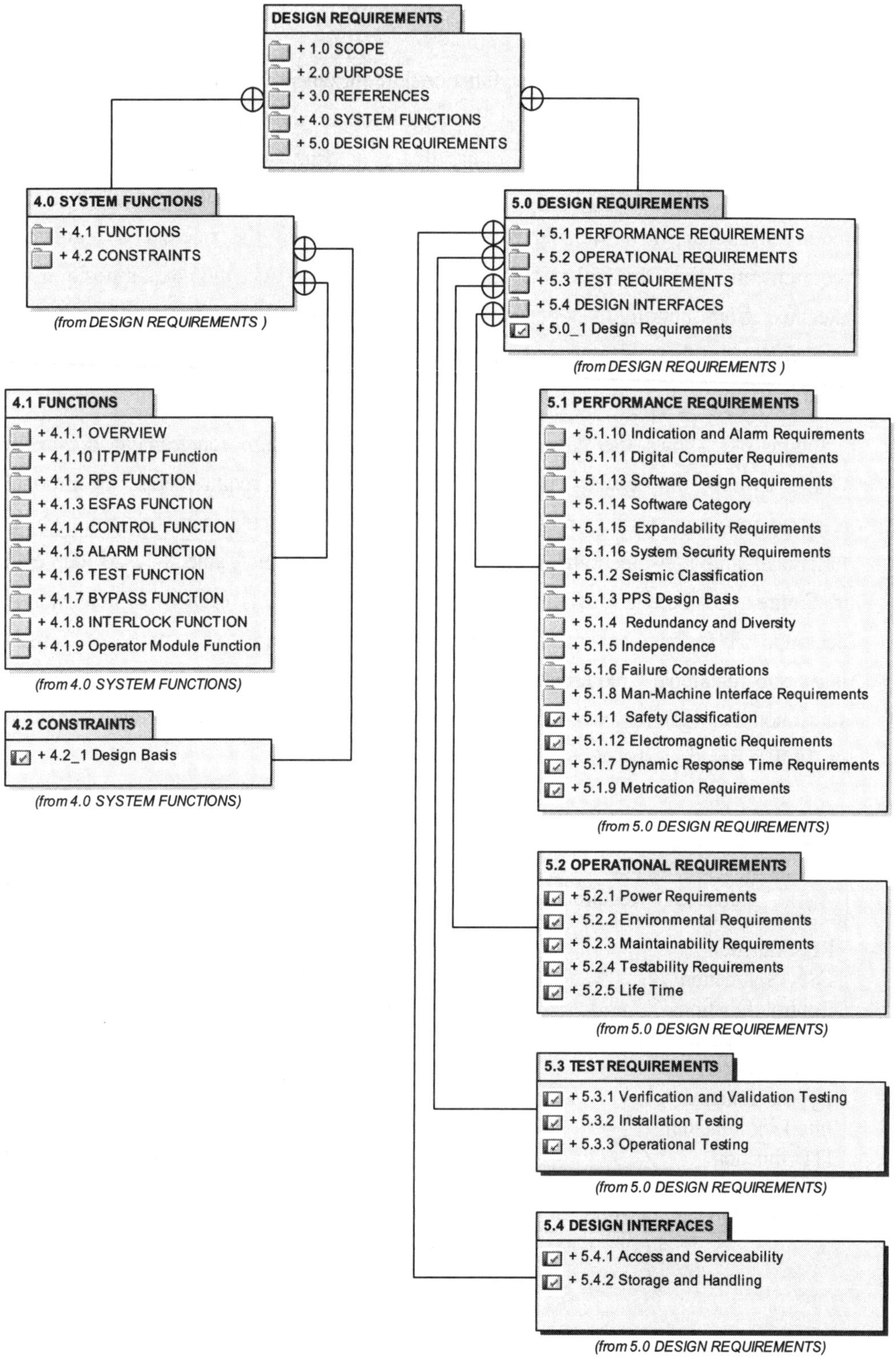


Figure 8-3: PPS Design Requirements Document Decomposition

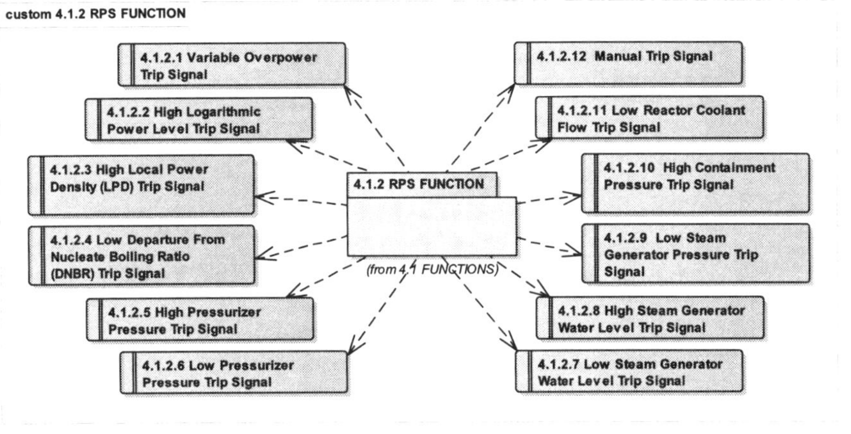


Figure 8-4: Decomposition of the RPS Functional Requirements

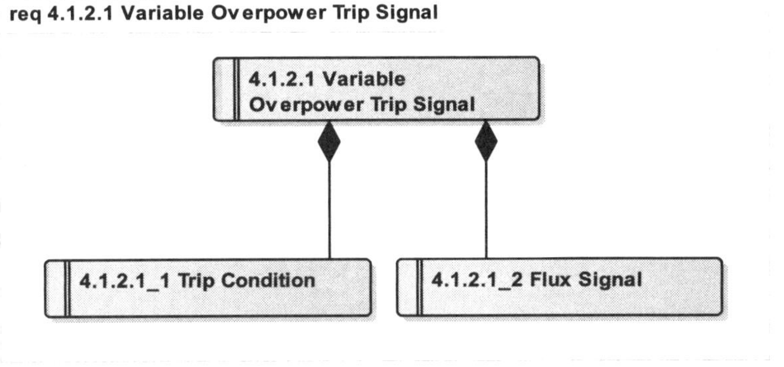


Figure 8-5: Decomposition of the VOPT Functional Requirements

It should be noted that in Figure 8-3, Chapter 5.3 of the requirements specification contains both the test requirements and the verification and validation requirements. These requirements are merged although INCOSE [10] defines the Verification and Validation requirements as an output of the Stakeholder Needs Definition process, while the test requirements are an output of the system requirements definition process.

Thus, for the case of the VOPT, the requirements are traced to the Safety Analysis Report (SAR), IEEE Std, and the USNRC Regulatory guides. Then the result is summarized in the RTM of Figure 8-6. In addition, Figure 8-7 shows the traceability diagram of the VOPT function to the applicable top-level standards and regulatory requirements.

| DR of PPS | | | | SAR 7.2 | | STD (IEEE STD 603.2009) | | SECY/RG/BTP | GDC (SRP 7.2) |
|---|---|---|---|---|---|---|---|---|---|
| 4.0 SYSTEM FUNCTIONS | 4.1 Functions | 4.1.1 | Overview (PPS Function) | 7.2.1.1 | System Description | Clause 4 | | RG 1.206, C.I.7.2 | 20 |
| | | 4.1.2 | RPS Function | 7.2.1.1.1 | Trips | Clause 6.1 & 7.1 | Automatic Control | | |
| | | 4.1.2.1 | VOPT | 7.2.1.1.1.1<br>7.2.2.2.1<br>7.2.2.5 | Variable Overpower Trip<br>Setpoint Methodology | Clause 6.8 and ANSI/ISA 67.04.01- 2006 | Setpoints | RG 1.105<br>BTP 7-12 | 20 |
| | | 4.1.6 | Test Function | 7.2.1.1.9<br>7.2.2.3.3 | Testing<br>Testing Criteria | Clause 5.7, 6.5<br>IEEE Std 338-2006 | Capability for testing and calibration | RG 1.22<br>RG 1.118<br>BTP 7-17 | 21 |
| 5.0 DESIGN REQUIREMENTS | 5.1 Performance Requirements | 5.1.3.1 | RPS Design Basis | 7.2.1.2 | Design Bases | Clauses 4.d, 4.k, 5.2, 7.3 | Completion of protective action | RG 1.153 | 20 |
| | | 5.1.4.1 | Redundancy | 7.2.1.1.7 | Redundancy | Clause 5.1 and IEEE Std. 379-2000 | Single-failure criterion | SECY-93-087 18: II Q<br>RG 1.53<br>BTP 7-19 | 21, 24 |
| | | 5.1.4.2 | Diversity | 7.2.1.1.8<br>7.2.1.1.11 | Diversity<br>System Arrangement | Clause 6.3 | Control protection interaction | RG 1.75 | 21, 22, 24, 29 |
| | | 5.1.5 | Isolation | 7.2.2.3.1 | General Design Criteria | Clauses 5.6.3 and IEEE Std 384-2008 | Independence between safety systems and other systems | RG 1.75<br>BTP 7-11 | 22, 24 |
| | | 5.1.6 | Failure Considerations | 7.2.2.3.2<br>7.2.2.4 | General Design Criteria<br>Failure Modes and Effects Analysis (FMEA) | Clause 5.5 and IEEE Std. 7-4.3.2-2003 | System Integrity | | 21, 23 |
| | 5.2 Operational Requirements | 5.2.4 | Testability Requirements | 7.2.1.1.9<br>7.2.2.3.3 | Testing<br>Testing Criteria | Clause 5.7, 5.10, 6.5 and IEEE Std 338-2006 | Capability for testing and calibration | RG 1.22<br>RG 1.118<br>BTP 7-17 | 21 |
| | 5.3 Test Requirements | | | | | | | | |

Figure 8-6: VOPT Requirement Traceability Matrix

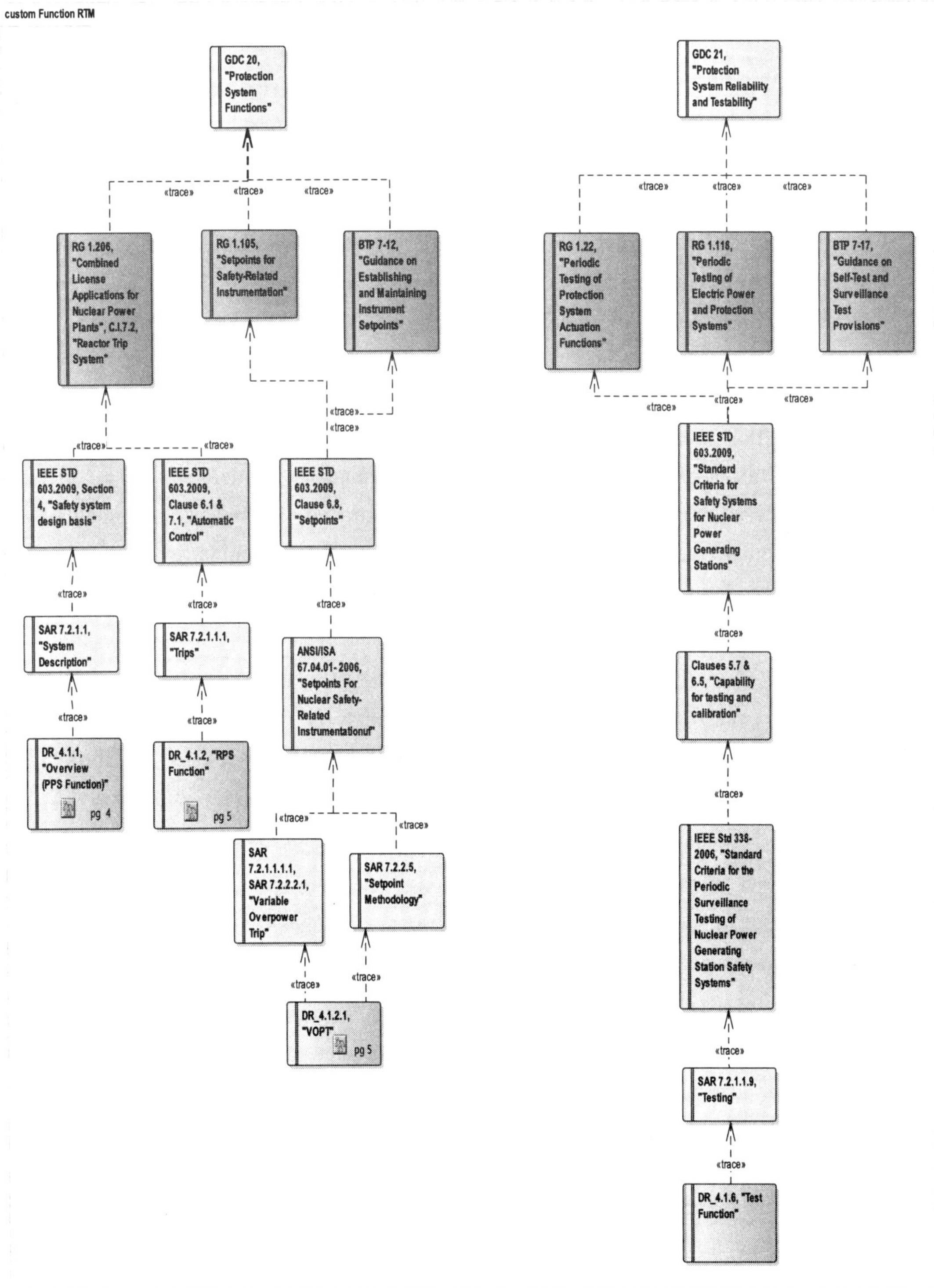


Figure 8-7: Functional & Performance Requirement Traceability diagrams

The goal of traceability analysis is to verify that the input requirements of an activity have been exhaustively taken into account by checking their trace in the resulting documentation; and to verify that all critical requirements have been identified and included through the development life cycle. The traceability is a top-down approach, which originates from the General Design Criteria (GDC). This approach is from the GDC to the Regulatory Guides (RG), which provides guidance to the licensees and applicants on implementing specific parts of the NRC's regulations, techniques used by the NRC staff in evaluating specific problems or postulated accidents, and data needed by the staff in its review of applications for permit or licenses. [30]. The RGs endorse relevant IEEE standards that establish specifications and procedures designed to maximize the reliability of the requirements or regulatory guidelines [31]. The functional requirement of the PPS can be traced to the safety analysis report, which can be further traced to the IEEE standards that were endorsed by the regulatory guides by the USNRC.

Figure 8-7 shows some of the requirements of the VOPT as traced to the top level. The pink box shows the design requirements, the yellow boxes show the traceability analysis to the safety analysis report, the green boxes are the standards the SARs and design requirements are traced to. The brown boxes and blue boxes are the regulatory guides and the GDCs respectively.

Figure 8-4 shows the decomposition of the PPS functions while Figure 8-5 is the decomposition of one of those PPS functions. The decomposed PPS functions is further traced to the top-level requirement as stated by USNRC. Figure 8-7 is the a further breakdown of some of the elements in the traceability matrix of Figure 8-6.

All the RPS functions shown in Figure 8-4 is activated by one of three set-point algorithms: Fixed set-point, Variable set-point and Contact signal from CPCS. It is generally required that the PPS shall have set-point algorithms for the selected plant parameters. Two types of set-point algorithms are used within PPS, namely: fixed set-point and variable set-point. The variable set-point can further be classified into: variable set-point with manual reset, and variable set-point with automatic rate limiting. The fixed set-point algorithm has a fixed set-point value for any process input value, while the variable set-point has no fixed value and it is varied between the floor (minimum) and ceiling (maximum) values.

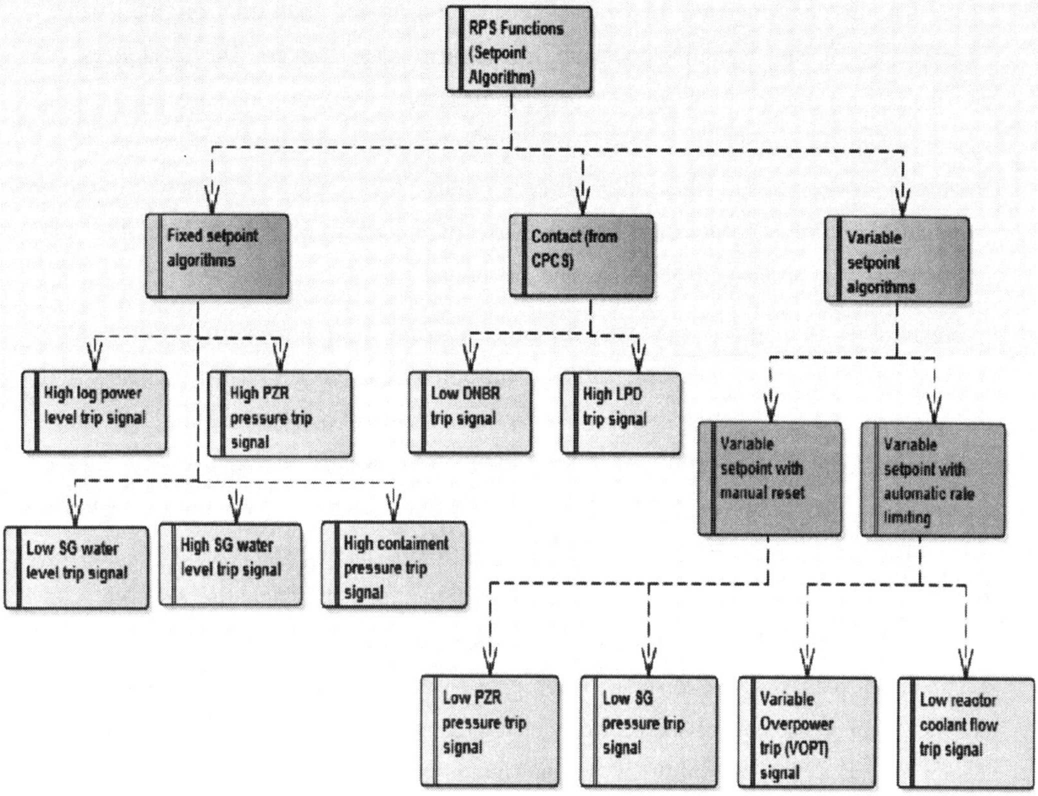


Figure 8-8: RPS bistable trip function set-points

Figure 8-8 shows the RPS bistable trip functions for the selected parameters and their set-point classifications. In the figure, there are also contact signals from the CPCS that represent low DNBR trip and high LPD trip, which are either logic 0 or logic 1.

Figure 8-9 shows three trip signal waveforms. In Figure 8-9(a), the waveform indicates a fixed set-point with a changing process variable (e.g. Pressurizer Pressure). The change in the process variable can either be increasing or decreasing. The fixed set-point with increasing process variable compares the process input and generates a pretrip and trip signals to the coincidence logic whenever the process input is greater than or equal to the set-point. The fixed set-point can be for both increasing and decreasing processes with hysteresis value. Hysteresis causes the output (trip/pre-trip state) to react less rapidly than it otherwise would, by considering recent history. The hysteresis value is typically equal to 1% of span. For a rising process variable, once the trip state is achieved, the hysteresis value is algebraically subtracted from the pre-trip and/or trip set-points causing the new set-points to be lower than the original value. This results in the system staying in trip/pre-trip state longer. The system is so configured to ensure that if the process variable were to increase again, the system state is stable and does not yoyo between trip and un-tripped. The hysteresis is removed when the process value falls below the new set-point (original set-point – hysteresis). For the case of a decreasing process variable, the hysteresis value is instead added to the set-point.

Figure 8-9 (b) shows the rate limiting variable set-point with an increasing process variable. This type of variable set-point permits automatic increase and decrease of the set-point based on the action (increase or decrease) of the input process variable (e.g. Steam Generator Level). The design attempts to maintain a fixed differential between the bistable input and the set-point. The design includes the ability to adjust the rate at which the set-point is allowed to change. If the input signal is changing at a rate greater than the rate at which the set-point can change, the differential between the two values eventually becomes zero, creating a condition such that the bistable logic generates the trip. When the bistable trip occurs, it prevents the set-point from changing until the bistable trip clears. The design includes the capability of having fixed upper and lower limits as well as hysteresis value. The design also provides a variable pre-trip set-point that is always related to the trip set-point by a fixed value. The maximum value for the trip set-point is referred to as the ceiling. Once the trip set-point reaches this value, it does not increase further, even if the process variable continues to increase. Similarly, below a minimum value, the floor, the trip set-point will stop decreasing even if the process variable continues to decrease.

Figure 8-9(c) shows the manual reset variable set-point for a decreasing process. Variable set-points with reset have a trip and pre-trip step offset values. When a reset is selected while the process variable is moving in the direction of trip, the new trip set-point is calculated by subtracting a step value from the process value (unless a floor or ceiling is reached first). When the process is moving in away from the trip, the trip set-point tracks the process value by a fixed step. As an example, for the Low Pressurizer Pressure Trip, during plant cooldown, as the process variable falls below normal, the set-point can be manually by pressing the reset button on the operator module. This is repeated until the floor is reached and the plant is shut down. During plant startup, the set-point automatically increases and remains at a fixed value below the process variable.

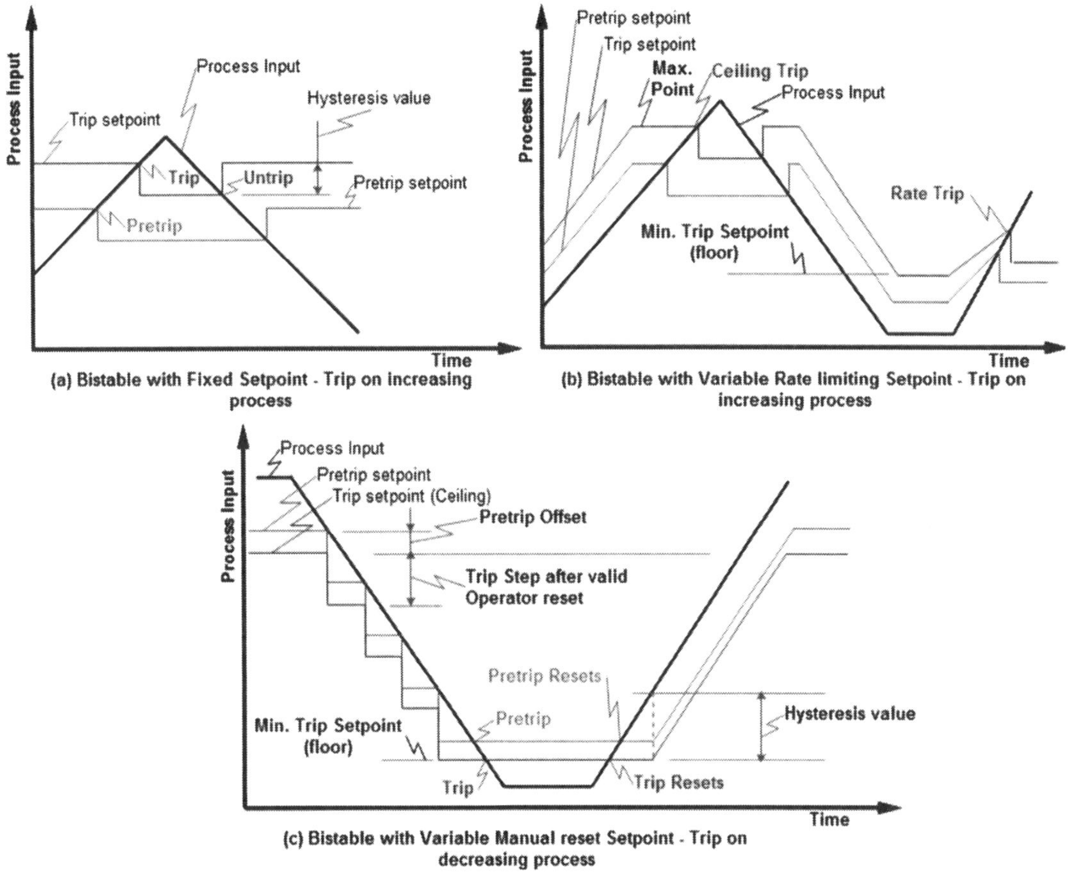


Figure 8-9: Reactor Bistable trip algorithm configurations

## 8.1.2 Performance Requirements

From a SE perspective, the technical processes involved in performance requirement analysis are the same as those for the functional requirement analysis discussed previously with the addition of identifying performance measures. It is an iterative and recursive process of decomposition, traceability, and analysis. As illustrated in Figure 8-10, the PPS performance requirements can be derived from the DR document. Each performance requirement artifact is decomposable to lower depth. An instance of this is the decomposition and traceability of the software design requirements illustrated in Figure 8-11 which is one of the performance requirement artifacts. The figure shows traceability to the Reg. guides and the IEEE Stds. which signifies that the software design requirements satisfy the top-level requirements.

There are other performance requirements, which must be decomposed such as the PPS design bases. The performance requirement for the VOPT example is shown also in Figure 8-6 and Figure 8-7, above.

Since performance requirements analysis must include performance measures, some specific measures for the PPS design in the case of VOPT and LPPT are tabulated as shown in Table 8-2. These measures benchmark the performance of new design.

Table 8-2: Performance requirement measures [30]

| REQUIREMENTS | VALUE | APPLICABILITY |
|---|---|---|
| Uncertainty | < 0.2% | Less than 0.2% of the selected full range value for a period of 39 days across the range of environmental conditions. |
| Accuracy of the software | >= 0.1% | The accuracy of the PPS software shall be equal to or better than 0.1% |
| Variable over power trip (VOPT) | 225 msec | PPS cabinet response time from the input of the BP to the output of initiation relays |
| Pressurizer pressure - Low | 225 msec | PPS cabinet response time from the input of the BP to the output of initiation relays |
| Steam generator water level - High | 225 msec | PPS cabinet response time from the input of the BP to the output of initiation relays |

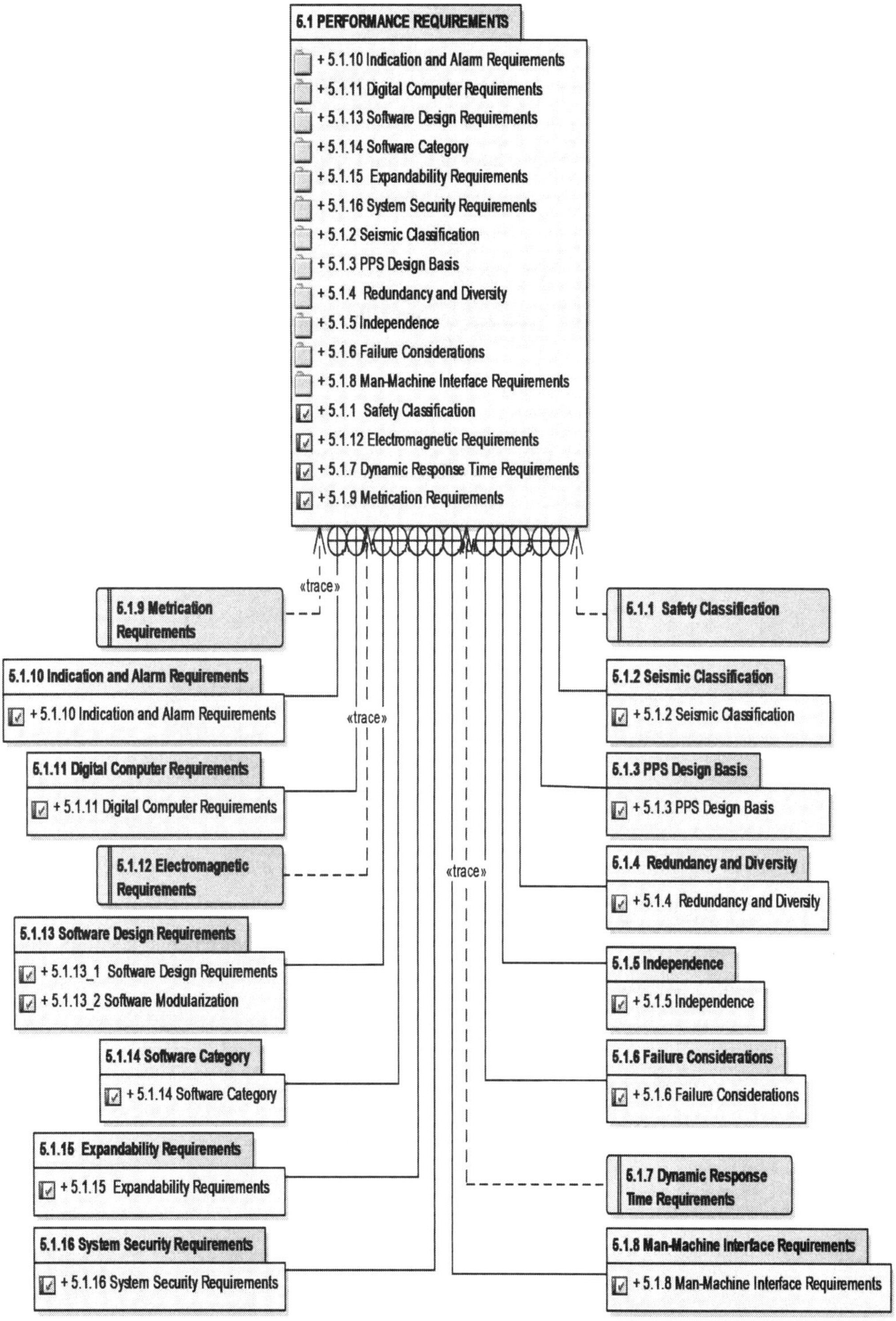


Figure 8-10: PPS Design Requirements (Performance Requirements) decomposition

### 8.1.3 Software Design Requirements

The PPS software consist of two types of categories as follows:

i. Bistable Processor and Coincidence Processor software
ii. Maintenance and Test Panel (MTP) and Interface and Test Processor (ITP)

The PPS software shall be designed to meet the requirements and recommendations as:

| Regulation Guide | Industrial Standard |
|---|---|
| USNRC RG 1.152 | IEEE Std. 7-4.3.2 |
| USNRC RG 1.168 | IEEE Std. 828 |
| USNRC RG 1.169 | IEEE Std. 829 |
| USNRC RG 1.170 | IEEE Std. 830 |
| USNRC RG 1.171 | IEEE Std. 1008 |
| USNRC RG 1.172 | IEEE Std. 1012 |
| USNRC RG 1.173 | IEEE Std. 1028 |
| | IEEE Std. 1047 |

The PPS software shall be modularized and the interface module shall be designed to be flexible to allow the design change in external systems with ease considering short project schedule. The software design requirements decomposition and traceability for the PPS is shown in Figure 8-11.

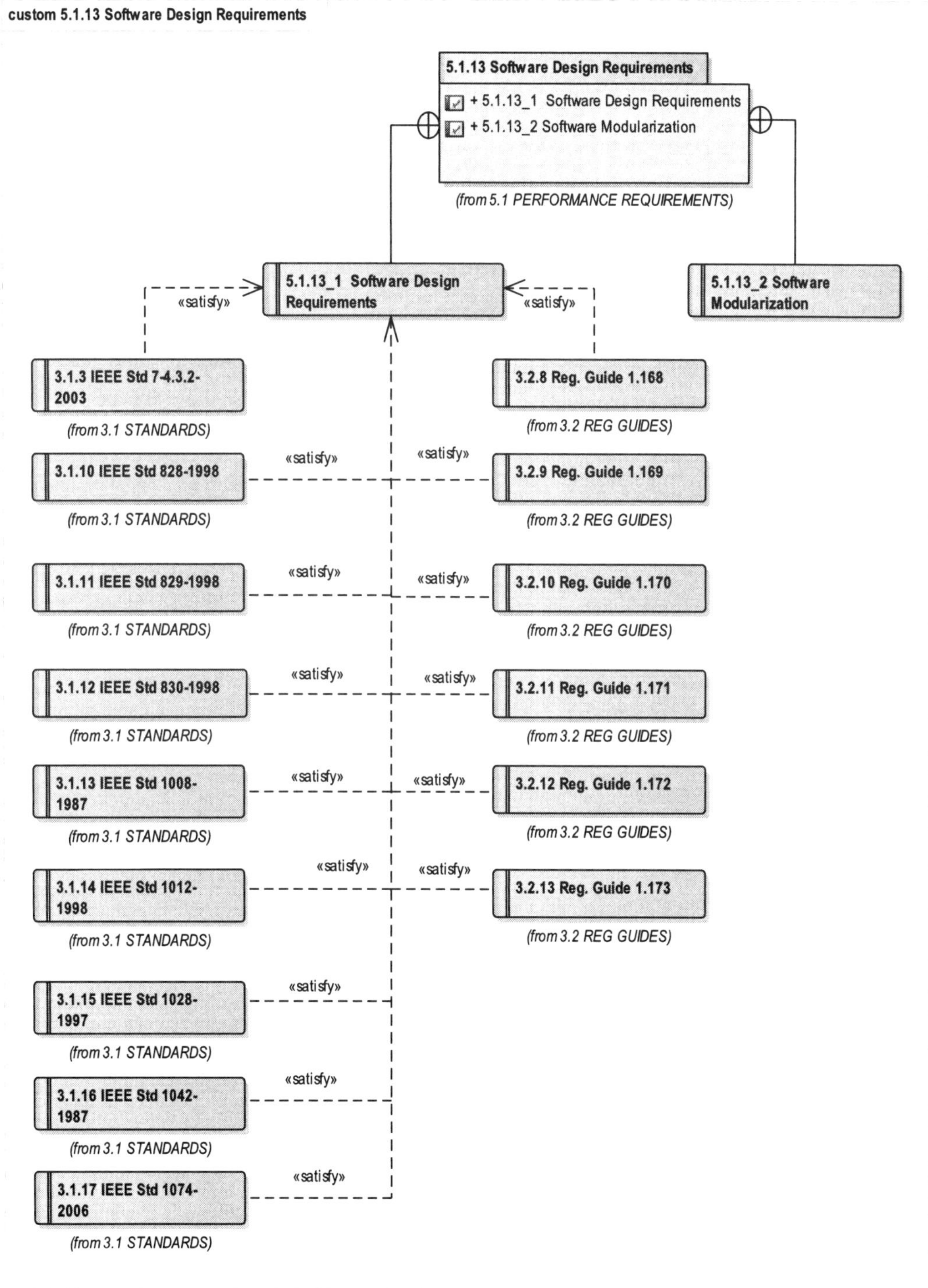


Figure 8-11: Decomposition and Traceability of Software Requirements

## 8.2 PPS Functional Architecture Definition and Analysis

Functional analysis is the process by which the functions, which the system must perform, are identified and the relationship between them defined. However, there are numerous methods of functional analysis, in this section use cases and functional flow block diagrams are used to model the PPS functions.

The modeling of any system's operation often starts with a scenario-based modeling. Typically, a use case analysis can be performed to identify and analyze the system users and interactions. A typical usecase analysis the PPS design is shown in Figure 8-12. The figure shows how different actors interact with the system in different operational and functional states. This provides useful insight into how the system's functions are to be performed and the mode of interaction between them. The operator interacts with the system to initiate the execution of PPS function. The engineer interacts with the system to initiate maintenance functions such as calibration and set-point operations. Since the PPS automatically monitors plant parameters and compares them to some set-point values before initiating a function, the PPS by itself is an actor in the use case analysis.

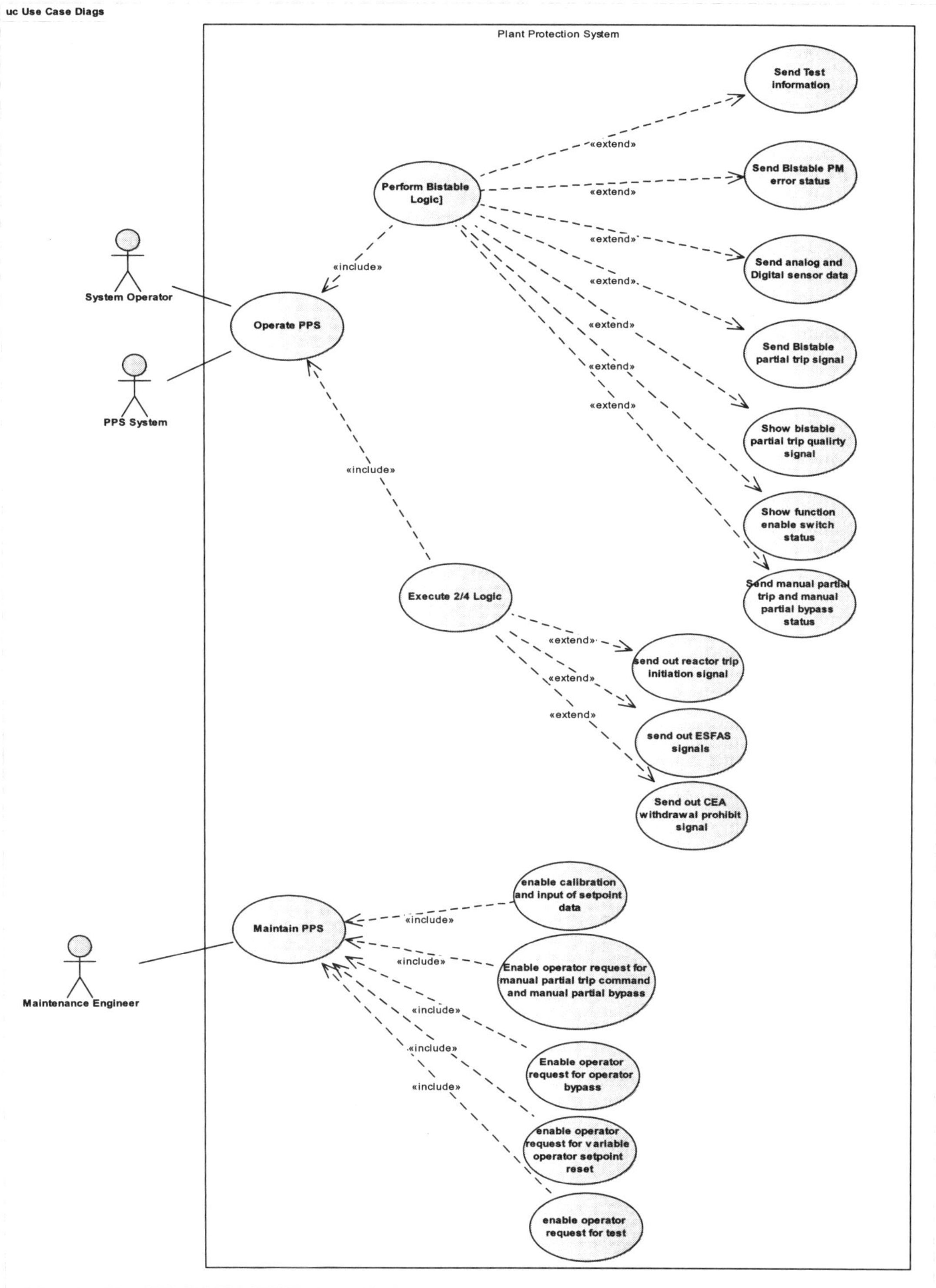


Figure 8-12: Use Case analysis for operational requirement modeling

To proceed with the system's functional modeling, a model based systems engineering tool such as the Functional Flow Block Diagram (FFBD) are typically used to model the system's functional algorithms. The FFBD arrange the system functions in a logical sequence so that any specified operational use of the system can be traced end-to-end [31].

The development of FFBD can be performed using system engineering tools. Modeling software enables the development of Enhanced Functional Flow Block Diagrams (EFFBD), which not only allow for the functional modeling of the system, but its simulation. Figure 8-13 and Figure 8-14 show the use of EFFBD for the PPS bistable Variable Overpower Trip (VOPT) and Low Pressurizer Pressure Trip (LPPT).

In figure 8-13 and 8-14, the rectangular blocks represent the functions to be performed while the ovals represent the data that is input to a given function. In these diagrams, the system consists of both parallel or independent functions as well as sequential functions. The ovals represent data that is input to or output from the functions. Of particular importance are the green oval blocks. These represent triggers, which are input signals that cause the execution of a function block.

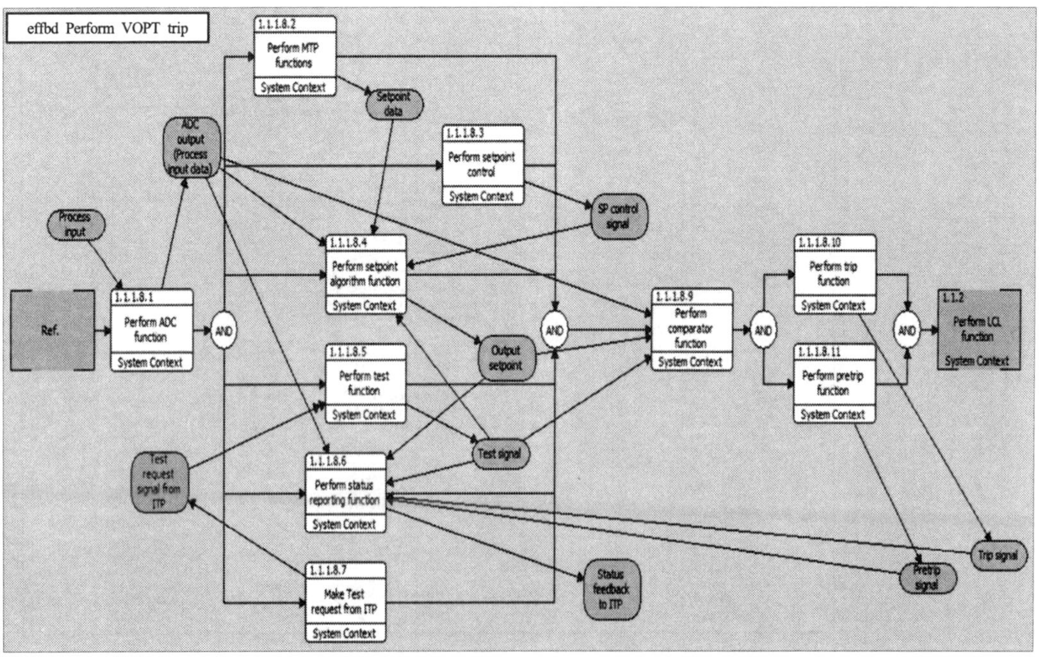


Figure 8-13: PPS bistable VOPT EFFBD model

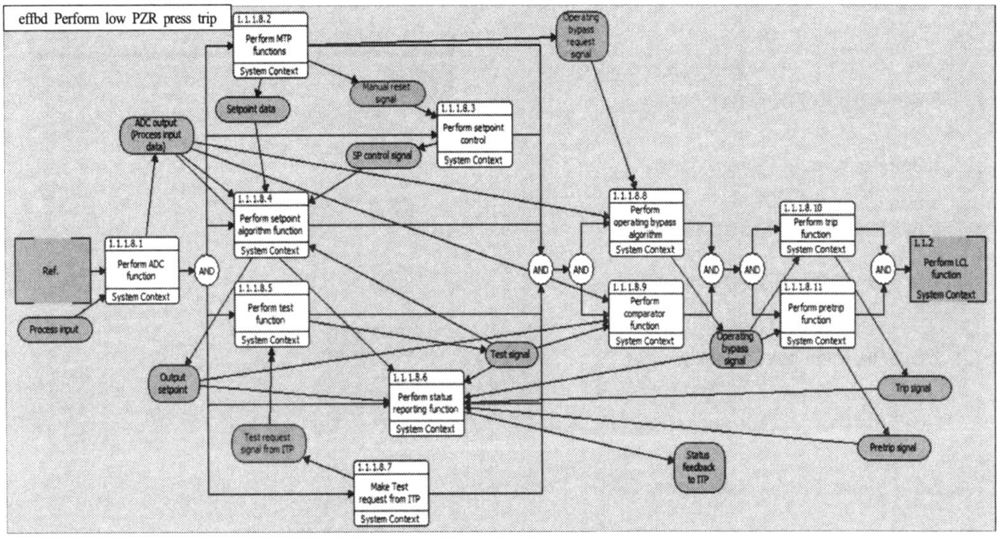


Figure 8-14: PPS bistable LPPT EFFBD model

Once the EFFBD is developed, it is simulated to observe its execution. Figure 8-15 and Figure 8-16 show the results of the simulation of the models of the VOPT and LPPT algorithms. Although the code is not automatically generated from the model, the results show the concurrent execution of the functional flow of the algorithm.

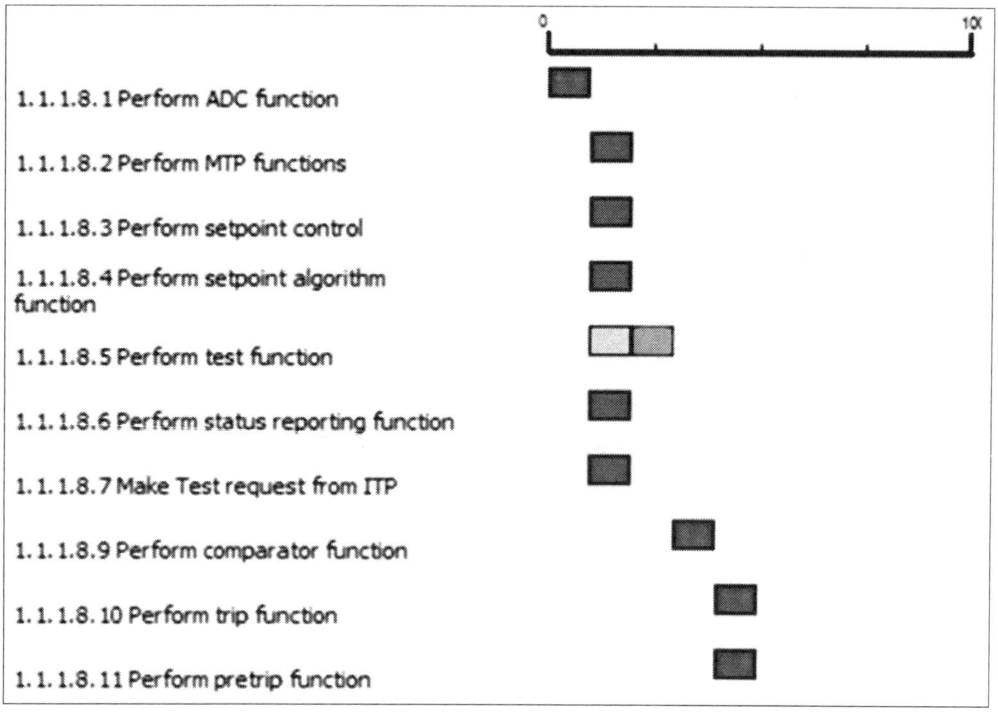


Figure 8-15: VOPT algorithms EFFBD model results from CORE 9

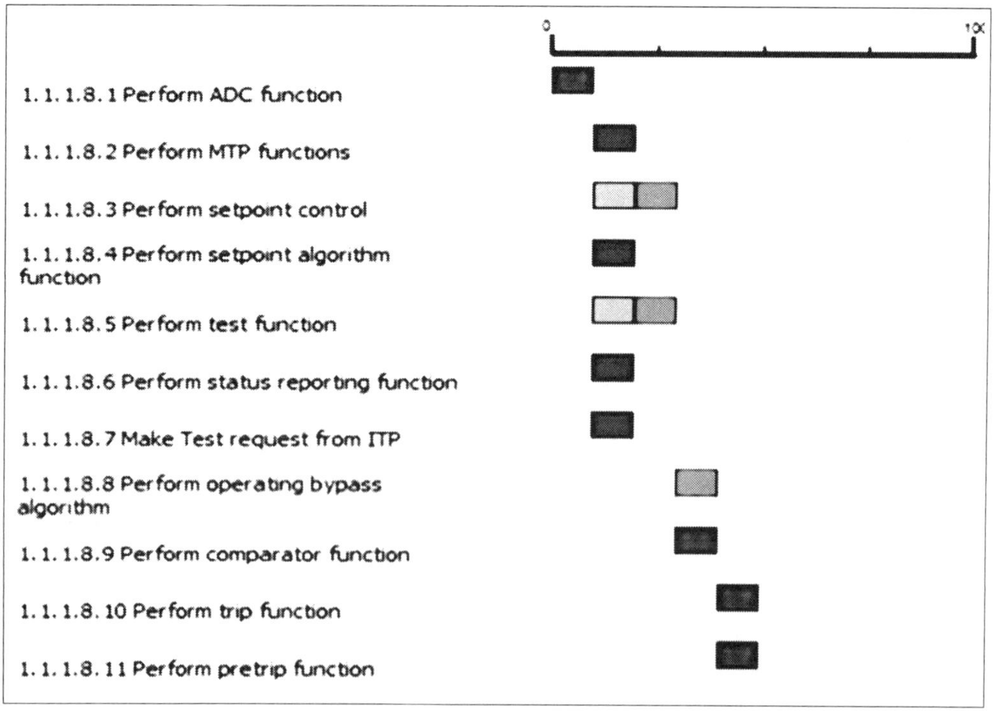


Figure 8-16: LPPT algorithms EFFBD model results from CORE 9

The dark (■) blocks signify the execution of a function that has received all the necessary resources, without having to wait for a trigger. The brightest (□) blocks are functions that are enabled but are waiting to receive a trigger. The bright (▩) blocks represents a function that is executing after receipt of a trigger. By simulating the EFFBD, the operation and interaction of the different functions can be observed, analyzed, and modified for optimal performance.

## 8.3 Design Architecture

After the completion of the system architecture, system development proceeds to the design architecture. Many methods can be used at this stage. One such method is the Finite State Machine with Data Path (FSMD) model. This method lends itself particularly to the development of digital circuits on platforms such as FPGA and ASIC.

An FSMD combines a Finite State Machine (FSM) and regular sequential circuits. An FSM is "used to model a system that transits among a finite number of internal states. The transitions depend on the current state and external input. Unlike a regular sequential circuit, the state transitions of an FSM do not exhibit a simple, repetitive pattern" [32]. In an FSMD, the FSM, which is sometimes known as a control path, examines the external commands and status and generates control signals to specify operation of the regular sequential circuits, which are known collectively as a data path [33]. FSMD is a structured design approach, which offers design simplification and can speed up the design procedure as well as enhances the design verification. For the case of VOPT and LPPT functions, the FSMD approach will involve the following design steps:

- Interface definition of the FSMD for the trip function
- Data path design
- FSM controller design

### 8.3.1 VOPT Design

#### (1) *Interface Definition of FSMD for VOPT Algorithm*

Figure 8-17 shows the interface definition of the VOPT algorithm which indicates the inputs and outputs to be used and generated in the VOPT function. In this case, the set-point is not fixed, it is determined by the reactor power input. However, there is need to input the set-point range limit (floor and ceiling) as well as hysteresis value.

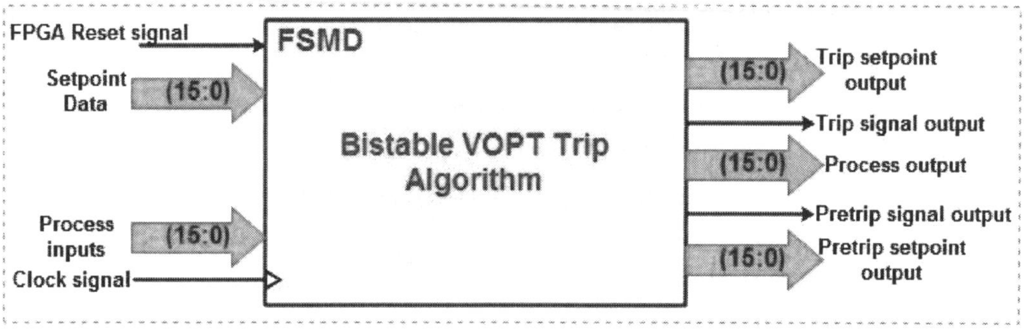


Figure 8-17: FSMD block diagram for VOPT set-point bistable trip function

#### (2) *VOPT Data path Design*

The data path shown in Figure 8-18 consists of registers as memory elements, buses, and multiplexers for storing and moving data from one place to another. The registers are designed to incorporate asynchronous reset signals and input-enable signals. The change in data output of the register depends on the enable signal, giving the FSM control over the data path. The data path compares the process variable input (PI) to the trip set-point. The data path is designed with registers, adders, comparators, and multiplexers. The registers are used to store the sepoints and the current PI value. The register enable signals receive from FSM controller are used to control the alteration of the registers output, should in case operator wants to change the set-point values. The adders are used for addition of hysteresis value to the set-points. The comparators continuously compare the PI with set-point and issue an output flag to the FSM controller whenever the PI is greater than the set-point or greater than the set-point minus hysteresis.

PART 2

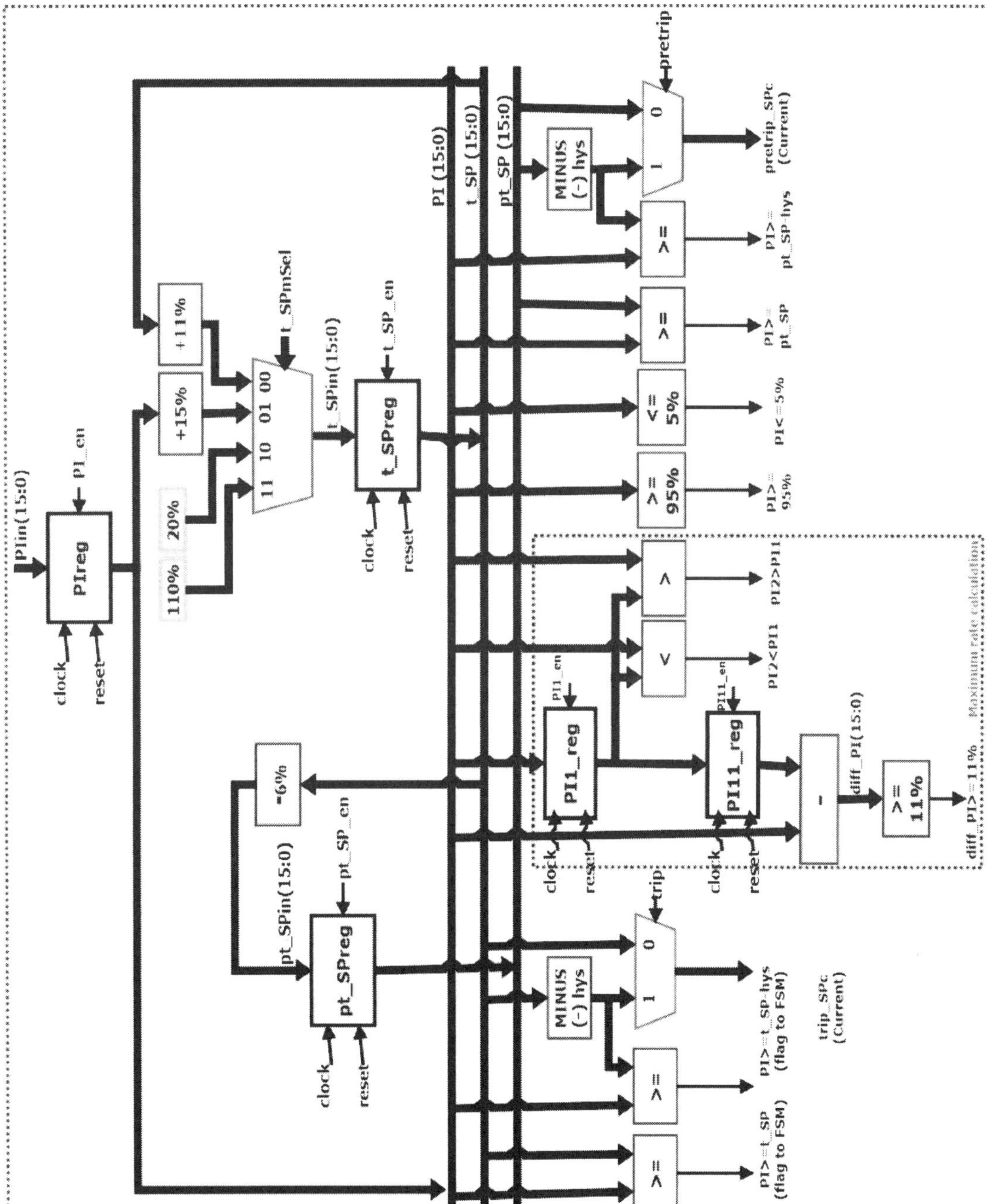


Figure 8-18: Data path for VOPT algorithm

There are additional comparators (>=95%) and (<=5%) which are 'active high' used to determine the ceiling and the floor conditions respectively. There are also some additional adders (+15%), (+11%), and (-6%) whose outputs go to 4-to-1 multiplexer, which are used to calculate the value of trip set-point, when PI is increasing at a rate less than 11%/min, increasing at a rate greater than 11%/min or equal to 11%/min respectively.

The addition of 4-to-1 multiplexer is to select the trip set-point at a particular point in time depending on the signal value of the multiplexer select line (t_SPmSel) from the FSM controller. From the data path diagram, if the "t_SPmSel = 00" command is received from the FSM controller; the 4-to-1MUX select t_SP+11% value for trip set-point (see Table 8-3). The pretrip set-point is calculated from trip set-point value at every value of trip set-point by the addition of '-6%' to the trip set-point value.

Table 8-3: 4-to-1 multiplexer use for VOPT trip set-point selection

| 4-to-1 MUX (Input lines) | Select line (t_SPmSel) from FSM Controller | 4-to-1 MUX (Output (t_Spin)) |
|---|---|---|
| t_SP + 11% | 00 | t_SP + 11% |
| PI + 15% | 01 | PI + 15% |
| 20% | 10 | 20% |
| 110% | 11 | 110% |

Within the red-dotted rectangular box in Figure 8-18 there are two registers, three comparators, and one subtractor in which their operation is being control by a different state machine transition diagram. The logic applied here is that when the system is initialized, the PI1_reg (PI1 register) gets the value of the reactor power and stored it as PI1 and then continually monitors any change in the reactor power using the two comparators (greater than and less than). Immediately a change is notice the system updates the output of PI11_reg (PI11 register) to get PI1 value stored in PI1_reg, in order to evaluate the difference between the two values, (PI2 – PI1) and at same time evaluate the comparator (>=11%) whose output is sent to state machine for necessary action decision. The interfacing diagram of this data path is shown in Figure 8-19. It clearly depicts the signal and data inputs and outputs of the data path interfacing parameters.

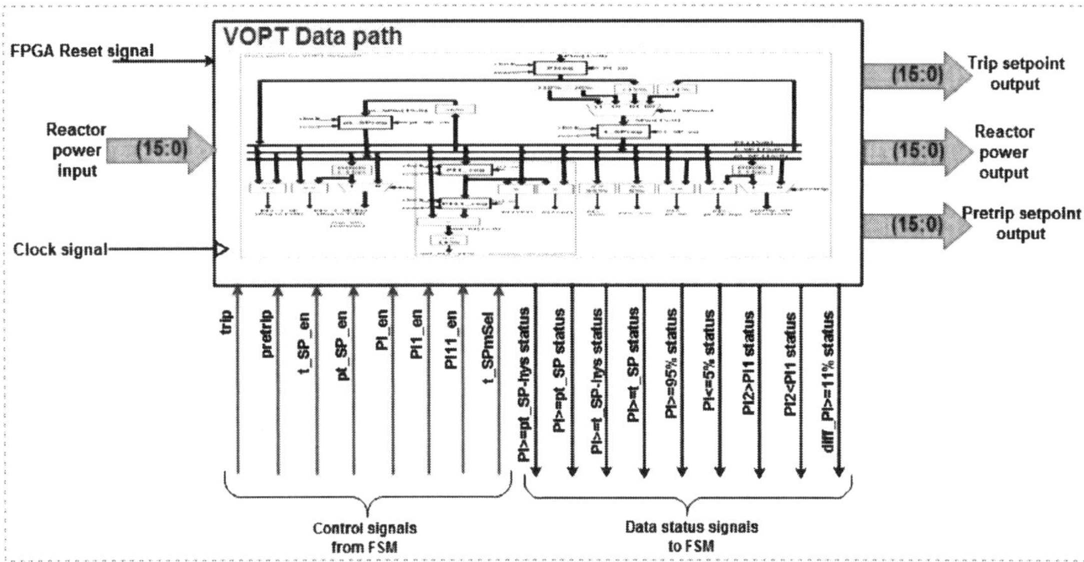


Figure 8-19: VOPT data Path interfacing diagram

### *(3) VOPT FSM Controller Design*

The FSM controller block diagram for both trip and pretrip, showing the signal interfaces is depicted in Figure 8-20. The diagram is a Moore type FSM in which the output is only the function of the current state of the machine.

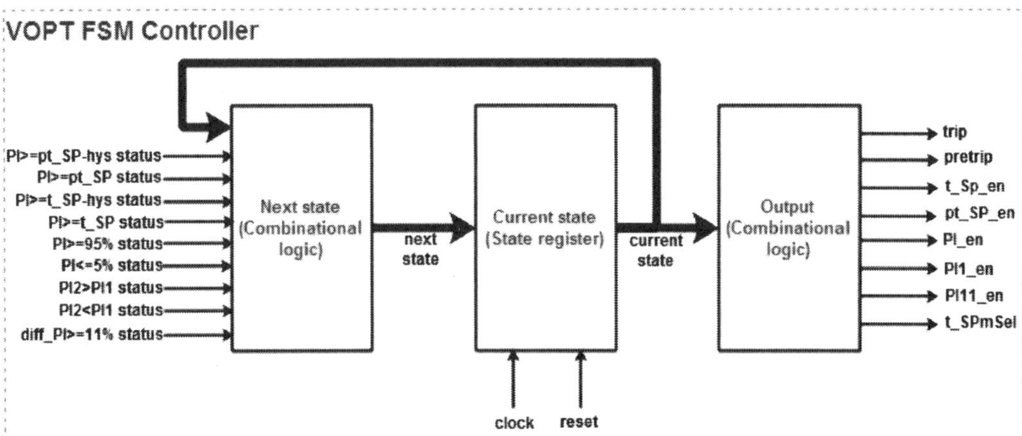


Figure 8-20: FSM controller block diagram for VOPT algorithm

In designing the state transition diagram for VOPT, it is important to take note of the point within the set-point limits where the trip is asserted. There are two conditions in which the trip signal will be transmitted to the Coincidence Logic (CL) in order to trip the reactor:

- If the power increase rate is greater than the pre-determined value (11%/min) AND the reactor power value is greater than or equal to trip set-point, the logic provides a trip signal to CL.
- If the reactor power is equal to the ceiling of 110% (the maximum allowable power set-point increase), the trip signal is provided to CL.

The transition diagram shown in Figure 8-21 is used to control the calculation of the reactor power increase rate, and it has six states: Start, Check1, Diff, Check2, Rate_S, and Adjust states.

- The Start state is the state at which the PI1 register (PI1_reg) in the data path is initialized to store the reactor power called PI1 when the reset signal is active high, and then transits to Check1 state when the reset signal is active low.
- The Check1 state is the state in which the state machine checks for the change in reactor power input called PI2. At this state, the state machine can transit to either Diff state or Adjust state if PI2>PI1 or PI2<PI1 respectively.
- The Diff is the state in which the PI11 register (PI11_reg) is enabled (PI11_en = '1') in order to allow the data path to compute the different (named diff_PI) between PI2 and PI1, and then transits to Check2 state.
- The Check2 state is the state where the state machine checks if diff_PI is greater than or equal to 11%/minute rate power. At this state, the state machine can transit to either Rate_S state or Start state if diff_PI>=11% or diff_PI<11% respectively.
- The Rate_S state is the state in which the maximum rate condition is energized (Rate = '1') and PI1_reg (the value of PI1) is updated to the current reactor power input PI2, and then transits to the Check1 state.
- The Adjust state is the state in which the state machine de-energized the maximum rate condition (Rate = '0') and PI1_reg (the value of PI1) is updated to the current reactor power input PI2, and then transit back to the Check1 state.

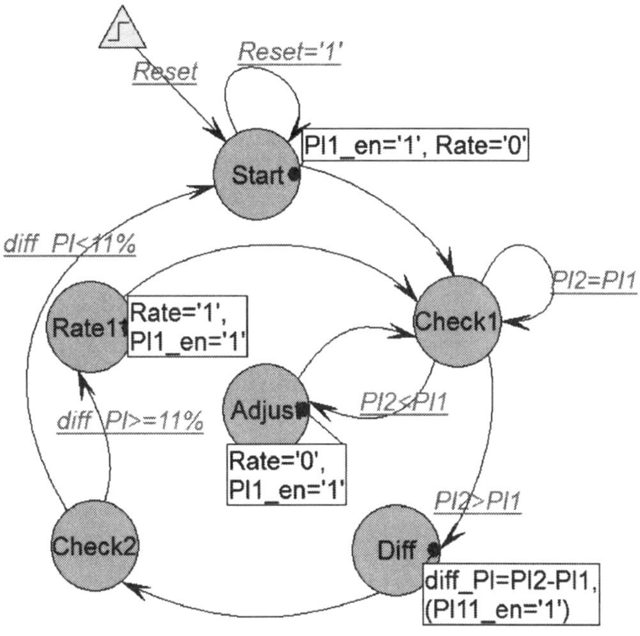


Figure 8-21: VOPT maximum rate determination state transition diagram

The transition diagram shown in Figure 8-22 is used to control the calculation of the trip set-point as well as trip signal determination, and it has seven states: Follow, Floor, MaxRate, Update, Ceiling, trip_S, and Untrip_S states.

- The Follow state is the state where the set-point tracks the reactor power using the difference of the amount of the band (15%), when the process input is greater than the minimum set-point minus the band (15%), and the rate of change of reactor power is less than the maximum allowable rate of 11%/min.
- The Floor state is the state where the set-point is fixed at a minimum set-point (20% in this case) when the reactor power is less than the value, MIN_SP – 15% (5%).
- The MaxRate state is the state in which the trip set-point increases at a maximum rate when the power changes at a rate greater than the maximum rate of 11%/min.
- The Update state is the state where the state machine updates the trip set-point for every change in reactor power. The Ceiling state is the state where the set-point is fixed at the maximum set-point of 110% when the power is greater than or equal to the value, 95%.

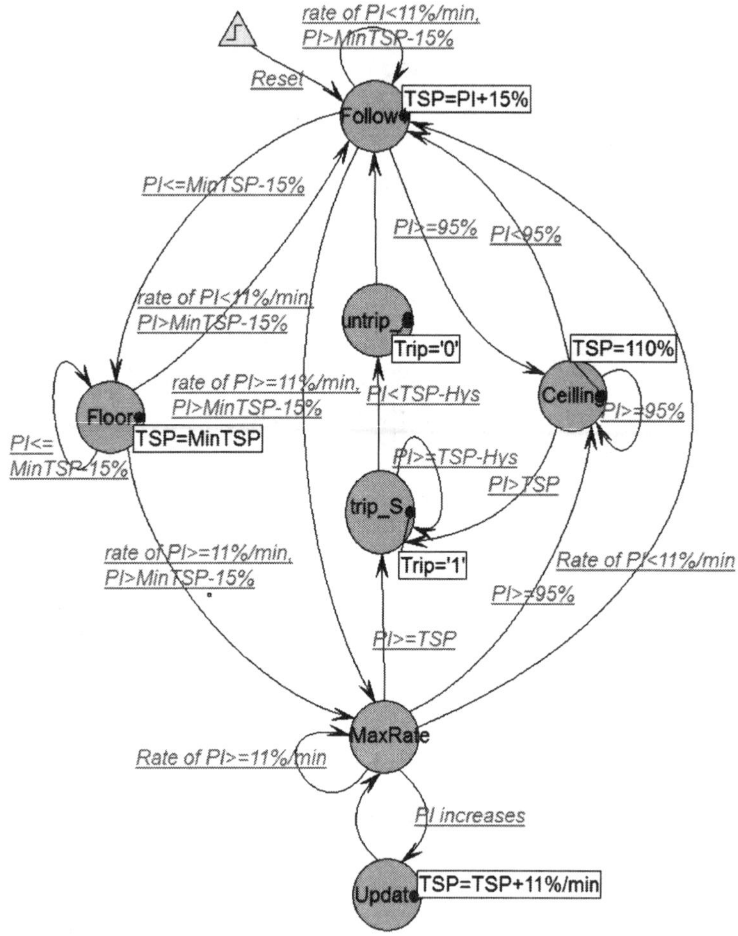


Figure 8-22: VOPT trip set-point calculation state transition diagram

- The trip_S state is the state in which the state machine issues a trip when the reactor power is greater than or equal to trip set-point.
- The Untrip_S state is the state in which the state machine is untrip when reactor power is less than or equal to the value, TSP-Hys (PI<TSP-Hys).

PART 2

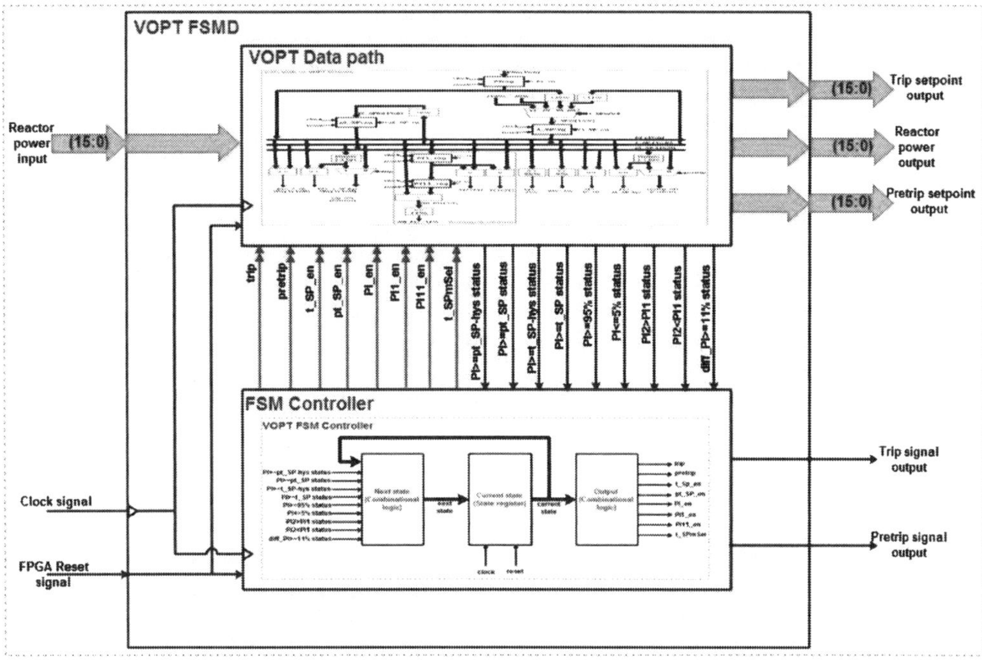


Figure 8-23: FSMD final design for VOPT configuration

Having designed both the data path and the FSM controller for VOPT, the final FSMD design structure for VOPT is designed as shown in Figure 8-23. This is done by integrating the designed data path and FSM controllers together to obtain the final design.

## 8.3.2 LPPT Design

### (1) *Interface Definition of FSMD for LPPT Algorithm*

Figure 8-24 shows the interface definition of the LPPT algorithm that indicates the expected inputs and outputs. In this case, the set-point is not fixed, it is determined by the pressurizer pressure input. However, there is a need to input the set-point range limit (floor and ceiling) as well as hysteresis value. Furthermore, there are other external signals to the FPGA which are manual reset and SOB (status of operating bypass request) that allow the operator to reset the set-point and to request for operating bypass respectively.

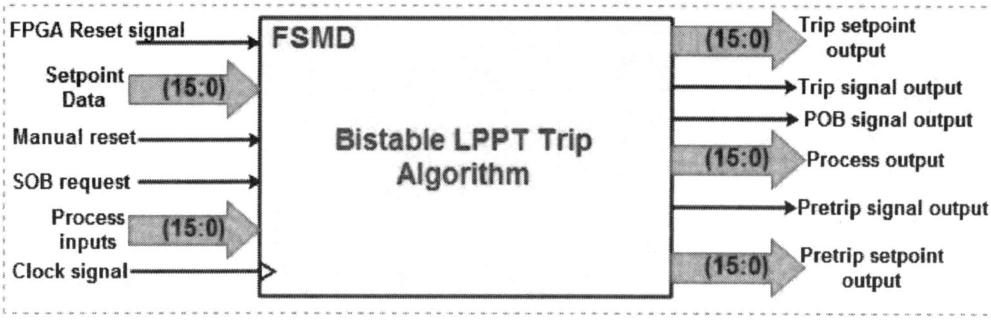


Figure 8-24: FSMD block diagram for LPPT set-point bistable trip function

## (2) *LPPT Data path Design*

The data path for the LPPT algorithm is shown in Figure 8-25, and it consists of registers multiplexers, adders, subtractors, and comparators. The LPPT data path contains the same components as that of VOPT algorithm (See Figure 8-18). The difference is the data contents.

The components inside the red dotted rectangular box of Figure 8-25 are used to evaluate the rate of change in the pressurizer pressure value at any point in time in order to allow the system controller to know whether the pressure is increasing or decreasing. The data path also contains two additional comparators (PI<=400 and PI>=500) inside the purple-dotted rectangular box which are used to check for the condition of operating bypass to respectively permit or automatically remove the operating bypass.

### *(3) LPPT FSM Controller Design*

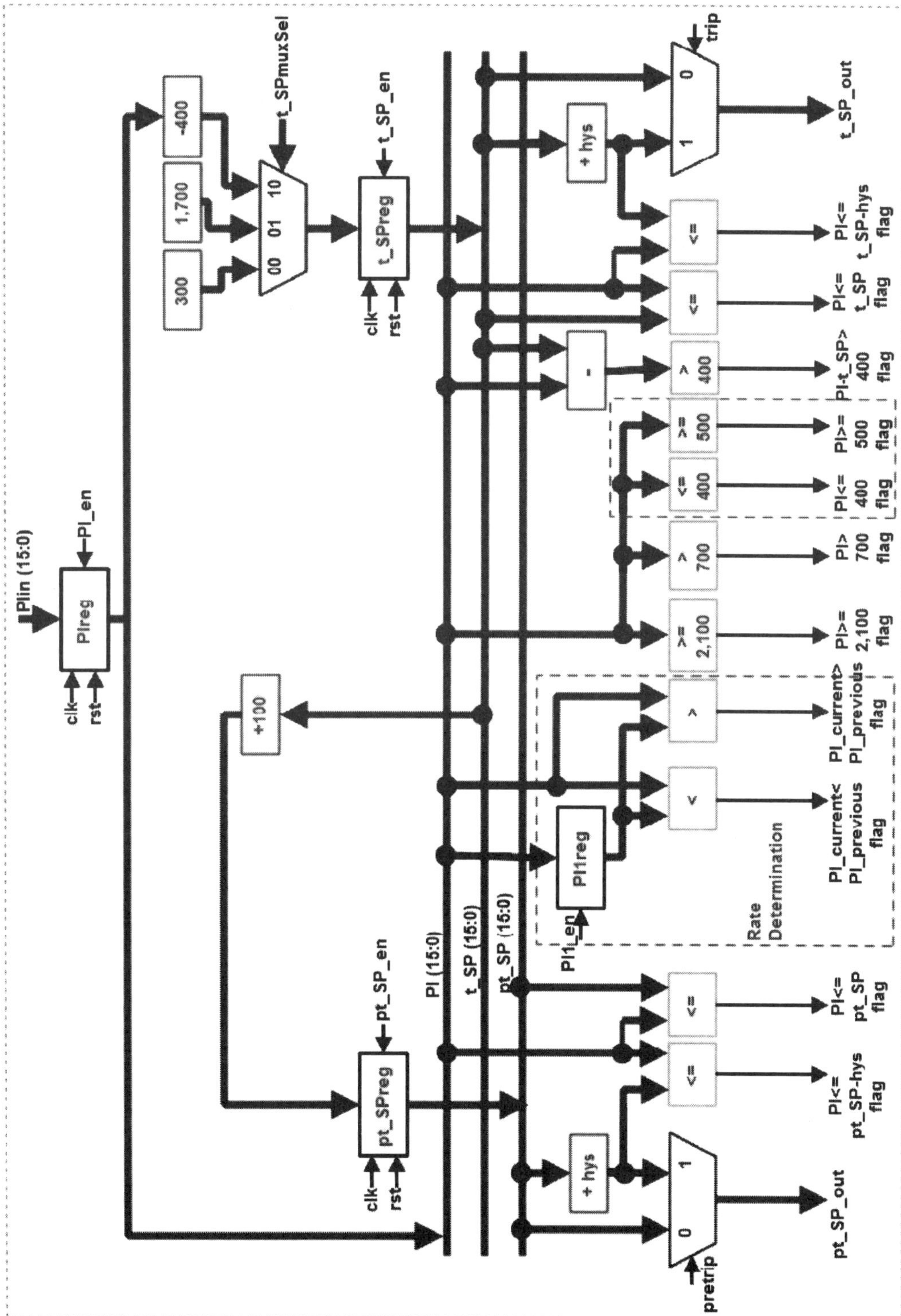


Figure 8-25: Data path for LPPT algorithm

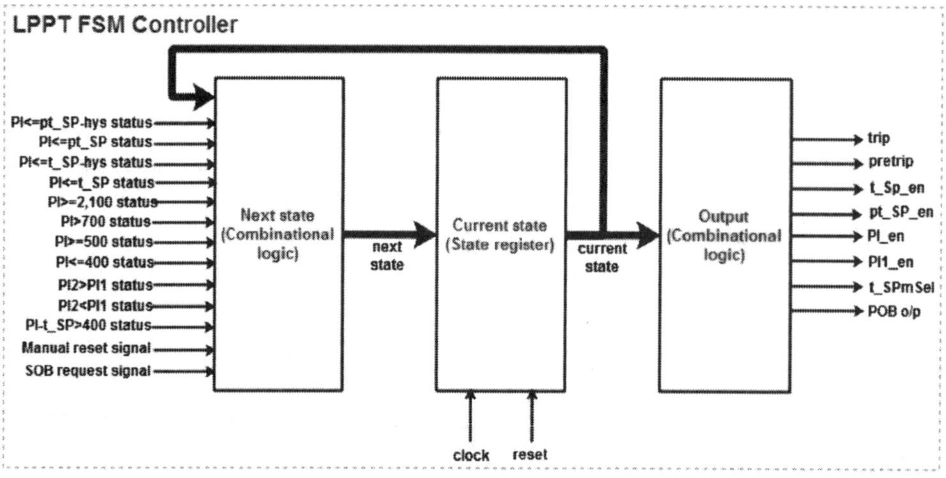


Figure 8-26: FSM controller block diagram for LPPT algorithm

The FSM controller block diagram of LPPT for both trip and pretrip showing the signal interfaces is depicted in Figure 8-26.

The FSM controller transition diagram for the LPPT algorithm in Figure 8-27 illustrates the control the LPPT pretrip set-point and pretrip signal. The transition diagram shown in Figure 8-27 has six states. It is used to control the change in pressure input value and the permission of operating bypass. The states are Start, Wait, Upd1, Upd2, Allow, and Remove states.

- The Start state is the state in which the PI1 register (PI1_reg) in the data path is initialized to store the pressure called PI1 when the reset signal is active high, and then transits to Wait state when the reset signal is active low.
- The Wait state is the state in which the state machine waits until there is change in pressure input value called PI2 or the condition for operating bypass is met. At this state, the state machine can transit to Upd1 state, Upd2 state, Allow state, or Remove state if PI2>PI1, PI2<PI1, PI<=400, or PI>=500 respectively.
- The Upd1 state is the state in which the PI1 register (PI1_reg) in the data path is enabled (PI1_en = '1') to update its output value to current pressure value and at same time sends output PI_rate>0 = '1' (pressure increasing alert) to third state machine shown in Figure 8-28 to calculate the trip set-point, and then transits to Wait state to waits until there is another change in pressure input value.

P A R T 2

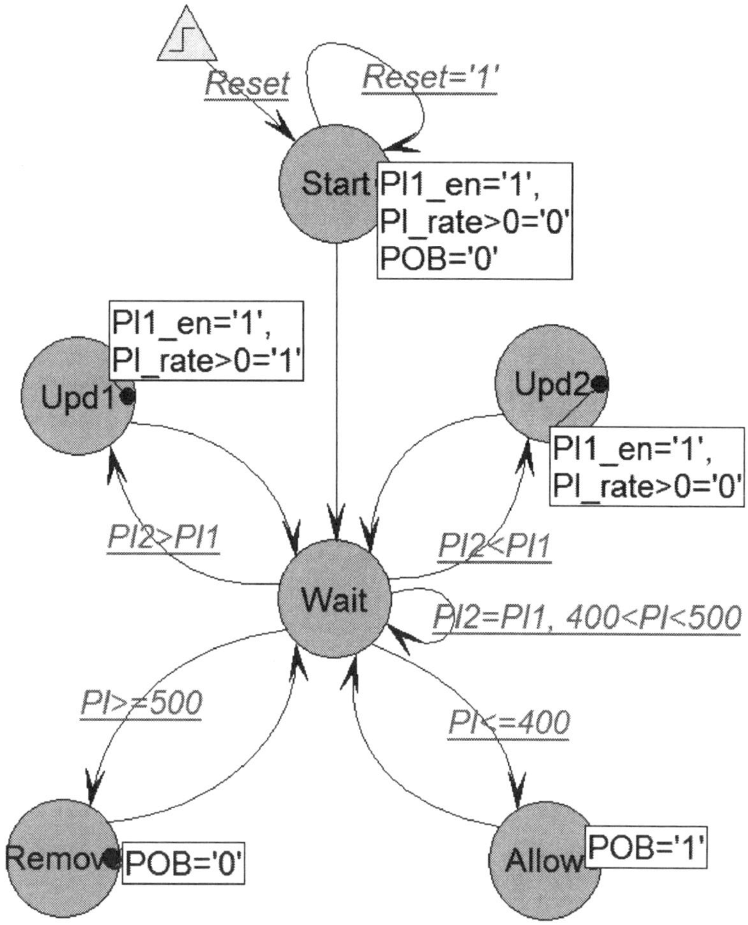


Figure 8-27: LPPT rate and Permission of operating bypass state transition diagram

- The Upd2 state executes the same actions as that of Upd1 state, only that it sends output PI_rate>0 = '0' (pressure decreasing alert) to third state machine shown in Figure 8-28, and then transits to Wait state.
- The Allow state is state where the operating bypass is permitted when the pressurizer pressure is less than or equal to 400psia (PI<=400). At this state, the permission of operating bypass status (POB = '1') is activated, and then the state machine transits back to Wait state.
- The Remove state is state where the operating bypass is automatically remove when the pressurizer pressure is greater than or equal to 500psia (PI>=500).

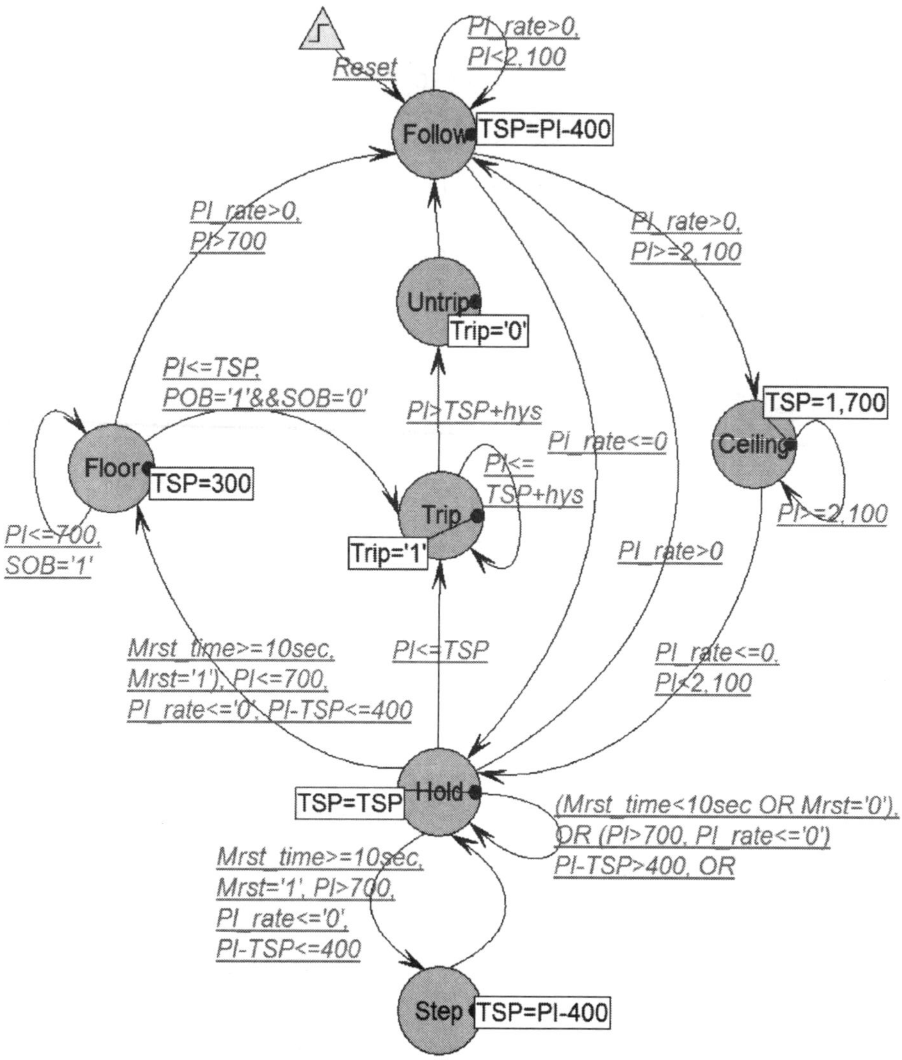


Figure 8-28: LPPT trip set-point calculation state transition diagram

The transition diagram shown in Figure 8-28 is used to control the calculation of the trip set-point as well as trip signal determination. It controls the operation of t_SPreg (trip set-point register) and trip set-point multiplexers shown in the data path. It has seven states: Follow, Floor, Ceiling, Hold, Step, trip_S, and Untrip_S states.

PART 2

- The Follow state is the state in which the trip set-point follows the pressurizer pressure using the difference of the margin, 400psia (TSP=PI-400), when the pressurizer pressure is less than 2,100psia (PI<2,100), and the rate of change of pressurizer pressure is greater than the zero value (PI_rate>0 = '1'). The Floor state is the state in which the set-point is fixed at minimum set-point value of 300psia (TSP=300) when the pressurizer pressure is less than or equal to 700psia (PI<=700) and the condition of operating bypass is activated.
- The Ceiling state is the state where the set-point is fixed at the maximum set-point of 1,700psia (TSP=1,700) when the pressurizer pressure is greater than or equal to 2,100psia (PI>=2,100).
- The Hold state is the state where the set-point is fixed at the previous set-point (TSP=TSP) when the pressurizer pressure is greater than 700psia (PI>700) and the rate of change of pressurizer pressure is less than or equal to zero (PI_rate>0 = '0').
- The Step state is state the where the the trip set-point reduces by maintaining a constant margin of the, 400psia using the pressurizer pressure value (TSP=PI-400) when the operator manual reset signal is activated.
- The trip_S state is the state in which the state machine issues a trip when the pressurizer pressure is less than or equal to trip set-point.
- The Untrip_S state is the state in which the state machine is deactivate the trip when pressurizer pressure is greater than the value, TSP+Hys (PI>TSP+Hys).

Figure 8-29 shows the final FSMD design structure for LPPT with internal and external interface signals. This is done by integrating the designed data path and FSM controllers together to obtain the final design.

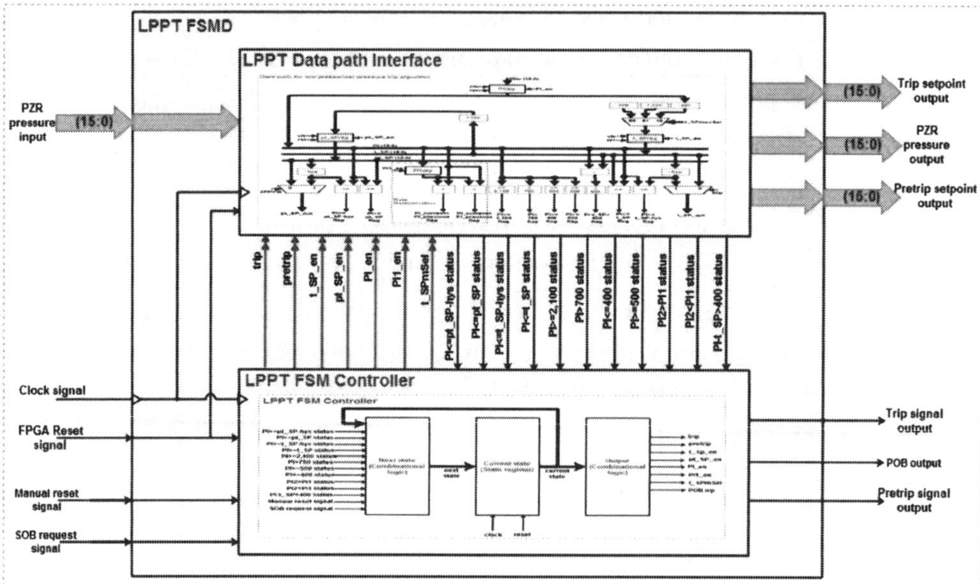


Figure 8-29: FSMD final design for LPPT configuration

Once the detailed design is complete, the next step is the implementation. The implementation of these functions on an FPGA will be discussed in Chapter 11.

## PROBLEMS

1. What the difference between functional requirement analysis and performance requirement analysis?
2. Where can you derive the requirements of PPS performance?
3. Define the Finite State Machine with Data Path. What are its advantages?
4. Using the FSMD, design the VOPT function.
5. Using EFFBD, draw the fixed set-point function.

## FURTHER READING

- KHNP, "APR1400 Design Control Document : Tier 2, Chapter 7," 2018
- Pong P. Chu, "RTL Hardware Design Using VHDL," Wiley, 2006

PART 2

Chapter 9

# ESF-CCS REQUIREMENTS & ARCHITECTURE DEFINITION

## SUMMARY

A description of the ESF-CCS is given in Chapter 4. The primary purpose of the ESF-CCS is to actuate the plant ESF components upon receipt of initiation signals from different control platforms including soft controls, manual switches and the plant protection system and radiation monitoring system. These components are provided to mitigate the consequences of the unlikely release of radioactive materials from the reactor pressure boundary. The detailed description and design bases of ESF-CCS can be found in several technical documents such as the Design Control Document (DCD), and should be read to have a full understanding of the system.

Similar to the previous chapter on PPS, this chapter presents some level of details around the ESF-CCS requirements and architecture definition using the systems engineering concept, which has been the focus of discussion in the preceding chapters. In this chapter, to demonstrate the concepts presented, only one ESF function, Safety Injection Actuation System, was dealt with.

The overall scheme of this chapter is illustrated in Figure 9-1 as mind-map.

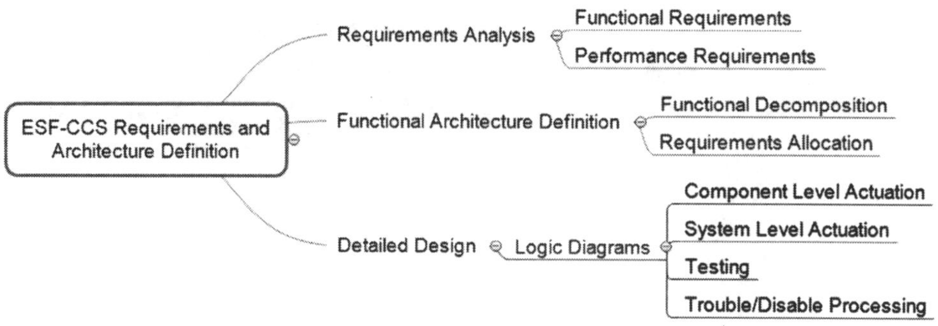


Figure 9-1: Mind Map of Chapter 9

**OBJECTIVES**

- Requirement Analysis of ESF-CCS including Functional and Performance requirements
- Definition of functional architecture in ESF-CCS
- Discussion of Detailed Design of ESF-CCS

## 9.1 ESF-CCS Requirement Analysis

Many requirements are common to both PPS and ESF-CCS. One commonality is that they are both safety-critical systems and are interconnected via the ESFAS. While top-level requirements for safety-critical systems, such as redundancy and diversity requirements, apply to both systems, there are specific functional and performance requirements that must be met in the development of an ESF-CCS design. Some of these requirements are discussed in this section. By decomposing the design specification document for the ESF-CCS, a bottom-up decomposition and traceability of the system's requirements can easily be performed.

### 9.1.1 Functional Requirements Analysis

In order to re-engineer the system for an FPGA platform, an analysis of the requirements specification of the existing system is required. Figure 9-2 shows the decomposition of the design specification of the ESF-CCS. The figure shows the functional requirements, the performance requirements, and the detailed design requirements.

PART 2

The primary function of the ESF-CCS is to actuate the ESF components so as to limiting the consequences of certain accident conditions. In particular, SIAS is automatically initiated by a low pressurizer pressure signal or a high containment pressure signal and manual initiation function is provided from the MCR. The ESF functions are:

- ESF actuation. Table 9-1 lists the actuation and control provisions of the system.
- Perform coincidence logic on system level initiation signals from PPS and MCR
- Status monitoring and testing from the OM and MTP/ITP
- Carry out priority logic for component actuation

Table 9-1: ESF Actuation Control Provisions [34]

| No. | Method | Description | Activated Via |
|---|---|---|---|
| 1 | Automatic | Automatic system level actuation | PPSs |
| 2 | ESCM and Manual Confirm Switches | Manual individual component control | Software controllers and hardware confirm switches located at the operator workstations in the MCR and RSR |
| 3 | Manual System Actuation Switches | Manual system level actuation | Hardware switches located at the:<br>RSR Workstations<br>MCR Safety Console |
| 4 | Manual Component Control Switches | Manual individual component control for the "Minimum Inventory (MI)" ESF components | Hardware switches located on the MCR safety console |
| 5 | Diverse Manual Actuation Switches | Manual system level actuation that is not dependent upon software | Hardware switches located at the MCR safety console |
| 6 | Local Manual Actuation Switches | Local manual switches for channelized ESF actuation. | Switches are located on the MTC |

The central functional requirements are the system actuation functions defined in 6.13 of Figure 9-2. Each of the actuation function requirements can be further decomposed and traced as illustrated in Figure 9-3. For instance, the requirements for the Safety Injection Actuation (SIA) functions can be decomposed and traced.

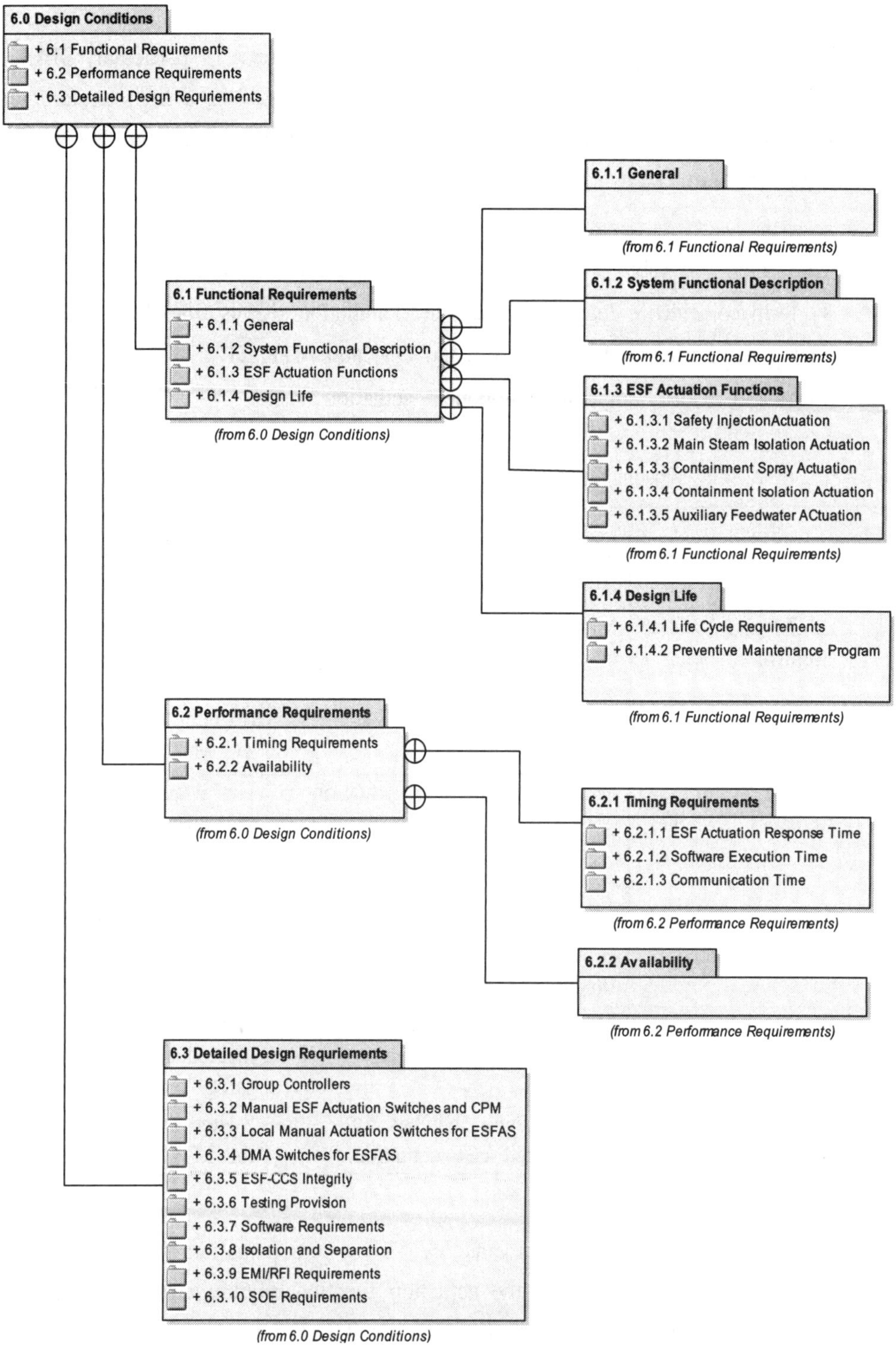


Figure 9-2: Decomposition of the ESF-CCS Design Specification

PART 2

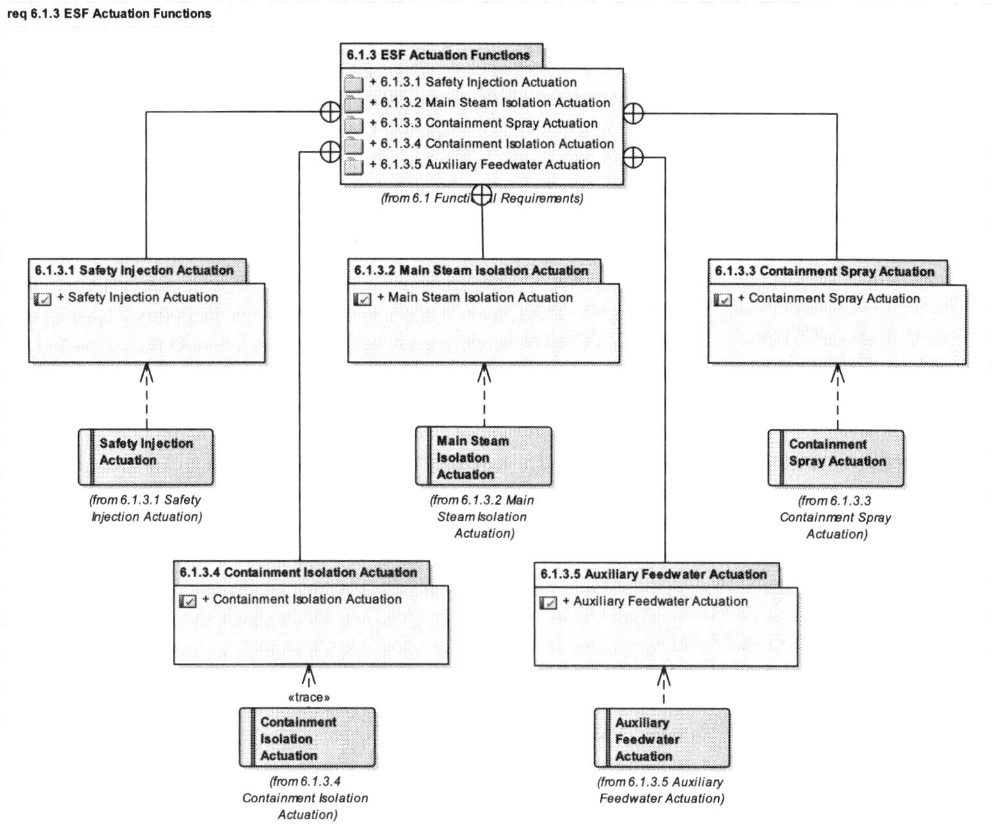


Figure 9-3: Further Requirement Decomposition of the Actuation Functions

It should be noted that the detailed design requirements artifact from the decomposition of the ESF-CCS design specification also contains requirement for functional and non-functional requirements at more detailed level.

### 9.1.2 Performance Requirement Analysis

Figure 9-4 illustrates the decomposition of the performance requirements. The major performance requirements are:

i. **Dynamic Response Time Requirements:** The response time between input of a GC and output of an LC due to PPS initiation signal shall be less than 240 milliseconds.

ii. **Digital Computer Requirements:** Processors and communication system are required to be fast enough to meet the response time requirements and to perform self-diagnostic tests. In addition, they are required to be of open architecture system and be available commercially. This means that they would necessarily be subjected to commercial grade item dedication. In addition, the contents of the memory should be non-volatile and expandable, to cater for any addition of functionality.

iii. **Software Design Requirements:** The software is required to be modularized. Modular programming emphasizes separation f code into independent sections that are capable of performing a particular function, or an aspect of the function. In addition, the interface modules should be designed so as to be flexible and expandable, to allow the design change in external systems with ease.

iv. **Redundancy and Diversity:** These principles are implemented as follows:

- The ESF-CCS is composed of 4 channels physically and electrically isolated channels
- Two out of four coincidence logic is provided for the actuation signals
- In each channel, there are redundant group controllers for the processing of ESFAS signals
- Capability for manual actuation that is independent of any software

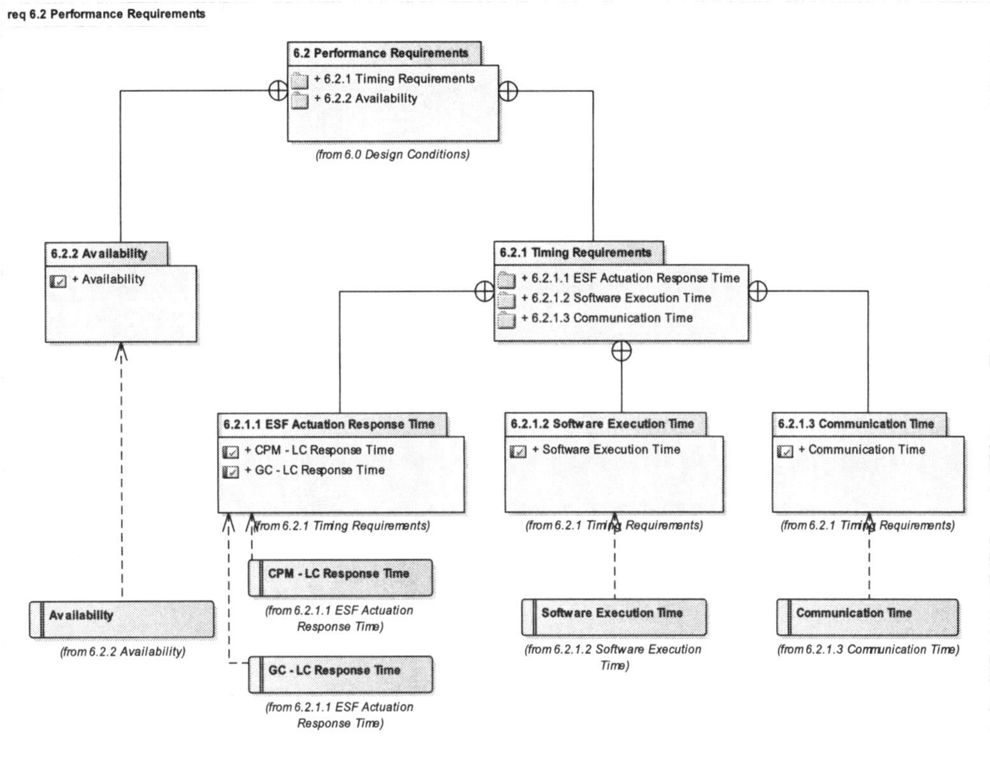


Figure 9-4: ESF-CCS Performance Requirements Decomposition

## 9.2 ESF-CCS Functional Architecture Definition

Functional Analysis/Allocation is an examination of a defined function to identify all the sub functions necessary for the accomplishment of that function. The sub functions are arrayed in a functional architecture to show their relationships and interfaces (internal and external). Upper-level performance requirements are flowed down and allocated to lower-level sub functions [3]. It does not describe either the hardware architecture or software architecture of the system. The functional architecture does not describe how the system shall operate, but what the system will do. Figure 9-5 illustrates the inputs, outputs, constraints, and enablers of this process, while Figure 9-6 depicts how functional decomposition is carried out.

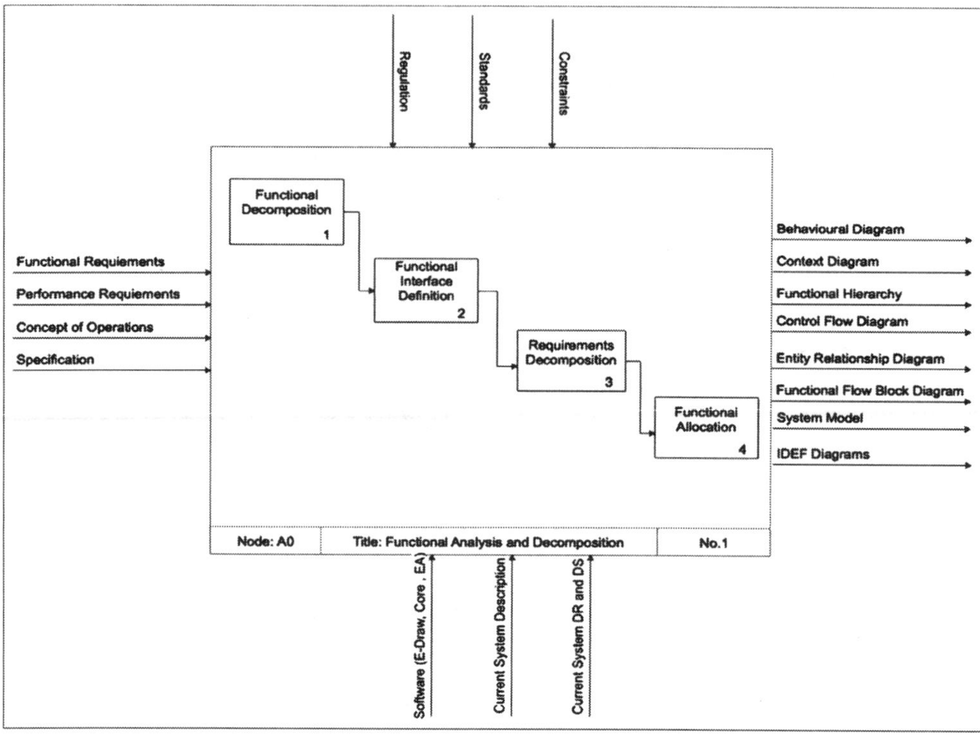


Figure 9-5: Functional Analysis and Decomposition IDEF0 Diagram

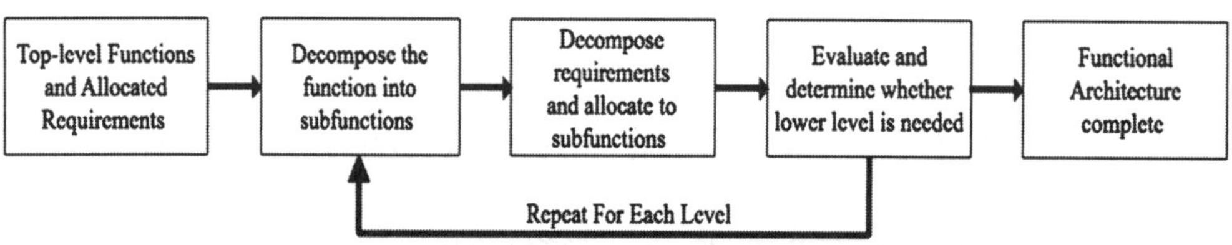


Figure 9-6: Process of Functional Analysis/Allocation Process [10]

In Figure 9-6 the first process of the functional analysis process is to allocate the requirements to the top-level functions.

Table 9-2: Allocation of Requirements

| Function | Top level Requirements |
|---|---|
| Process Input Signals | 1.1<br>2.1<br>2.3<br>2.6<br>2.5<br>4 |
| Component Level Command Processing | 2.1<br>2.3<br>2.5<br>2.6 |
| System Level Command Processing | 2.3<br>2.5<br>2.6 |
| Component Command Control | 2.3<br>2.5<br>2.6 |
| Component Command Control | 2.3<br>2.5<br>2.6 |
| Indication and Alarm Processing | 1.5.1<br>1.5.2<br>1.6.2<br>2.3<br>2.5<br>2.6 |

Next, the functions are decomposed into sub functions. In Figure 9-2, each of the functions that had been defined in Table 9-1 above, is decomposed into its constituent functions.

Requirements are then allocated to the sub-functions. In order to do this a simple table such as the example shown below in Table 9-3 can be utilized. By so doing, the traceability of the system functions to the requirements is ensured. The designer can then be confident that all the requirements have been implemented in the design.

Table 9-3: Sample Table for the Allocation of Requirement to Functions

| Function | Top-level Requirements (index) | Sub-functions | Low level requirements (index) |
|---|---|---|---|
| Process input signals | 2.1 | Process ESCM signals | 2.1.2.1.2 |
| | | Process PPS signal | 2.1.1.3 |
| | | Process MI signal | 2.1.1.4.1<br>2.1.1.2 |
| | | Process Local Actuation Signals | 2.1.1.4.1 |
| | | Process MTP Signals | 2.1.1.4.1<br>2.1.1.5.1<br>2.1.1.6.1<br>2.1.2.1.1.3 |
| | | Process DMA Signals | 2.1.1.4.1<br>2.1.1.4.2<br>2.1.1.4.3<br>2.1.2.1.1.2<br>2.1.2.1.2.2 |

Lastly, the designer must determine whether further decomposition is required. From Figure 9-7, most of the functions are decomposed only once, while some functions (e.g., Validation of ESCM commands) are decomposed further. Once the process is completed, the functional architecture is completely defined and the detailed design can proceed.

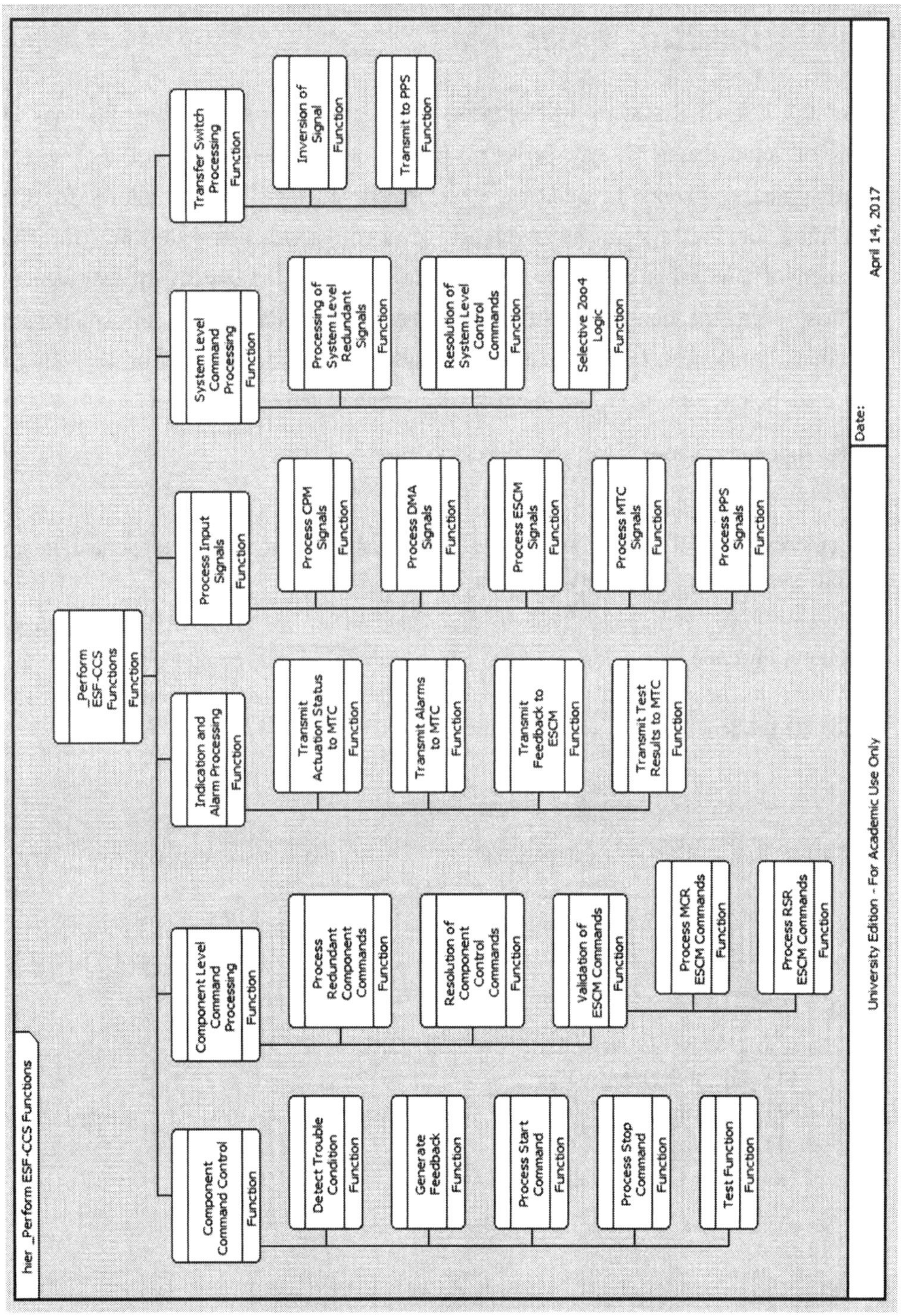


Figure 9-7 Functional Decomposition

## 9.3 Detailed Design

For the ESF-CCS system development, representation of the system functions in terms of logic diagrams may serve to give a clear understanding of the system function and operation. In addition, development of logic models allows for the simulation of the system at a higher level of abstraction, allowing for the detection of any errors or faulty logic, before beginning the system implementation. As has been mentioned, the ESF-CCS has two main modes of component actuation: component level actuation and system level actuation. For the design presented below, the logic was developed and simulated in Matlab.

The functions whose logic was developed are:

1. Component Level Actuation: This is the actuation of a single component in an ESF system e.g. the Safety Injection Pump
2. System Level Actuation: This is the actuation of an entire ESF system e.g. Safety Injection Actuation System, which causes all the components within the system to be actuated. The table 9-4 shows all the components that are actuated when the SIA system is actuated.

Table 9-4 : SIAS Components and Actuated State

| Component | Actuated State |
|---|---|
| Letdown Isolation Valve | Close |
| Shutdown Cooling Pump | Start |
| SI Line Isolation Valve | Open |
| SI Pump | Start |
| Check Valve Leakage Line Isolation Valve | Close |
| SIT Fill and Drain Valve | Close |
| SIT Fill Line Isolation Valve | Close |
| SIT Isolation Valve | Open |
| Hot Leg Check Valve Leakage Line Isolation Valve | Close |

3. Test Processing: The testing of the ESC-CCS subsystems and components
4. Trouble/Disable Processing: The detection of a trouble/disable state in the component and the generation of the associated indication and alarm signals

### 9.3.1 Component Level Actuation

There are 3 ways of achieving component control:

1. MCR ESCM
2. RSR ESCM
3. Minimum Inventory Switches

In the design, as indicated in Figure 9-8, the command switches are represented by two boxes, one for the start position, and the other for the stop position. To start the component, the start signal must be set to logic 1. Similarly, to stop the component, the stop signal must be set to logic 1. In order for a command from the MCR ESCM to be processed, the following signals must be present:

- Component Command (Start or Stop set to logic 1)
- ESCM Enable (Logic 1)
- Confirmation Switch (Logic 1)
- HS7, RSR Transfer Switch (Logic 0)

In order to determine if there is a component command, a comparator is used(Table 9-5). If both start and stop signals are set to 0, there is no command. In addition, if both start and stop commands are set to 1, this state is considered invalid, and no command should be implemented. Therefore, there is only a valid command when the start and stop signals are not equal.

Table 9-5: Determination of the Presence of a Command

| Start | Stop | Comparator Output (ESCM Command) | Explanation |
|---|---|---|---|
| 0 | 0 | 1 | No command |
| 1 | 0 | 0 | Valid Command (Start) |
| 0 | 1 | 0 | Valid Command (Stop) |
| 1 | 1 | 1 | Invalid Command, take no action |

The output of the comparator is then inverted. This inverted signal is combined with the ESCM Enable, Confirmation Switch, and Inverted RSR Transfer signal. If the RSR Transfer Signal is set to true, this indicates that the control of the system has been transferred from the MCR to the RSR, the system should therefore ignore commands from the MCR. In addition, if the ESCM is disabled or confirmation switch signal lacking, commands from the ESCM should be ignored.

The truth table for this logic is listed in Table 9-6.

Table 9-6: ESCM Validation (When there is a valid Command)

| INPUT | | | | OUTPUT |
|---|---|---|---|---|
| ESCM Command | ESCM Enable | Confirmation Switch | Inverted RSR Transfer | Validation Signal |
| 1 | 0 | 0 | 0 | 0 |
| 0 | 0 | 0 | 1 | 0 |
| 1 | 0 | 1 | 0 | 0 |
| 1 | 0 | 1 | 1 | 0 |
| 1 | 1 | 0 | 0 | 0 |
| 1 | 1 | 0 | 1 | 0 |
| 1 | 1 | 1 | 0 | 0 |
| 1 | 1 | 1 | 1 | 1 |

This combined signal indicates that all conditions necessary for a signal from the MCR to be considered valid, are true. The signal is then combined (logical AND) with the start command, as well as with the stop command. The truth table is defined in Table 9-7.

Table 9-7: Truth Table for ESCM Start and Stop Commands

| INPUT | | | OUTPUT | |
|---|---|---|---|---|
| Start | Stop | Validation Signal | MCR Start | MCR Stop |
| 1 | 0 | 1 | 1 | 0 |
| 0 | 1 | 1 | 0 | 1 |
| 1 | 0 | 0 | 0 | 0 |
| 0 | 1 | 0 | 0 | 0 |

The RSR carries out the same logic as described for the MCR. The only difference is that, a valid signal is received when the RSR transfer switch (HS 7) is set to logic 1, indicating transfer of control to the RSR from the MCR. The MI (HS 2) is located in the safety console, and is used to start or stop the component. There is no additional validation carried out on this signal.

The outputs of these three means of initiation are combined (using OR logic), so that if there is a valid command from any of the sources, the component is actuated. Figure 9-8 shows the component level actuation logic.

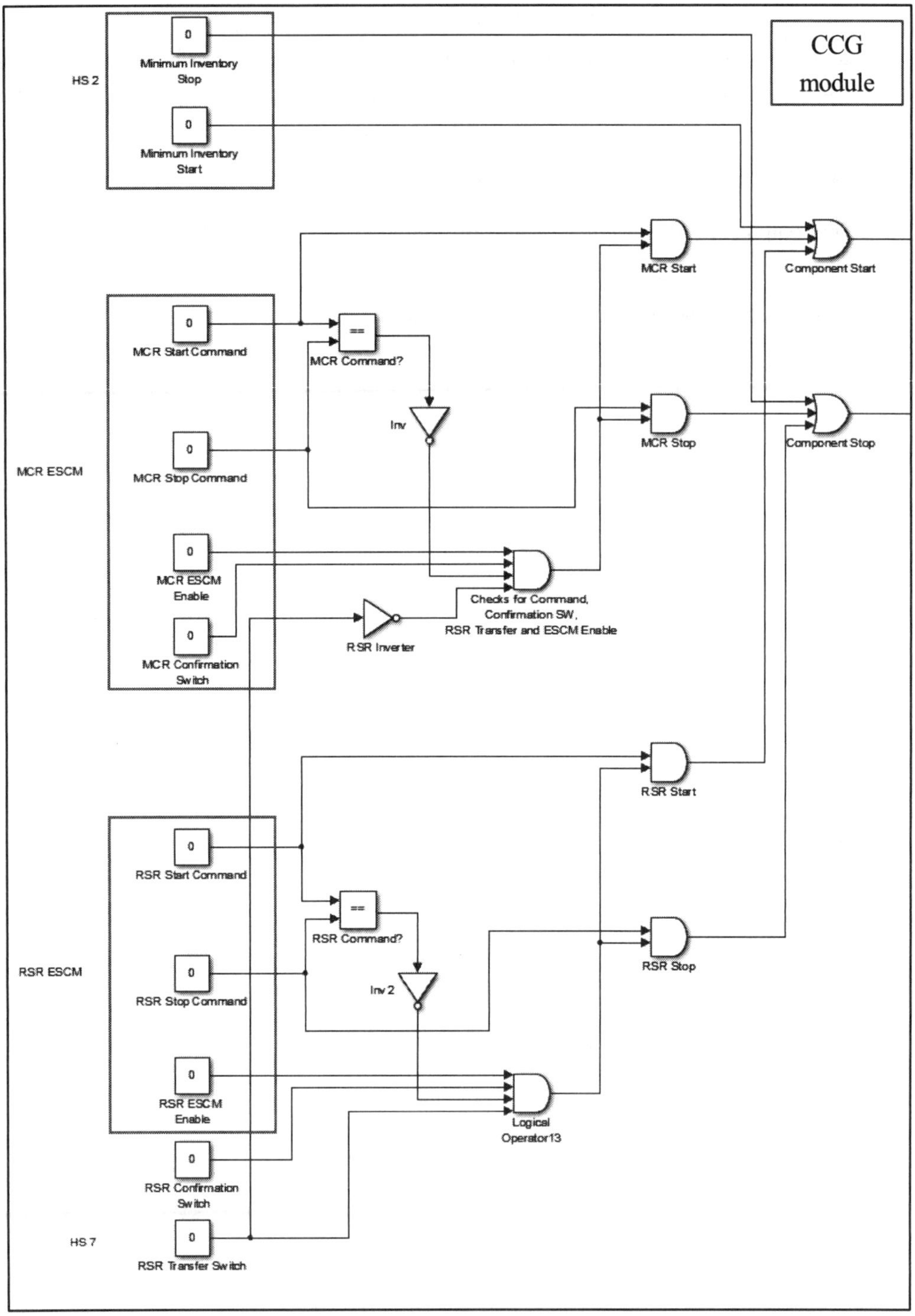


Figure 9-8: Component Level Actuation Logic

## 9.3.2 System Level Actuation

The system level actuation logic is indicated in Figure 9-9. There are 4 ways system level actuation of the ESFAS is achieved:

- Automatic generation by the PPS, when plant parameters exceed the set-point.
- Manual Actuation by switches on MTC
- Manual Actuation by switches on MCR Safety Console (HS 3)
- Diverse Manual Actuation Switches on MCR Safety Console.
- The DMA switches are not shown in Figure 9-9, because these signals are processed by the (Component Interface Module) CIM. Selective 2 out of 4 logic (indicated in Table 9-8) is carried out on the inputs from the PPS, MTC, and MI. This logic is defined as:

(A OR B) AND (C OR D)

Table 9-8: Selective 2oo4 logic, truth table

| INPUT (From PPS/ MTC/ MI) | | | | OUTPUT (ESFAS Signal) |
|---|---|---|---|---|
| A | B | C | D | |
| 0 | 0 | 0 | 0 | 0 |
| 0 | 0 | 0 | 1 | 0 |
| 0 | 0 | 1 | 0 | 0 |
| 0 | 0 | 1 | 1 | 1 |
| 0 | 1 | 0 | 0 | 0 |
| 0 | 1 | 0 | 1 | 0 |
| 0 | 1 | 1 | 0 | 1 |
| 0 | 1 | 1 | 1 | 1 |
| 1 | 0 | 0 | 0 | 0 |
| 1 | 0 | 0 | 1 | 1 |
| 1 | 0 | 1 | 0 | 0 |
| 1 | 0 | 1 | 1 | 1 |
| 1 | 1 | 0 | 0 | 1 |
| 1 | 1 | 0 | 1 | 1 |
| 1 | 1 | 1 | 0 | 1 |
| 1 | 1 | 1 | 1 | 1 |

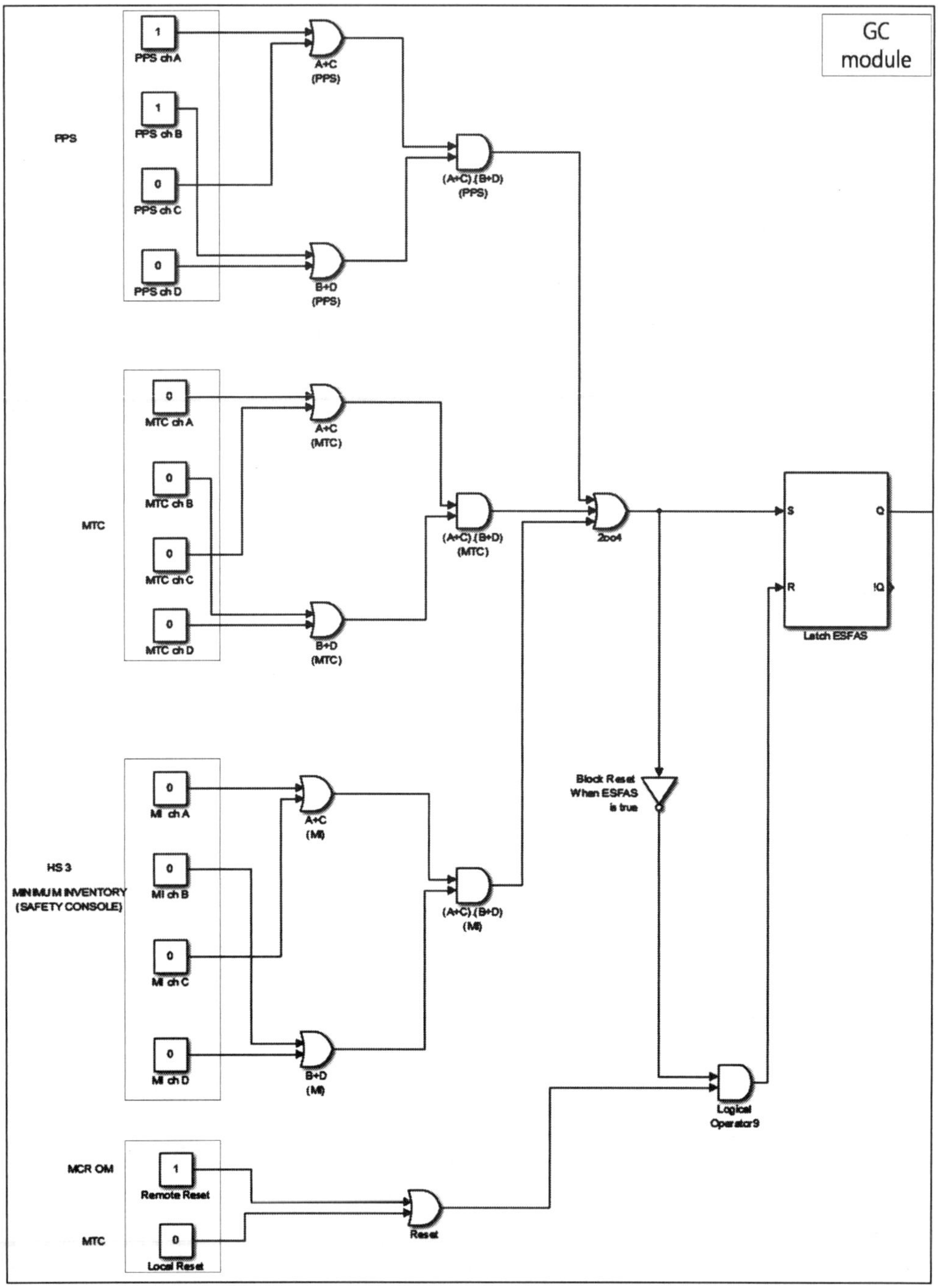


Figure 9-9: System Level Actuation Logic

The outputs of this logic are then passed through an OR gate, so that the system is actuated if any one of these three input sources. In addition, the ESFAS signal is latched. This is implemented using an S-R flip-flop. By latching this signal, the ESFAS remains in the actuate state, even if the signals from the PPS go back to 0. In order to reset the signal, an operator must give reset command either from the Operator Module (OM) in the MCR, or from the MTC. However, the operator should not be able to issue a reset command if there is an ESFAS signal, therefore the inverted ESFAS signal is used to inhibit the Reset command(Logical AND between inverted ESFAS and reset command)

The Truth table for the SR flip-flop is detailed in Table 9-9. $Q_{N-1}$ represents the current output while $Q_{N-1}$ represents the output in the previous clock cycle.

Table 9-9: SR Flip Flop (Latch Logic) Truth Table

| INPUT | | OUTPUT, (QN) |
|---|---|---|
| S (ESFAS) | R(Reset, Local or Remote) | |
| 0 | 0 | $Q_{N-1}$ |
| 0 | 1 | 0 |
| 1 | 0 | 1 |

### 9.3.3 Test Processing

The inputs and outputs of the test logic are listed in Table 9-10.

Table 9-10: Inputs and Outputs of Test Logic

| INPUT | SOURCE |
|---|---|
| Test Key Switch | MTC |
| Test ID | MTC |
| ESFAS | System Level Actuation Logic |
| **OUTPUT** | **DESTINATION** |
| ESF Demand | Start Command Processing Logic |
| ESF Received Test Mode | MTC |
| Misc. Disable (ESF Blocked Test Mode) | Trouble/Disable Processing Logic<br>MTC |

In order to initiate a test as shown in Figure 9-10, the MTC send a test ID to the system. The system compares this ID to the stored ID value. If the two are the same, and the Test Key Switch is in the test position (Logic 1), then the system is tested. The Misc. Disable is set to true.

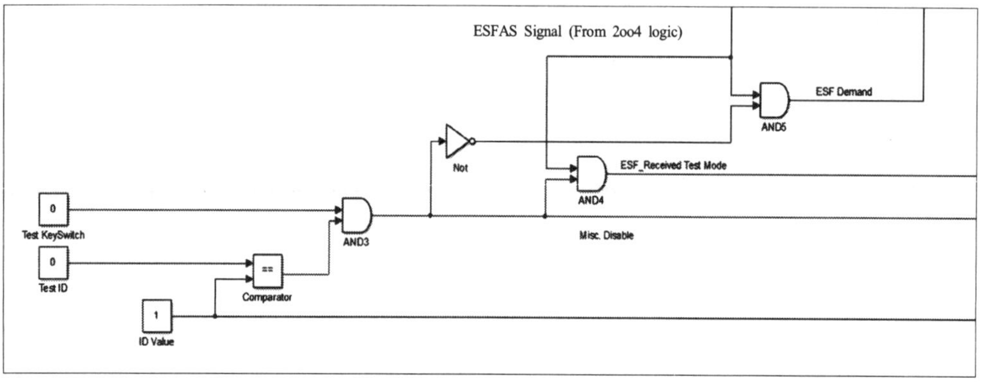


Figure 9-10: Test Logic

In addition, if there is a valid test demand, the ESF demand is disabled. ESF Received Test Mode indicates that both a valid test demand and ESFAS signal exist. Start and Stop Command Processing.

The Start logic contains an ESF override block, this is because the ESFAS signal is of type ESF-2, and can be over-ridden by an operator command. The output of the over-ride block or the presence of an operator command can generate a Start Demand. In addition, the start signal may be inhibited by an opposite command block from the stop logic

The stop logic is simple and consists of a logical AND operation between the Stop command from the Component Actuation Logic, and an inverted ESF-Status Signal. The ESF-Status signal acts as an inhibitor, preventing the actuation of a Stop command, if a start command exists.

The outputs of this block are the Start and Stop Demands, which are transmitted to the CIM, for component actuation as illustrated in Figure 9-11.

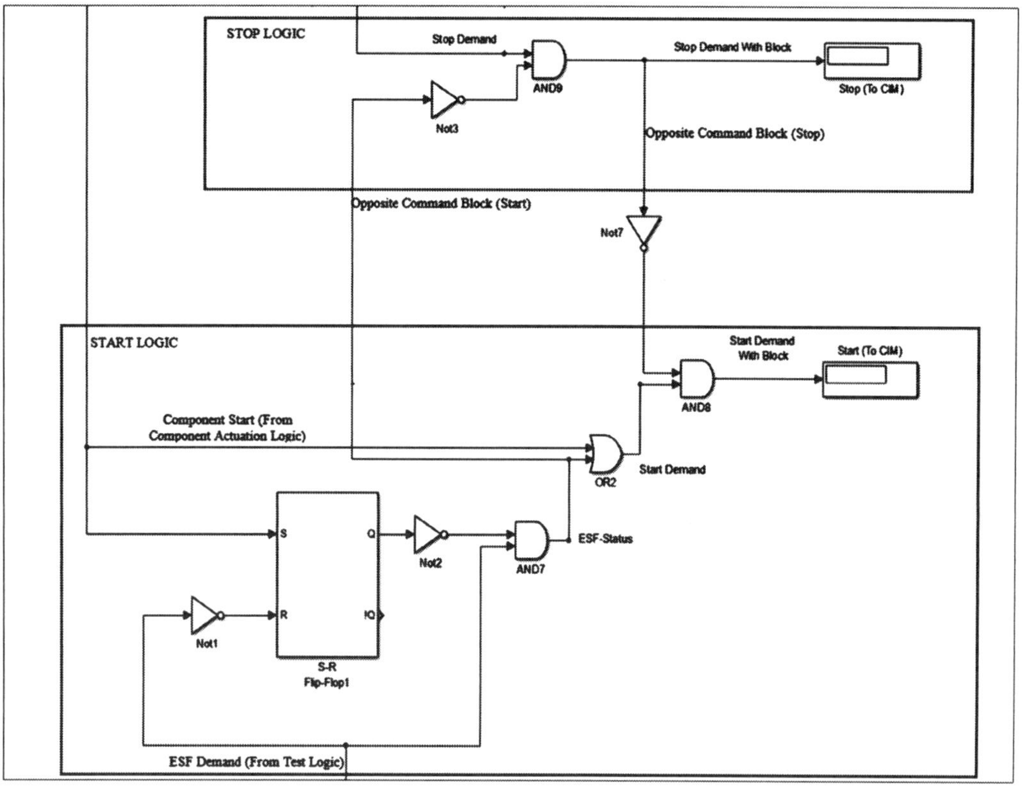


Figure 9-11: Start and Stop Command Processing Logic

## 9.3.4 Trouble / Disable Processing

This logic is used to generate Trouble and Disable Alarms based on received commands as well as feedback from the CIM. In addition, the alarms generated are latched and a Trouble/Disable acknowledges signal from the operator is required to reset the alarms. Conditions that require the trouble output to be true are:

- Start failure or unexpected stop as detected by a circuit breaker not closed after a start command is given
- Stop failure or unexpected start as detected by a circuit breaker closed after a stop command is given
- Motor overload as detected by electrical protection

For the conditions above, the trouble alarm is generated in the presence of the signals listed below:

- Start command from the system, and the lack of a corresponding start status form the CIM after a time delay (0.5 seconds)
- Stop command from the system and the presence of a start status from the CIM after a time delay (0.5 seconds)
- Electrical Protection signal from the CIM

Conditions that require the disable output to be true:

- Loss of trip/closing circuit control power as detected by a control power monitoring circuit.
- Absence of a start permissive
- Loss of breaker trip or closing continuity as detected by a coil continuity monitoring circuit.
- Existence of a mechanical protection signal that would block starting of the equipment (e.g. lockout relay)
- Module disables condition such as caused by a self-diagnostic test.
- Circuit breaker not in position as detected by a cell switch (This switch is used to indicate the draw out circuit breaker position)

A disable alarm is generated in the presence of the following signals:

- Absence of a start status from the CIM, accompanied by a Close Coil Open signal from the CIM
- Start signal and a Trip coil open signal from the CIM
- Absence of a CIM Ready signal
- Port X Blocked signal from the CIM
- CIM Fault signal
- Misc. Disable from test circuit
- I/O Trouble Diagnostic Signal from the CIM
- Cell Switch Signal

## PROBLEMS

1. Explain the functions of each component of the ESF-CCS?
2. Why does the ESF-CCS include both system-level and component-level actuation?
3. Define the functions of ESF-CCS are decomposed
4. Describe the detailed functions of Component Interface Module.
5. Explain how the trouble/disable logic is processing and alarms are generated?

## FURTHER READING


- KHNP, APR1400 Design Control Document, Chapter 7
- A. F. Kana, Digital Logic Design

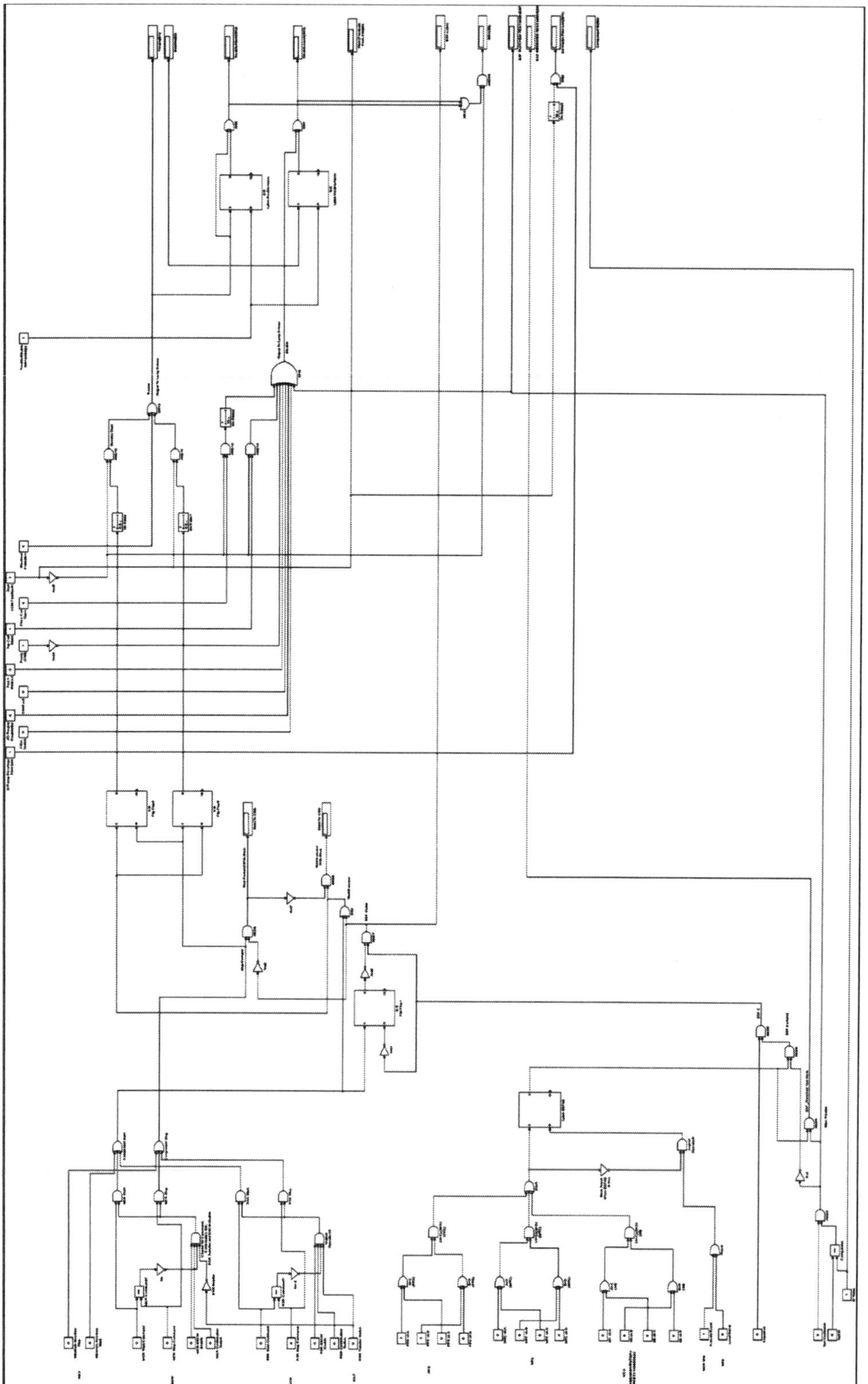

Figure 9-12 Safety Injection Actuation System(Pump A)

Chapter 10

# CPCS REQUIREMENTS AND ARCHITECTURE DEFINITION


## SUMMARY

In PART 1, Chapter 4 gives an overview of the CPCS. The purpose of the Core Protection Calculator System (CPCS) is to protect the reactor core from exceeding the safety limit by monitoring pertinent Reactor Core conditions and to generate Departure from Nucleate Boiling Ratio (DNBR) and Local Power Density (LPD) trips during plant operation.

Unlike the PPS and ESF-CCS whose logic is fairly simple - comparison of a process variable to a trip set-point – the CPCS carries out complicated arithmetic in order to compute the thermos-hydraulic properties needed to ascertain the safety of the fuel and the core. In addition, as this system is responsible for the protection of the core, it has very stringent response time requirements, to enable the detection of an unsafe condition and trip the reactor in a timely manner. For this reasons, the algorithms used in the computation of the CPCS are complex and do not easily lend themselves to implementation on an FPGA platform.

In this Chapter, the CPCS is re-engineered in order to reduce the current algorithm complexity and increase the dynamic performance for execution of the algorithm. The simplification approach is based on hot channel fluid properties calculation such as mass flux and enthalpy distribution. These concepts, as they are critical to understanding the system modification, will be presented in this chapter.

The chapter is structured as shown in the mind-map in Figure 10-1.

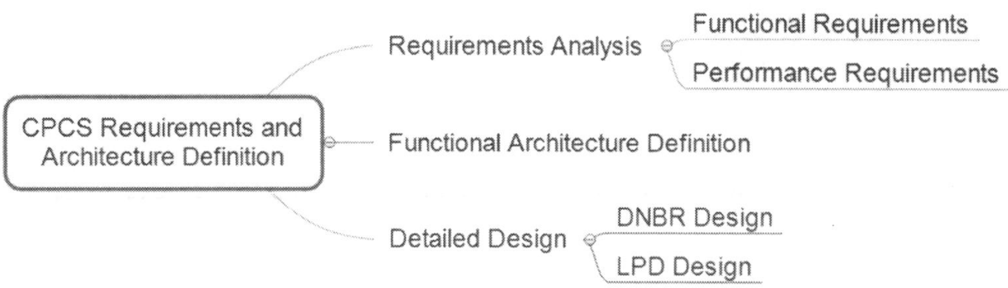


Figure 10-1: Mind Map of Chapter 10

**OBJECTIVES**

- Requirement Analysis of CPCS including Functional and Performance requirements
- Definition of functional architecture in CPCS
- Discussion of Detailed Design of CPCS

## 10.1 CPCS Requirements Analysis

The CPCs generate DNBR and LPD pre-trip and trip signals when the calculated DNBR or LPD values exceeds their respective set-points, or when certain auxiliary condition are met. The CPCs channel Pre-trip and Trip outputs are used by the PPS logic to trip the reactor.

The requirements of the CPCS take into consideration Anticipated Operational Occurrences (AOOs) and Postulated Accidents (PA). In the case of AOOs, the reactor trip signal should be generated so as not to exceed the Specified Acceptable Fuel Design Limits (SAFDL). In the event of certain postulated accident, the CPCS should be designed to generate appropriate reactor trip signal to mitigate the consequences. The hierarchy of CPCS requirements is depicted in

Figure 10-2. The following documents should be reviewed in order to elicit the system requirements:

- 10 CFR 50
- SRP Sections 7.2 and 7.3
- Applicable Regulatory Guides and Standards
- Design Control Document
- Engineering Documents

## 10.1.1 Functional Requirements

### (1) Trip Functions

The CPCS generates DNBR trip, LPD trip, and Auxiliary trip signals when the process variables exceed the pre-defined set-points [18] :

- DNBR Trip: the low DNBR trip is provided to trip the reactor when the calculated DNBR approaches a preset value. The DNBR trip set-point is 1.29.
- LPD Trip: The high LPD is provided to trip the reactor when the core peak LPD reaches a preset value (689 W/cm or 21 kW/ft.)
- Auxiliary Trips: when the core conditions are out of the allowed operating conditions, the auxiliary trip is generated. These conditions are:
    - VOPT provides protection for sudden power increases.
    - Asymmetric Steam Generator Transient (ASGT) is a trip that provides protection for instantaneous closure of the Main Steam Isolation Valves (MSIVs) to a single steam generator.
    - Low Pressure and Low DNBR Trip (LPLDT) provides protection from events that decrease the RCS inventory.
    - Range trip on several parameters, which ensures the core conditions are within the analyzed operating space.
    - Pump trip precludes operation with less than two reactor coolant pumps running.
    - Hot leg saturation trip prevents operation with substantial voiding in the hot leg fluid.
    - Internal processor failure trip provides a trip signal whenever the CPC is in test, initialization, or when an internal processor fault is detected.
    - Low reactor coolant pump shaft speed trip is generated when the RCP shaft speed is below a predefined value

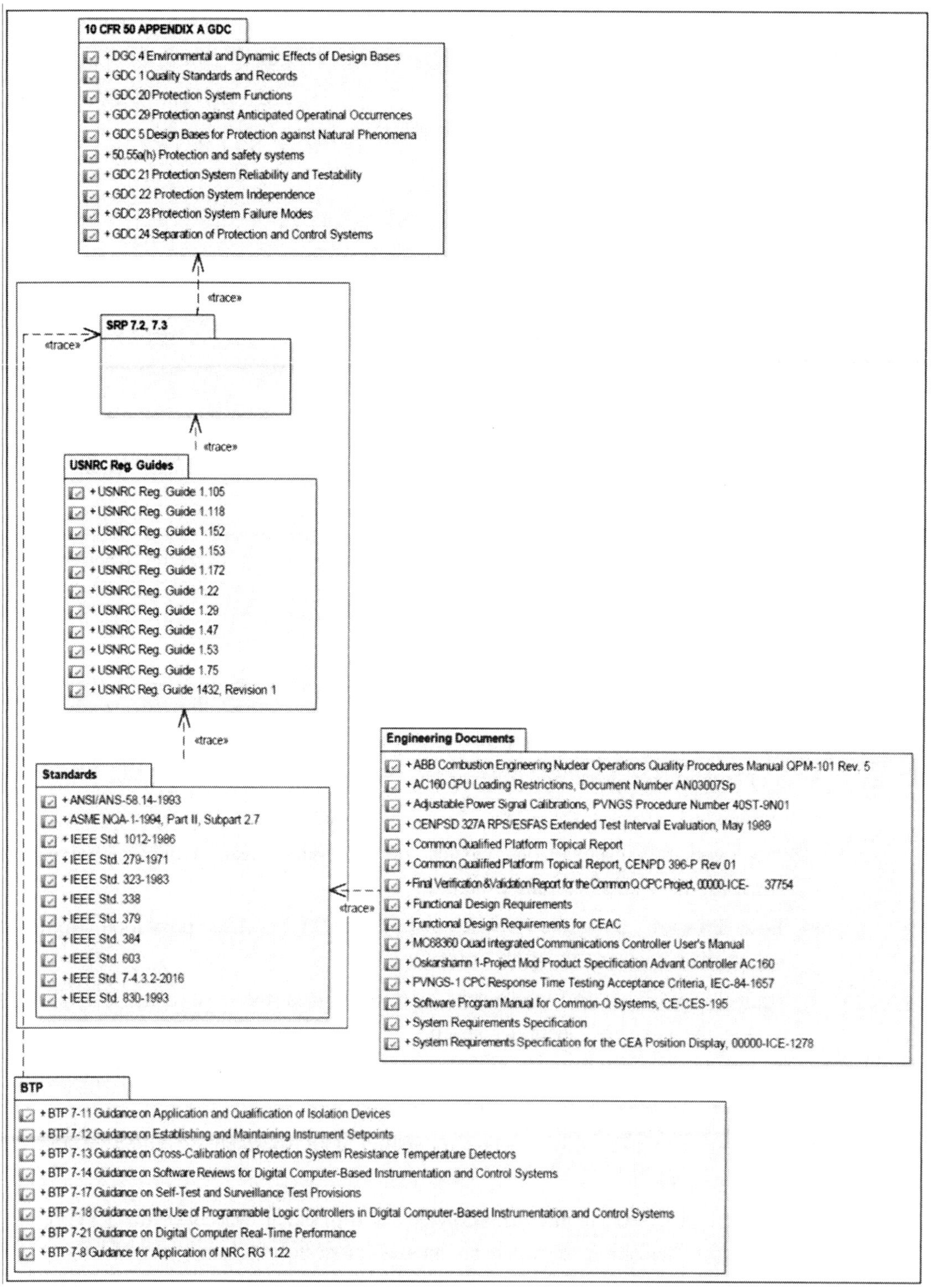


Figure 10-2: Requirements hierarchy of CPCS

PART 2

#### (2) CEA Withdrawal Prohibit (CWP)

The CPC generated CWP signal to prevent any abnormal condition in the core due to CEA withdrawal in the following situations:

- Reactor power cutback
- DNBR pre-trip
- LPD pre-trip
- CEA deviations within a subgroup
- Regulating groups out of sequence or deviation between subgroups

### 10.1.2 Performance Requirements

The CPCS is re-engineered to meet the original system performance requirements defined in this section.

#### (1) Response Time Requirements

The CPCS has six interdependent programs of which five are resident in the CPC processor and the sixth in the CEAC processor,

- Coolant mass flow program (FLOW)
- DNBR and power density update program (UPDATE)
- Power distribution program (POWER)
- Static DNBR and power density program (STATIC)
- Trip sequence program (TRIPSEQ)
- CEAC penalty factor program is implemented in each of the two CEACs.

The execution times of these programs are scheduled on a priority basis. The execution time of each protection program shall be fixed and the more frequently executed programs shall be given a higher priority. Any failure in the CPC programs on the specified schedule would result in a channel trip on Low DNBR and High LPD. CEAC Fail will occur in the associated CEAC in the event of failure to execute within the scheduled time. The Table 10-1 shows the program execution intervals and input sampling rates of the CPCS software.

### (2) Accuracy

The accuracy requirements are based on the following uncertainties:

- Loading effects
- Reference voltage supply regulation
- Electrical noise
- Linearity
- A/D converter power supply sensitivity
- quantization

The CPCS hardware shall meet the accuracy requirements as specified in Table 10-2.

Table 10-1: Program Execution Intervals and Input Sampling Rates

| **Program/ Processor** | **Inputs used** | **Execution interval** | **Remarks** |
|---|---|---|---|
| FLOW (CPC) | RCP Shaft Speed | 50 ms | • Input signal processing must update input value at least once per 50 ms for speeds above 70% rated speed.<br>• FLOW to be combined with UPDATE and TRIPSEQ, and executed every 50milliseconds, in the order FLOW-UPDATE-TRIPSEQ |
| UPDATE (CPC) | Cold Leg Temp.<br>Hot Leg Temp.<br>RCS Pressure<br>CEA Deviation Penalty Factors<br>Ex-core Detector Signals | 50 ms (100 ms in the legacy CPCs). | • CEA Deviation PFs must be updated by CEACs at least once per 100 ms.<br>UPDATE to be combined with FLOW and TRIPSEQ, and executed every 50milliseconds, in the order FLOW-UPDATE-TRIPSEQ |
| POWER (CPC) | CEA Positions | 250 ms (1.000 seconds in the legacy CPCs). | |
| STATIC (CPC) | None | 2 sec. | |

| Program/ Processor | Inputs used | Execution interval | Remarks |
|---|---|---|---|
| TRIP SEQUENCE (TRIPSEQ) | STATIC, POWER, FLOW and UPDATE programs | 50 msec | • TRIPSEQ to be combined with FLOW and UPDATE and executed every 50milliseconds, in the order FLOW-UPDATE-TRIPSEQ<br>(TRIPSEQ is executed immediately following FLOW execution in the legacy CPCs). |
| CEAC (CEAC 1) | CEA Positions from CPPs | 100 msec | • CEAC 1 uses RSPT-1 on each CEA.<br>• Produces CEA deviation penalty factors |
| | | 200 msec | • Provides target CEA position to CPC |
| CEAC (CEAC 2) | CEA Positions from CPPs | 100 msec | • CEAC 2 uses RSPT-2 on each CEA.<br>• Produces CEA deviation penalty factors |
| | | 200 msec | • Provides target CEA position to CPC |

Table 10-2: CPC Process Input Signals [17]

| Signal | Number per CPC Channel | Description | Representative Range | Signal Type | Accuracy Required |
|---|---|---|---|---|---|
| Reactor Coolant Pump Speed | 4 | Reactor coolant pump shaft speed | 10～100% of rated speed | digital | 0.25% of rated speed |
| Cold Leg Temperature | 2 | Temperature in primary coolant cold legs, 1 of 2 for each steam generator | 446°F～626°F | analog | 1.0°F |
| Hot Leg Temperature | 2 | Temperature in primary coolant hot legs 1 and 2 | 482°F～662°F | analog | 1.0°F |
| Pressure | 1 | Pressurizer pressure | 1500～2500Psia | analog | 6.00Psia |
| Ex-Core Neutron Flux | 3 | Excore neutron detector signals | 0～200% Power | analog | 0.5% power |
| Deviation Penalty Factor | 2 | CEA deviation penalty factor from CEACs | 0～8.0 | digital | 10～2 |

| Signal | Number per CPC Channel | Description | Representative Range | Signal Type | Accuracy Required |
|---|---|---|---|---|---|
| CEA Position | 23 | Target CEA position | 0~100% withdrawal (0-381cm) | analog | 0.5% |
| CEA Position | 70/CEAC | CEA position from CEA Position Processor (CCP) | 0~100% withdrawal (0~381cm) | analog | 0.5% |

Table 10-3: CPC Output Signals Type and Range

| Signal | Type | Range |
|---|---|---|
| Low DNBR Trip | Contact Output | 0, 1 (logical) |
| Low DNBR Pretrip | Contact Output | 0, 1 (logical) |
| High LPD Trip | Contact Output | 0, 1 (logical) |
| High LPD Pretrip | Contact Output | 0, 1 (logical) |
| CEA Withdrawal Prohibit | Contact Output | 0, 1 (logical) |
| CPC Sensor Failure | Contact Output | 0, 1 (logical) |
| DNBR Margin | Analog | 0~2 (unitless) |
| LPD Margin | Analog | 0~25 (KW/Ft) |
| Calibrated Neutron Flux Power | Analog | 0~200 (% of rated power) |
| Core Coolant Mass Flow Rate | Analog | 0~2 (fraction of design flow) |

### (3) Design Requirement Formulation

Low Departure from Nucleate Boiling Ratio (DNBR) and high local power density trips:

- Assure that the Specified Acceptable Fuel Design Limits (SAFDL) on departure from nucleate boiling and centreline fuel melting are not exceeded during Anticipated Operational Occurrences (AOO)
- Assist the ESF in limiting the consequences of certain postulated accidents.

① **Specified Acceptable Fuel Design Limit (SAFDL)**

The acceptable SAFDLs for COLSS/CPC analysis and Non-LOCA transient analysis are:

DNBR SAFDL: The DNBR in the limiting coolant channel of the reactor core shall be maintained greater than or equal to the value that precludes DNB on the limiting fuel pin at a 95% probability with a 95% confidence.

Fuel centerline melt SAFDL: The peak linear heat rate in the limiting fuel pin of the reactor core shall be maintained less than or equal to that value which precludes centerline fuel melting

② **Anticipated Operational Occurrences (AOOs)**

Figure 10-3 shows the AOOs considered for the design requirements:

### (4) Postulated Accidents

The postulated accidents that are used to determine the design requirements for the CPCS are as follows:

- Reactor coolant pump shaft seizure
- Steam generator tube rupture
- Steam line break containment

The CPC's are designed to provide a reactor trip when required for the above AOO and postulated accidents when initiated from a power level greater than the CPC operating bypass power set-point.

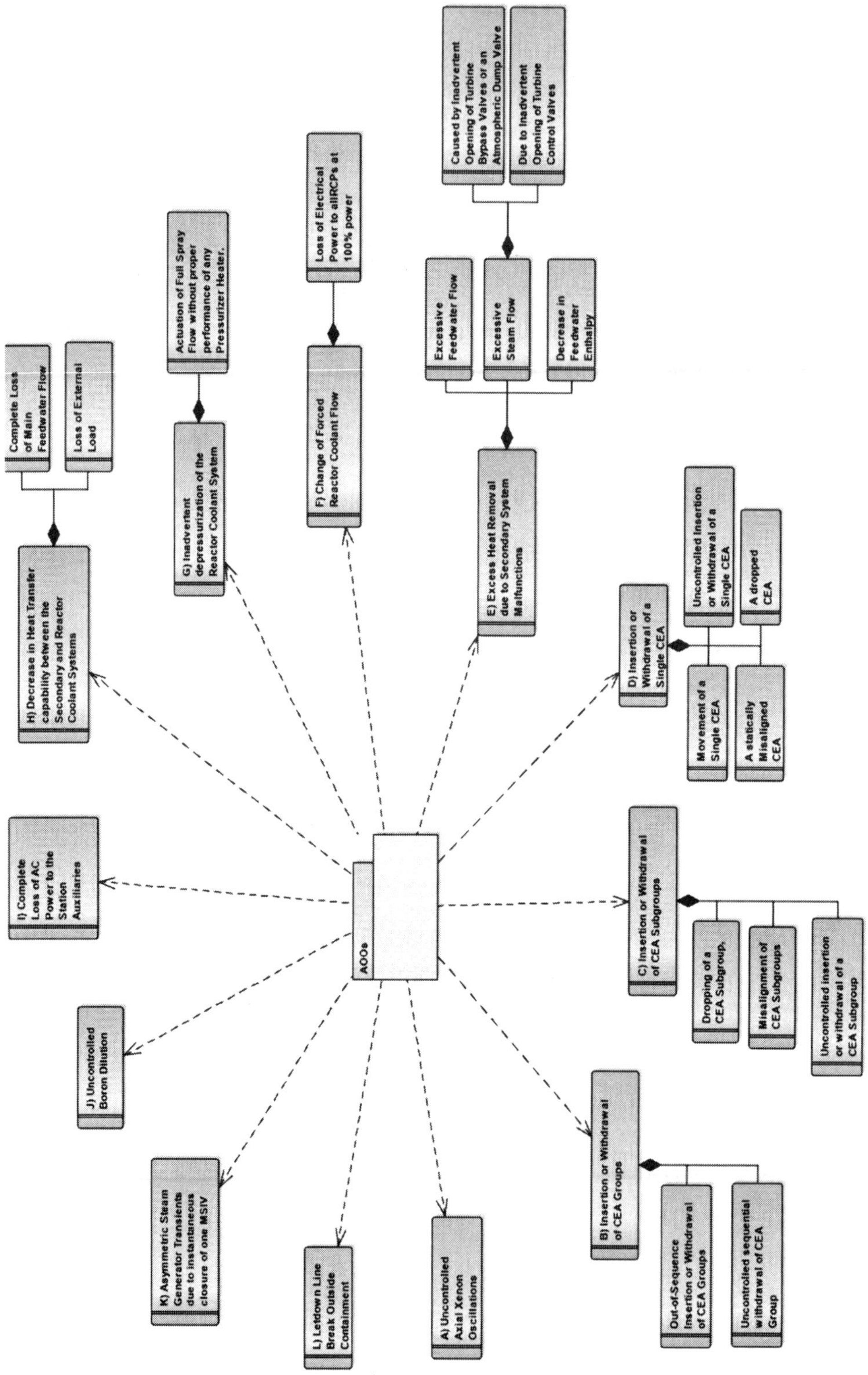


Figure 10-3 Design Bases Events/Anticipated Operational occurrences

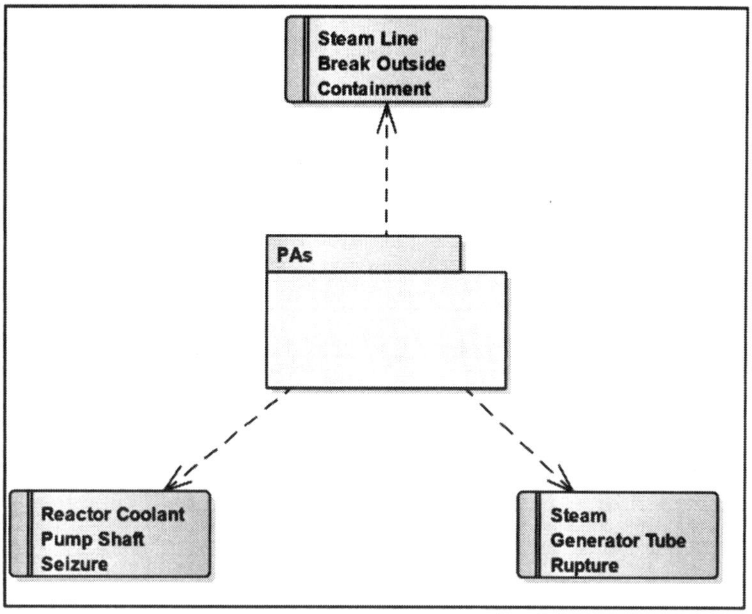


Figure 10-4: Postulated Accidents (PA)

### (5) Requirements Traceability Matrix

The reason for review of the requirements is better to understand the source of the requirements of the current system for the new system development. The traceability of requirement hierarchy is illustrated in Figure 10-5.

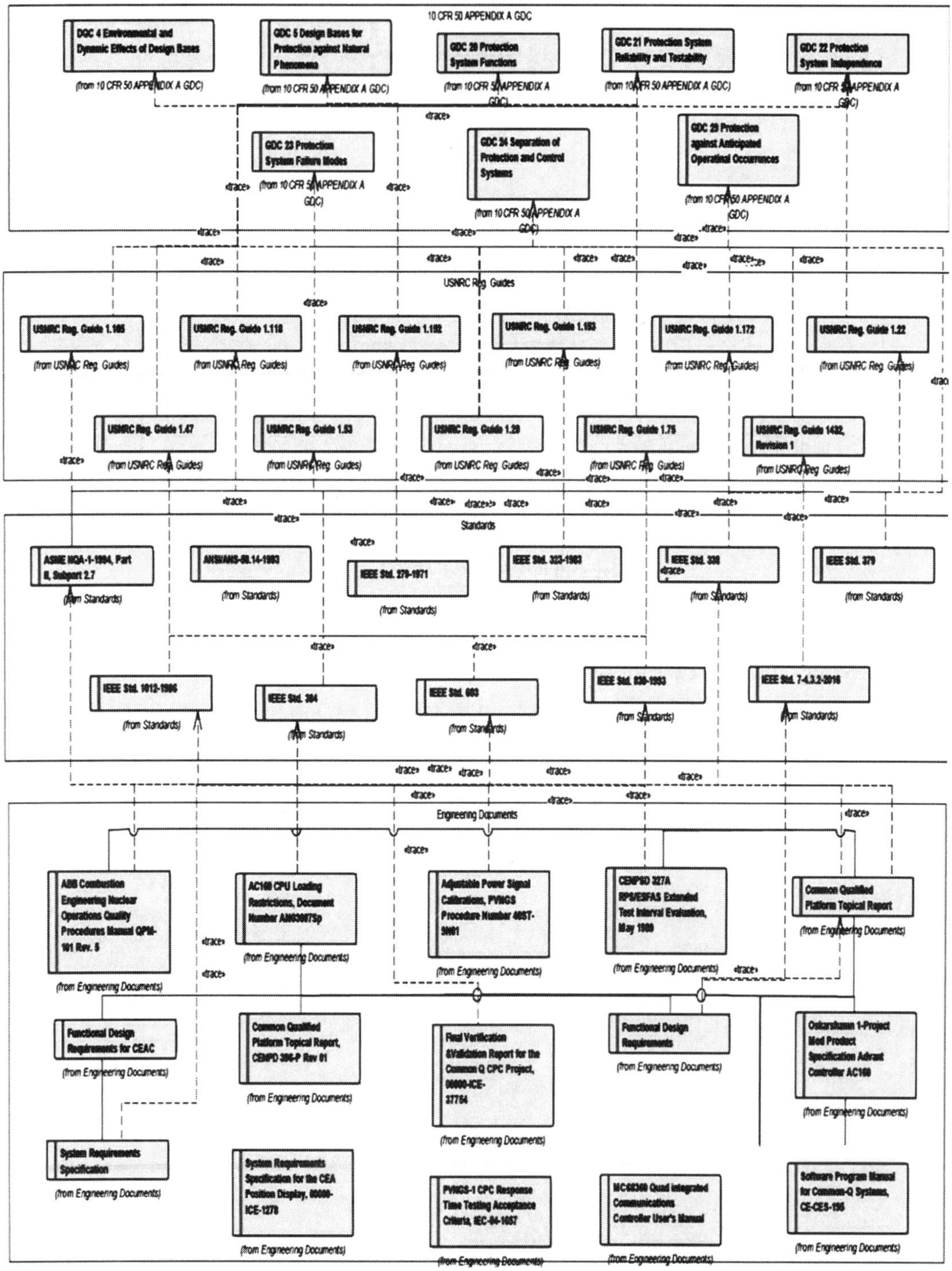


Figure 10-5: Requirements Traceability Matrix of the CPCS

PART 2

## 10.2 CPCS Functional Architecture

In order to enable the computation of the LPD and DNBR, the CPC algorithm is divided into 6 main computational blocks, each implemented on its own processor. The Figure 10-6 shows the CPCS program modules.

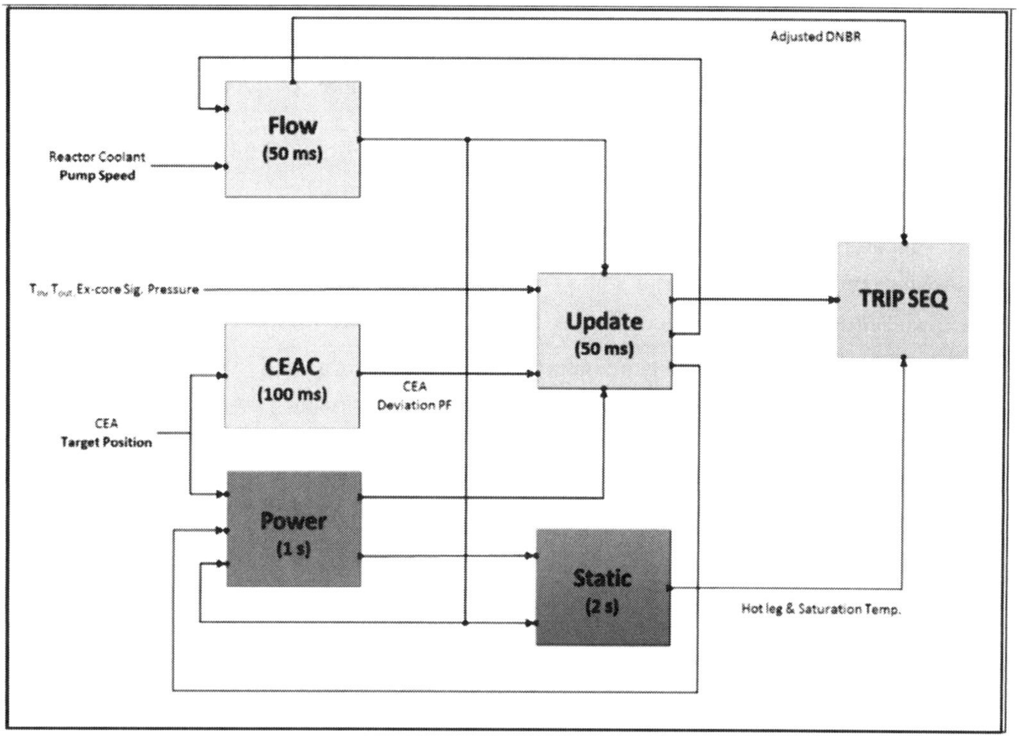


Figure 10-6: CPCS Modules

The legacy CPCS consists of the FLOW, CEAC, POWER, STATIC, UPDATE, and TRIPSEQ program modules. Each of the above-mentioned program modules performs the following functions [35] :

1. The FLOW program computes a normalized flow rate in each leg of the primary coolant system and in the reactor core, in order to compute an adjusted value of DNBR based on the number of RCPs running.
2. The function of the CEAC is to scan all CEA positions and, based on any single CEA deviation detected within a CEA subgroup, to calculate the single CEA position-related penalty factors necessary to ensure that the CPCs calculate conservative approximations to the actual core peak LPD and DNBR during

single CEA-related AOOs. The CEAC must also be capable of detecting a reactor power cutback event.

3. The purpose of the POWER program is to compute the core average axial power distribution, pseudo hot pin power distribution, and the three-dimensional power peak from the ex-core detector signals and target CEA positions. The ex-core detector signals, CEA positions and the temperature shadowing factor are also used to compute a power normalization factor required for calculation of calibrated neutron flux power in UPDATE.
4. The UPDATE program computes an updated value of DNBR, quality margin and LPD based on temperature, pressure, core power, flow, and power distribution.
5. The STATIC program computes the static values of DNBR, hot channel quality, primary thermal power, and maximum hot leg temperature. In addition, this program establishes static values of the process variables that, in turn, constitute the baseline conditions for DNBR update.
6. The TRPSEQ issues trip outputs when the computed variables within the program structure violate predetermined set-point values. Also, core conditions outside the analyzed parameter space, less than two reactor coolant pumps running, an AGST, a rapid power rise or the CPC in an other than normal operating mode will cause a trip output to be issued.

### 10.2.1 Simplified DNBR and LPD algorithm

The simplified DNBR and LPD algorithms consist of eight modules. All modules work with the same cycle execution time. The simplified algorithm uses the inputs signal collected from the sensors to calculate the minimum DNBR and LPD. The new simplified algorithm assumes that there are no deviations between Control Elements Assemblies (CEA) so the penalty factor is equal one. The input signals are listed in Table 10-4 below.

Table 10-4: Input Signals for DNBR and LPD Algorithm

| Inputs | Description | Representative Range | Signal Type |
|---|---|---|---|
| Reactor Coolant Pump Speed | Reactor coolant pump shaft speed | 10~100% of rated speed | Digital |
| Cold Leg Temperature | Temperature in primary coolant cold legs, 1 of 2 for each steam generator | 446°F~626°F | Analog |
| Hot Leg Temperature | Primary coolant hot legs 1 and 2 | 482°F~662°F | Analog |
| Primary Pressure | Pressurizer pressure | 1500~2500 Psia | Analog |
| Ex-Core Neutron Flux | Ex-core neutron detector signals | 0~200% Power | Analog |
| CEA Position | Target CEA position | 0~100% withdrawal | Analog |

The functional block diagram shown in Figure 10-7 illustrates the modules as follow:

- Input signals collection and normalization
- Thermal Hydraulic coolant condition
- Power Distribution Synthesis (Axial and Radial)
- Heat Flux Compensation
- Linear heat rate distribution
- Hot channel coolant condition
- Minimum DNBR calculation
- Local Power Density Calculation

Integration Definition for Functional Modeling (IDEF0) was used as the high-level design diagram of the simplified system. IDEF0 involves a preliminary or high-level design of the main sub-modules with an overall picture algorithm to show how the sub-modules fit together. IDEF0 was used to represent functional interfaces between the sub-modules. It was used to systematically identity, define, tabulate, design, and analyze the functional interface. Figure 10-8 show the IDEF0 diagram for the algorithm and the inputs and outputs to and from each sub-module. A description of the inputs and outputs to each functional block is given in Table 10-5

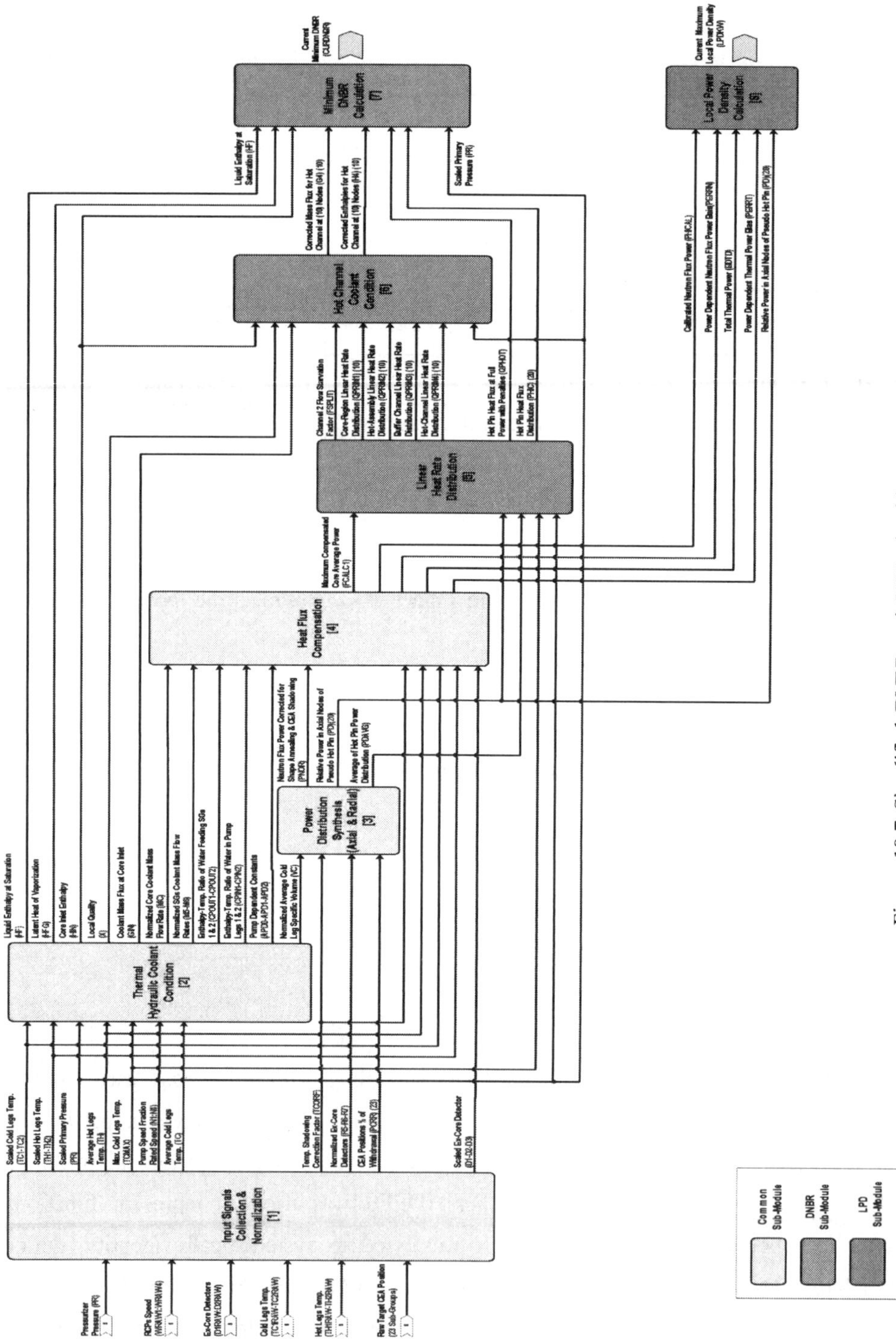


Figure 10-7 Simplified DNBR and LPD algorithm

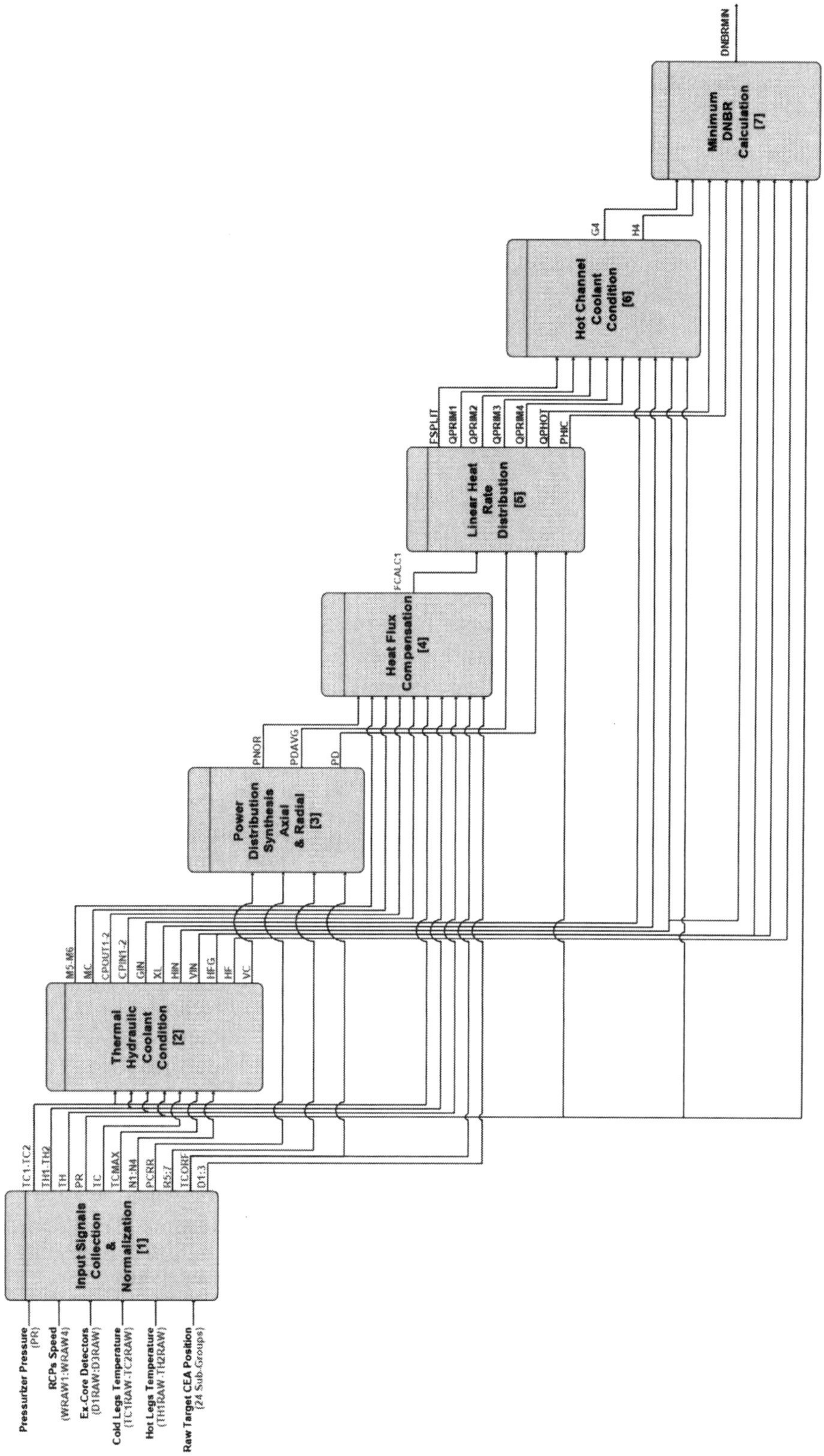


Figure 10-8 IIDEFO for the simplified CPCS algorithm

Table 10-5: Inputs and Outputs to Modules

| Sub-Module | Inputs To Sub-Module | Outputs From Sub-Module |
|---|---|---|
| Input Signals Collection & Normalization | • RCP SSSS Signals<br>• Raw Cold Legs Temperature<br>• Raw Hot Legs Temperature<br>• Raw Ex-Core Detectors<br>• Raw CEA Subgroup Position<br>• Raw Pressurizer Pressure | • Scaled Cold Legs Temperature<br>• Scaled Hot Legs Temperature<br>• Scaled Primary Pressure<br>• Scaled Ex-Core Detector<br>• Average Hot Legs Temperature<br>• Average Cold Legs Temperature<br>• Max. Cold Legs Temperature<br>• Temp. Shadowing Correction Factor<br>• Pump Speed Fraction Rated Speed<br>• CEA Positions % of Withdrawal<br>• Normalized Ex-Core Detectors Response |
| Thermal Hydraulic Coolant Condition | • Max. Cold Legs Temperature<br>• Scaled Primary Pressure<br>• Scaled Hot Legs Temperature<br>• Scaled Cold Legs Temperature<br>• Average Hot Legs Temperature<br>• Pump Speed Fraction Rated Speed<br>• Average Cold Legs Temperature | • Core Inlet Enthalpy<br>• Latent Heat of Vaporization<br>• Liquid Enthalpy at Saturation<br>• Inlet Core Coolant Specific Volume<br>• Coolant Mass Flux at Core Inlet<br>• Normalized SGs Coolant Mass Flow Rates<br>• Normalized Core Coolant Mass Flow Rate<br>• Local Quality<br>• Enthalpy-Temp. Ratio of Water Feeding SGs 1 & 2<br>• Enthalpy-Temp. Ratio of Water in Pump Legs 1 & 2<br>• Normalized Average Cold Leg Specific Volume |
| Power Distribution Synthesis (Axial & Radial) | • CEA Positions % of Withdrawal<br>• Normalized Ex-Core Detectors Response<br>• Temperature Shadowing Correction Factor<br>• Normalized Average Cold Leg Specific Volume | • Relative Power in Axial Nodes of Pseudo Hot Pin<br>• Neutron Flux Power Corrected for Shape Annealing & CEA Shadowing<br>• Average of Hot Pin Power Distribution |

| Sub-Module | Inputs To Sub-Module | Outputs From Sub-Module |
|---|---|---|
| Heat Flux Compensation | • Scaled Cold Legs Temperature<br>• Scaled Hot Legs Temperature<br>• Average Hot Legs Temperature<br>• Temp. Shadowing Correction Factor<br>• Scaled Ex-Core Detector<br>• Normalized Core Coolant Mass Flow Rate<br>• Normalized SGs Coolant Mass Flow Rates<br>• Enthalpy-Temp. Ratio of Water Feeding SGs 1 & 2<br>• Enthalpy-Temp. Ratio of Water in Pump Legs 1 & 2<br>• Neutron Flux Power Corrected for Shape Annealing & CEA Shadowing | • Maximum Compensated Core Average Power |
| Linear Heat Rate Distribution | • Scaled Primary Pressure<br>• Maximum Cold Legs Temp.<br>• Average of Hot Pin Power Distribution<br>• Relative Power in Axial Nodes of Pseudo Hot Pin<br>• Maximum Compensated Core Average Power | • Hot Pin Heat Flux Distribution<br>• Channel 2 Flow Starvation Factor<br>• Hot Pin Heat Flux at Full Power with Penalties<br>• Core-Region Linear Heat Rate Distribution<br>• Hot-Assembly Linear Heat Rate Distribution<br>• Buffer Channel Linear Heat Rate Distribution<br>• Hot-Channel Linear Heat Rate Distribution |
| Hot Channel Coolant Condition | • Scaled Primary Pressure<br>• Core Inlet Enthalpy<br>• Coolant Mass Flux at Core Inlet<br>• Local Quality<br>• Channel 2 Flow Starvation Factor<br>• Core-Region Linear Heat Rate Distribution<br>• Hot-Assembly Linear Heat Rate Distribution<br>• Buffer Channel Linear Heat Rate Distribution<br>• Hot-Channel Linear Heat Rate Distribution | • Corrected Mass Flux for Hot Channel at (10) Nodes<br>• Corrected Enthalpies for Hot Channel at (10) Nodes |

| Sub-Module | Inputs To Sub-Module | Outputs From Sub-Module |
|---|---|---|
| Minimum DNBR Calculation | • Scaled Primary Pressure<br>• Latent Heat of Vaporization<br>• Core Inlet Enthalpy<br>• Liquid Enthalpy at Saturation<br>• Inlet Core Coolant Specific Volume<br>• Hot Pin Heat Flux at Full Power with Penalties<br>• Hot Pin Heat Flux Distribution<br>• Corrected Mass Flux for Hot Channel at (10) Nodes<br>• Corrected Enthalpies for Hot Channel at (10) Nodes | • Minimum DNBR |
| Local Power Density Calculation | • Calibrated Neutron Flux Power<br>• Power Dependent Neutron Flux Power Bias<br>• Total Thermal Power<br>• Power Dependent Neutron Flux Power Bias<br>• Relative Power in Axial Nodes of Pseudo Hot Pin | • Current Maximum Local Power Density |

## 10.3 Detailed Design

The detailed design for the DNBR and LPD trip algorithms was performed using Flowcharts and EFFBDs. For each of the functions defined in Section 10.2.1, the associated flowcharts and EFFBD are presented here.

### 10.3.1 DNBR Design

#### (1) Input signals collection and *normalization*

In this module, the input signals coming from the sensors are calibrated and normalized for use in the next calculation. These input signals are raw pressurizer pressure signal, two raw temperature signals from hot leg and other two for cold leg, the reactor coolant pumps shaft speed sensors, raw ex-core neutron flux detector signals for upper, middle and lower detectors, raw target CEA subgroup position (23 subgroups).

The main functions in this module are shown in Figure 10-9 and can be divided into six parallel functions as follow:

- Calculate the primary cycle pressure using the pressurizer pressure raw signal,
- Convert the raw target CEA subgroups positions to percentage of CEA subgroups withdrawal from the core,
- Calculate pump speed as a function of fraction of rated speed,
- Normalize ex-core detectors values and response,
- Calculate scaled values for two hot legs temperature and the average hot leg temperature,
- Calculate scaled values for two cold legs, the average cold legs temperature, maximum cold legs temperature, and the temperature shadowing correction factor.

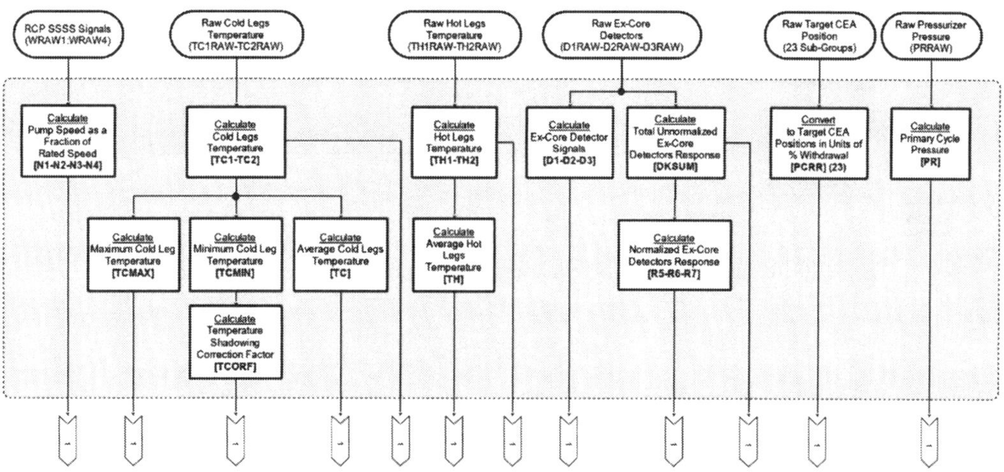


Figure 10-9: Input Signals Collection and Normalization Module Flow Chart

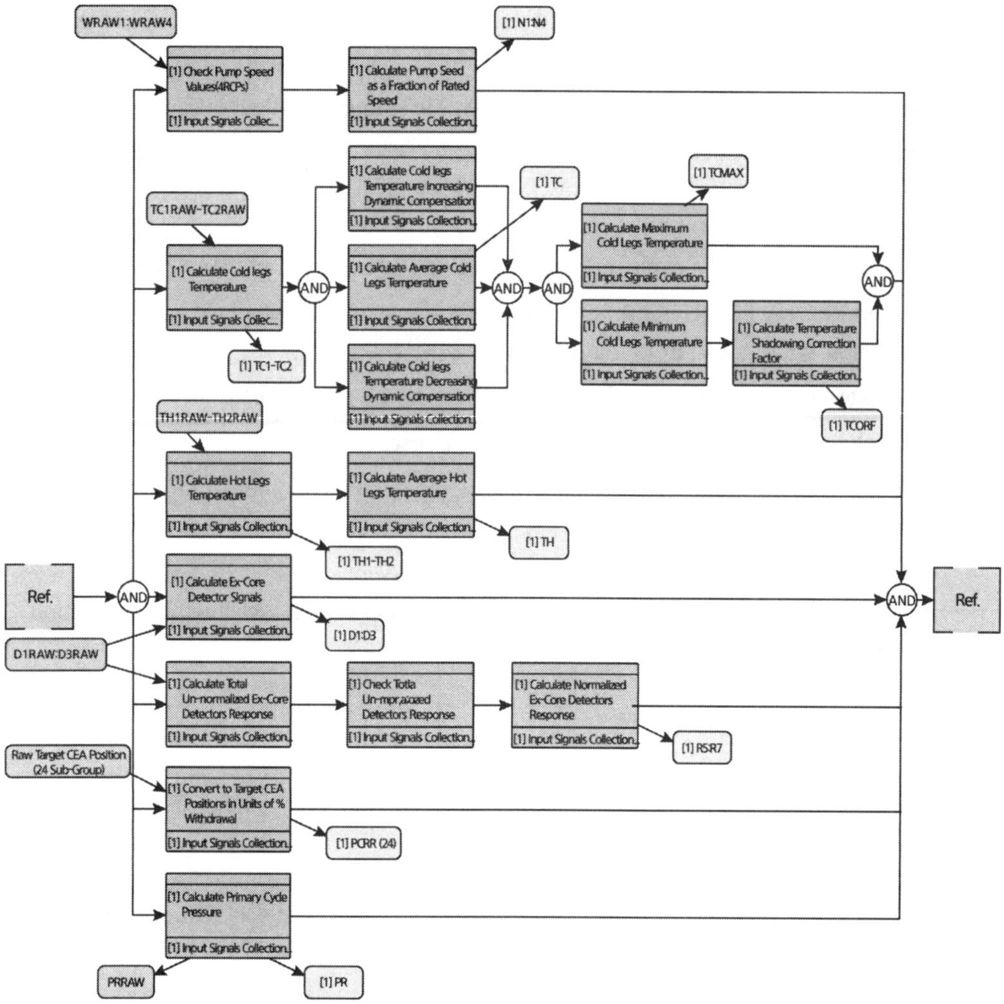


Figure 10-10: Input Signals Collection and Normalization Module EFFBD

## (2) Thermal Hydraulic coolant condition

This module receives the calculated reactor coolant pumps speed as a fraction of rated speed, scaled cold legs temperature values, maximum cold legs temperature, scaled hot legs temperature values, average hot legs temperature, scaled primary cycle pressure as illustrated in Figure 10-12 to calculate the following parameters:

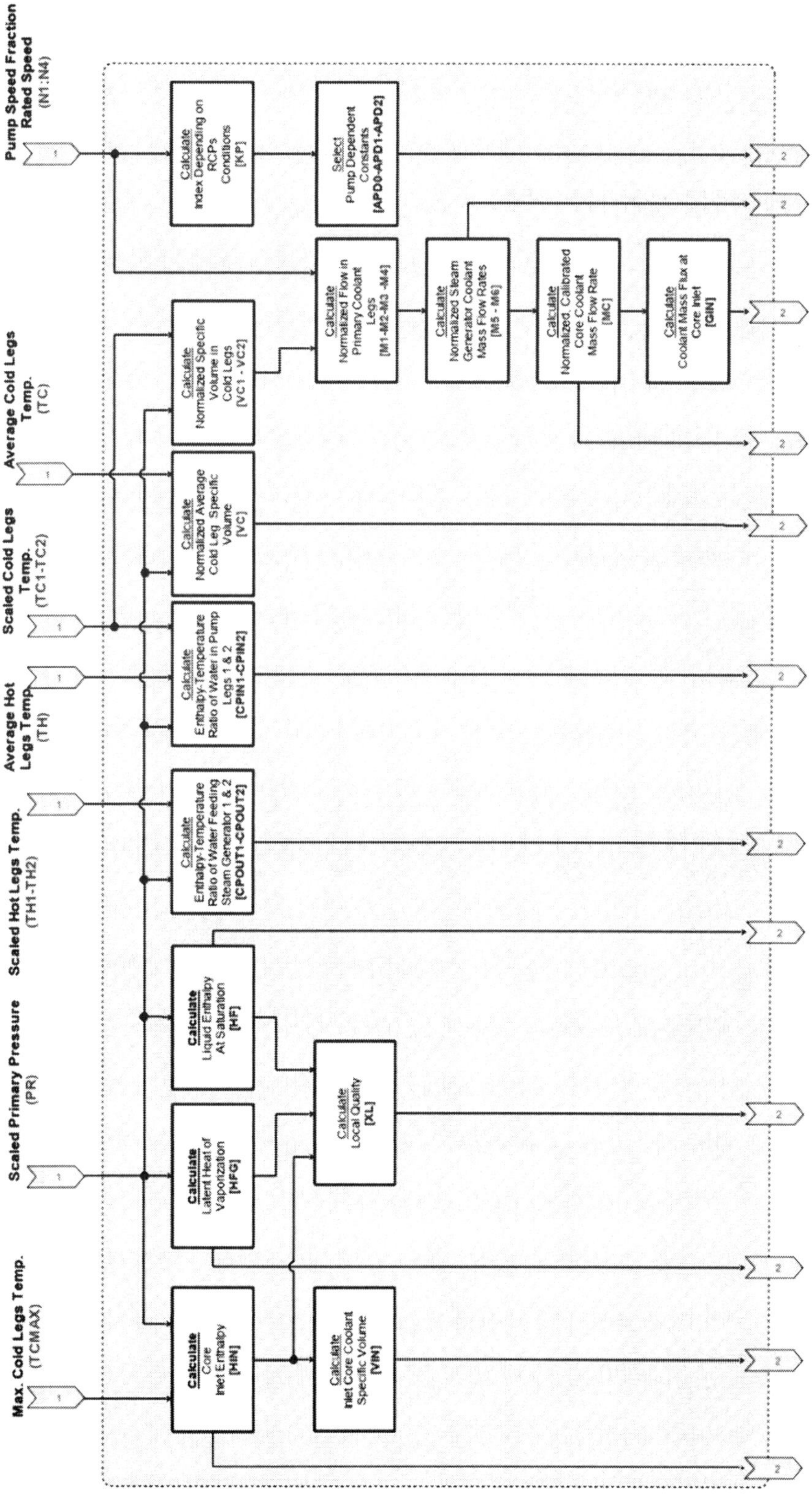


Figure 10-11: Thermal Hydraulic Coolant Condition Module Flow Chart

- Normalized steam generators coolant mass flow rates,
- Normalized core coolant mass flow rate,
- Coolant mass flux at core inlet
- Enthalpy-Temperature ratio of water in reactor coolant pump Legs and of water feeding steam generators,
- Liquid enthalpy at saturation,
- Latent heat of vaporization,
- Core inlet enthalpy,
- Inlet core coolant specific volume.

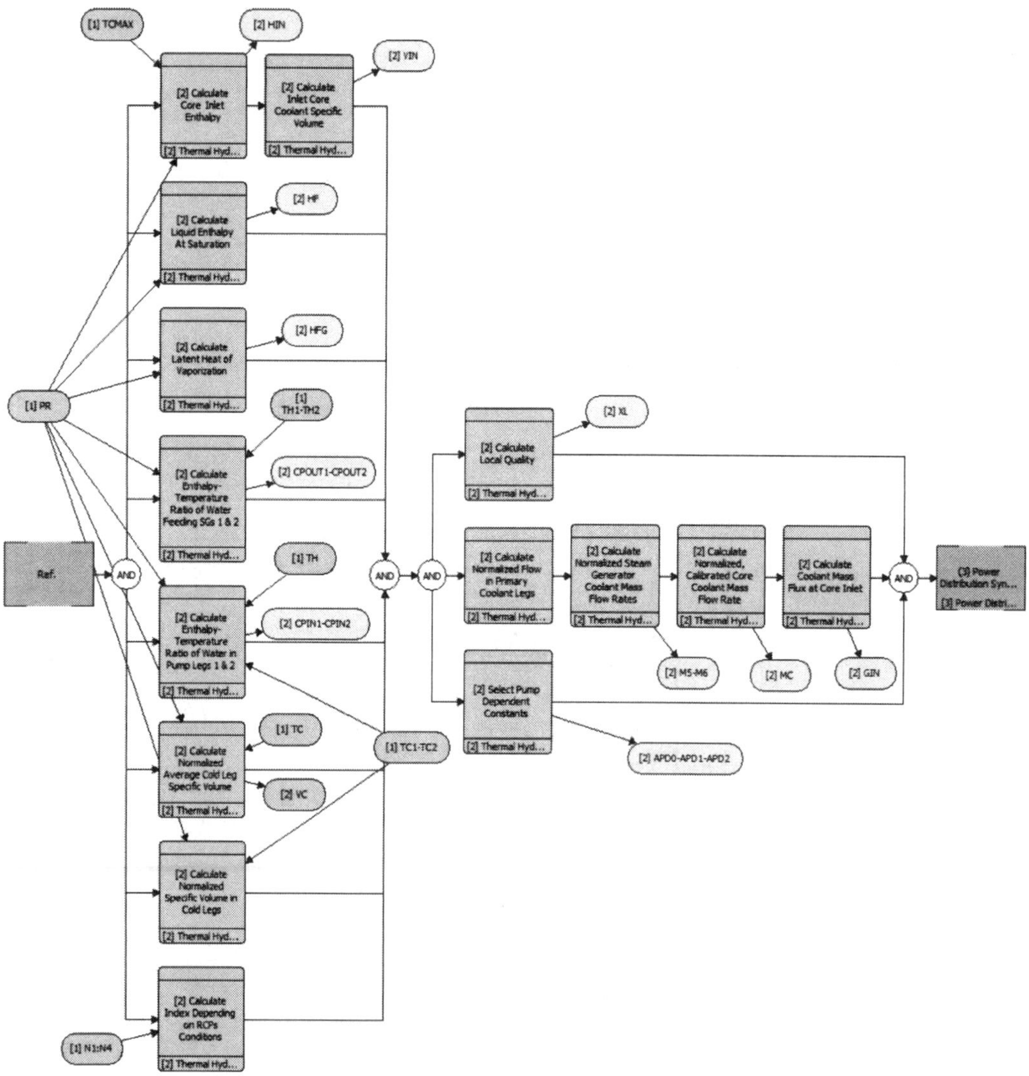


Figure 10-12: Thermal Hydraulic Coolant Condition Module EFFBD

### (3) Power Distribution Synthesis (Axial and Radial)

The purpose of this module is to compute the core average axial power distribution, pseudo hot pin power distribution, and the three-dimensional power peak from the ex-core detector signals and target CEA positions. The ex-core detector signals, CEA positions, and the temperature shadowing factor are also used to compute a power normalization factor required for calculation of calibrated neutron flux power [36].

This module receives its inputs from Input signals collection and normalization module as illustrated in Figure 10-13. Figure 10-14 is EFFBD of this module.

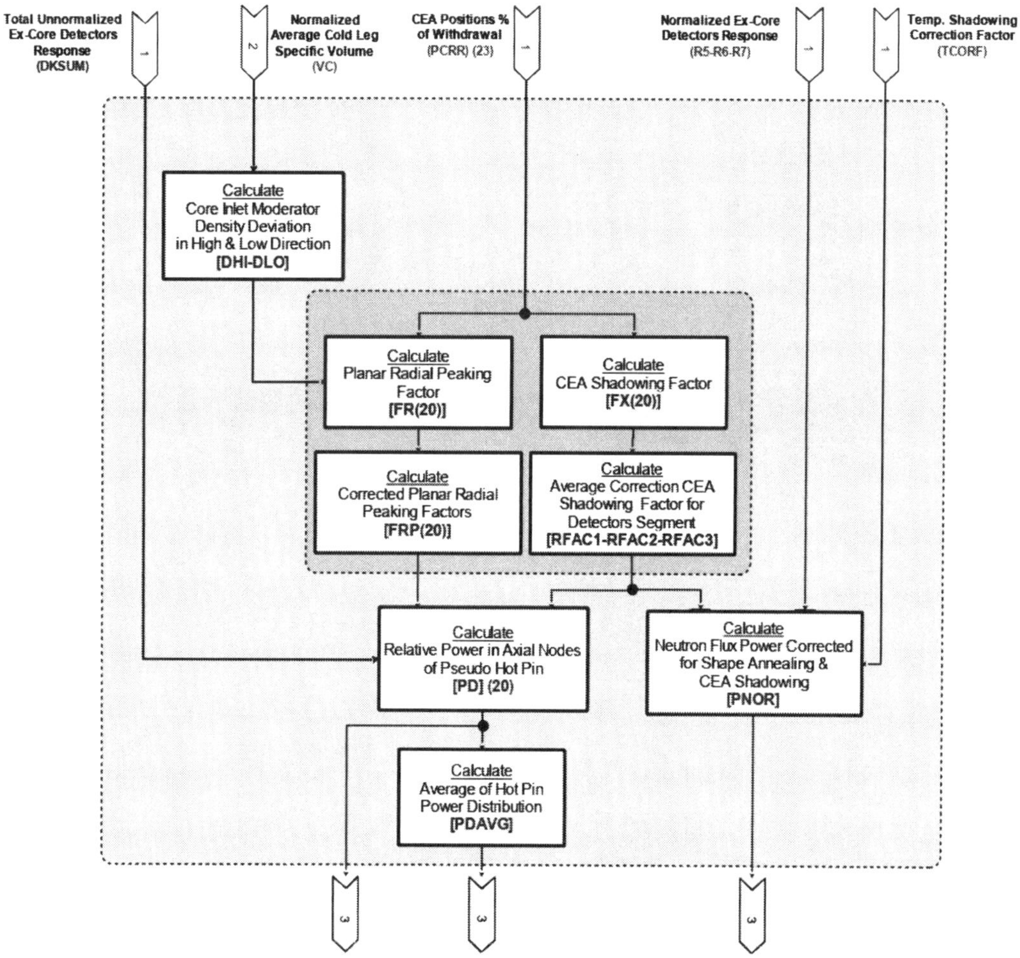


Figure 10-13: Power Distribution Synthesis (Axial and Radial) Module Flow Chart

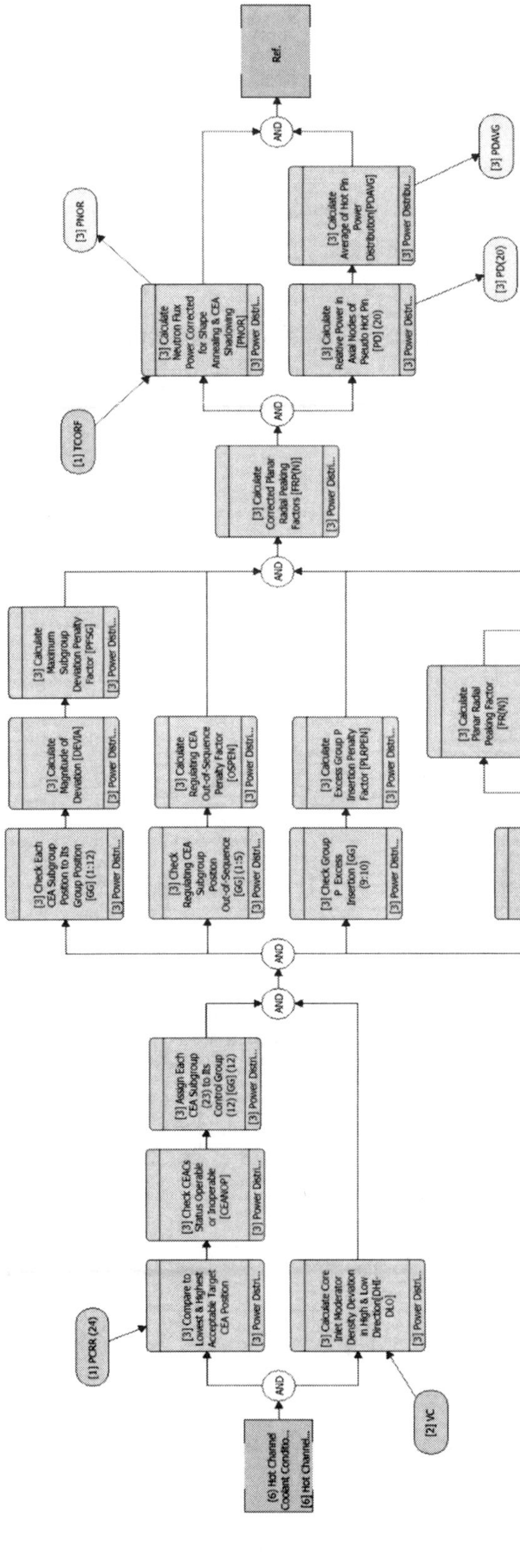

Figure 10-14: Power distribution synthesis(Axial and Radial) Module EFFBD

### (4) Heat Flux Compensation

The purpose of this module is to compute the maximum compensated core average power. The compensated thermal power and the compensated neutron flux power are calculate from the total thermal power and calibrated neutron flux power taking in consecration the fuel dynamics. The core average heat flux is selected as the maximum of the compensated thermal power and compensated neutron flux power values the algorithm of this module illustrated in Figure 10-15. Figure 10-16 is EFFBD for this module.

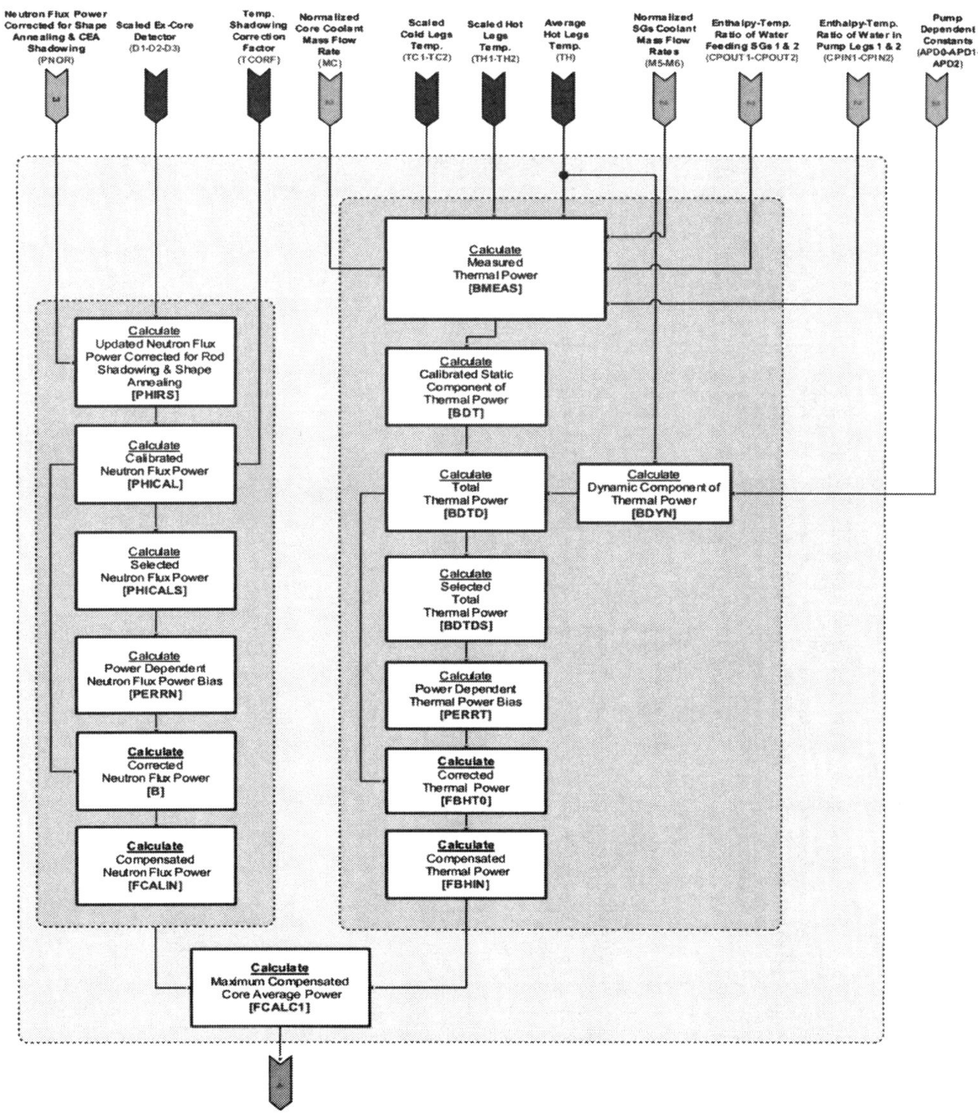


Figure 10-15: Heat Flux Compensation Module Flow Chart

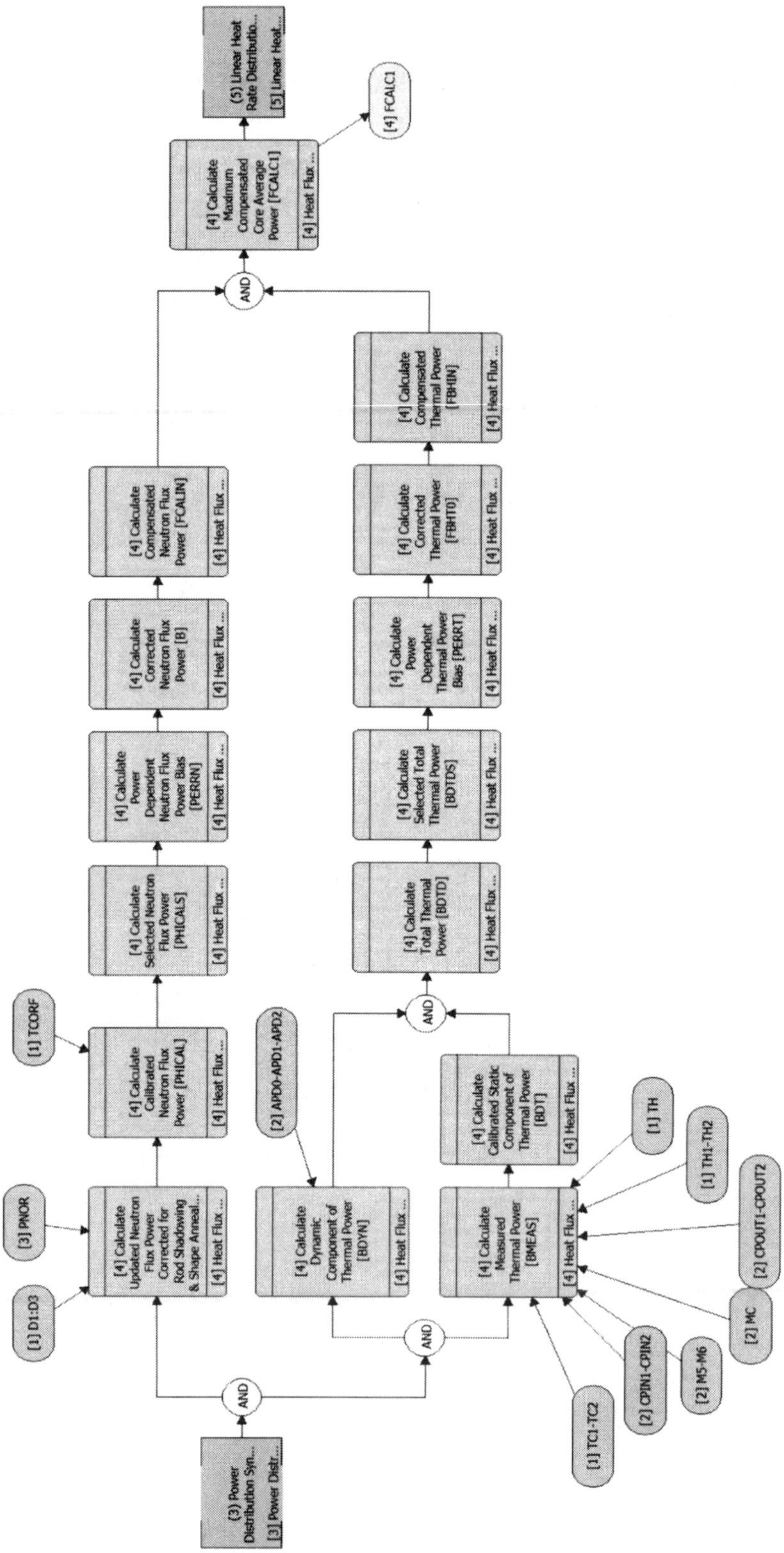


Figure 10.16: Heat Flux Compensation Module EFFBD

## (5) Linear Heat Rate Distribution

The linear heat distribution are computed for four channel modeling as illustrated in Figure 10-17. These channels are:

- Channel one is the reactor core,
- Channel two is the hot assembly,
- Channel three is the buffer channel,
- Channel 4 is the hot channel.

The 20-element hot pin axial heat flux distribution is combined with the integrated one pin radial peak, and collapsed to 10-element distributions.

EFFBD of this module is shown in Figure 10-18.

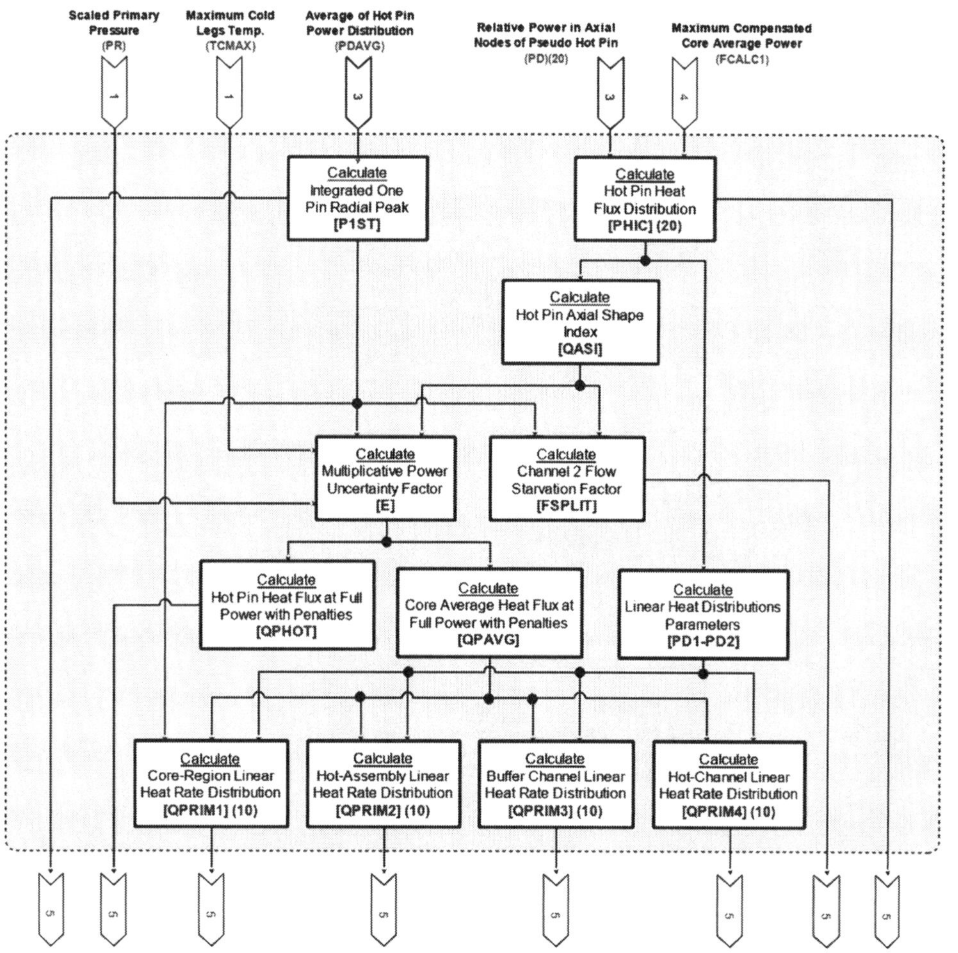


Figure 10-17: Linear Heat Rate Distribution Module Flow Chart

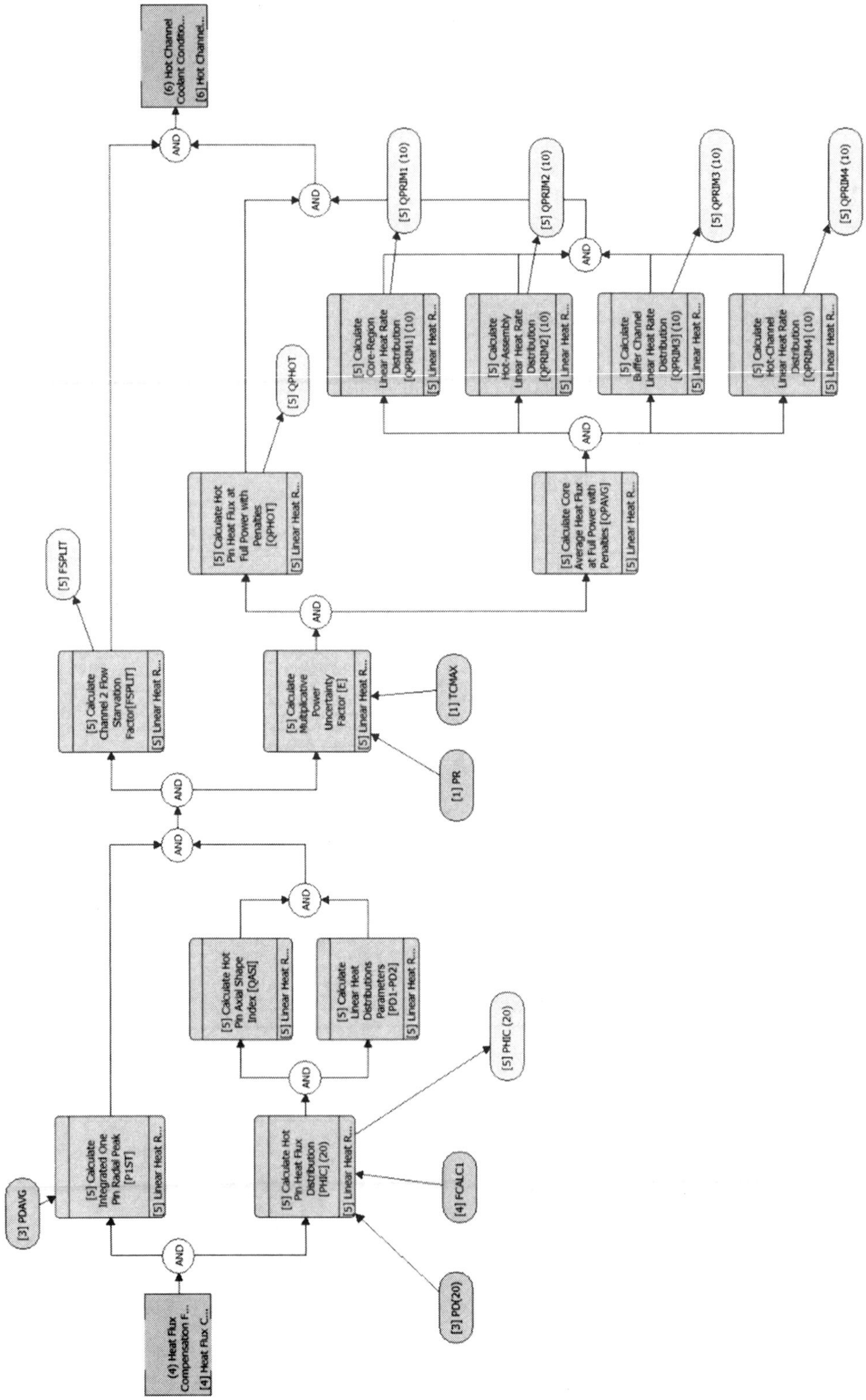


Figure 10-18: Linear Heat Rate Distribution Module EFFBD

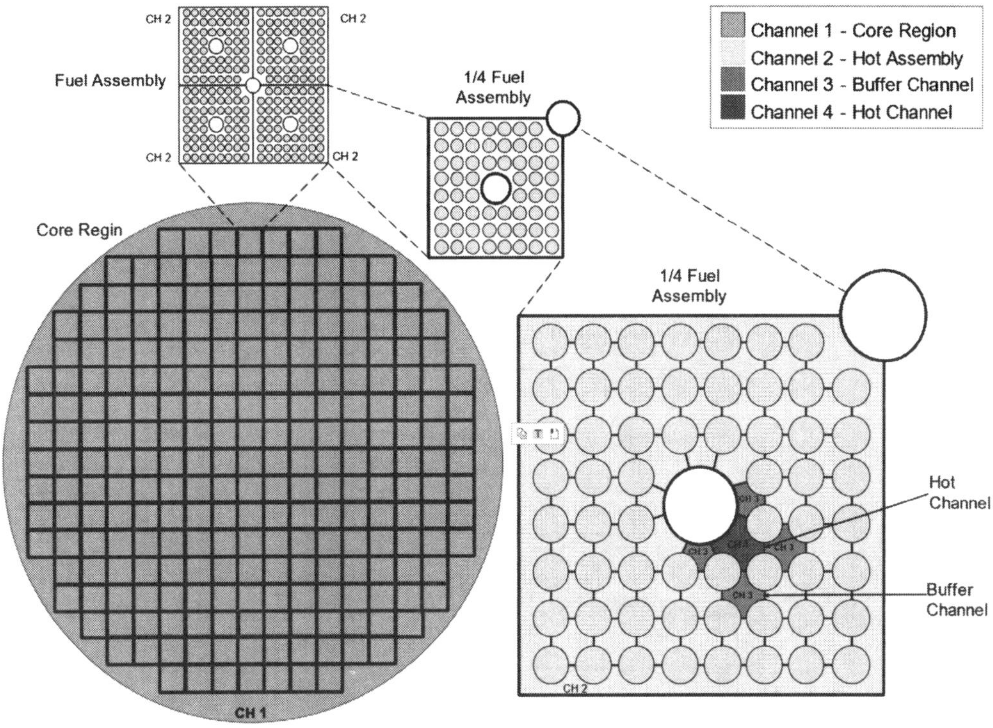


Figure 10-19: Four Channel Modeling for Linear Heat Distribution

## (6) Hot Channel Coolant Condition

First, the core and hot assembly channels fluid properties such as enthalpy, mass flux, cross flow and pressure drop axial distributions are calculated for ten nodes depends on the properties of the upstream and downstream nodes. Second the calculated fluid properties for channel one and two are used to calculate the fluid properties for channel three and four (buffer and hot) channels. The hot channel fluid properties such as enthalpy and mass flux distributions will be the output of this module. The hot channel distributions (channel 4) will be used in the critical heat flux calculations. Four-channel modeling is shown in Figure 10-17 and the module flow chart is shown in Figure 10-20. Figure 10-21 depicts EFFBD for the module.

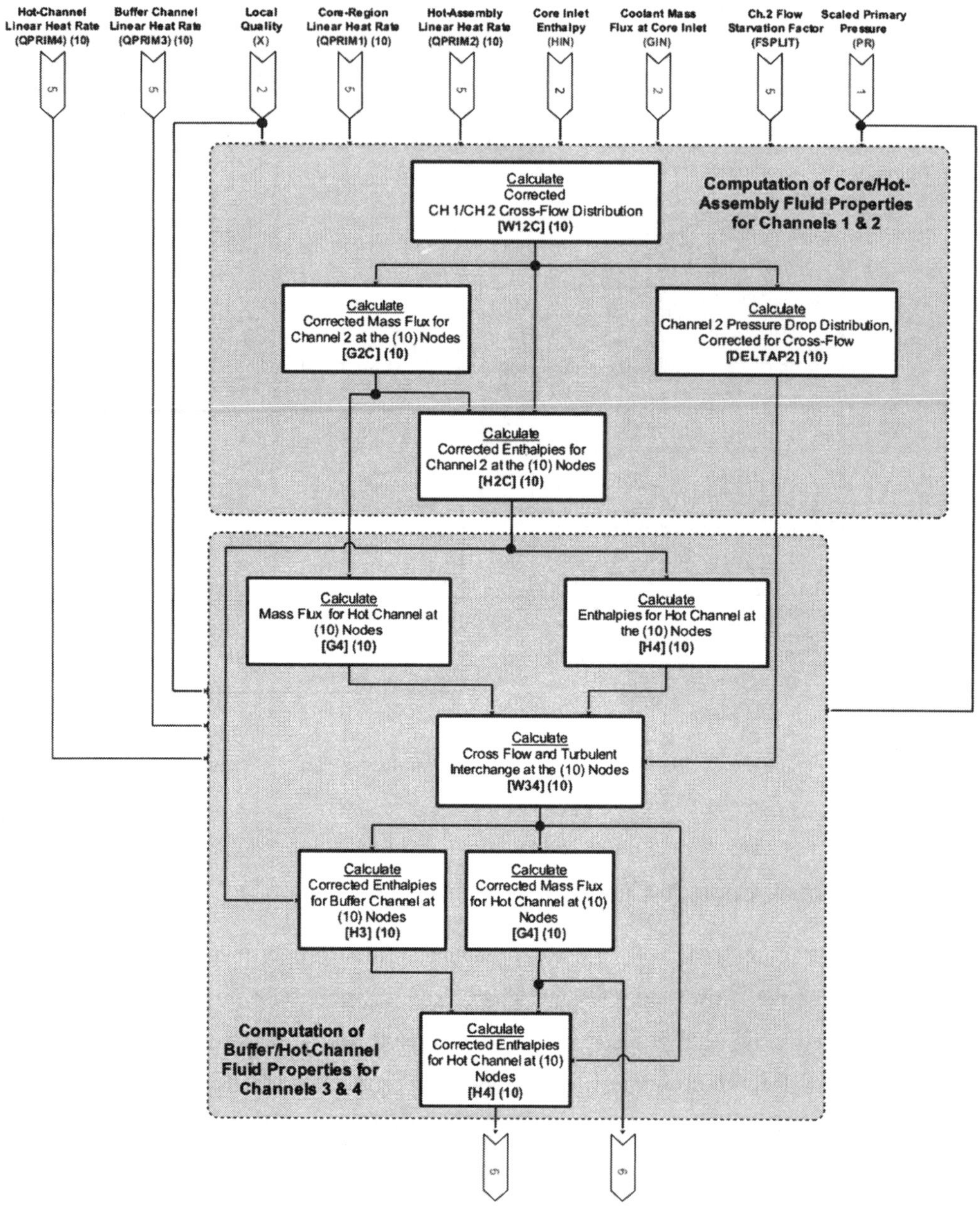


Figure 10-20: Hot Channel Coolant Conditions Module Flow Chart

P A R T 2

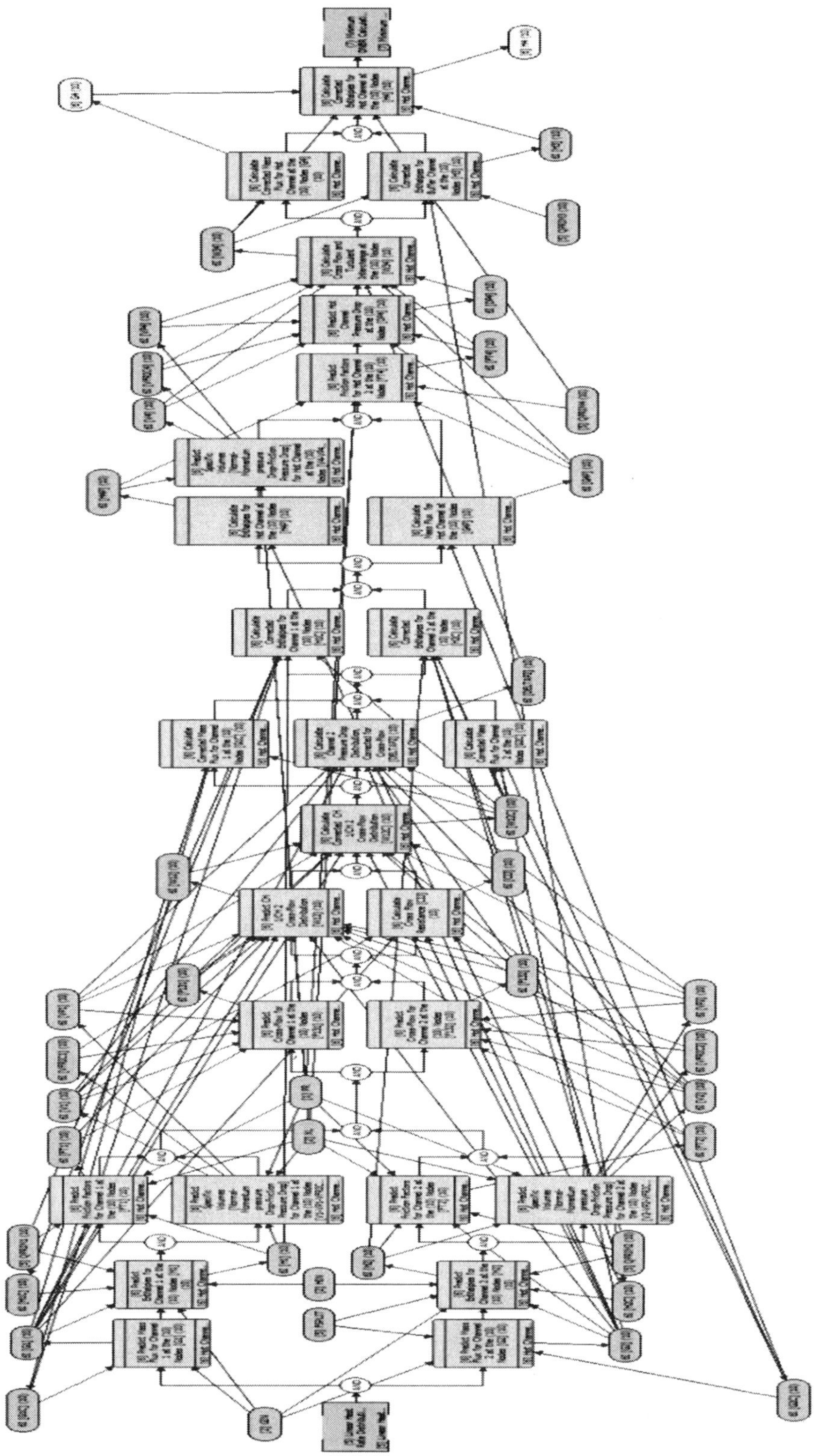

Figure 10-21: Hot Channel Coolant Conditions Module EFFBD

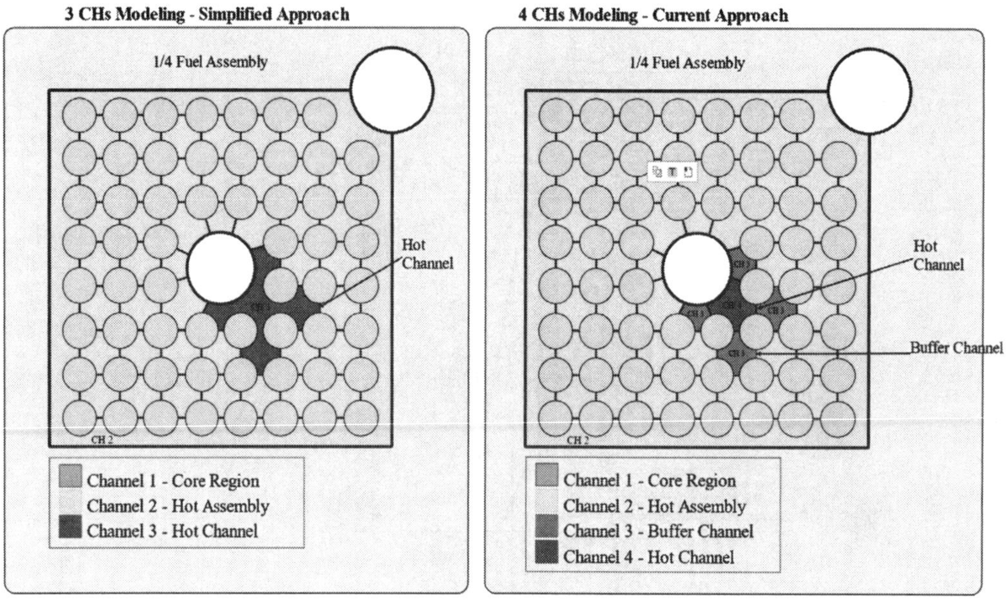


Figure 10-22: Four and three channels modeling for linear heat rate

- **Simplified Hot Channel Fluid Properties and Heat Flux Calculations**

  The current DNBR algorithm uses four channels modeling to calculate linear heat distributions, which are used to predict the fluid properties in the hot channel. The simplified DNBR algorithm was developed using three modeling channels, which includes the core region, hot assembly, and hot channel as illustrated in Figure 10-22. Linear heat distributions were calculated in the core region and hot assembly channels to predict the mass flux and enthalpy distributions for 11 nodes in the hot channel.

## (7) Minimum DNBR Calculation

The purpose of this module is to calculate the DNB ratio in hot channel for twenty nodes after that select the minimum value as illustrated in Figure 10-23 and 10-24.

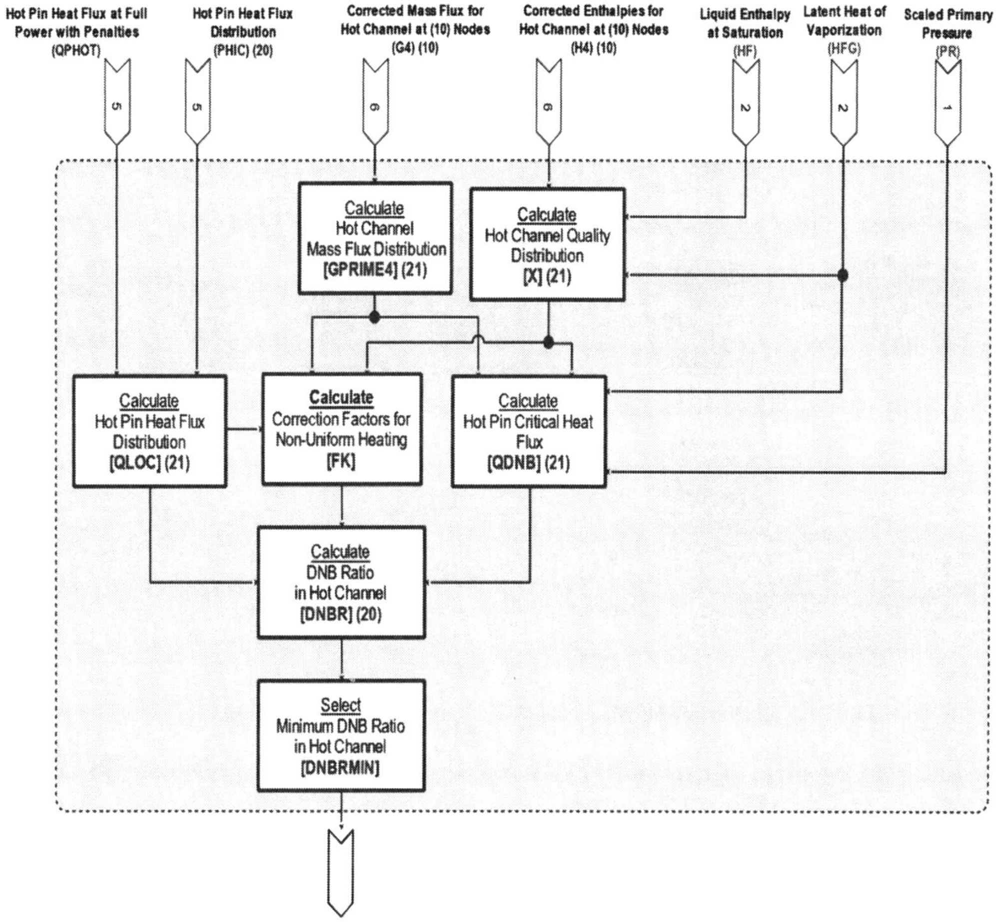


Figure 10-23: Minimum DNBR Calculation Module Flow Chart

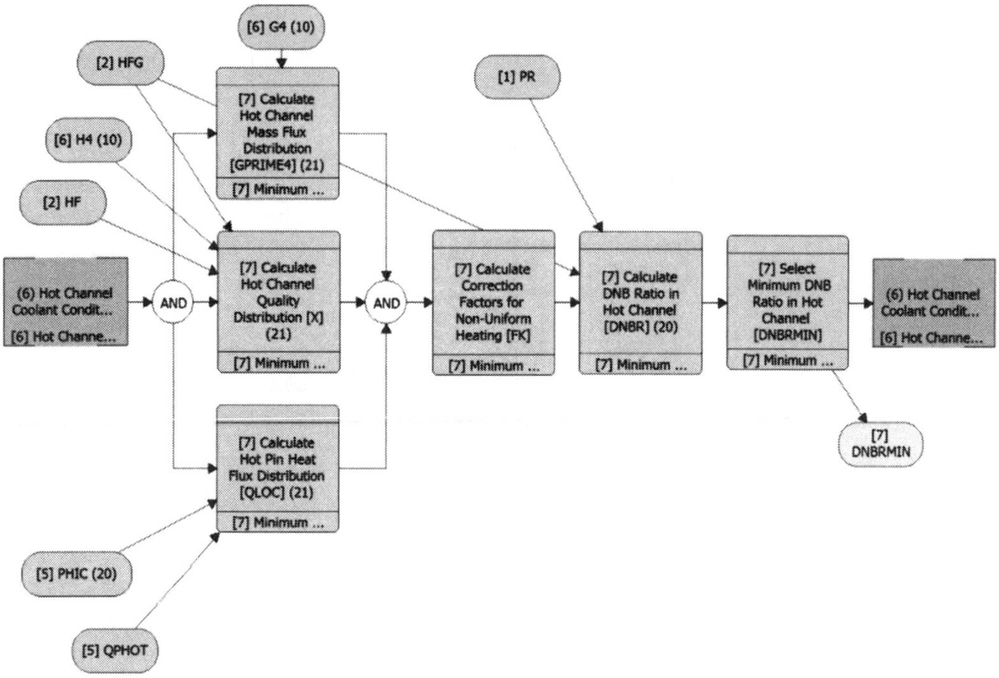


Figure 10-24: Minimum DNBR Calculation Module EFFBD

## 10.3.2 LPD Design

In order to compute the LPD, only a subset of the functional blocs defined in Section 10.2 are used. LPD functional flow is shown in Figure 10-25. Figure 10-26 illustrates the EFFBD for the LPD algorithm.

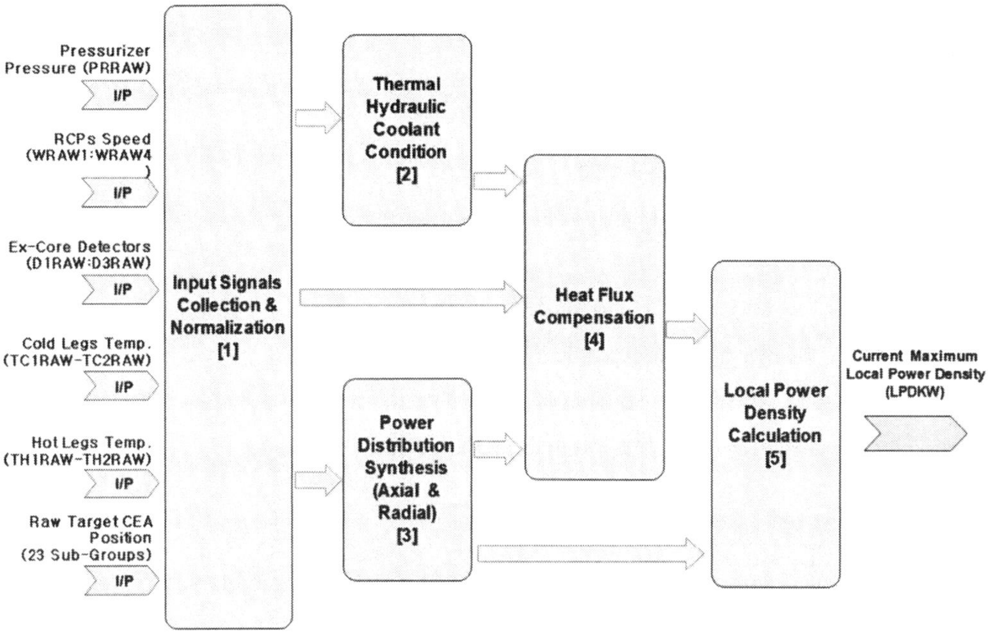


Figure 10-25: LPD Function Flow Diagram

The LPD Calculation module calculates hot pin heat flux distribution. It selects maximum 3-D peaking factor, calculate corrected core average power, and generates local power density (LPDKW). Input signals, output signals, and functional flow of module 5 are described in the EFFBD in Figure 10-26.

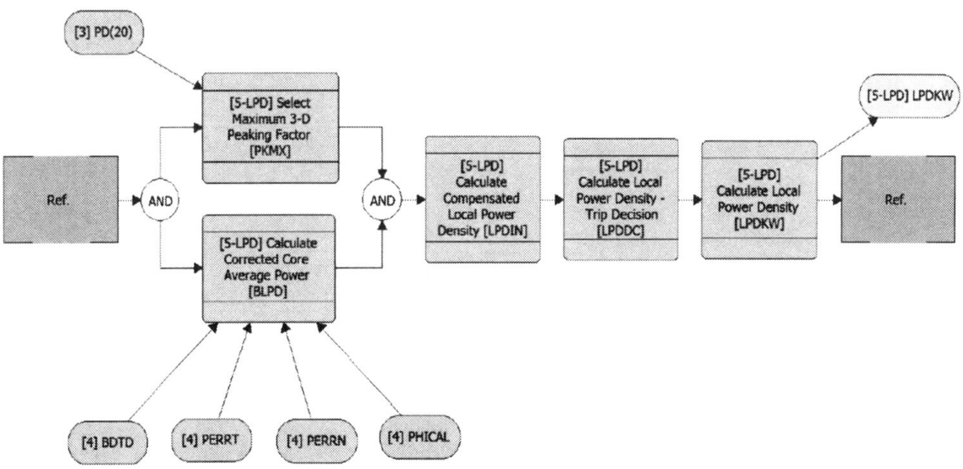


Figure 10-26: LPD Calculation Module EFFBD

## PROBLEMS

1. What is AOO. Gve an example
2. Explain how the CPCS works.
3. What is DNBR?
4. What is LPD?
5. Discuss the hot channel and buffer channel.
6. Explain the specified acceptable fuel design limit.
7. List all the prohibition situations of CEA withdrawal
8. Discuss why we need the minimum DNBR and high LPD

## FURTHER READING


- KHNP, "APR1400 Design Control Document Document : Tier-2, Chapter 7," 2018
- John R. Lamarsh and Anthony J. Baratta, "Introduction to Nuclear Engineering" Prentice Hall, 2001
- Wang-Kee In, Young-Ho Park and Seung-Yeob Baeg, "Development and Assessment of Advanced Reactor Core Protection System," Journal of Power and Energy System, 2012.

# Chapter 11 IMPLEMENTATION OF FPGA BASED SYSTEMS

In PLC based systems, during the implementation phase, the software code is developed and then integrated with the hardware. The final product consists of both software and hardware. In the case of FPGA based systems, the implementation phase consists of writing/embedding the description of the circuit (HDL code) onto the FPGA board, the result is hardware only. Thus, while most of the FPGA development lifecycle is similar to that of software development, extra considerations must be made during the implementation. In this chapter, the FPGA implementation process is defined and an example of an FPGA based MMIS (whose design was discussed in the previous chapters) is given. Further, the development of the HDL code can be done manually or using automatic code generators. The writing of HDL has been covered extensively by multiple texts. In this chapter, the use of MATLAB to generate code is addressed. The structure of the chapter is as shown in the mind map below(Figure 11-1).

FPGA based systems can be used to perform functions that are safety critical and require high reliability, also referred to as HDL Programmable Devices (HPDs); they can also run without an operating system and the design architecture is inherently parallel. A Programmable Logic Controller (PLC) can be replaced by FPGA in order to reduce the Common Cause Failure (CCF) probability.

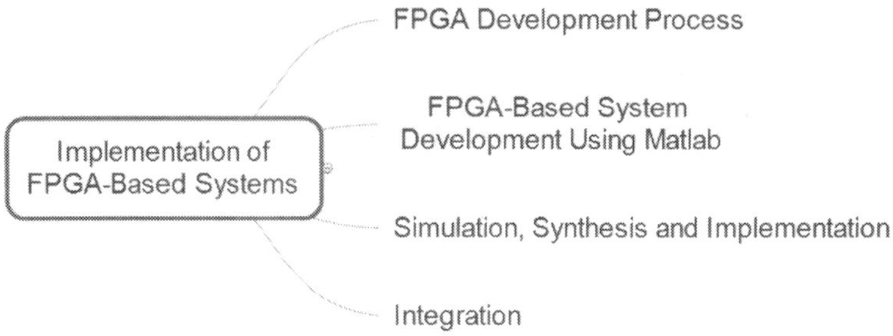


Figure 11-1: Mind Map of Chapter 11

## OBJECTIVES

- Define the development process of FPGAs
- Discuss the implementation of a Safety Critical System on FPGA

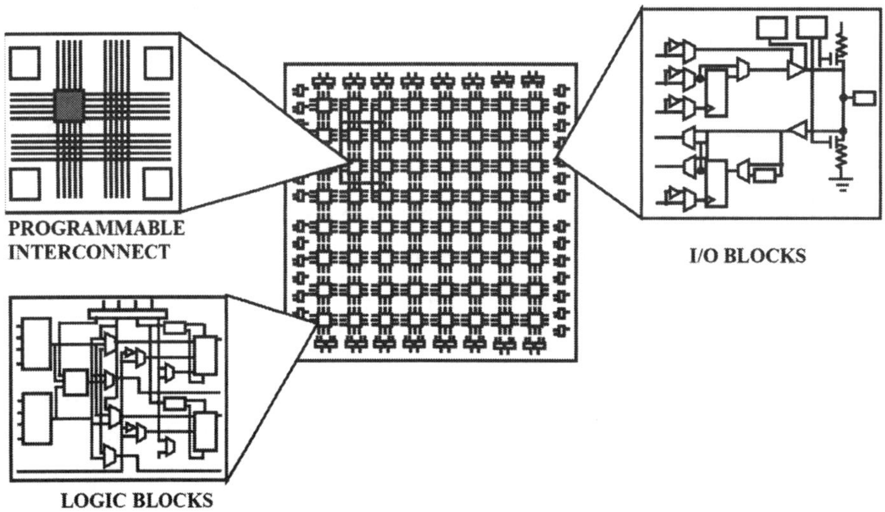


Figure 11-2: Internal Structure of FPGA

The field-programmable gate array (FPGA) is an integrated circuit that consists of internal hardware blocks with user-programmable interconnects to customize operation for a specific application. Interconnects can readily be reprogrammed, allowing an FPGA to accommodate changes to a design or even support a new application during the lifetime of the part. A single FPGA can replace thousands of discrete components by incorporating millions of logic gates in a single integrated circuit (IC) chip. The internal resources of an FPGA chip consist of a matrix of configurable logic blocks (CLBs) surrounded by a periphery of I/O blocks shown in Figure 11-2. Signals are routed within the FPGA matrix by programmable interconnect switches and wire routes [12].

FPGA technology has been applied in the satellite, military, aerospace, aircraft, transportation industries, and in other industries where the reliability of applications is of primary importance. Considering the benefits and favorable experience with the use of FPGAs in other industries, the nuclear power industry has started to introduce FPGAs in nuclear power plants for safety, safety related and non-safety applications.

## 11.1 Field Programmable Gate Array Development Process

The FPGA development process, as defined by the IAEA [12], comprises four main development phases, as shown in the V-shaped diagram in Figure 11-3.

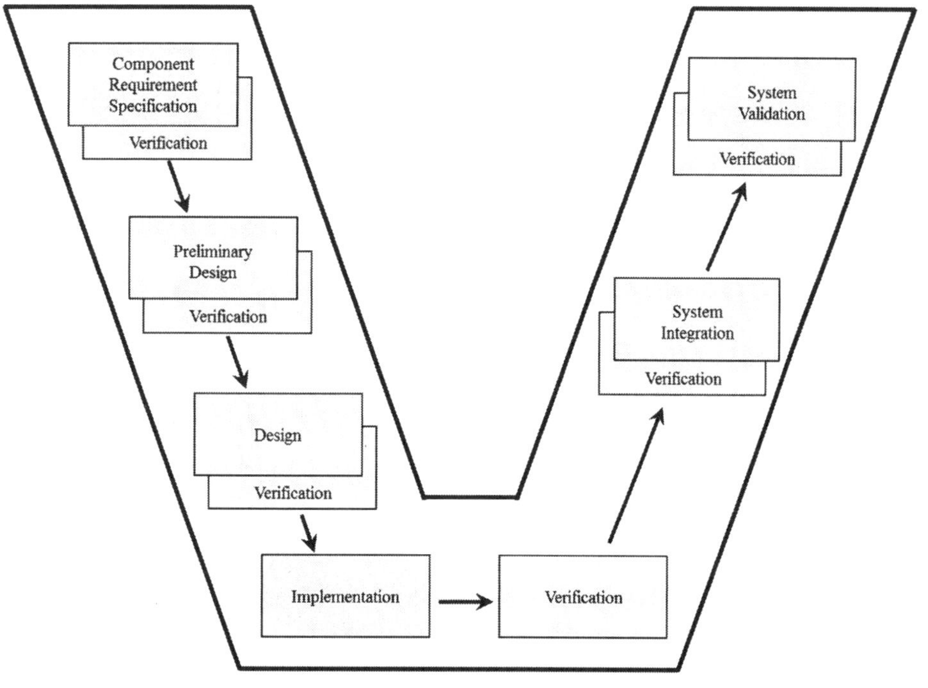


Figure 11-3: V-shaped field programmable gate array (FPGA) programming model

### 11.1.1 System Implementation

The generic design flow of an FPGA includes following steps as shown in the Figure 11-4. Design Entry is the first step is design entry and description. This step describes design functionality either by using schematic editors or by using one of the various Hardware Description Languages (HDLs) like Verilog or VHDL. The second crucial step in the design process is FPGA simulation. Designs of all sizes should require Functional and Timing simulation. Simulation can be done using waveforms and providing stimulus or using a test bench.

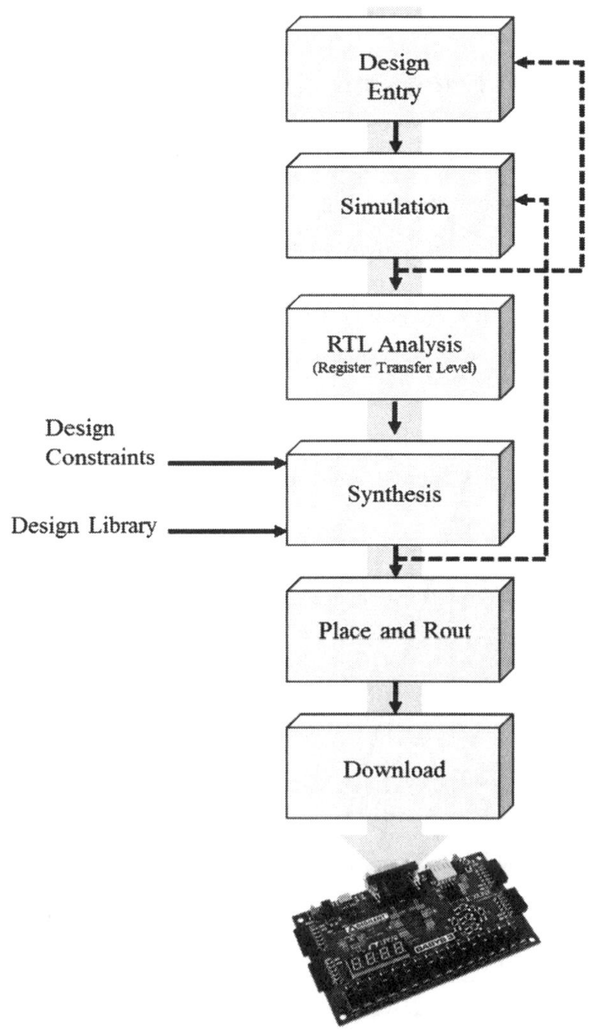


Figure 11-4: FPGA Design Flow Process

Register Transfer Level (RTL) analysis process the required hardware is described by implying logic defining flip-flops, latches, and how data is transferred between them. Synthesis Process succeeds the RTL and includes generic optimization, slack optimizations, power optimizations followed by placement and routing. Implementation includes Partition, Place and route. The output of design implementation phase is bit-stream file.

Place and Route is composed of two steps, placement and routing. The first step, placement, involves deciding where to place all electronic components, circuitry, and logic elements in a generally limited amount of space. This is followed by routing, which decides the exact design of all the wires needed to connect the placed components. This step must implement all the desired connections while following the rules and limitations of the manufacturing process. The final process is the download step where the design is written onto the FPGA board.

## 11.2 FPGA Based System Implementation Using MATLAB

MATLAB programming language is a high-level language, so it is easy to use it to develop any complex functions. Using MATLAB give the advantages to save the time and effort for developing a complex algorithms and functions. The new approach using MATLAB as a first step before go through the FPGA development process has many features such as: it is easy to modify the code and change it at any time of developing the software also during the verification phase the MATLAB Simulink can be used to simulate the real time simulation for the code.

The approach includes developing the MATLAB code that describes the functions of the system after that the code is verified and tested using different test cases once the code is tested and verified now it is the step to convert the code from MATLAB m file to VHDL code using HDL Coder application in MATLAB. In this step all the inputs, variables and outputs are converted from floating point domain to fixed point domain to define the required HDL resources which are required to implement the code in the hardware domain. Figure 11-5 shows the development process using MATLAB approach.

Before developing the code for the system an Integration Definition for Functional Modeling (IDEF0) diagram and N Square Diagram ($N^2$) are used to describe the code and give an overall guidance to the architecture of the algorithm in order to move the software coding phase.

IDEF0 involves a preliminary or high-level design of the main sub-modules with an overall picture of the code. The diagram shows the interfaces between all the inputs and the outputs. $N^2$ diagram can be used to represent functional interfaces between the system functions. It can be used to systematically identify, define, tabulate, design, and analyze functional interfaces for the system.

In the coding phase, the actual code was programmed using MATLAB software but before starting the programming of the algorithm each main function of the system should be decomposed into sub functions. For main function, an Enhanced Functional Flow Block Diagram (EFFBD) should be developed to show the functions tasks identified through functional decomposition and display them in their logical, sequential relationship. EFFBDs played a key role to minimize the time required to develop the code and organize the responsibilities for the programming team.

Each box in the EFFBD diagram represents a sub-function or a task under each main function and it has inputs and outputs. To perform the main function all the sub-functions that included in the EFFBD shall be performed.

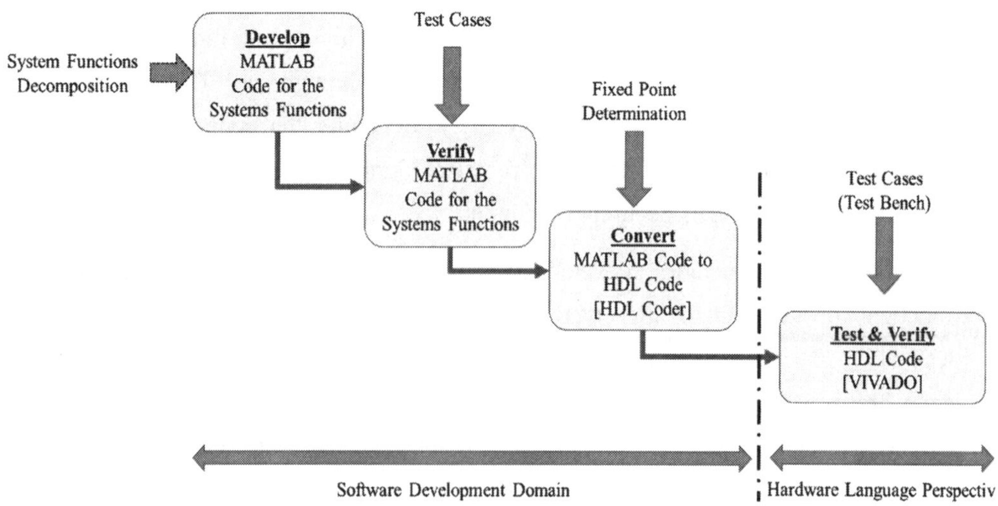


Figure 11-5: Development process using MATLAB approach

P A R T 2

From the programming point of view to perform each sub-function, there were many mathematical equations. These equations are programmed using MATLAB M file code. Using the EFFBDs are important during testing the code because it made the verification process easier to find the error and solve it.

## 11.3 Simulation, Synthesis and Implementation

After verifying the code in MATLAB domain and converting it to HDL code, verification of the HDL is performed using an Integrated Development Environment for FPGA Development. FPGA vendors have their own custom software for the FPGA development. Lebero IDE is one od software tool set to design Microsemi FPGA products. VIVADO is a software suite produced by Xilinx for synthesis, implementation, and analysis of HDL designs. In this work, the VIVADO is chosen to simulate, synthesis and implement FPGA based CPCS system. The processes are as follow;

- The converted VHDL code can be simulated in VIVADO software to verify the correct operation of the HDL code.
- Simulation is performed using the test-benches developed by Matlab and observing the system operation as waveforms.

An example is illustrated in Figure 11-6.

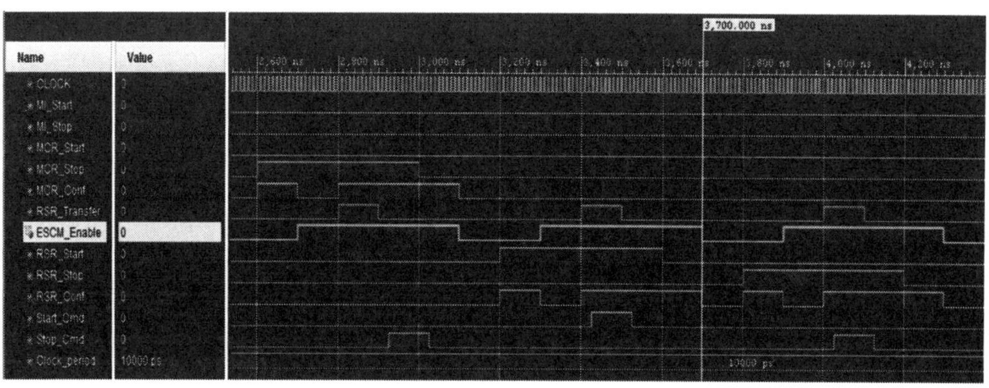


Figure 11-6: HDL Simulation

Once the design operation has been verified by simulation, synthesis is performed. Synthesis translates the HDL code into logical resources (AND, OR gates) that will perform the required function. In Figure 11-7, the sample VHDL code that performs priority logic for a Solenoid Operated Valve (SOV) in the CIM is shown. When Synthesis is performed, the code is translated into discrete components, as illustrated in Figure 11-7. In this figure, the VHDL code has been translated into AND gates, OR gates and Registers.

After synthesis, place and route is performed. In this phase, the post synthesis netlist is mapped onto actual resources on the FPGA board. Figure 11-9 illustrates the post-implementation netlist for the CIM priority logic defined in Figure 11-8. In this figure, the system is represented using the actual Registers and Look-up Tables on the FPGA board on which it shall be implemented. The implementation process generates a bitstream file that is used to program the FPGA board.

```
library IEEE;
use IEEE.STD_LOGIC_1164.ALL;

entity Priority_P01 is
    Port ( Clk              : in STD_LOGIC;
           Rst              : in STD_LOGIC;
           Selector         : in STD_LOGIC;
           Open_X           : in STD_LOGIC;
           Close_X          : in STD_LOGIC;
           Disable_RMC_X    : in STD_LOGIC;
           Open_Z           : in STD_LOGIC;
           Close_Z          : in STD_LOGIC;
           Open_Local       : in STD_LOGIC;
           Close_Local      : in STD_LOGIC;
           Open_Q           : out STD_LOGIC;
           Close_R          : out STD_LOGIC;
           X_Blocked        : out STD_LOGIC;
           Z_Blocked        : out STD_LOGIC);
end Priority_P01;

architecture Behavioral of Priority_P01 is

    --===================================================================================
    -- Internal Signals
    --===================================================================================
    Signal S_Selector        : STD_LOGIC;
    Signal S_Open_X          : STD_LOGIC;
    Signal S_Close_X         : STD_LOGIC;
    Signal S_Disable_RMC_X   : STD_LOGIC;
    Signal S_Open_Z          : STD_LOGIC;
    Signal S_Close_Z         : STD_LOGIC;
    Signal S_Open_Local      : STD_LOGIC;
    Signal S_Close_Local     : STD_LOGIC;
```

Figure 11-7: Sample VHDL code for SOV priority logic(Partially implemented)

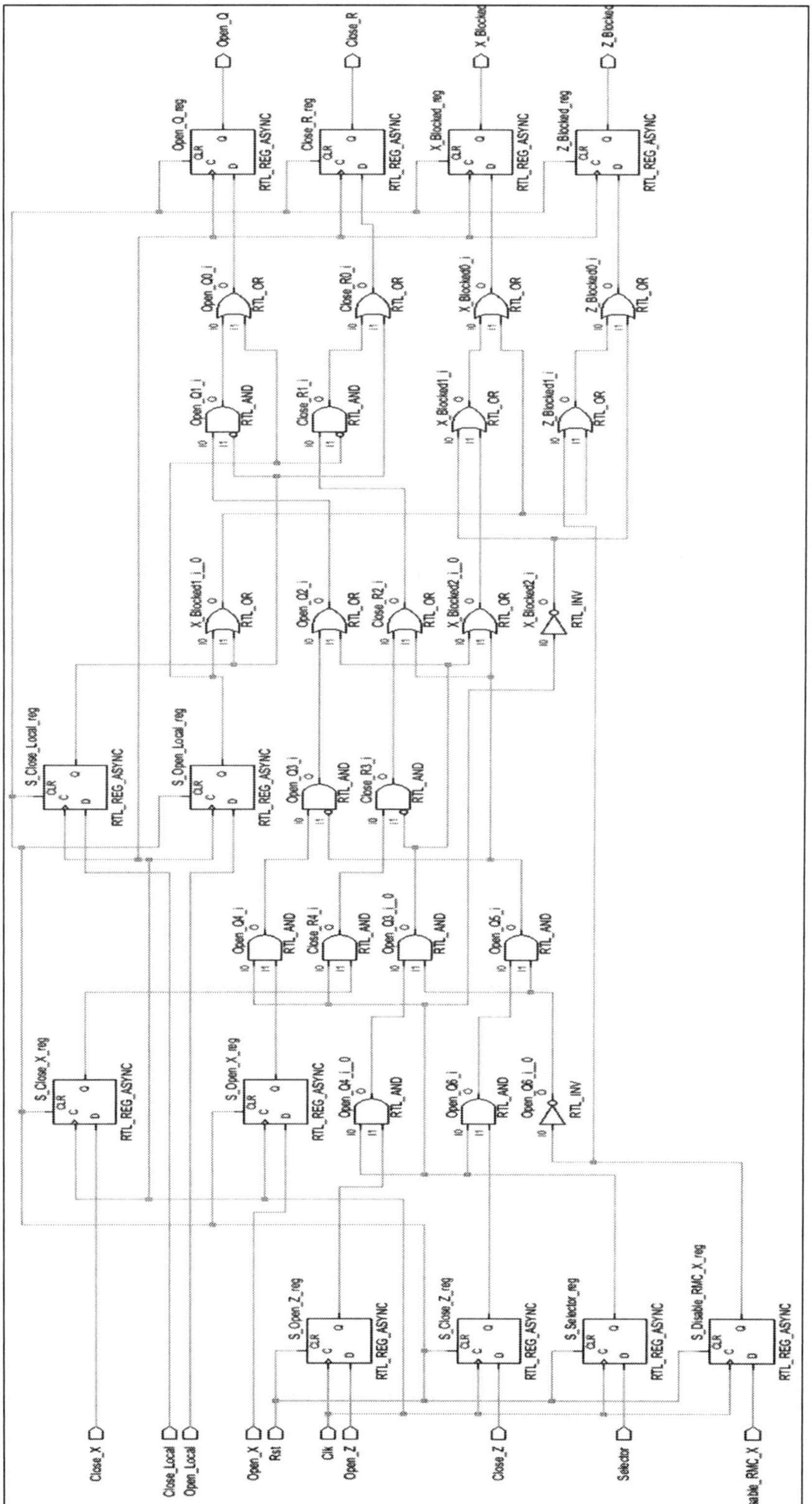


Figure 11-8: Post-synthesis netlist of the CIM priority logic

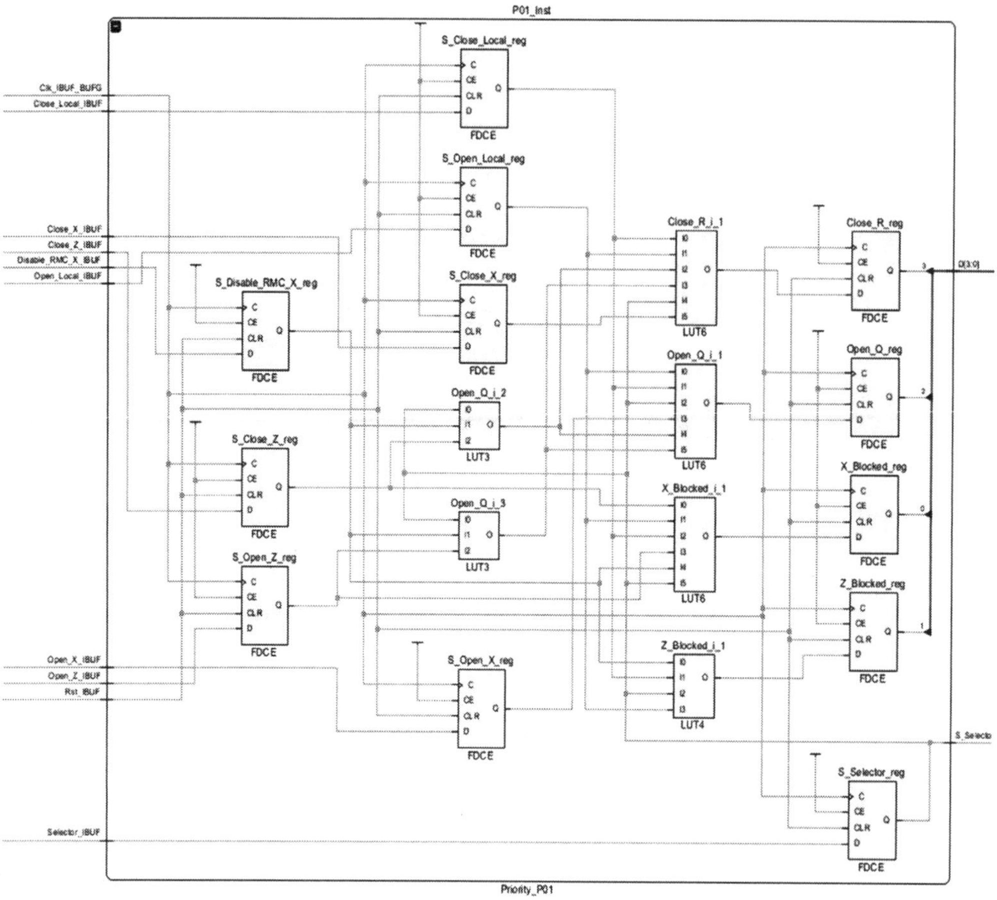


Figure 11-9: Post-Implementation Netlist of the CIM priority logic

## 11.4 Integration

The implementation of a system using FPGAs may be achieved using a single FPGA board or multiple boards. If multiple boards are utilized, integration of these boards into a single system or sub-system is performed, following the implementation phase. For the safety MMIS described in the preceding chapters, the system was implemented on multiple boards, which were subsequently integrated.

The LPD function of the CPCS is implemented on 4 Nexys Video boards (one for each channel) as illustrated in Figure 11-10. These components are implemented on the Nexys Video board. This FPGA board is different from that chosen for the implementation of the PPS and ESF-CCS due to the fact that the CPCS implements numerous mathematical computations and therefore requires extra resources, not available on the Basys 3 board. Due to the fact that the CPCS receives inputs from a multitude of sensors, for the current implementation, these signals were combined into two states, normal and trip. The Normal input gives the board all the process inputs at values that represent the normal operating condition. The Trip Condition gives a combination of process inputs that cause the generation of a trip signal.

As the CPCS receives inputs from a multitude of sensors, for the current implementation, these signals were combined into two states, normal and trip. The Normal input gives the board all the process inputs at values that represent the normal operating condition. The Trip Condition gives a combination of process inputs that cause the generation of a trip signal.

Each channel of the PPS was implemented on 5 hardware modules: 3 bistable processors and two local coincidence logic processors, as illustrated in Figure 11-11. The input to the BP modules is a process variable signal, which in the NPP is provided by a sensor. In this implementation, as the use of sensors would not be practical, switches were used to provide 16 bit and 8 bit process variable inputs. In addition, in order to display the output of the modules a 4 digit seven segment display and 16 Light Emitting Diodes (LED). The Seven Segment display is used to display the trip set-point while the LEDs are used toindicate the pre-trip and trip states. The LCL modules receive signals from the

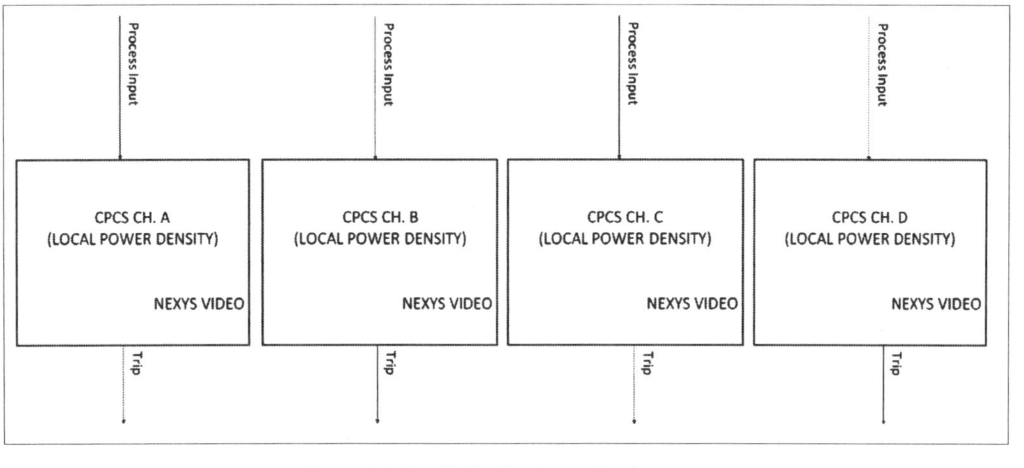


Figure 11-10: CPCS Hardware Configuration

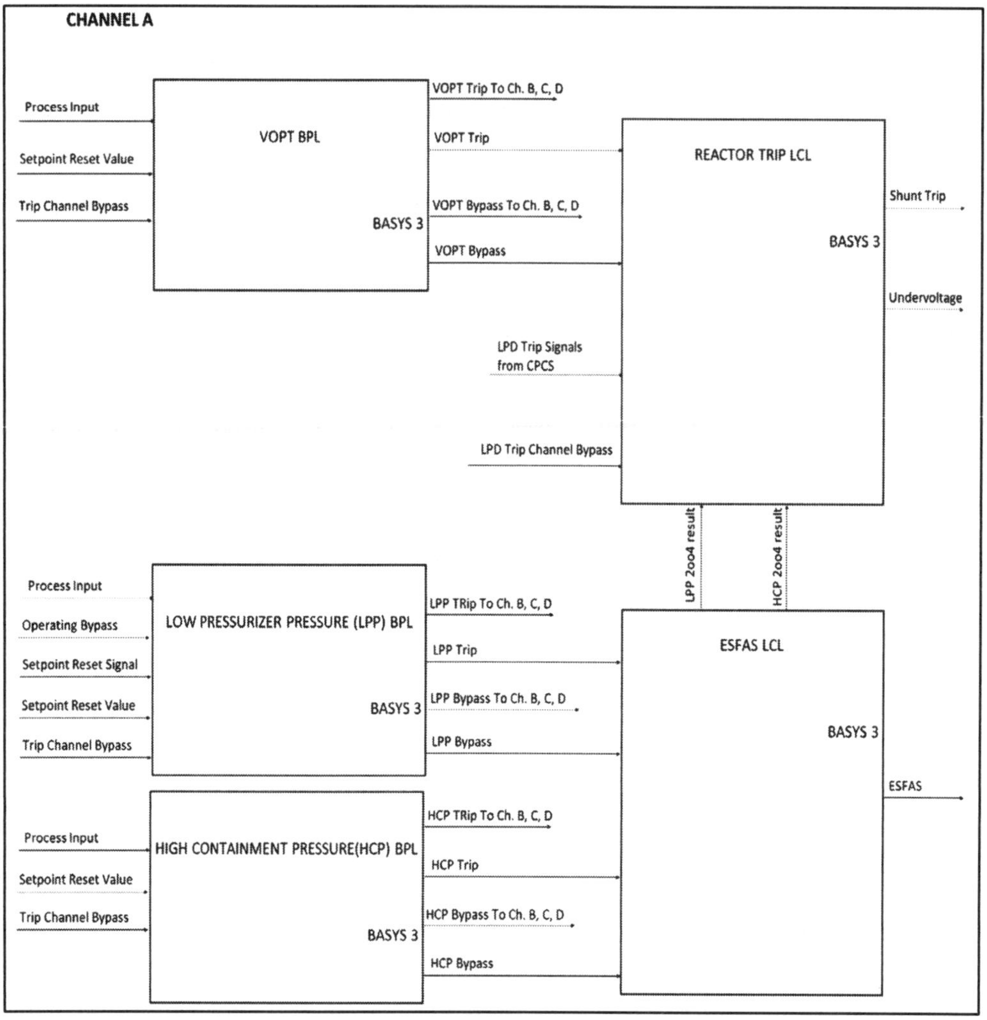


Figure 11-11: PPS Hardware Components

BPs and do not require the use of switches. For the display, it utilises the seven segment display to indicate the presence of trip signals.

Each channel of the ESF-CCS is composed of two components, the group controller and loop controller, each implemented on an independent FPGA board, as illustrated in Figure 11-12. The group controller is responsible for the validation and resolution of the system-level and component-level actuation signals. It therefore receives signals from the PPS LCL as well as the operator module and safety console.The loop controller is responsible for transmitting the actuation signals to the ESF component via the Component Interface Module, as well as generating feedback and alarm signals.

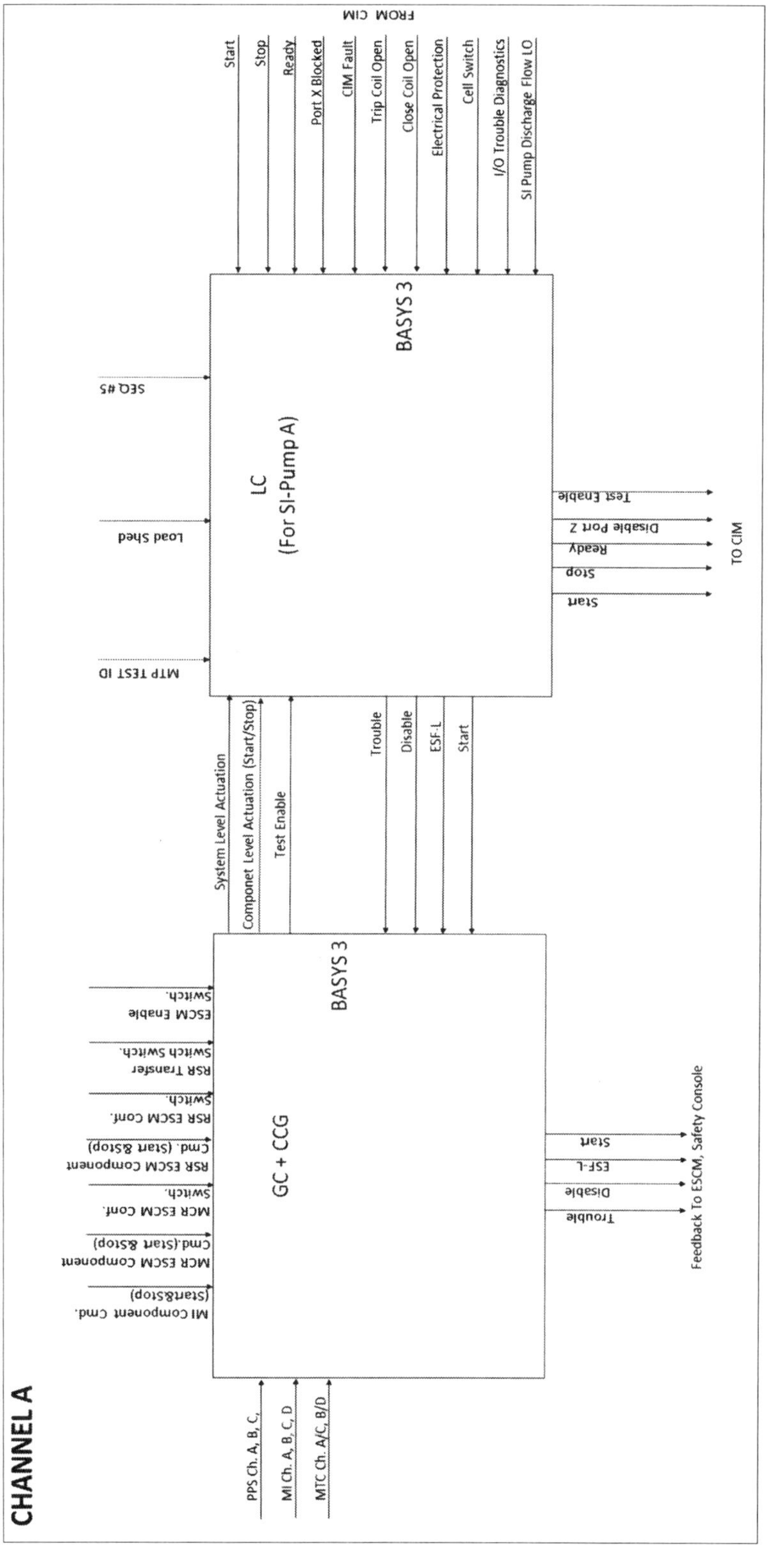


Figure 11-12: ESF-CCS Hardware Diagram

The three sub-systems (PPS, CPCS and ESF-CCS) were combined into a single system. Figure 11-13 illustrates all four channels of the system while Figure 11-14 illustrates a single channel of the integrated system.

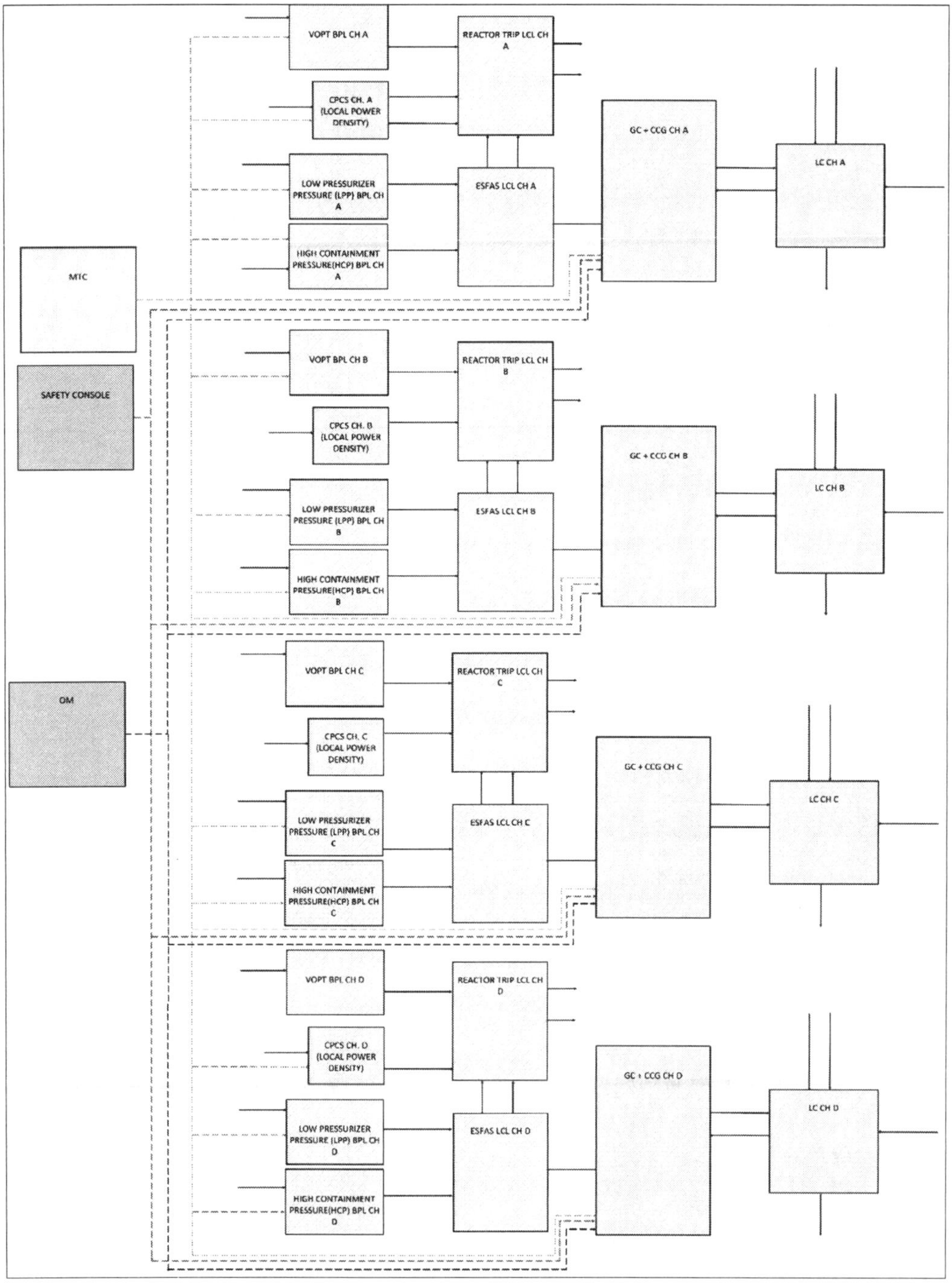


Figure 11-13: Four Channels of the Integrated System

Figure 11-14: Integrated FPGA-based MMIS

Figure 11-15 shows the safety MMIS system which is intended to check the feasibility of the system implementation using FPGA hardwares. So that satisfy this objective, only small portion of functions are chosen.

Followings indicate what kind of FPGA modules are used for implementing chosen MMIS functions;

- five (5) Basys-3 FPGA modules for PPS,
- one (1) Nexus-video module for CPCS,
- two Basys-3 modules for ESF-CCS,
- one (1) Basys-3 for OM,
- one (1) Basys-3 for Safety Console

## PROBLEMS

1. What is an FPGA? How are FPGAs different from PLCs?
2. Explain the FPGA development processes defining in this book.
3. In this chapter, MATLAB was given as an entry method for the HDL code development. What other methods can you think of?
4. What is Synthesis?
5. What is FPGA implementation?
6. What do you think are the challenges in FPGA-based design and implementation?

## FURTHER READING


- IAEA, "Application of Field Programmable Gate Arrays in Instrumentation and Control Systems of Nuclear Power Plants," Energy Series No. NP-T-3.17, 2016.
- R. Fink, C. Killian and T. Nguyen, "Recommended Approaches and Design Criteria for Application of Field Programmable Gate Arrays in Nuclear Power Plant Instrumentation and Control Systems," EPRI, 2011.
- B. Fink, C. Killian, T. Nguyen, A. Druilhe, and Frédéric Daumas, "Guidelines on the Use of Field Programmable Gate Arrays (FPGAs) in Nuclear Power Plant I&C Systems," EPRI, Technical Report 1019181, 2009.
- USNRC, NUREG/CR-7006, ORNL/TM-2009/20, "Review Guidelines for Field-Programmable Gate Arrays in Nuclear Power Plant Safety Systems," 2010.
- Pong P. Chu, "RTL Hardware Design Using VHDL," Wiley 2006

Chapter 12

# VERIFICATION AND VALIDATION

## SUMMARY

The FPGA has mixed characteristics of both hardware and software. As a result, the development V&V activities differ from purely software based systems. Currently, regulators endorse the use of IEEE 1012-2004 for the verification of software systems. However, IEEE-1012-2004 specifies the minimum tasks in supporting all system, software, and hardware lifecycle, therefore, it cannot be directly applied to V&V process for FPGA-based systems. Even though the top level V&V processes and underlying activities are generic and can be used for FPGAs, the low level tasks are software specific, and not directly applicable to FPGAs." [37]. Therefore, methods and processes other than those defined for computer-based systems must be used. IEC62566 gives a detailed outline for the V&V tasks to be performed at each step in the development process for FPGA based systems. In this chapter, some of the V&V activities defined by IEC-62566 are introduced and examples given. The chapter is organized as shown in Mind-map of the Figure 12-1.

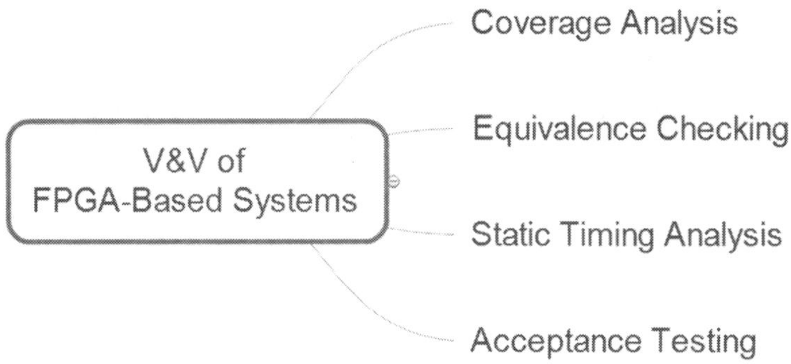


Figure 12-1: Mind Map of Chapter 12

**OBJECTIVES**

- Introduce the verification methods for FPGA-based systems
- Overview of Static Timing Analysis and Acceptance Testing method

## 12.1 Overview of FPGA Verification and Validation

Digital I&C systems can share code, data transmission, data, and process equipment to a greater degree than analog systems [14]. Although this sharing of information and resources leads to numerous advantages, it also increases the consequence of a single failure within the system. Therefore, greater care must be taken in the development of digital systems, to ensure their reliability. In order to ensure this, BTP 7-14, requires that reviewers of digital I&C systems:

- Confirm that acceptable plans were prepared to control software development activities,
- Obtain evidence that the plans were followed in an acceptable software development life cycle (SDLC), and
- Obtain evidence that the process produced acceptable design outputs.

To achieve those goals, the SDLC processes must be clearly defined: the method of carrying out the processes, relevant standards, outputs, and associated validation and verification (V&V). Current regulation endorses IEEE 1012 as the standard to be followed for the verification of safety critical systems. However, this is a software standard and as such "the low level tasks are software specific, and not directly applicable to FPGAs." [37]. Thus, for the verification of FPGA-based systems, multiple standards/ guidance must be followed:

- IEEE Std. 1012 : Standard for Software Verification and Validation
- IEC 62566 : Nuclear power plants – Instrumentation and control important to safety – Development of HDL-programmed integrated circuits for systems performing category A functions

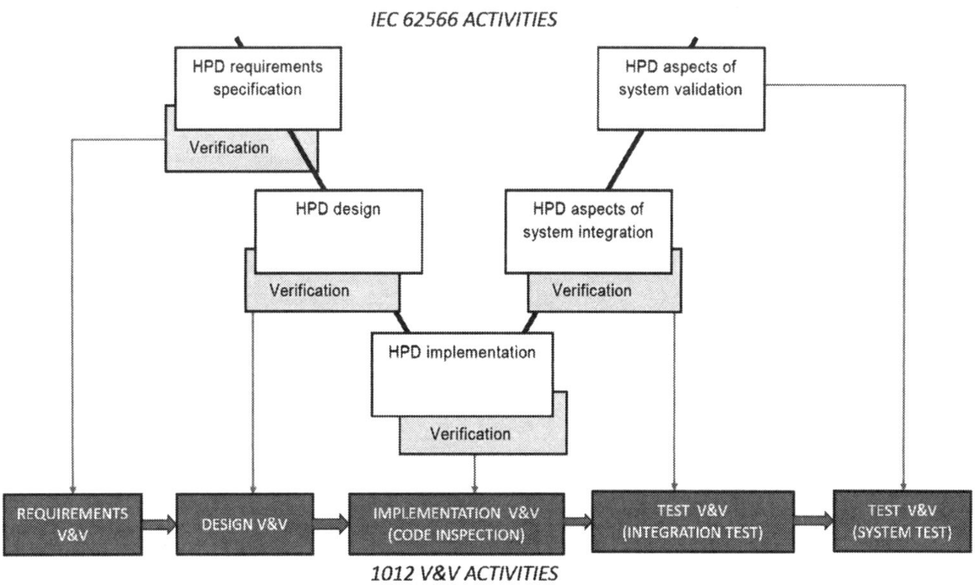


Figure 12-2: IEC 62566 and IEEE 1012 Verification activities

Figure 12-3 outlines the verification activities to be performed at each step of the FPGA development process.

Verification provides independent design reviews, and checks to ensure that the products produced at each phase of the development meet the requirements for that phase. It is performed by a team that is independent of the design team. This independence is aimed at ensuring that the verification will not be influenced by the design approaches and mind-set of the design team, and can therefore perform its task without any bias.

In this section, Coverage Test, Equivalence Checking, and Timing Analysis are discussed.

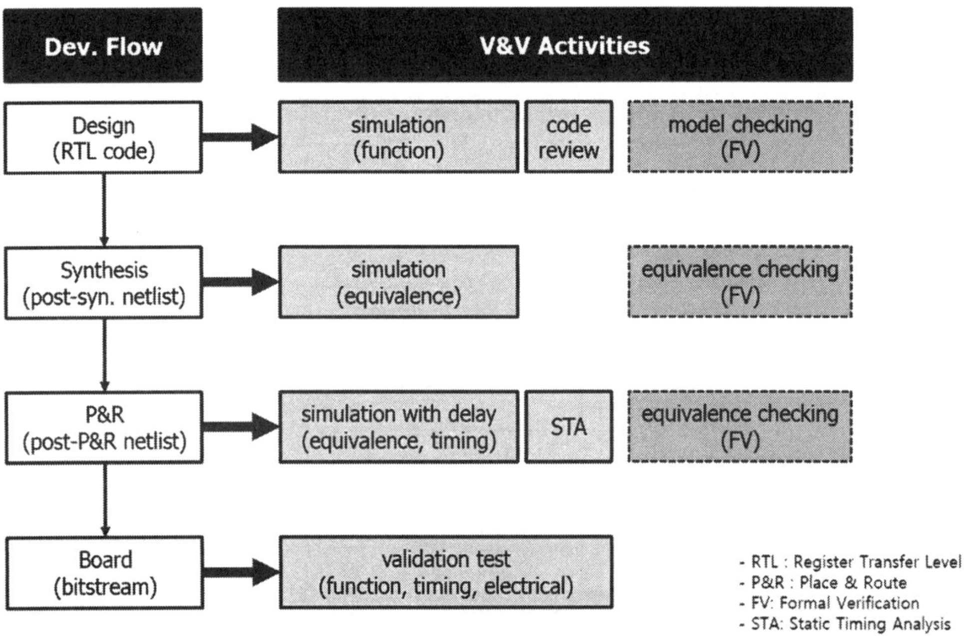


Figure 12-3: Verification Activities for FPGA based Systems

P A R T 2

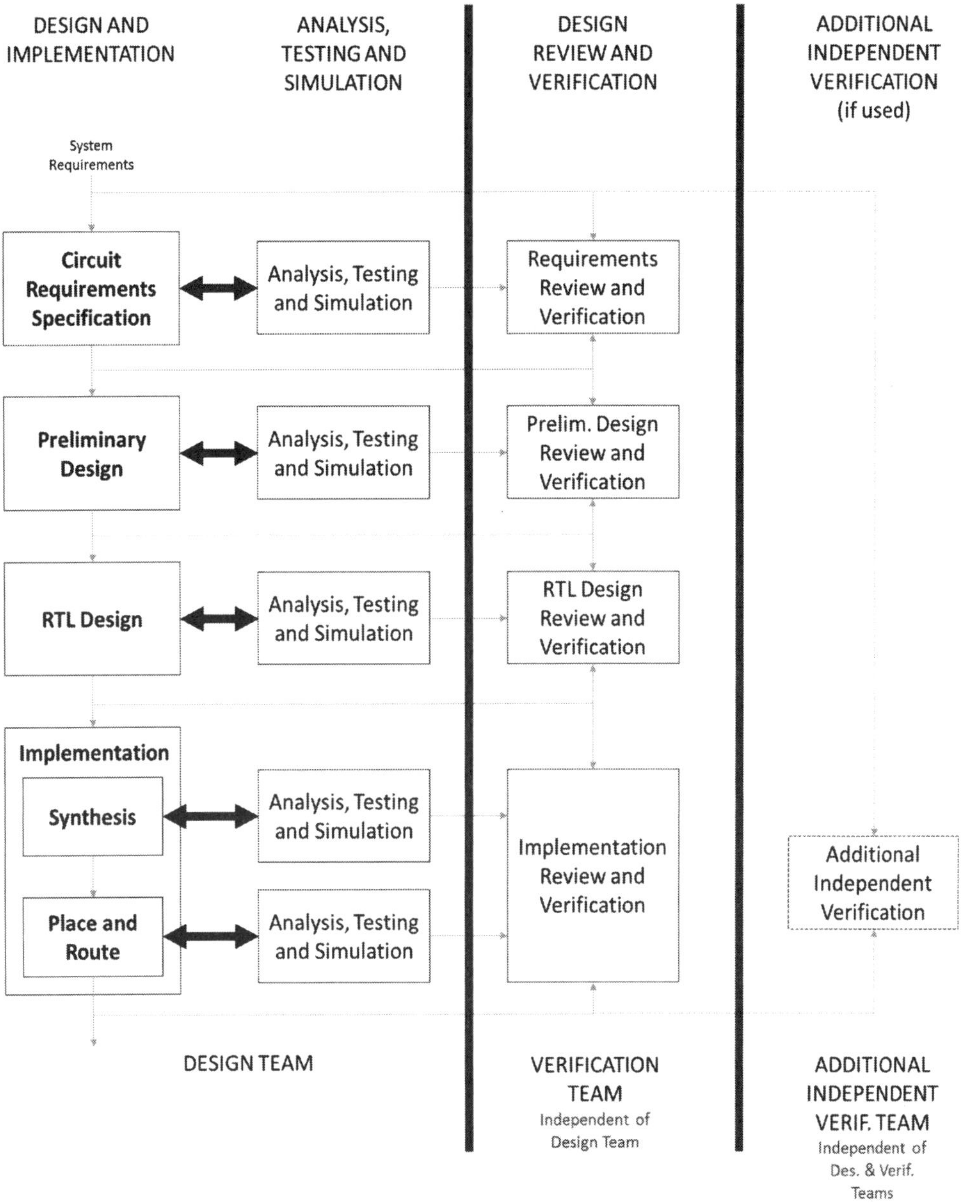


Figure 12-4: Independence of Design and Verification activities [11]

## 12.2 Coverage Analysis

Code coverage is a technique that allows engineers to collect the statistics on the execution of each line of HDL code, and evaluate the quality of their test. Code coverage can be roughly divided into statement coverage and branch coverage. Statement coverage provides information on which statements inside the VHDL or Verilog code were executed during simulation and how many times. Branch coverage examines the execution of conditional statements. It provides the data on which branches were executed during the simulation, how many times each branch was executed, and how many times the branch condition evaluated to true or false.

Code coverage and functional coverage are extensively used during validation to evaluate the effectiveness of testing. Ideally, designs must reach 100% statement, branch and functional coverage, however exceptions are made if it is known that a given coverage point is unreachable or not important.

### 12.2.1 Types of Coverage

In this section, we will briefly discuss seven types of coverage that are often used in simulation:

i. ***Statement Coverage:*** reports what fraction of HDL statements are executed. In most coding styles, each line corresponds with a statement. However, this is not strictly required. An example of statement coverage is illustrated in Section

ii. ***Branch Coverage:*** reports what fraction of branches created by decision statements such as if à elsif à else à end if and case blocks are executed (note that an if à end if block has an implied empty else branch). An example of branch coverage is shown in Section 3.6.2.

iii. ***Condition Coverage:*** reports what fraction of boolean sub-expressions in a decision statement (e.g., if) that have evaluated to both true and false values. For example, "if (var_a = '1' and var_b = '2') then" has two sub-expressions: "var_a = '1'" and "var_b = '2'".

iv. ***Path Coverage:*** reports what fraction of combinations of branches have been

PART 2

executed. For example, a block with two if statements, one after another has four paths since each if statement has two possible values, true and false.

v. Functional Coverage: reports what fraction of user-configured signal values, statements, properties, or cross products of these coverage points are executed. It is used extensively in industry. However, to be effective it requires careful implementation by a designer with deep domain knowledge into the module under test.

vi. ***Mutation Coverage:*** reports on the percentage of tests, which detect an error that is intentionally inserted into RTL [2]. For example, a mutation of the original RTL may include referencing the wrong variable or modifying operators. A test is said to have 'killed' the mutation if it successfully detects the change. A test that detects many mutations (i.e., one that has high mutation coverage) is likely more comprehensive and thorough than one with low mutation coverage. Unfortunately, mutation coverage relies heavily on the quality of the mutations generated and it is complicated to implement effectively.

vii. ***Tag Coverage:*** reports what fraction of signal assignments have an effect on output signals (i.e., how many signal assignments are observable at the output). For every signal assignment, a tag signal corresponding to that signal is also updated, allowing the propagation of signal assignments to be traced. A test with high tag coverage would cause many signal assignments to be detected at the module's output. This type of coverage, while similar to statement coverage, is much more difficult to attain because the effect of many statements will only be reflected in a module's output under very limited circumstances.

### Example of Statement and Branch Coverage

For example, in Table 12-1, if a = 0 and b = 0, statement coverage would be 4/6 = 67% because statements 1, 2, 3, and 6 were executed. The process for branch coverage is similar but instead of counting statements, it counts branches created by conditionals (such as if and case). The branch coverage for the code in Table 12-1: Example illustrates statement and branch coverage techniques Table 12-1 would only be 2/4 = 50% since branch 1 (the main program branch) and branch 2 were the only branches executed.

Table 12-1: Example illustrates statement and branch coverage techniques

```
setup ( ) ; // Statement 1 & Branch 1
 if ( a == 0) { // Statement 2 & Branch 2
do_something ( ) ; // Statement 3
 } else if ( b == 1) { // Statement 4 & Branch 3
do_something_else ( ) ; // Statement 5
} // There is an implied else branch here
 // else { } Branch 4

 teardown ( ) ; / / Statement 6
```

For PPS Low Pressurizer Pressure Trip, VHDL test cases are generated to meet the 100% code coverage. ALDEC software tool is used to perform the code coverage test. Figure 12-5: PPS Low Pressurizer Pressure Trip Coverage Test shows coverage testing result.

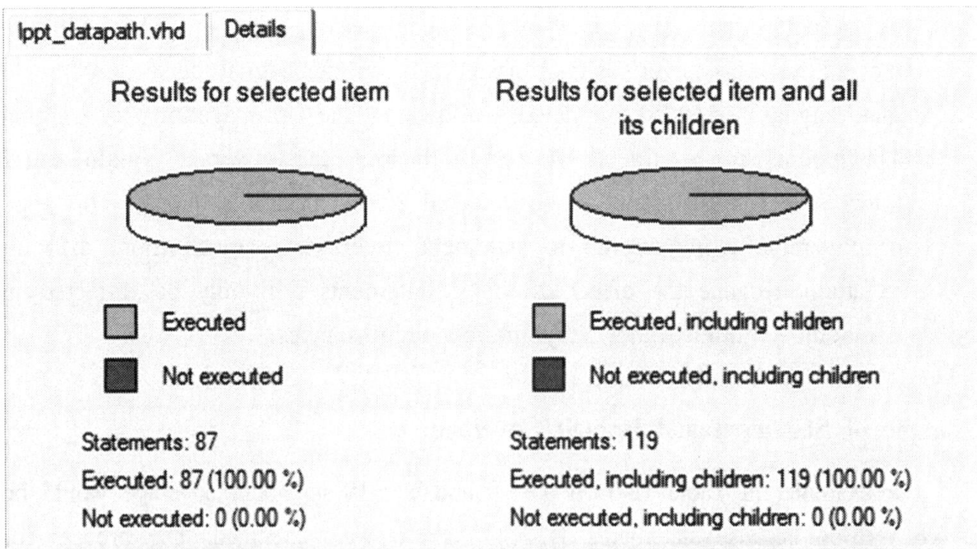


Figure 12-5: PPS Low Pressurizer Pressure Trip Coverage Test

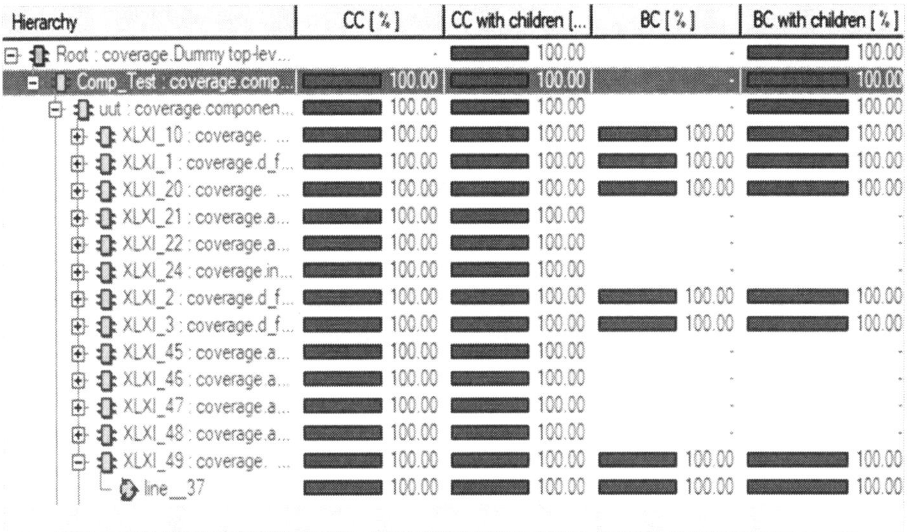

| Hierarchy | CC [ % ] | CC with children [... | BC [ % ] | BC with children [ % ] |
|---|---|---|---|---|
| Root : coverage.Dummy top-lev... | - | 100.00 | - | 100.00 |
| Comp_Test : coverage.comp... | 100.00 | 100.00 | - | 100.00 |
| uut : coverage.componen... | 100.00 | 100.00 | - | 100.00 |
| XLXI_10 : coverage. ... | 100.00 | 100.00 | 100.00 | 100.00 |
| XLXI_1 : coverage.d_f... | 100.00 | 100.00 | 100.00 | 100.00 |
| XLXI_20 : coverage. ... | 100.00 | 100.00 | 100.00 | 100.00 |
| XLXI_21 : coverage.a... | 100.00 | 100.00 | - | - |
| XLXI_22 : coverage.a... | 100.00 | 100.00 | - | - |
| XLXI_24 : coverage.in... | 100.00 | 100.00 | - | - |
| XLXI_2 : coverage.d_f... | 100.00 | 100.00 | 100.00 | 100.00 |
| XLXI_3 : coverage.d_f... | 100.00 | 100.00 | 100.00 | 100.00 |
| XLXI_45 : coverage.a... | 100.00 | 100.00 | - | - |
| XLXI_46 : coverage.a... | 100.00 | 100.00 | - | - |
| XLXI_47 : coverage.a... | 100.00 | 100.00 | - | - |
| XLXI_48 : coverage.a... | 100.00 | 100.00 | - | - |
| XLXI_49 : coverage. ... | 100.00 | 100.00 | 100.00 | 100.00 |
| line__37 | 100.00 | 100.00 | 100.00 | 100.00 |

Figure 12-6 ESF-CCS Code Coverage Analysis

For each of sub-systems, tests were carried out using the test cases previously developed. A test bench was developed to cover these cases. In order to achieve the goals, the test bench must be analyzed and the code coverage determined. The results of the coverage test are illustrated in Figure 12-6.

## 12.3 Equivalence Checking

The validation process of a design mainly consists of three tasks [5]:

- The designers have to give a formal specification, which describes the desired behavior of the circuit. The designers have to assure themselves of the completeness of their specification.
- After having designed a first implementation of the circuit, which is usually given at the register-transfer level, it has to be proven that the specified properties
  are satisfied by the implementation, i.e., that the design is logical correct.
- Starting from a register-transfer-level description, a design is synthesized to a gate level netlist. This netlist is then subject to layout synthesis. After all, there is almost no design where not some manual corrections need to be done.

Equivalence checking is the problem of checking whether two circuit descriptions specify the same behavior. Figure 12-7 shows the overall equivalence checking

Equivalence checking is solutions for systematic bugs. Synthesis and PnR are performed by software tools. These tools perform aggressive optimizations in order to ensure that the developed conforms not only to the RTL but meets the timing requirements. Therefore it is essential to ensure that the software while performing both the synthesis and PnR does not introduce any systematic errors.

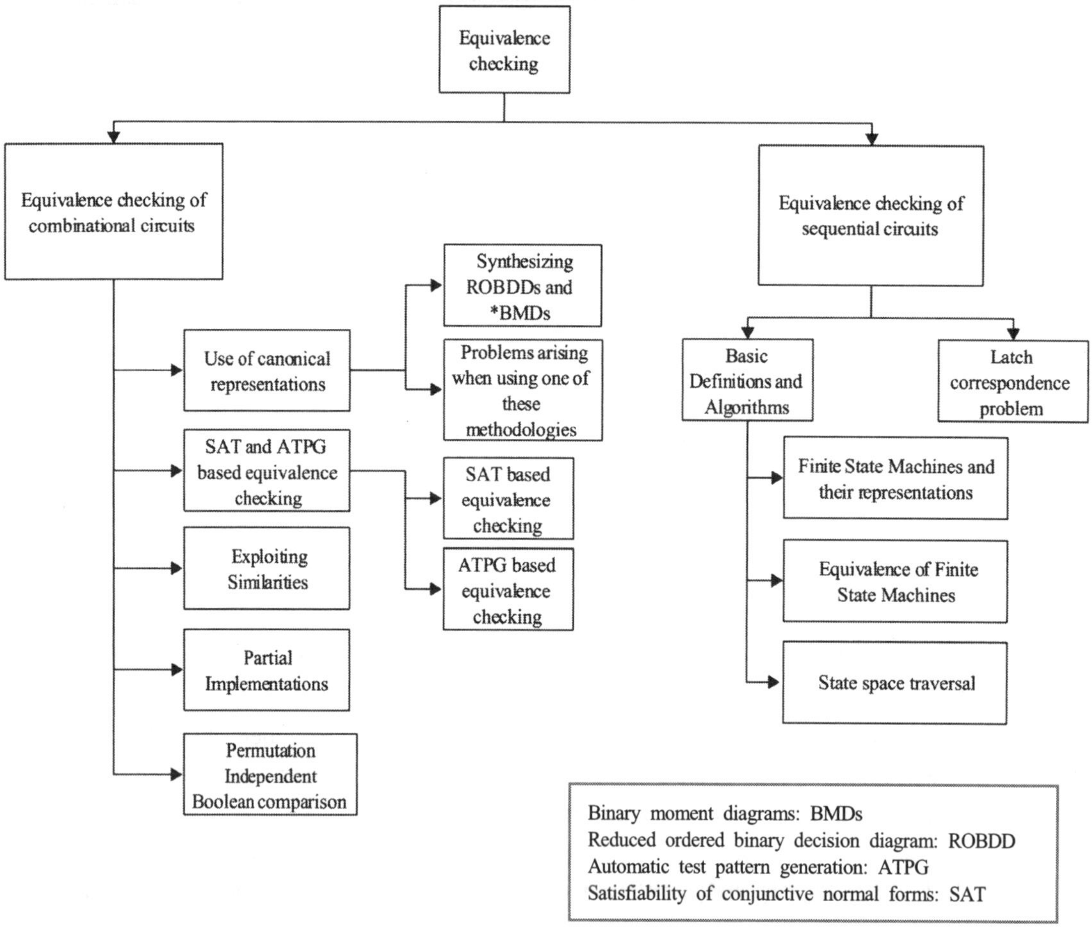


Figure 12-7 Equivalence Checking

The Systematic errors "can occur anywhere in the FPGA with little relationship to the design section under development. They often produce unexpected behavior or are triggered by unusual, corner-case scenarios, making the creation of verification tests complex and time consuming" [38]. To identify these errors, Formal Equivalence Checking (EC) is used. EC checks that the netlist developed is equivalent to the RTL design. As shown in Figure 12-8 [38].

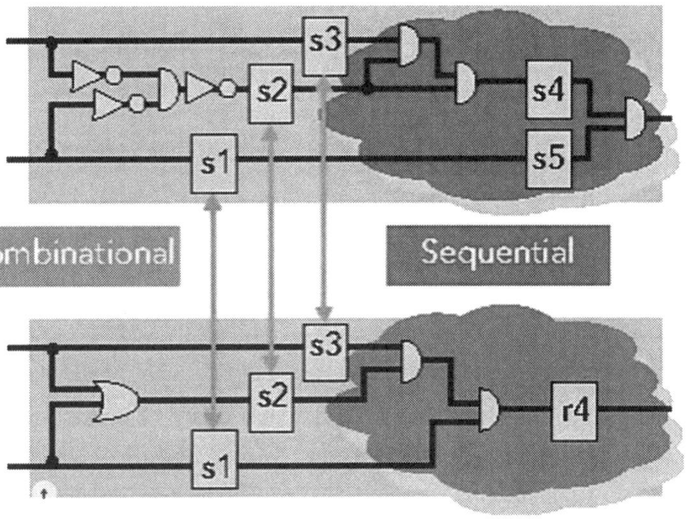


Figure 12-8 Equivalence checking by sequential optimizations

The use of EC in the FPGA flow has the following benefits:

- It eliminates the probability that any faults in the design are related to the development tool (synthesis and place and route). This eases the verification process as any errors/faults observed can be traced to the design (RTL) and solved.
- 'Confidence that no systematic, corner case bugs exist in the final design, ensuring consistency between the verified RTL code and gate-level final design" [38].
- Confidence to leverage the most aggressive optimizations available without concern for the introduction of errors, leading to the highest quality design.

## 12.4 Static Timing Analysis

Real time systems are characterized by reliability and predictability. Run time instrumentation is required for the dynamics of run time behavior .By performing Static Timing Analysis (STA) the time delay between signal input and output can be analyzed. STA can prove that the software module or task will meet the specification of time constraints. STA should prove that a safety critical command message will generate correctly and meet the given time constraints when it generated.

The input delay defines the external system-level path delay on a primary input port relative to a clock edge at the interface of the design [39]. The input delay value is specified in nanoseconds (ns), and can be positive or negative, depending on the clock and data relative phase at the interface of the device. The output delay specifies the output path delay of an output port relative to a clock edge at the interface of the design. Similarly, the output delay value can be positive or negative, depending on the clock and data relative phase outside the FPGA device. Similarly, the output delay value can be positive or negative, depending on the clock and data relative phase outside the FPGA device. Once the constraints are defined, static timing analysis can then be performed.

The Design Timing Summary reports timing slack values in order to determine if the design satisfies the timing constraints. Any negative slack values, or a number of failing endpoints, are an indication of a timing problem. Slack is therefore the difference between the clock period or data required time and the total path delay through the circuit. A positive slack, s, implies that the signal may be delayed by s without affecting the overall delay of the circuit. Conversely, negative slack implies that a path is too slow, and the path must be sped up (or the reference signal delayed) if the whole circuit is to work at the desired speed.

Figure 12-9 illustrates the timing summary for the loop controller module. In this summary all slack is seen to be positive, in addition, there are no failing endpoints, which indicates that the module meets the set timing requirements.

A detailed analysis of the module timing is also generated, as illustrated in Figure 12-10.

**Design Timing Summary**

| Setup | | Hold | | Pulse Width | |
|---|---|---|---|---|---|
| Worst Negative Slack (WNS): | 77.556 ns | Worst Hold Slack (WHS): | 0.117 ns | Worst Pulse Width Slack (WPWS): | 49.500 ns |
| Total Negative Slack (TNS): | 0.000 ns | Total Hold Slack (THS): | 0.000 ns | Total Pulse Width Negative Slack (TPWS): | 0.000 ns |
| Number of Failing Endpoints: | 0 | Number of Failing Endpoints: | 0 | Number of Failing Endpoints: | 0 |
| Total Number of Endpoints: | 376 | Total Number of Endpoints: | 376 | Total Number of Endpoints: | 176 |

**All user specified timing constraints are met.**

Figure 12-9 Timing Summary for the Loop Controller Component

```
------------------------------------------------------------------------------------
| Tool Version : Vivado v.2017.2 (win64) Build 1909853 Thu Jun 15 18:39:09 MDT 2017
| Date         : Mon Dec 11 09:45:08 2017
| Host         : IC_LAB-HP running 64-bit Service Pack 1  (build 7601)
| Command      : report_timing
| Design       : LC
| Device       : 7a35t-cpg236
| Speed File   : -1  PRODUCTION 1.17 2017-05-11
------------------------------------------------------------------------------------

Timing Report

Slack (MET) :              77.556ns  (required time - arrival time)
  Source:                  Display/q_reg_reg[16]/C
                             (rising edge-triggered cell FDRE clocked by sys_clk_pin  {rise@0.000ns fall@50.000ns period=100.000ns})
  Destination:             Disp[6]
                             (output port clocked by sys_clk_pin  {rise@0.000ns fall@50.000ns period=100.000ns})
  Path Group:              sys_clk_pin
  Path Type:               Max at Slow Process Corner
  Requirement:             100.000ns  (sys_clk_pin rise@100.000ns - sys_clk_pin rise@0.000ns)
  Data Path Delay:         7.269ns  (logic 4.173ns (57.413%)  route 3.096ns (42.587%))
  Logic Levels:            2  (LUT6=1 OBUF=1)
  Output Delay:            10.000ns
  Clock Path Skew:         -5.139ns (DCD - SCD + CPR)
    Destination Clock Delay (DCD):    0.000ns = ( 100.000 - 100.000 )
    Source Clock Delay      (SCD):    5.139ns
    Clock Pessimism Removal (CPR):    0.000ns
  Clock Uncertainty:       0.035ns  ((TSJ^2 + TIJ^2)^1/2 + DJ) / 2 + PE
    Total System Jitter     (TSJ):    0.071ns
    Total Input Jitter      (TIJ):    0.000ns
    Discrete Jitter          (DJ):    0.000ns
    Phase Error              (PE):    0.000ns

    Location             Delay type                Incr(ns)  Path(ns)    Netlist Resource(s)
  -------------------------------------------------------------------    -------------------
                         (clock sys_clk_pin rise edge)
                                                      0.000     0.000 r
    W5                                                0.000     0.000 r  CLOCK (IN)
                         net (fo=0)                   0.000     0.000    CLOCK
    W5                   IBUF (Prop_ibuf_I_O)         1.458     1.458 r  CLOCK_IBUF_inst/O
                         net (fo=1, routed)           1.967     3.425    CLOCK_IBUF
    BUFGCTRL_X0Y0        BUFG (Prop_bufg_I_O)         0.096     3.521 r  CLOCK_IBUF_BUFG_inst/O
                         net (fo=175, routed)         1.618     5.139    Display/CLOCK_IBUF_BUFG
    SLICE_X64Y25         FDRE                                           r  Display/q_reg_reg[16]/C
  -------------------------------------------------------------------    -------------------
    SLICE_X64Y25         FDRE (Prop_fdre_C_Q)         0.518     5.657 r  Display/q_reg_reg[16]/Q
                         net (fo=12, routed)          0.853     6.511    Display/sel[0]
    SLICE_X65Y26         LUT6 (Prop_lut6_I4_O)        0.124     6.635 r  Display/Disp_OBUF[6]_inst_i_1/O
                         net (fo=1, routed)           2.242     8.877    Disp_OBUF[6]
    U7                   OBUF (Prop_obuf_I_O)         3.531    12.408 r  Disp_OBUF[6]_inst/O
                         net (fo=0)                   0.000    12.408    Disp[6]
    U7                                                                  r  Disp[6] (OUT)
  -------------------------------------------------------------------    -------------------

                         (clock sys_clk_pin rise edge)
                                                    100.000   100.000 r
                         clock pessimism              0.000   100.000
                         clock uncertainty           -0.035    99.965
                         output delay               -10.000    89.965
  -------------------------------------------------------------------
                         required time                         89.965
                         arrival time                         -12.408
  -------------------------------------------------------------------
                         slack                                 77.556
```

Figure 12-10 Detailed Timing report for the LC Module

Timing

Intra-Clock Paths - sys_clk_pin - Setup

| Name | Slack | Levels | High Fanout | From | To | Total Delay | Logic Delay | Net Delay | Requirement | Source Clock | Destination Clock | Exception |
|---|---|---|---|---|---|---|---|---|---|---|---|---|
| Path 1 | 79.950 | 1 | 1 | Elec_Pro | S_Elec_Pro_reg/D | 1.443 | 0.405 | 1.037 | 100.0 | sys_clk_pin | sys_clk_pin | |
| Path 2 | 79.965 | 1 | 1 | Seq | S_Seq_reg/D | 1.463 | 0.409 | 1.054 | 100.0 | sys_clk_pin | sys_clk_pin | |
| Path 3 | 80.010 | 1 | 1 | T_D_Ack | S_T_D_Ack_reg/D | 1.380 | 0.417 | 0.962 | 100.0 | sys_clk_pin | sys_clk_pin | |
| Path 4 | 80.011 | 1 | 1 | ESFAS | S_ESFAS_reg/D | 1.416 | 0.426 | 0.990 | 100.0 | sys_clk_pin | sys_clk_pin | |
| Path 5 | 80.013 | 1 | 1 | Start_Cmd | S_Start_Cmd_reg/D | 1.463 | 0.421 | 1.042 | 100.0 | sys_clk_pin | sys_clk_pin | |
| Path 6 | 80.077 | 1 | 1 | Stop_Cmd | S_Stop_Cmd_reg/D | 1.402 | 0.413 | 0.989 | 100.0 | sys_clk_pin | sys_clk_pin | |
| Path 7 | 80.261 | 1 | 1 | CIM_Start | S_CIM_Start_reg/D | 1.107 | 0.415 | 0.693 | 100.0 | sys_clk_pin | sys_clk_pin | |
| Path 8 | 80.372 | 1 | 1 | Test_ID | S_Test_ID_reg/D | 1.127 | 0.404 | 0.724 | 100.0 | sys_clk_pin | sys_clk_pin | |
| Path 9 | 80.544 | 1 | 1 | I_O_Trbl | S_I_O_Trbl_reg/D | 0.861 | 0.409 | 0.452 | 100.0 | sys_clk_pin | sys_clk_pin | |
| Path 10 | 80.556 | 1 | 1 | CIM_Fault | S_CIM_Fault_reg/D | 0.843 | 0.424 | 0.418 | 100.0 | sys_clk_pin | sys_clk_pin | |

Figure 12-11 Delay Information

Figure 12-11 shows no warning information related the time delay. By this analysis, it can be reported that for the static analysis are expected to meet the criteria that recommended by IEC 62566 to develop static verification.

## 12.5 Acceptance Testing

The Acceptance test process involves determining whether the system meets it's functional requirements in an environment which is real, or as close to real as can reasonably be created. The major validation areas are:

- Functional operation,
- System level performance,
- External and Internal interfaces.

Test procedures include, validation of operational and maintenance procedures that will be contained in documentation which will be provided to the end user upon delivery of the system. Any User documentation required for a specific test must be developed prior to the start of testing and reviewed for sufficiency (with respect to the test) as part of test procedure development.

In addition, the following types of tests shall may be conducted on first kind of systems with all elements that are part of the system connected together:

- The input signals used for the system acceptance test are simulated.
- The system is exercised using both steady state and dynamic inputs

## PROBLEMS

1. What is coverage analysis? Define the different types of coverage analysis.
2. What are the coverage requirements of IEC 62566?
3. What are the coverage requirements of NUREG/CR 7006?
4. Why is equivalence checking of FPGAs necessary? At what phase in the development cycle is equivalence checking performed?
5. What is the difference between static testing and dynamic testing?
6. Why is timing analysis important for Safety Systems?
7. Define acceptance testing.

## FURTHER READING


- USNRC, "Review Guidelines for Field-Programmable Gate Arrays in Nuclear Power Plant Safety Systems," NUREG/CR-7006, ORNL/TM-2009/20, 2010.
- IEEE-1012, "IEEE Standard for System, Software, and Hardware Verification and Validation," 2016
- B. Fink, C. Killian, T. Nguyen, A. Druilhe, and Frédéric Daumas, "Guidelines on the Use of Field Programmable Gate Arrays (FPGAs) in Nuclear Power Plant I&C Systems," EPRI, Technical Report 1019181, 2009.
- K. J. Hayhurst, D. S. Veerhusen, J. J. Chilenski and L. K. Rierson, "A Practical Tutorial on Modified Condition/ Decision Coverage," NASA/TM-2001-210876, 2001.
- Xilinx, "Vivado Design Suite User Guide, Using Constraints," June, 2017.

# REFERENCES

[1] USNRC, "NRC Regulations Title 10, Code of Federal Regulations," [Online]. [Accessed 7 June 2019].

[2] USNRC, "Appendix A to Part 50—General Design Criteria for Nuclear Power Plants," [Online]. [Accessed 7 June 2019].

[3] USNRC, "NRC Regulatory Guides," [Online]. [Accessed 7 June 2019].

[4] IEEE-603, "Standard Criteria for Safety Systems for Nuclear Power Generating Stations," 2009.

[5] IEEE-7-4.3.2, "Standard Criteria for Digital Computers in Safety Systems of Nuclear Power Generating Stations," IEEE Power and Energy Society, 1993.

[6] IEC-61226, "Nuclear Power Plants - Instrumentation and Control Important to Safety - Classification of Instrumentation and Control Functions"

[7] IEC-60880, "Nuclear Power Plants – Instrumentation And Control Systems Important To Safety – Software Aspects For Computer-Based Systems Performing Category A Functions"

[8] IEC-60987, "Nuclear Power Plants – Instrumentation And Control Important To Safety – Hardware Design Requirements For Computer-Based Systems"

[9] J. Xiong and L. Li, "Transparent-Box: Efficient Software Testing Method Combining Structural and Functional Testing together," *International Journal of Computers,* vol. 7, no. 3, pp. 69-90, 2013.

[10] INCOSE, "Systems Engineering Handbook," 4th ed., Wiley, 2015.

[11] R. Fink, C. Killian and T. Nguyen, "Recommended Approaches and Design Criteria for Application of Field Programmable Gate Arrays in Nuclear Power Plant Instrumentation and Control Systems," 2011.

[12] IAEA, "Application of Field Programmable Gate Arrays in Instrumentation and Control Systems of Nuclear Power Plants," January 2016. [Online]. [Accessed 15 January 2017].

[13] Agarwal, "Buy Electronics & Electrical Projects in India," 2017. [Online]. [Accessed 10 01 2019].

[14] USNRC, BTP 7-19, "Guidance for Evaluation of Computer-based I&C Systems Review Responsibilities," [Online].

[15] RPC Radiy, "FPGA-Paved Road to Business Success," September 2017. [Online].

[16] I. Lab, "Performance Requirement Ananlysis of the Safety Injection Actuation System," I&C Lab, 2017.

[17] KHNP, APR1400 DCD Chapter 7.

[18] KEPCO and KHNP, "Safety I&C System," [Online]. [Accessed 5 June 2019].

[19] Y. Liu, M. Yang, F. Li, C. Zhang and M. Hashim, "A Survey on the Quantitative Evaluation Methods of Software Reliability of Digital I&C Systems at NPPs," in *21st International Conference on Nuclear Engineering,* 2013.

[20] T. L. Chu, M. Yue, G. Martinez-Guridi and J. Lehner, "Review of Quantitative Software Reliability Methods," 2010.

[21] ISO/IEC/IEEE 24765, "Systems and Software Engineering - Vocabulary," 2010.

[22] U.S Nuclear Regulatory Commission, "Design Specific Review Standard for mPower iPWR Design," May 2013. [Online]. [Accessed 25 March 2017].

[23] T. J. McCabe and A. H. Watson, "Structured Testing: A Testing Methodology Using the Cyclomatic Complexity Metric," 1996.

[24] G. Miller, "The Magical Number Seven, Plus or Minus Two: Some Limits on Our Capacity for Processing Information," *The Psychological Review,* March 1956.

[25] T. J. M. and C. W. Butler, "Design complexity measurement and testing," *Communications of the ACM,* vol. 32, no. 12, pp. 1415-1425, December 1989.

[26] G. Stark, R. Durst and C. Vowell, "Using Metrics in Management Decision Making," *IEEE Computer,* vol. 27, pp. 42 - 48, September 1994.

[27] S. R. Chidamber and C. Kemerer, "A Metrics Suite for Object Oriented Design," *IEEE Transactions on Software Engineering,* vol. 20, pp. 476- 493, June 1994.

[28] G. G. Preckshot, "Reviewing Real-Time Performance of Nuclear Reactor Safety Systems," Washington, DC, 1993.

[29] C. S. Smidts and M. Li, "Software Engineering Measures for Predicting Software Reliability in Safety Critical Digital Systems," Washington, DC, 2000.

[30] J. C. Jung, "Improved design architecture to minimize functional complexity of plant protection system for nuclear power plant," *Nuclear Engineering and Design,* vol. 309, pp. 97 - 103, 2016.

[31] NASA, "Space Systems Engineering: Functional Analysis Module," [Online]. [Accessed 24 April 2019].

[32] P. P. Chu, Embedded SoPC Design with Nios II Processor and Verilog Examples, O'Reilly.

[33] P. P. Chu, FPGA Prototyping by VHDL Examples, John Wiley & Sons, Inc, 2008.

[34] J. C. Jung and J. Mayaka, "Complexity reduction of the Engineered Safety Features Component Control System," *Nuclear Engineering and Design,* vol. 331, pp. 194-203, 2018.

[35] J. C, Jung, D. H. Jeon, "Design architecture of simplified core protection calculator system algorithm," *Nuclear Engineering and Design,* vol. 331, pp. 1-10, 2018.

[36] S. Han and S. J. Park, "A Benefit Assessment of Using In-Core Neutron Detector Signals in Core Protection Calculator System," in *In-core instrumentation and core assessment, Proceedings of a Specialists' Meeting.*

[37] USNRC, "Review Guidelines for Field-Programmable Gate Arrays in Nuclear Power Plant Safety Systems," 2010.

[38] Embeded Computing Design, "Strategies for verifying an FPGA design," November 2014. [Online]. [Accessed 10 June 2019].

[39] Xilinx, "Vivado Design Suite User Guide, Using Constraints," June, 2017.

[40] K. J. Hayhurst, D. S. Veerhusen, J. J. Chilenski and L. K. Rierson, "A Practical Tutorial on Modified Condition/ Decision Coverage," 2001.

# PART 3

# NON-SAFETY SYSTEMS

# Chapter 13 NUCLEAR STEAM SUPPLY SYSTEM (NSSS)

**SUMMARY**

This chapter aims at describing the characteristics of the Nuclear Steam Supply System (NSSS) of the APR 1400 Nuclear Power Plant. In this chapter, we do not cover the Balance of Plant (BOP) Control, and Turbine/generator control system. As shown in Figure 13-1, this chapter covers the analysis of NSSS control system, regulations, codes and standards, as well classification of the NSSS structure. Additionally, basic control functions, the characteristics of specific systems, and control system Common Cause Failures (CCFs) analysis are also discussed in this chapter.

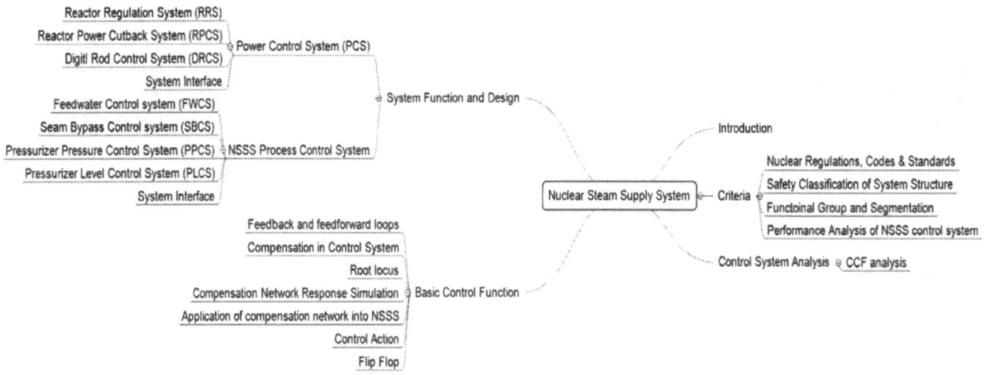


Figure 13-1: Mind Map for Chapter 13

**OBJECTIVES**

- Describe the Nuclear Steam Supply System (NSSS)
- Introduce the regulatory requirements related to the NSSS
- Describe the interface between the systems
- Define the CCF analysis of non-safety control sysstems

## 13.1 Introduction

There are several systems, such as the Power Control System (PCS), the Process-Component Control System (P-CCS), in the non-safety control system in the APR 1400. The PCS consists of the Reactor Regulating System (RRS), the Digital Rod Control System (DRCS), and the Reactor Power Cutback System (RPCS). In addition, the P-CCS encompasses the Nuclear Steam Supply System (NSSS) Process Control System (NPCS) and Balance of Plant (BOP) control systems. The NPCS includes the Feedwater Control System (FWCS), Steam Bypass Control System (SBCS), Pressurizer Pressure Control System (PPCS), Pressurizer Level Control System (PLCS), and other miscellaneous NSSS control functions. Figure 13-2 depicts the structure of non-safety control systems in the APR 1400 [1].

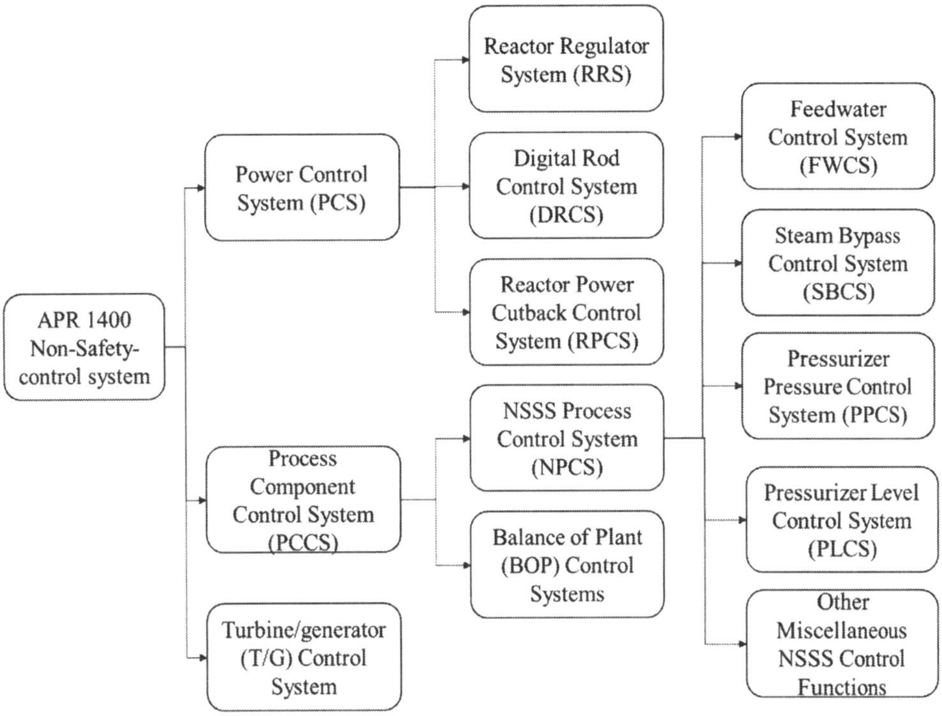


Figure 13-2: Structure of Non-Safety Control System

## 13.2 Nuclear Regulations, Codes & Standards

The purposes of providing design criteria, bases, or standards are to ensure adequate protection of public health and safety, to promote the common defense and security, to protect the environment, and to keep the radiation doses to plant personnel reasonably low.

Nuclear power plant design is governed by the law. If the design does not meet the law, plant construction and operation cannot be permitted. Each country has Atomic Laws and develops code and standards. In Korea, Ministry of Science and Technology (MOST) provides rules for nuclear power plants and review the application of plants. Also, code and standards for Korean industry has been made and named as KEPIC (Korea Electric Power Industry Code). In this document, U.S. rules and regulatory practice are introduced as guidance. U.S. laws and regulatory rules developed by Nuclear Regulatory Committee (NRC) are described hereunder. These documents are available in USNRC website.

Acceptance criteria for the design of the Control and Monitoring System are based on meeting the relevant requirements of the Standard Review Plan (SRP) Chapter 7.5, 7.6, and 7.7. The criteria apply to all Control and Monitoring equipment, including hardware, software, and firmware, from sensors to actuation devices that are required to initiate reactor shutdown and auxiliary supporting features and other auxiliary features.

### 13.2.1 Code of Federal Regulations (CFR)

Code of Federal Regulation law 10 is the law for energy, and Sections 50 and 52 are the laws for power plant design. The 10 CFR binds all person and organization who receive a license from the regulation agency. As top-tier regulation requirements for important to safety information systems, the 10 CFR 50.55a (h), "Protection and Safety Systems," requires compliance with IEEE Std 603-1991, "IEEE Standard Criteria for Safety Systems for Nuclear Power Generating Stations" as well as 10 CFR 50.55a(a)(1), "Quality Standards" and 10CFR 50.34(f), "Additional TMI-Related Requirements". Table 13-1 illustrates the applicable 10 CFR for Control and Monitoring system design and development.

Table 13-1: Applicable 10 CFR for Control and Monitoring system design and development

| Basis | Title | Related System |
|---|---|---|
| 10 CFR Part 50 | | |
| 50.55a(h) | Criteria for Protection Systems for Nuclear Power Generating Stations (IEEE Standard 603) | Information system- important to safety<br>Control system-non safety |
| 50.55a(a)1 | Quality Standards | Information system- important to safety<br>Control system-non safety |
| 50.34(f) | Additional TMI-Related Requirements | Information system- important to safety |
| 50.34(f)(2)(v) | Item I.D.3: Bypass and Inoperable Status Indication | Information system- important to safety |

#### 13.2.1.1 General Design Criteria

10CFR50 Appendix A provides General Design Criteria (GDC) that plant design shall meet. Table 13-2 shows the applicable GDC in control and monitoring system specifically.

Table 13-2: Applicable General Design Criteria

| Basis | Title | Related System |
|---|---|---|
| 10 CFR Part 50, App. A | | |
| GDC1 | Quality Standards and Records | Information system-important to safety |
| GDC 2 | Design Basis for Protection Against Natural Phenomena | Information system-important to safety |
| GDC 4 | Environmental and Missile Design Basis | Information system-important to safety |
| GDC 10 | Reactor Design | Control system |
| GDC 13 | Instrumentation and Control | Information system-important to safety<br>Control system-non safety |
| GDC 15 | Reactor Coolant System Design | Control system |
| GDC 19 | Control Room | Information system-important to safety<br>Control system-non safety |
| GDC 24 | Separation of Protection and Control Systems | Information system-important to safety<br>Control system-non safety |
| GDC 28 | Reactivity Limits | Control system |
| GDC 29 | Protection Against Anticipated Operational Occurrences | Control system |
| GDC 44 | Cooling Water | Control system |

#### 13.2.1.2 Regulatory Guide (Reg. Guide) and Industrial Standards

Reg. Guide provides NRC staffs position that they agrees the design meets the regulation rules. Safety-related post-accident monitoring instrumentation for Regulatory Guide 1.97 applicable Category 1, 2, and 3 variables is provided to monitor plant variables and systems during and following an accident, in accordance with the intent of TMI Action Item II.F.3.

Each of the control and monitoring system functions has been analyzed to determine the characteristics it must have to meet the required sections of regulatory guides and standards that apply to that system (Reg. Guide 1.97, NUREG-0696, NUREG-0737, IEEE Std. 497, etc.) as well as desired reliability and performance characteristics.

## 13.3 Safety Classification of System Structures

Non-Class 1E systems which interface with Class 1E safety systems are designed such that credible failures in the control and monitoring systems will not impact the operation of Class 1E safety systems. Interfaces between these systems employ isolation devices to maintain electrical independence between channels. Isolation devices are qualified with the Class 1E requirements.

In general, the NSSS control systems use dedicated transmitters and signal conditioning devices separated from those used by plant safety systems. Safety related devices provide parameters for control and/or monitoring, signal isolation is applied between the safety systems and the control and monitoring systems.

The QIAS and IPS receive data from both Class 1E and Non-Class 1E systems via qualified fiber-optic network interfaces to maintain isolation and perform signal validation.

Main Control Room and the Remote Shutdown Room are designed to support man-machine interface devices for each Class 1E safety channel as well as Non-Class 1E control channels. Low energy circuits are used and all panel mounted control and monitoring devices are isolated such that credible electrical faults originating in one channel will not affect other channels. Control panels are designed to maintain structural integrity, such that no control room missile hazard will occur as a result of a seismic event.

## 13.4 Functional Group and Segmentation

According to EPRI URD Chapter 10 Section 3.5.3, “Design Requirements for Availability and Reliability”, it is necessary to separate major controls logically to make redundancy and hence, the availability and reliability of control system will be increased as shown in Table 13-3.

Table 13-3: Control and monitoring functions requiring separation

| | |
|---|---|
| Reactivity Control | Automatic and manual control and monitoring of rod positions and manual control and monitoring of boron concentrations and neutron flux. |
| Steam Generator Inventory Control | Control and monitoring of main feedwater flow, main feed pumps and steam generator inventory. |
| Turbine Steam Demand Control | Control and monitoring of turbine throttle valves and bypass valves. |
| Pressurizer Pressure Control | Control and monitoring of the pressurizer heaters and spray and Reactor Coolant System pressure |
| Reactor Coolant System Inventory Control | Control and monitoring of makeup and letdown, Reactor Coolant System inventory and CVCS inventory. |

## 13.5 Performance Analysis of NSSS Control System

The conceptual design of NSSS control systems contains system design, workflow of performance analysis, Performance Related Design Bases Events (PRDBE), and Safety Related Design Bases Event (SRDBE). The system design of NSSS is shown in Figure 13-3. The work flow of performance analysis is performed in Figure 13-4.

The Design Bases Event (DBE) describes an occurrence that shall be directly considered in the design of the plant. It also shows the anticipated consequences of such events that must fall within acceptable limits. The DBE is separated into two categories: PRDBE and SRDBE [1].

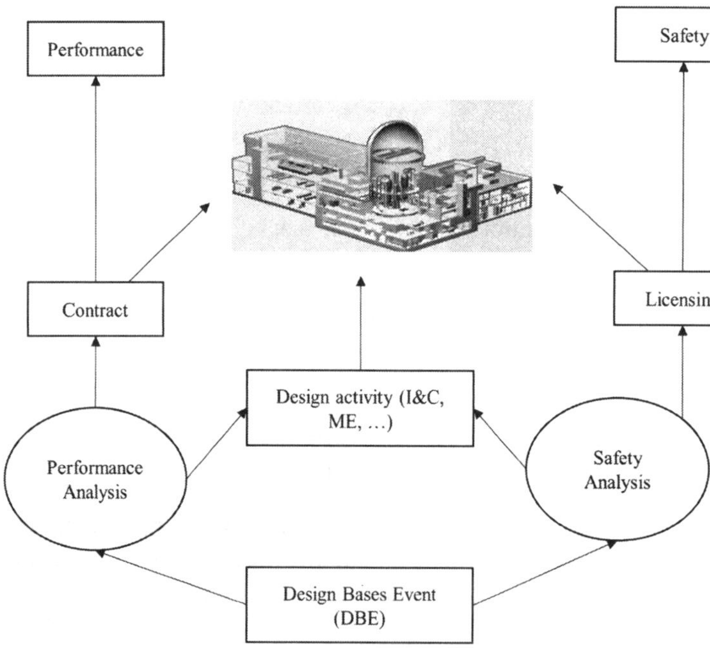


Figure 13-3: System design of NSSS

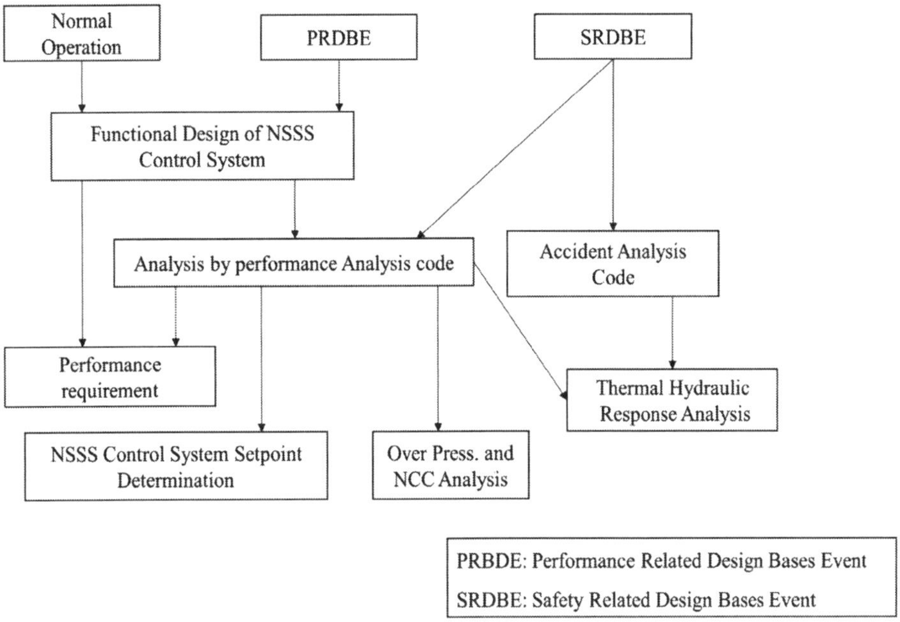


Figure 13-4: Work flow of performance analysis

In terms of the SRDBE, postulated occurrences that shall be accommodated by the plant design. There are two major event frequency categories: Anticipated operational occurrence (AOO), and Postulated Accident (PA). The AOO is expected to occur in the plant lifetime; however, the Events of Moderate Frequency can occur in a year, and the Infrequent Incidents may occur once in a plant lifetime. Nonetheless, the PA is not expected to occur in the plant lifetime, but the plant is designed to accommodate the event within the established acceptance criteria (Limiting Faults). The Plant Event Categorization is shown in Table 13-4. AOOs and PAs for the APR1400 fall into one of the following event categories [2]:

- Increase in heat removal by the secondary system
- Decrease in heat removal by the secondary system
- Decrease in reactor coolant system (RCS) flow rate
- Reactivity and power distribution anomaly

Table 13-4: Plant Event Categorization

| R.G 1.70 | 10 CFR | Safety Analysis | Design/ASME |
|---|---|---|---|
| Normal | Normal | Normal | level A (Normal)<br>> 60 over 60 year plant life |
| Incidents of moderate frequency:<br>which may occur during a calendar year | AOO<br>Fuel failure is not allowed. | AOO<br>Fuel failure is not allowed | Level B<br>(Upset)<br>> 1, < 60 |
| Infrequent Incidents:<br>which may occur once during the lifetime of plant | AOO<br>Fuel failure is not allowed. | AOO<br>Fuel failure is allowed.<br>Radiological A.C: small fraction of 10 CFR 100 | Level C<br>(Emergency)<br>> 0.06, < 1 |
| Limiting Faults:<br>which are not expected to occur, but are postulated | Accident | Accident:<br>Fuel failure is allowed.<br>Radiological A.C : within 10 CFR 100 | Level D<br>(Faulted)<br>< 0.06 |

- Increase in RCS inventory
- Decrease in RCS inventory
- Radioactive release from a subsystem or component

In terms of the PRDBEs, normal transient occurrences that shall be accommodated by the design of the plant. Consequences would be no more than normal service effects i.e., no equipment replacement, inspection or test. The general event category, including 6 modes: Power Operation, Startup, Hot Standby, Hot Shutdown, Cold Shutdown, Refueling.

The major PRDBEs related to the control system when the reactor power greater than 5%, or lower than 100% are expressed as follow:

- Steady state operation
- Turbine power step change of + 10% power
- Turbine power step change of - 10% power
- Turbine power ramp changes up to 5 percent/minute
- Turbine power step change of + 1% power
- Turbine power step change of - 1% power

- Turbine power ramp changes of + 1 percent/minute
- Turbine load rejection of up to 50% of rated power
- Turbine generator runback to house load
- Daily load cycle of 100-50-100% power
- Loss of a main feedwater pump
- Loss of a condensate pump when all are operating
- Turbine trip following a reactor trip
- Feedwater flow transfer from the downcomer to economizer

Additionally, the acceptance criteria for the PRDBE are expressed as follow:

- Without activating the Reactor Protection System
- Without activating the Diverse Protection System
- Without activating the Engineered Safety Features Actuation System (ESFAS)
- Without violating Technical Specification Limiting Condition for Operation (LCO)
- Without lifting the Pressurizer and/or Main Steam Safety Valves
- Without violating any Plant Operating Limit
- Without exceeding Normal Service Effects on any NSSS component or system. However, additional actions: component replacement, inspection, or test are not required after the PRDBE.

Next, the design of NSSS control algorithms and set-point should determinate so as to analyze the activities for the PRDBE.

- Conceptual Design of Pressurizer Pressure Control System (PPCS), Pressurizer Level Control System (PLCS), Reactor Regulating System (RRS), Feed Water Control System (FWCS), Steam Bypass Control System (SBCS), Reactor Power Cutback System (RPCS) should be analyzed.
- General requirements should be performed.
- The design, analysis and evaluation of the control algorithm iterate in order to meet the performance requirements.
- With the help of best-estimate analysis methodology, analyses of sensitivity to generate the minimized set-point of the control system which are performed.
- After the validation, the set-points are adjusted during the Power Ascension Test (PAT) if necessary.

In terms of Safety Analysis, the objective of Safety Analysis is to show that Design Basis Events are accommodated within the plant nuclear safety criteria. In order to achieve this objective, the Safety Analysis shall [2]:

- Confirms the Design of Structure, System, Components (SSC)
- Confirms the operator action and the action time
- Confirms the safety function within the design limit
- Confirms Safety Margin
- Provide the bases of the Limiting Condition of Operation of technical specification
- Provide the Thermal/Hydraulic data of the components

## 13.6 Basic Control Functions

### 13.6.1 Feedback and Feed Forward Loops

#### (1) Feedback loop

In a control system, there are two sorts of feedback loops: positive feedback and negative feedback [3]. As shown in Figure 13-5, the positive feedback has the reference input, $R(s)$, and the feedback output. The transfer function of the positive feedback is expressed:

$$T = \frac{G}{1 - GH} \qquad \textbf{Eq. 13-1}$$

Where T is the transfer function or overall gain of positive feedback control system, G is the open loop gain, and H is the gain of feedback path.

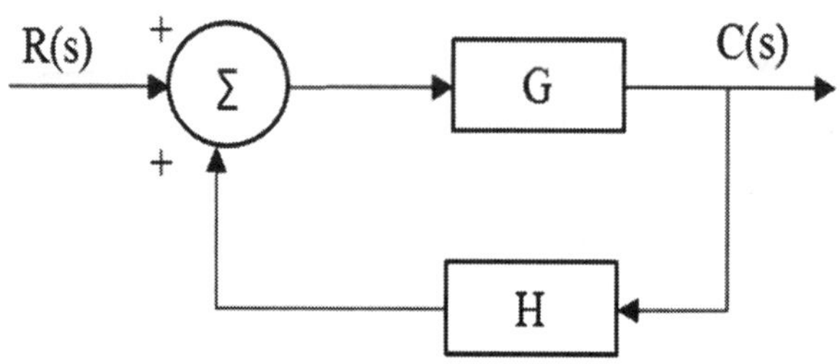


Figure 13-5: Positive feedback control system

The negative feedback loop is shown in Figure 13-6. This loop mitigates the error between the output of the system and the input, $R(s)$. The transfer function of the negative feedback is expressed:

$$T = \frac{G}{1+GH}$$ **Eq. 13-2**

Where T is the transfer function or overall gain of negative feedback control system, G is the open loop gain, and H is the gain of feedback path.

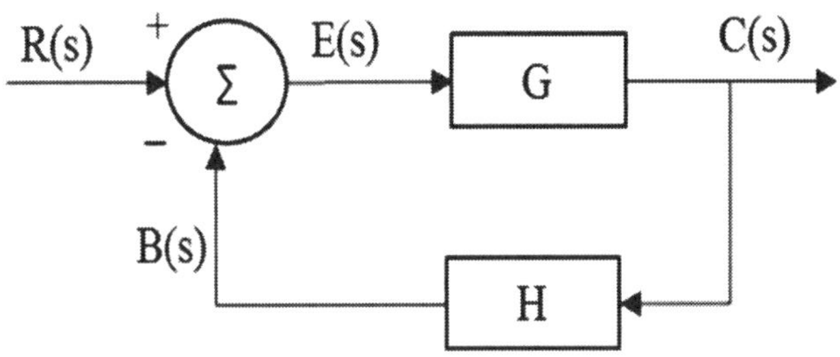


Figure 13-6: Negative feedback control system

### (2) Open-loop transfer function and feed-forward transfer function

As shown in Figure 13-6, the open-loop of the control system is considered as the ratio of the feedback signal B(s) to the actuating error E(s) [3].

The open-loop transfer function is expressed:

$$\frac{B(s)}{E(s)} = GH$$ **Eq. 13-3**

The feed-forward transfer function, which is considered as the ratio of the output C(s) to the error signal E(s) is shown as follow:

$$\frac{C(s)}{E(s)} = G$$ **Eq. 13-4**

If the feedback transfer function H is taken into account as unity, the open-loop transfer function and the feed forward transfer function will be the same.

### (3) Effects of feedback on overall gain

Based on Eq. 13-2, it is straightforward to recognize that the overall gain of the negative feedback control system is the ratio between G and (1+GH). Therefore, depending on the value of (1+GH), this overall gain can increase or decrease. If the value of (1+GH) is less than 1, the overall gain will rise. In this circumstance, the value of GH is negative as the gain of feedback path is negative. On the other hand, if (1+GH) value is greater than 1, the overall gain will decrease. In this circumstance, the value of GH is positive since the gain of the feedback path is positive. Because G and H are considered as frequency functions, the feedback will increase the overall gain of the system in one frequency range and diminish in the other frequency range [4].

### (4) Effect of feedback on sensitivity

The definition of the sensitivity of the overall gain of the negative feedback closed loop system (T) to the variation in open loop gain (G) is expressed:

$$S_G^T = \frac{\frac{\partial T}{T}}{\frac{\partial G}{G}} = \frac{Percentage\ change\ in\ T}{Percentage\ change\ in\ G} \qquad \textbf{Eq. 13-5}$$

Where  is the incremental change in T because of the incremental change in G. The Eq. 13-5ㅓ is rewritten as:

$$S_G^T = \frac{\partial T}{\partial G}\frac{G}{T} \qquad \textbf{Eq. 13-6}$$

Based on Eq. 13-2, the implementation of partial differentiation of both sides is expressed:

$$\frac{\partial T}{\partial G} = \frac{\partial}{\partial G}\left(\frac{G}{1+GH}\right) = \frac{(1+GH)-G(H)}{(1+GH)^2} = \frac{1}{(1+GH)^2} \qquad \textbf{Eq. 13-7}$$

The Eq. 13-2 can be rewritten as follow:

$$\frac{G}{T} = 1+GH \qquad \textbf{Eq. 13-8}$$

Substituting Eq. 13-7, and Eq. 13-8 into Eq. 13-6,

$$S_G^T = \frac{1}{(1+GH)^2}(1+GH) = \frac{1}{1+GH} \qquad \textbf{Eq. 13-9}$$

Therefore, the sensitivity of the overall gain of the closed loop control system is considered as the ratio of 1/(1+GH). Based on the value of (1+GH), the sensitivity will increase or decrease. If the value of (1+GH) is less than 1, the sensitivity will increase. In this scenario, the value of GH is negative because the gain of the feedback path is negative. However, if the value of (1+GH) is greater than 1, the sensitivity will reduce. In this scenario, the value of GH is positive because the gain of the feedback path is positive. Since G and H are considered as frequency functions, the feedback will increase the overall gain of the system in one frequency range and diminish in the other frequency range. Consequently, we are able to select the values of GH in such a way that the system is insensitive or less sensitive to parameter variations [4].

### (5) Effect of feedback on stability

It is important to properly select the feedback so as to make the control system stable. It is due to the fact that a system is considered to be stable if its output is under control. Otherwise, this system is considered to be unstable. In Eq. 13-2, if the value of denominator is zero, the output of the control system will be infinite. Therefore, the control system becomes unstable [4].

### (6) Effect of feedback on noise

The closed loop control system with noise signal is shown in Figure 13-7. Its transfer function is considered as Eq. 13-10 when making the input R(s) becomes zero.

$$\frac{C(s)}{N(s)} = \frac{G_b}{1+G_aG_bH} \qquad \textbf{Eq. 13-10}$$

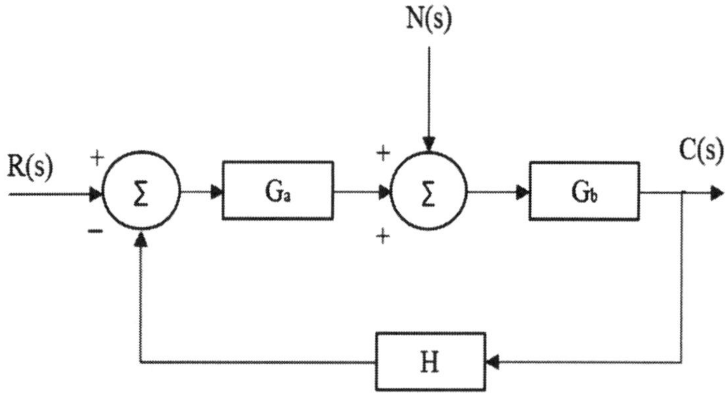


Figure 13-7: Closed loop control system with noise signal

The open loop control system with noise signal is displayed in Figure 13-8. Its transfer function is considered as Eq. 13-11 when making the input R(s) becomes zero.

$$\frac{C(s)}{N(s)} = G_b \qquad \textbf{Eq. 13-11}$$

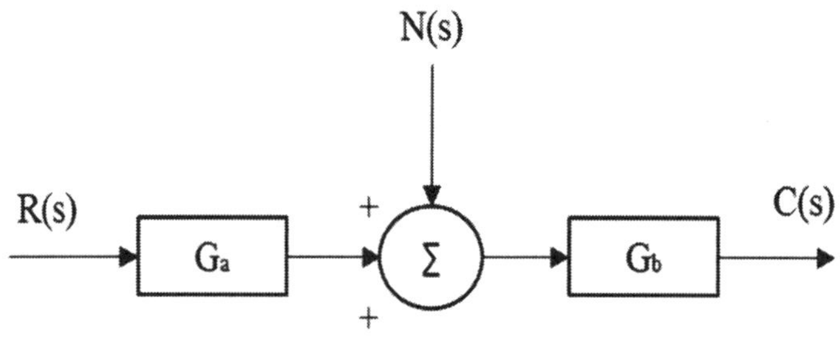


Figure 13-8: Open loop control system with noise signal

From Eq. 13-10 and Eq. 13-11, the gain due to noise signal is decreased in the closed loop control system, as the factor of $(1+G_aG_bH)$ is greater than 1 [4].

### 13.6.2 Compensation in Control system

There are many benefits as to why compensation network is employed in the control system. The compensating networks are used in order to obtain the desired performance of the system. Compensating networks are applied to the system in the form of feed forward path gain adjustment. The compensation compensates an unstable system to make it stable. A compensating network is also utilized to optimum overshoot. Moreover, these compensating networks increases the steady state accuracy of the system. It is important to note that the increase in the steady state accuracy brings instability to the system. The compensating networks also illustrates poles and zeros in the system thereby causes changes in the transfer function of the system. This leads to change performance specifications of the system. There are several methods of compensation, such as Series Compensator, Feedback Compensator, and Load Compensator which are shown in Figure 13-9, Figure 13-10, and Figure 13-11 [5].

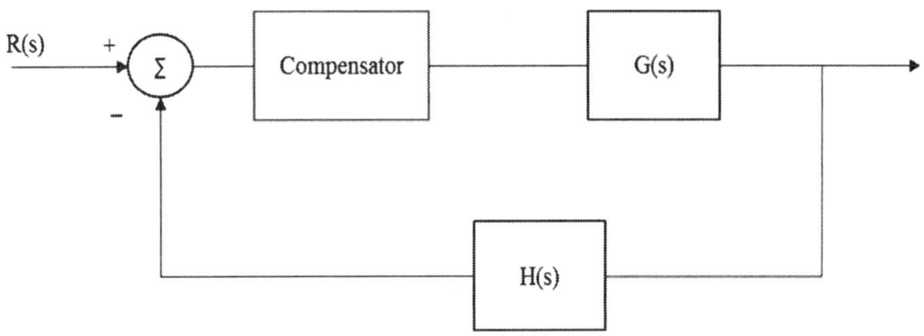


Figure 13-9: Series Compensator

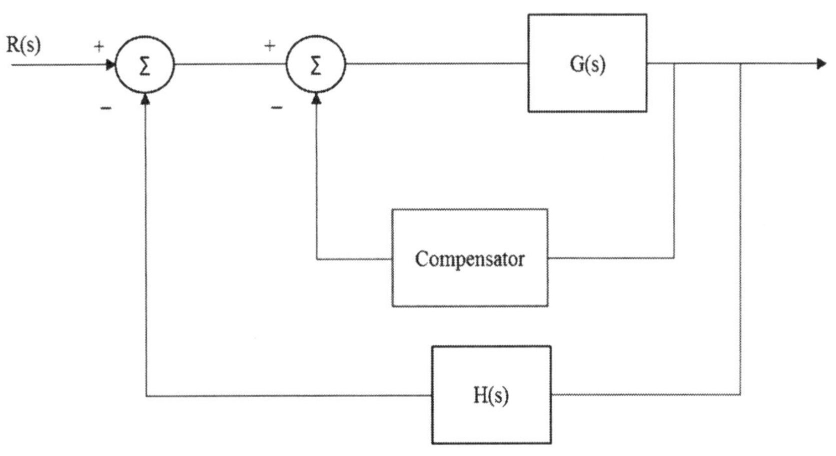


Figure 13-10: Feedback Compensator

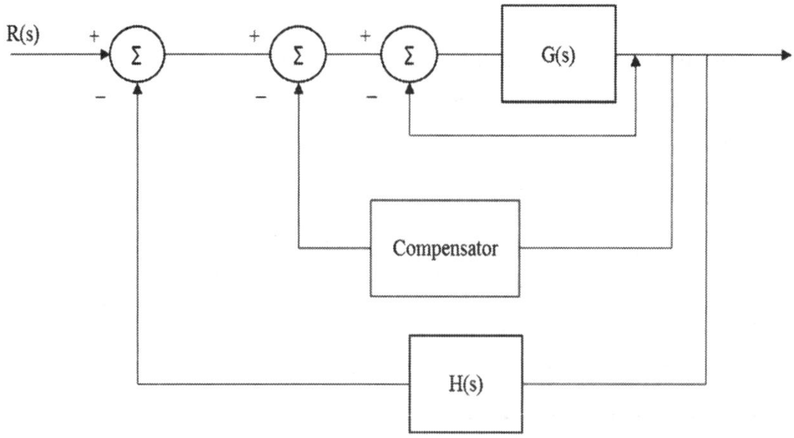


Figure 13-11: Load Compensator

### (1) Electrical system mathematic modeling

According to [3], the operational amplifier circuit illustrated in Figure 13-12 is considered as an electrical model in this section, the circuit is shown as:

$$\frac{E_i(s) - E'(s)}{Z_1} = \frac{E'(s) - E_o(s)}{Z_2} \qquad \textbf{Eq. 13-12}$$

As we consider  is approximately zero, so we have:

$$\frac{E_o(s)}{E_i(s)} = -\frac{Z_2(s)}{Z_1(s)} \qquad \textbf{Eq. 13-13}$$

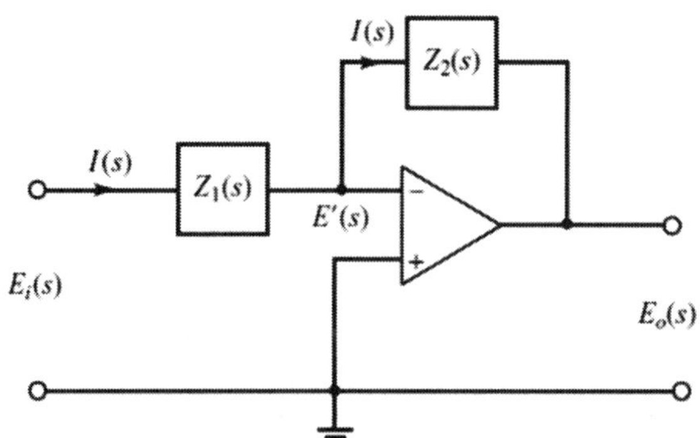


Figure 13-12: Operational amplifier circuit

Based on Figure 13-12, the operational amplifier circuit consists of capacitors and resistors shown in Figure 13-13. The transfer function of this circuit is accomplished as follows:

$$Z_1 = \frac{R_1}{R_1 C_1 s + 1}, \quad Z_2 = \frac{R_2}{R_2 C_2 s + 1} \qquad \textbf{Eq. 13-14}$$

Where $Z_1$, and $Z_2$ are the input impedance and the feedback impedance, respectively.

Based on Eq. 13-13, we have:

$$\frac{E(s)}{E_i(s)} = -\frac{Z_2}{Z_1} = -\frac{R_2}{R_1}\frac{R_1 C_1 s + 1}{R_2 C_2 s + 1} = -\frac{C_1}{C_2}\frac{s + \frac{1}{R_1 C_1}}{s + \frac{1}{R_2 C_2}} \qquad \textbf{Eq. 13-15}$$

We can see the minus sign in Eq. 13-15, so the circuit is sign inverting. In the actual application, a sign inversion is not convenient, so a sign inverter is added to either the input or the output of the circuit of Figure 13-12. Figure 13-14 shows an electronic circuit using operational amplifiers. The transfer function of the sign inverter is given:

$$\frac{E_o(s)}{E(s)} = -\frac{R_4}{R_3} \qquad \textbf{Eq. 13-16}$$

The value of $-R_4/R_3$ is the gain of the sign inverter.

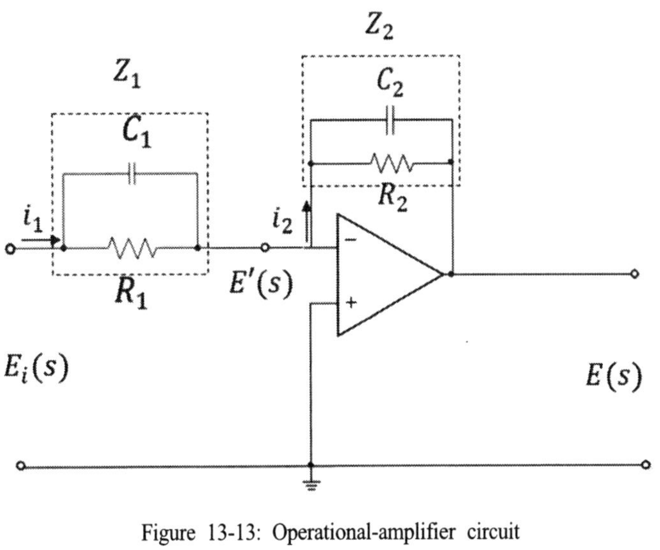


Figure 13-13: Operational-amplifier circuit

### (2) Phase Lead or Lag Compensation

There are several ways to perform the lead compensator or the lag compensator. However, we select an electronic circuit using operational amplifiers which is shown in Figure 13-14. If the value $R_1C_1 > R_2C_2$ or $\alpha < 1$, then the electronic circuit is the lead compensator. If the value $R_1C_1 < R_2C_2$, then the electronic circuit is the lag compensator.

The transfer function of the network illustrated in Figure 13-14 is obtained:

$$\frac{E_o(s)}{E_i(s)} = \frac{R_2R_4R_1C_1s+1}{R_1R_3R_2C_2s+1} = \frac{R_4C_1}{R_3C_2}\frac{s+\frac{1}{R_1C_1}}{s+\frac{1}{R_2C_2}}$$

$$= K_c\alpha\frac{Ts+1}{\alpha Ts+1} = K_c\frac{s+\frac{1}{T}}{s+\frac{1}{\alpha T}} \qquad \textbf{Eq. 13-17}$$

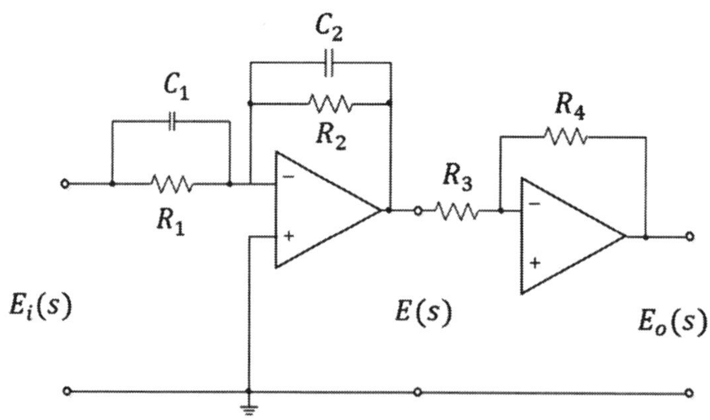


Figure 13-14: Electronic circuit using operational amplifiers

Where

$$T = R_1C_1,\ \alpha T = R_2C_2,\ K_c = \frac{R_4C_1}{R_3C_2}$$

$$\alpha K_c = \frac{R_4C_1}{R_3C_2}\frac{R_2C_2}{R_1C_1} = \frac{R_2R_4}{R_1R_3},\ \alpha = \frac{R_2C_2}{R_1C_1}$$

The value $\alpha K_c$ is taken into account as a dc gain of the network.

The lead network is considered when a system has one pole and one dominating zero. The zero, which is closer to the origin than all over zeros, is called the dominating zero. If the dominating zero is added for compensation in a control system, the lead compensation network will be selected. In the phase lead network, the fundamental requirement is all poles and zeros of the transfer function of the network must lie on the negative part of a real axis with a zero located at the origin of the nearest origin. These poles and zero interlace each other.

There are many influences of the phase lag compensation. Whereas the gain crossover frequency increases, the bandwidth decreases. Owing to the decreasing bandwidth, the response will be slower, the rise time and the settling time become larger. Additionally, the phase margin will be increased.

The phase lead compensation has several advantages when adding in the control system. First, the speed of the system increases as it shifts gain crossover frequency to a higher value. Second, the maximum overshoot of the system decreases. However, the steady state error of the system is not improved.

The lag network is considered when a system has one zero and one dominating pole. The pole, which is closer to the origin than all over poles, is called the dominating pole. If the dominating pole is added for compensation in a control system, the lag compensation network will be selected. In the phase lag network, the fundamental requirement is all poles and zeros of the transfer function of the network must lie on the negative part of a real axis with a pole located at the origin of nearest origin. These poles and zero interlace each other.

There are many influences of the phase lag compensation. Whereas the gain crossover frequency increases, the bandwidth decreases. Owing to the decreasing bandwidth, the response will be slower, the rise time and the settling time become larger. Additionally, the phase margin will be increased.

The phase lag compensation has several advantages when adding in the control system. First, the phase lag network enables low frequencies and high frequencies are attenuated. Second, the steady state accuracy increases. However, the speed of the system decreases. Figure 13-15 shows the pole-zero configuration of the lead and the lag networks.

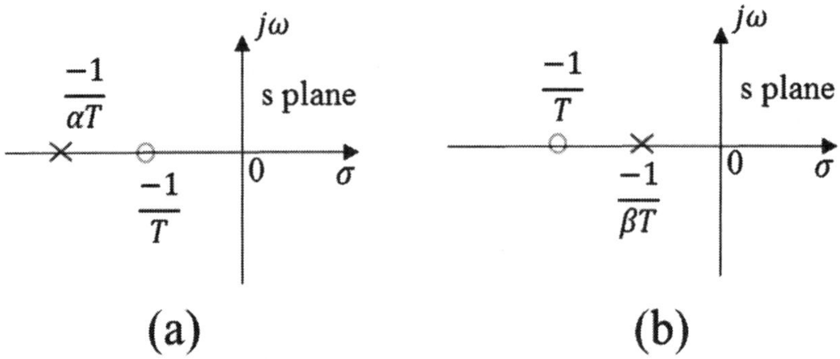


Figure 13-15: Pole-zero configurations: (a) lead network; (b) lag network

### (3) Phase Lag-Lead Compensation

In several situations, the single lead or lag compensation is not able to be satisfied the design specifications. For an unstable uncompensated system, the lead compensation provides fast response but does not provide enough phase margin whereas the lag compensation stabilize the system but does not provide enough bandwidth. Therefore, it is necessary to produce the lag-lead compensator network. Because of presenting this network, the speed of the system increases because it shifts gain crossover frequency to a higher value, and the steady state accuracy is improved. Similarly, we use the electronic circuit using operational amplifiers to perform the Lag-Lead network. Figure 13-16 shows the lag-lead compensating network. Figure 13-17 shows the Pole Zero Plot Lag Lead Network. The transfer function of the lag-lead compensating network is obtained as follow:

$$\frac{1}{z_1} = \frac{1}{R_1 + \frac{1}{C_{1s}}} + \frac{1}{R_3}, \text{ or } Z_1 = \frac{(R_1 C_1 s + 1) R_3}{(R_1 R_3) C_1 s + 1} \quad \textbf{Eq. 13-18}$$

$$Z_2 = \frac{(R_2 C_2 s + 1) R_4}{(R_2 + R_4) C_2 s + 1} \quad \textbf{Eq. 13-19}$$

Where, $Z_1$, $Z_2$ are the complex impedances.

Then, we have:

$$\frac{E(s)}{E_i(s)} = -\frac{Z_2}{Z_1} = -\frac{R_4}{R_3} \frac{(R_1 + R_3) C_1 s + 1}{R_1 C_1 s + 1} \frac{R_2 C_2 s + 1}{(R_2 + R_4) C_2 s + 1} \quad \textbf{Eq. 13-20}$$

Transfer function of the sign inverter is

$$\frac{E_o(s)}{E(s)} = -\frac{R_6}{R_5} \qquad \textbf{Eq. 13-21}$$

Thus the transfer function of the compensator is:

$$\frac{E_o(s)}{E_i(s)} = \frac{E_o(s)}{E(s)}\frac{E(s)}{E_i(s)}$$
$$\frac{E_o(s)}{E_i(s)} = \frac{R_4R_6}{R_3R_5}\left[\frac{(R_1+R_3)C_1s+1}{R_1C_1s+1}\right]\left[\frac{R_2C_2s+1}{(R_2+R_4)C_2s+1}\right] \qquad \textbf{Eq. 13-22}$$

$$\Rightarrow G_{lag-lead} = K_c\frac{\beta}{\gamma}\left(\frac{T_1s+1}{\frac{T_1}{\gamma}s+1}\right)\left(\frac{T_2s+1}{\beta T_2s+1}\right) = K_c\frac{\left(s+\frac{1}{T_1}\right)\left(s+\frac{1}{T_2}\right)}{\left(s+\frac{\gamma}{T_1}\right)\left(s+\frac{1}{\beta T_2}\right)} \qquad \textbf{Eq. 13-23}$$

Where

$$T_1 = (R_1+R_3)C_1, \quad \frac{T_1}{\gamma} = R_1C_1, \quad T_2 = R_2C_2, \quad \beta T_2 = (R_2+R_4)C_2$$

$$\gamma = \frac{R_1+R_2}{R_1} > 1, \ \beta = \frac{R_2+R_4}{R_2} > 1, \ K_c = \frac{R_2R_4R_6R_1+R_3}{R_1R_3R_5R_2+R_4}$$

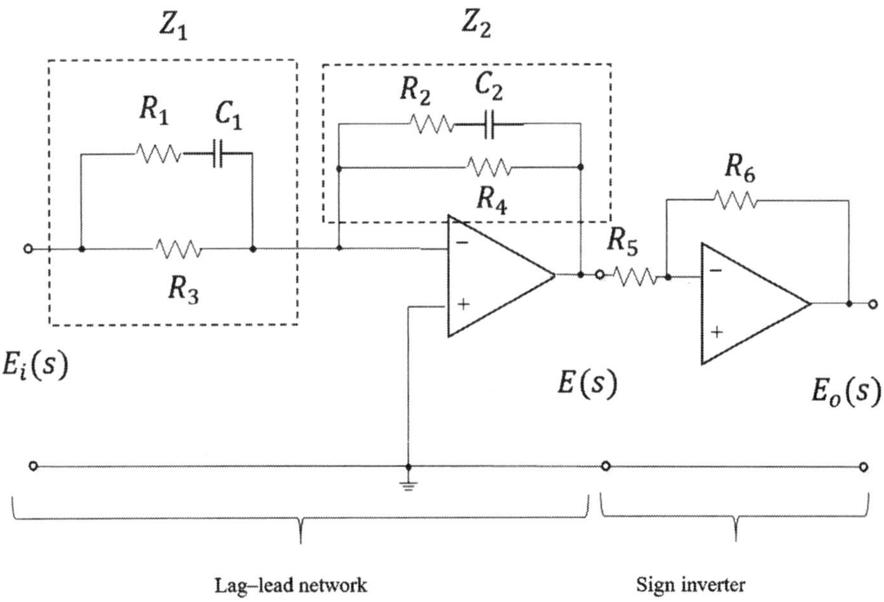


Figure 13-16: Lag-Lead Compensating Network

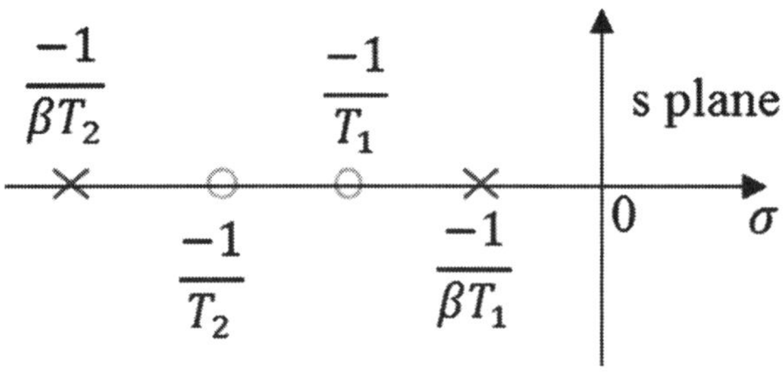


Figure 13-17: Pole Zero Plot Lag Lead Network

### 13.6.3 Stability Analysis Methods

There are various meanings of stability, but in general two types of statbility method are shown in Table 13-5.

Table 13-5: Stability Analysis Methods

| | Absolute Stability | Relative Stability |
|---|---|---|
| Time Domain Analysis | Routh-Hurwitz | Root Locus |
| Frequency Domain Analysis | Nyquist Plot | Bode Plot |

Table 13-6 shows the description of stability analysis methods.

Table 13-6: Description of Stability Analysis Methods

| Type | Description | Analytic Method | Remark |
|---|---|---|---|
| Absolute Stability | A numerical method is said to be absolutely stable for a given if all the roots of lie within the unit circle. A region of the complex plane is said to be a region of absolute stability if the method is stable for all in | - Routh-Hurwitz<br>- Nyquist Plot | Absolute stability means whether system is stable or unstable. |
| Relative Stability | The relative stability of a system is defined as the property measured by the relative real part of each root or pair of roots. | - Root Locus<br>- Bode Plot<br>- Zigler-Nichols | Relative Stability gives the degree of stability or how close it is to instability. |

In control system, there are also two kinds of methods to analyse the stability of a system as shown in Table 13-7.

Table 13-7: Analytic Methods

| | Absolute Stability | Description |
|---|---|---|
| Indirect Analytic Method | Routh-Hurwitz | - The Routh-Hurwitz stability criterion gives a simple algorithm to decide whether or not the zeros of a polynomial are all in the left half of the complex s-plane.<br>- A Hurwitz polynomial is an important requirement for a linear continuous-time invariant to be stable. |
| | Nyquist Plot | - A Nyquist plot is a frequency response plot utilized in control engineering and signal processing.<br>- Nyquist plots are prevalently employed to assess the stability of a system with feedback. |
| Direct Analytic Method | Root Locus | - The root locus of a feedback system is the graphical representation in the complex s-plane of the possible locations of its closed-loop poles for varying values of a certain system parameter.<br>- The points that are part of the root locus satisfy the angle condition.<br>- The value of the parameter for a certain point of the root locus can be accomplished utilizing the magnitude condition. |
| | Bode | Bode plots illustrates the gain and phase of a system as a function of frequency. |

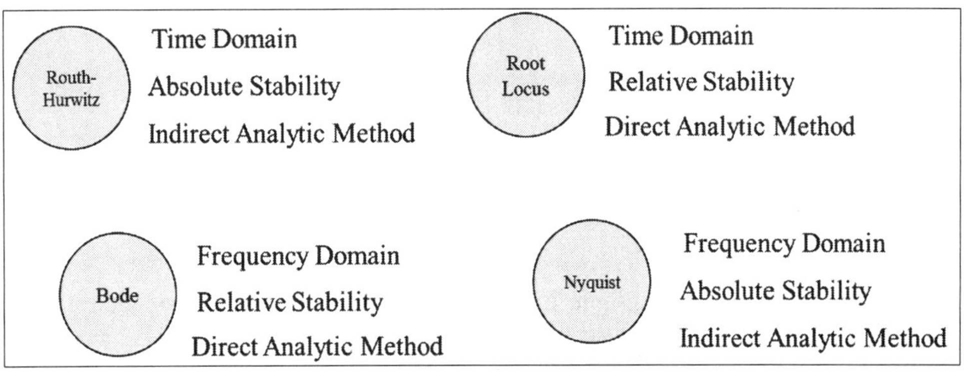


Figure 13-18 Characteristic of stability analysis methods

Figure 13-18 shows the characteristic of stability analysis methods.

### 13.6.4 Root Locus Plots

#### (1) Root Locus Plots

Figure 13-19 shows the control system with the negative feedback.

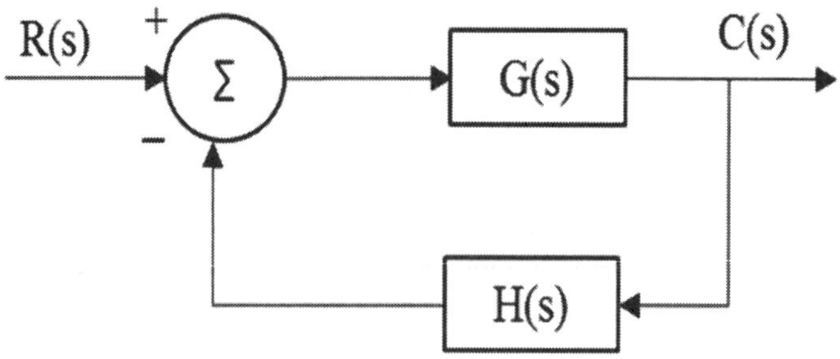


Figure 13-19: Control system

The closed-loop transfer function as shown in Eq. 13-2 is expressed again herein:

$$\frac{C(s)}{R(s)} = \frac{G(s)}{1+G(s)H(s)} \qquad \textbf{Eq. 13-21}$$

The angle condition is expressed:

$$\angle G(s)H(s) = \pm 180^\circ (2k+1) \qquad \textbf{Eq. 13-22}$$

The magnitude condition is expressed:

$$|G(s)H(s)| = 1 \qquad \textbf{Eq. 13-23}$$

According to [3], $G(s)H(s)$ is given:

$$G(s)H(s) = \frac{K(s+z_1)}{(s+p_1)(s+p_2)(s+p_3)(s+p_4)} \qquad \textbf{Eq. 13-24}$$

In order to gain insight into the root locus approach, the control system is considered in Figure 13-20.

PART 3

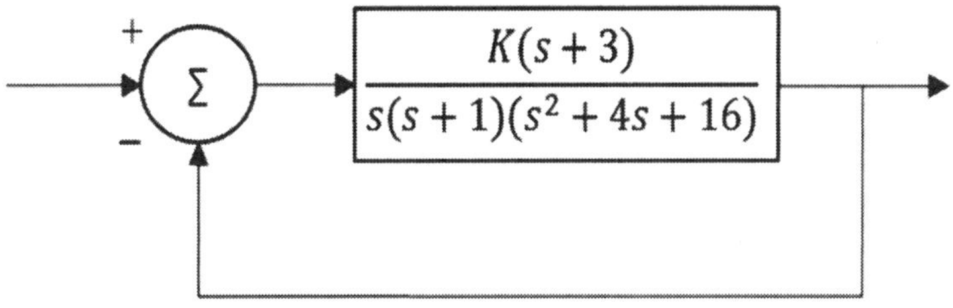


Figure 13-20: Control system

Matlab function will do a simulation of the aforementioned control system to observe the generation of root-locus plots and its related information. The root-locus plot is illustrated in Figure 13-21.

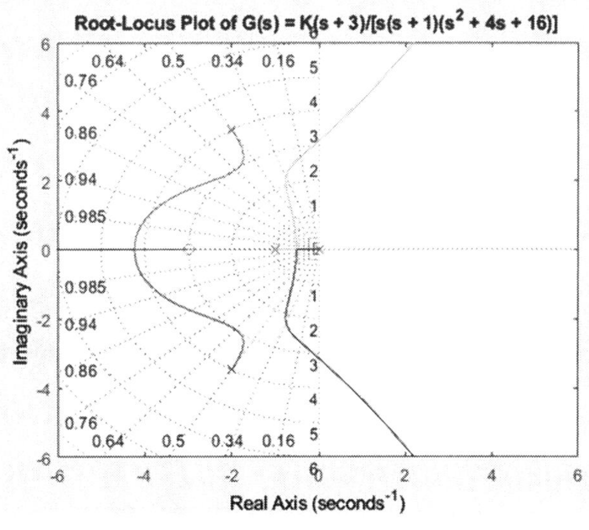


Figure 13-21: Root-locus plot

In order to be familiar with Matlab approach for the plotting root loci, another control system with the open loop transfer function of $G(s)H(s) = \frac{K}{s(s+0.5)(s^2+0.6s+10)}$, and the root-locus plot is shown in Figure 13-22.

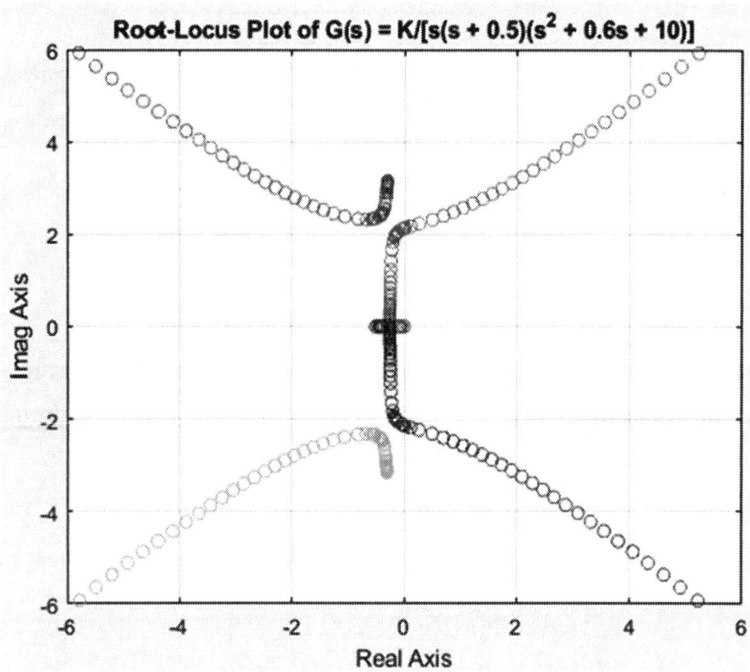


Figure 13-22: Root-locus plot

Next, the control system is illustrated in Figure 13-23, and the root-locus plot is shown in Figure 13-24. The system is stable in several limited ranges of the value of K: 0<K<12 and 73<K<154. The system is unstable in other ranges such as 12<K<73 and 154<K.

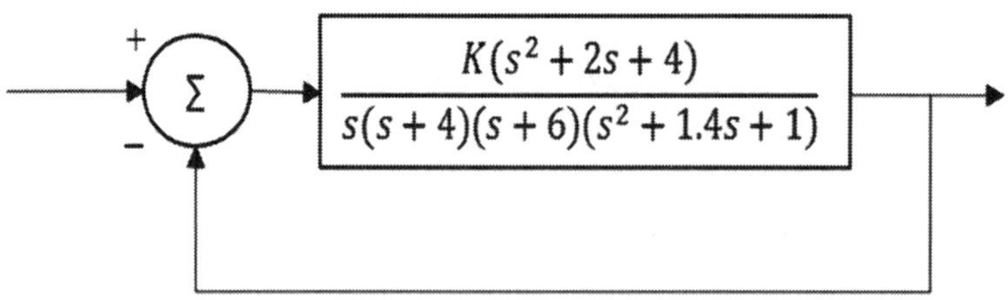


Figure 13-23: Control system

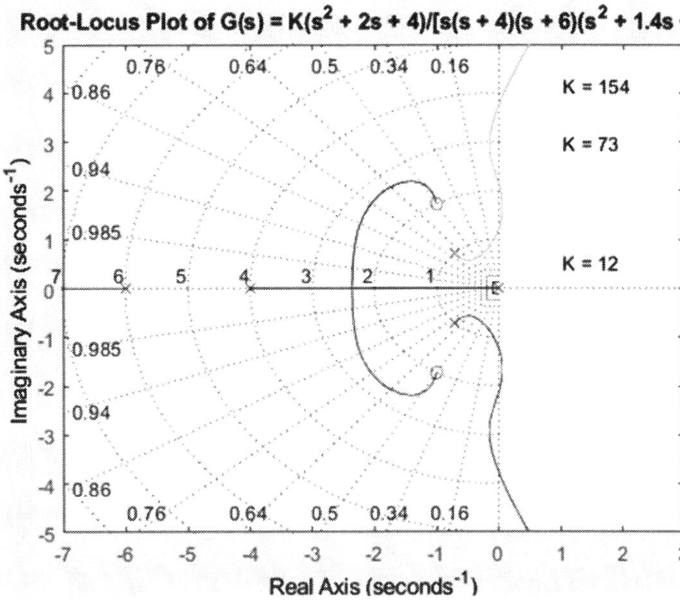


Figure 13-24: Root-locus plot of conditionally stable system

## (2) Lead compensation is designed using root locus approach

The control system is shown in Figure 13-25. The feed forward transfer function is

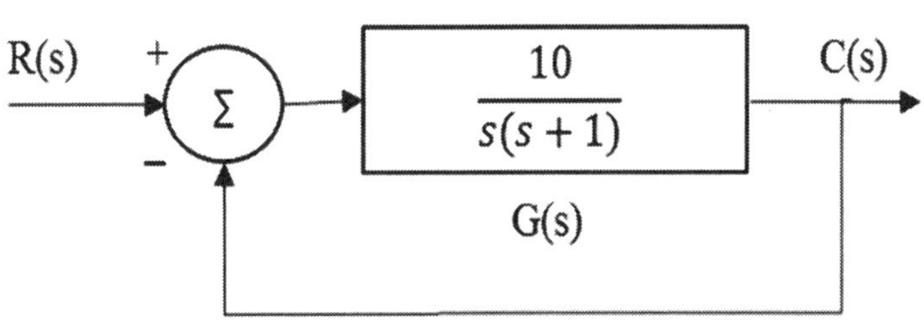


Figure 13-25: Control system

The closed loop transfer function of the system is:

$$\frac{C(s)}{R(s)} = \frac{10}{s^2+s+10} = \frac{10}{(s+0.5+j3.1225)(s+0.5-j3.1225))} \quad \textbf{Eq. 13-25}$$

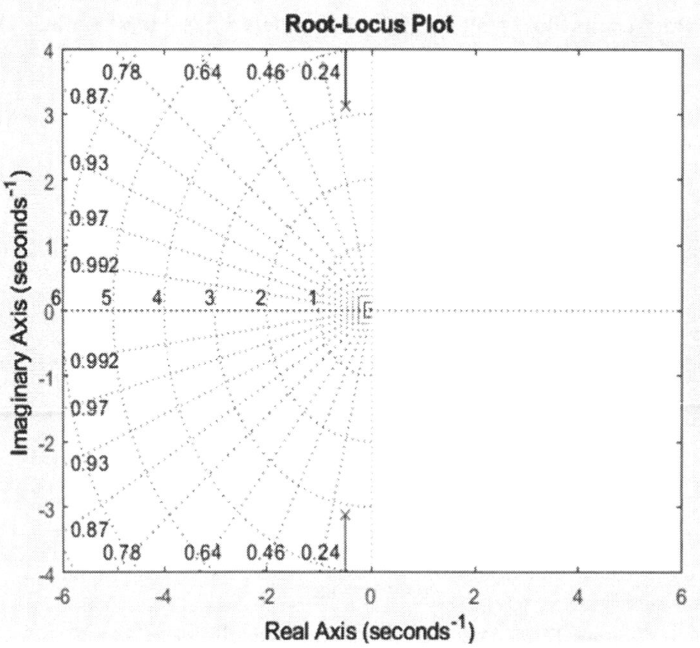


Figure 13-26: Root-locus plot

Using Matlab, the root locus of this system is illustrated in Figure 13-26.

The damping ratio of the closed-loop poles is $\zeta = \frac{\frac{1}{2}}{\sqrt{10}} = 0.1581$. Compared with the undamped natural frequency of the closed-loop system ($\omega_n = \sqrt{10} = 3.1623$), the damping ratio is small. This system has a large overshoot in the step response. Therefore, the lead compensator network is added to improve the stability of the system illustrated in Figure 13-27. In order to obtain this objective, the dominant closed-loop poles have the damping ratio ($\zeta = 0.5$) and the undamped natural frequency ($\omega_n = 3\, rad/s$). The expected location of the dominant closed-loop poles is determined:

$$s^2 + 2\zeta\omega_n + \omega_n^2 = s^2 + 3s + 9 = (s + 1.5 + j2.5981)(s + 1.5 + j2.5981) \quad \textbf{Eq. 13-26}$$

After the computation, the lead compensator $G_c(s)$ just designed is expressed:

$$G_c(s) = 1.2287\left(\frac{s + 1.9432}{s + 4.6458}\right) \quad \textbf{Eq. 13-27}$$

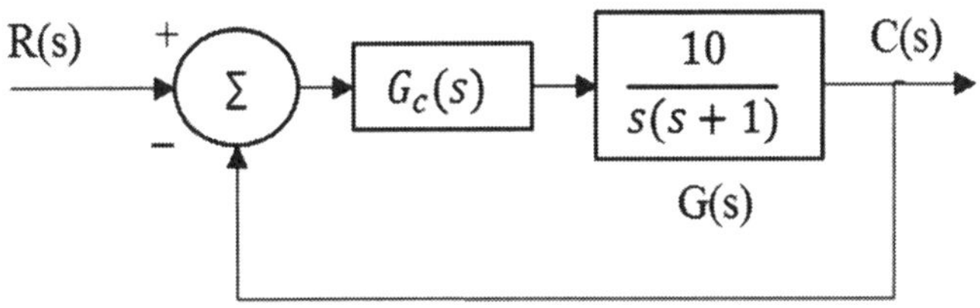


Figure 13-27: Compensated system

The open loop transfer function of the designed system is:

$$G_c(s)G(s) = 1.2287\left(\frac{s+1.9432}{s+4.6458}\right)\frac{10}{s(s+1)} \quad \textbf{Eq. 13-28}$$

The closed-loop transfer function is:

$$\frac{C(s)}{R(s)} = \frac{12.287(s+1.9432)}{s(s+1)(s+4.6458)+12.287(s+1.9432)} \quad \textbf{Eq. 13-29}$$

The root-locus plot of the designed system is shown in Figure 13-28.

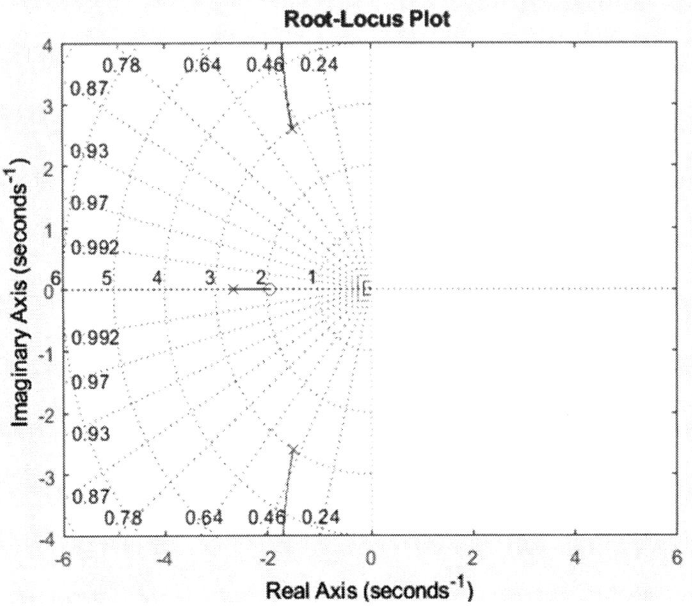


Figure 13-28: Root-locus plot of the designed system

The step responses between the compensated system and the uncompensated system is shown in Figure 13-29. The unit-ramp response curves of designed systems is illustrated in Figure 13-30.

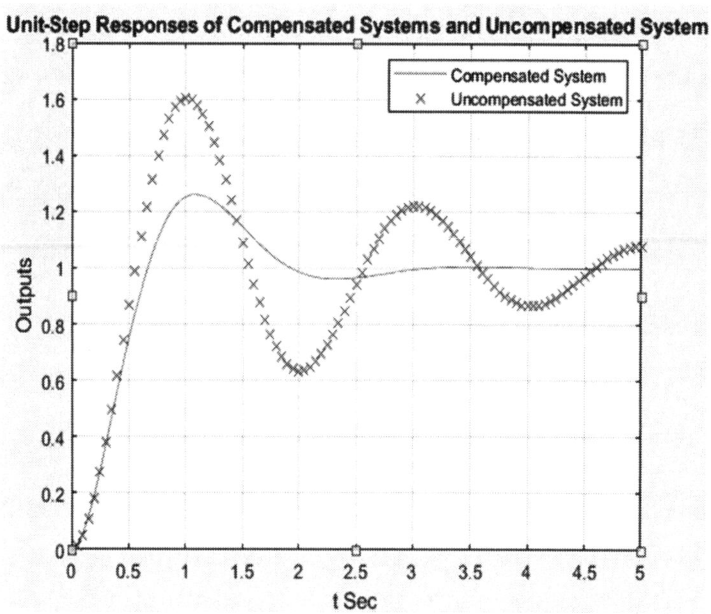


Figure 13-29: Unit-step response curves of designed systems and original uncompensated system

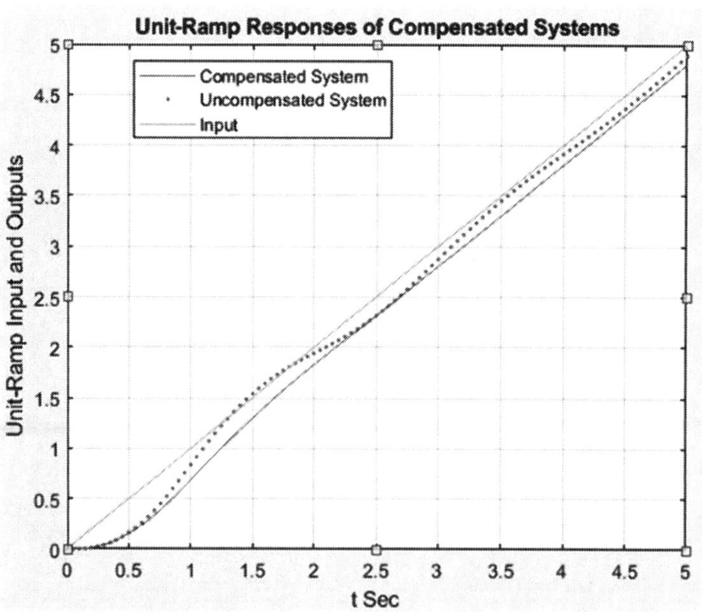


Figure 13-30: Unit-ramp response curves of designed systems

### (3) Lag compensation is designed using root locus approach

The system is considered in Figure 13-31. The feed forward transfer function is:

$$G(s) = \frac{1.06}{s(s+1)(s+2)} \quad \textbf{Eq. 13-30}$$

The closed-loop transfer function of this system is:

$$\frac{C(s)}{R(s)} = \frac{1.06}{s(s+1)(s+2)+1.06} \quad \textbf{Eq. 13-31}$$

The root locus of this system is shown in Figure 13-32.

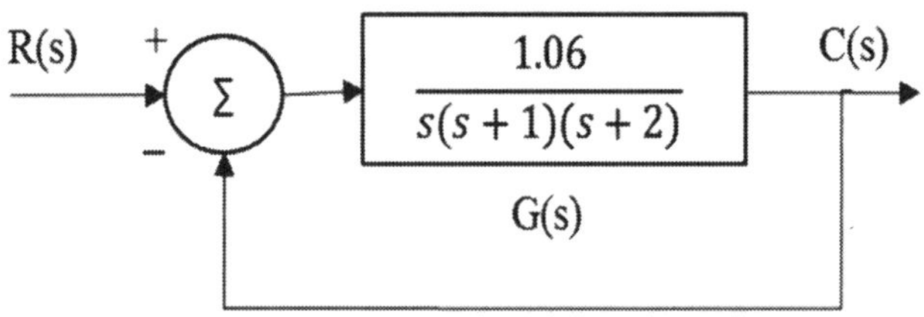


Figure 13-31: Control system

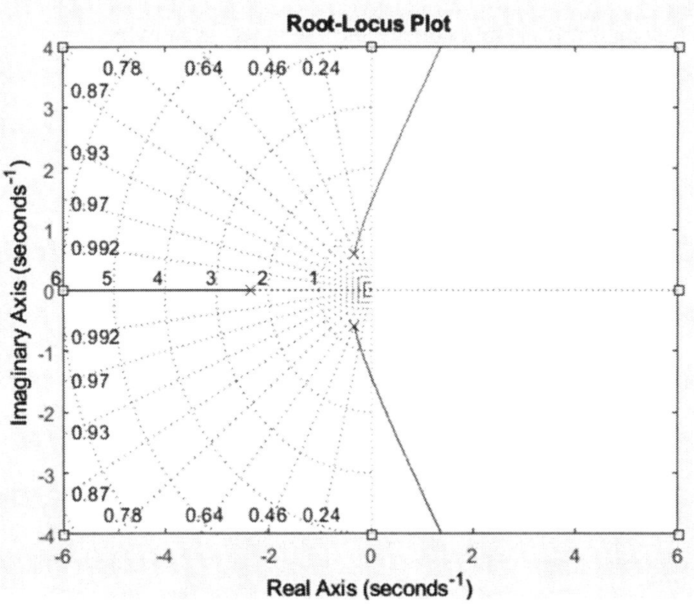


Figure 13-32: Root-locus plot

The damping ratio of the dominant closed-loop poles, $\zeta = 0.491$. The undamped natural frequency of the dominant closed-loop poles, $\omega_n = 0.673$ rad/sec. The static velocity error constant is 0.53 $\sec^{-1}$. The lag compensator is added to increase the static velocity error constant $K_v$ to approximately 5 $\sec^{-1}$ without changing the location of the dominant closed-loop poles. The expected closed-loop pole location is shown in Figure 13-33. The root-locus plot of the designed system is shown in Figure 13-34.

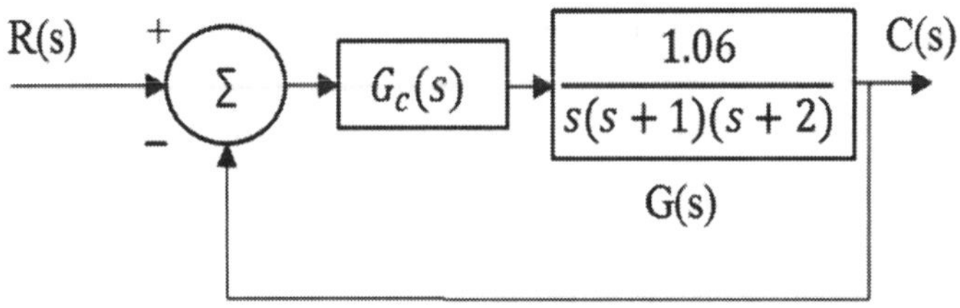


Figure 13-33: Compensated system

After the computation, the lead compensator $G_c(s)$ is expressed:

$$G_c(s) = 9.656\left(\frac{20s+1}{200s+1}\right) \quad \text{Eq. 13-32}$$

The open loop transfer function of the designed system is:

$$G_c(s)G(s) = 9.656\left(\frac{20s+1}{200s+1}\right)\frac{1.06}{s(s+1)(s+2)} \quad \text{Eq. 13-33}$$

$$= \frac{204.7s+10.24}{200s^4+601s^3+403s^2+2s}$$

$$= \frac{1.0235(s+0.05)}{s(s+0.005)(s+1)(s+2)}$$

The closed-loop transfer function of the lag compensated system is:

$$\frac{C(s)}{R(s)} = \frac{1.0235(s+0.05)}{s(s+0.005)(s+1)(s+2)+1.0235(s+0.05)} \quad \text{Eq. 13-34}$$

PART 3

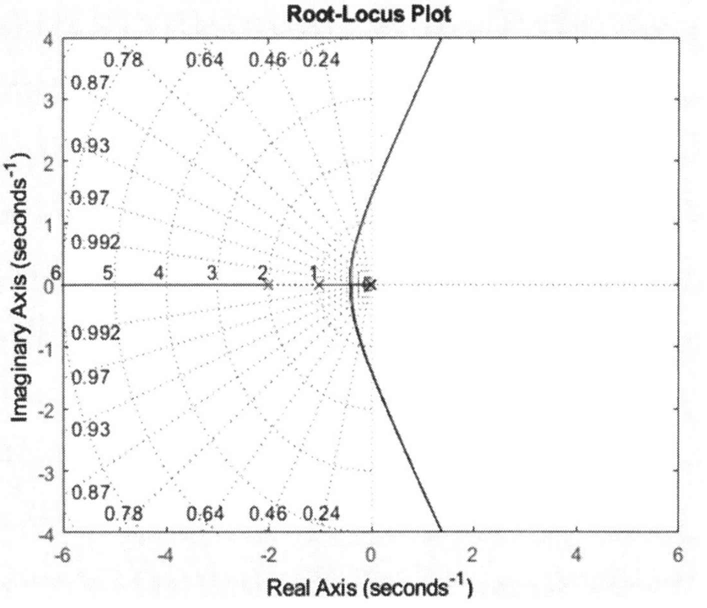


Figure 13-34: Root locus of designed system

The step responses between the compensated system and the uncompensated system is shown in Figure 13-35. The unit-ramp response curves of designed systems is illustrated in Figure 13-36.

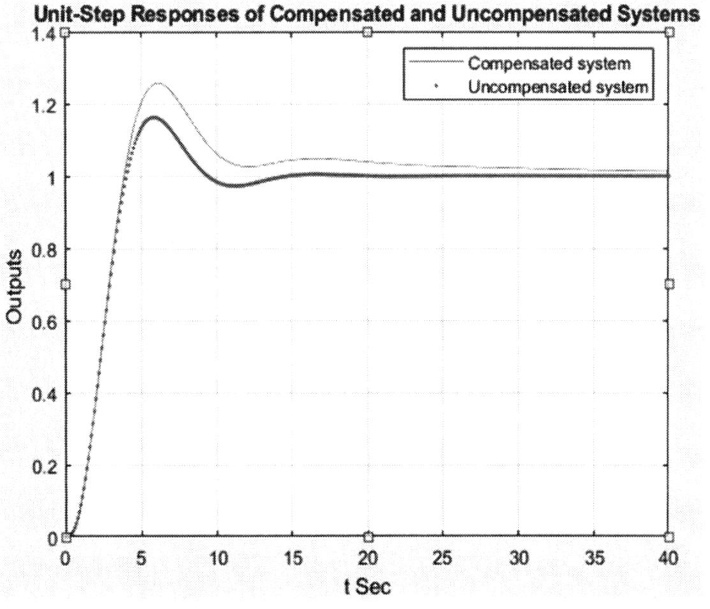


Figure 13-35: Unit-step responses of compensated and uncompensated systems

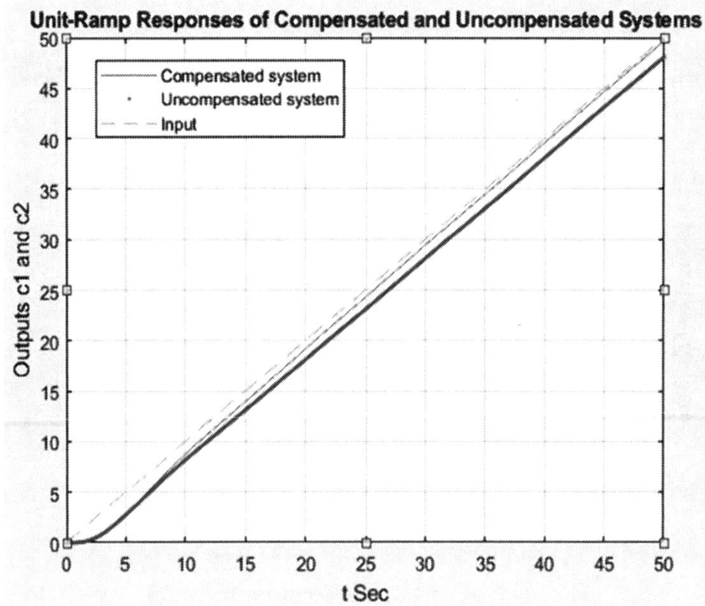


Figure 13-36: Unit-ramp responses of compensated and uncompensated systems

### (4) Lead-Lag compensation is designed using root locus approach

The control system is considered shown in Figure 13-38. The feed forward transfer function is:

$$G(s) = \frac{4}{s(s+5)} \quad \text{Eq. 13-35}$$

The closed-loop transfer function of this system is:

$$\frac{C(s)}{R(s)} = \frac{4}{s^2+5s+4} \quad \text{Eq. 13-36}$$

Using Matlab, the root locus plot of this system is shown in Figure 13-37.

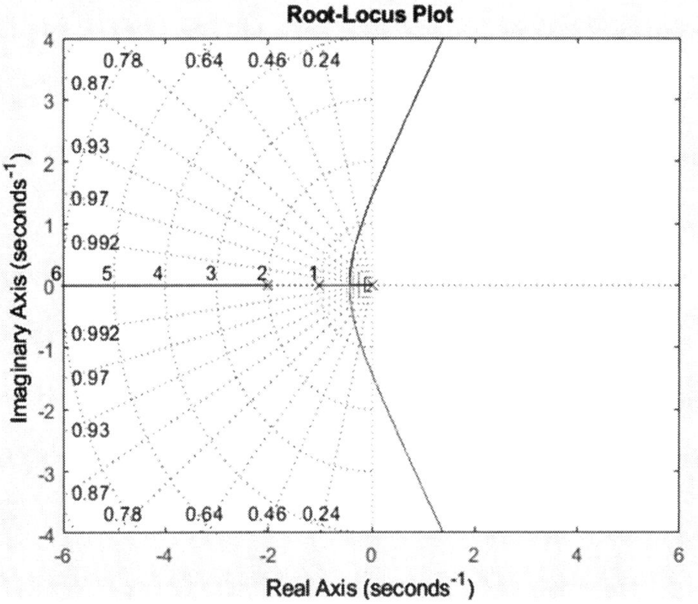


Figure 13-37: Root locus plot

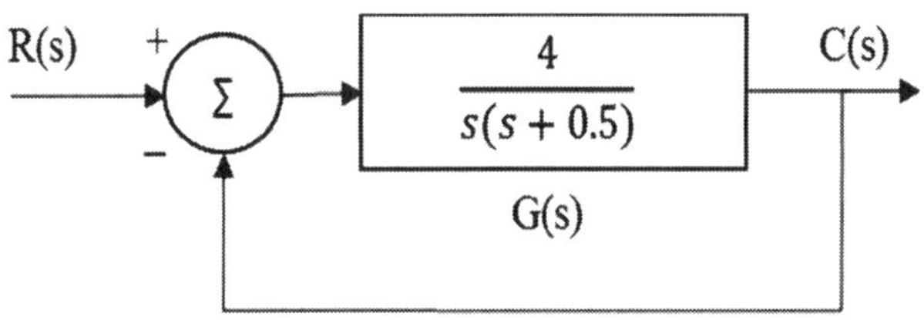


Figure 13-38: Control system

The damping ratio is 0.125, the undamped natural frequency is 2 rad/sec, and the static velocity error constant is 8 $sec^{-1}$. The lag-lead network is added into the control system to increase the damping ratio of the dominant closed-loop poles to about 0.5 and to make the undamped natural frequency to 5 rad/sec and the static velocity error constant to 80 $sec^{-1}$. The compensated system is shown in Figure 13-39.

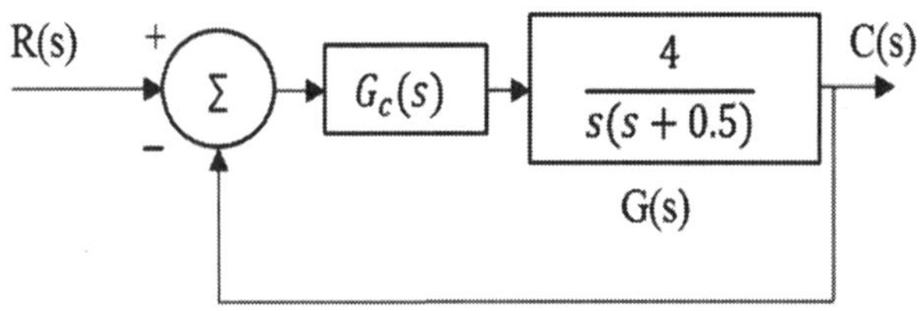


Figure 13-39: Lead-lag compensated system

After the computation, the lag-lead compensator $G_c(s)$ is expressed:

$$G_c(s) = \frac{10(2s+1)(5s+1)}{(0.1992s+1)(80.19s+1)} \quad \textbf{Eq. 13-37}$$

The open-loop transfer function of the lag-lead compensated system is:

$$G_c(s)G(s) = \frac{25.04(s+0.2)}{s(s+5.02)(s+0.01247)} \quad \textbf{Eq. 13-38}$$

The closed-loop transfer function of the lag-lead compensated system is

$$\frac{C(s)}{R(s)} = \frac{25.04(s+0.2)}{s(s+5.02)(s+0.01247)+25.04(s+0.2)} \quad \textbf{Eq. 13-39}$$

The root-locus plot of the lag-lead compensated system is shown in Figure 13-40.

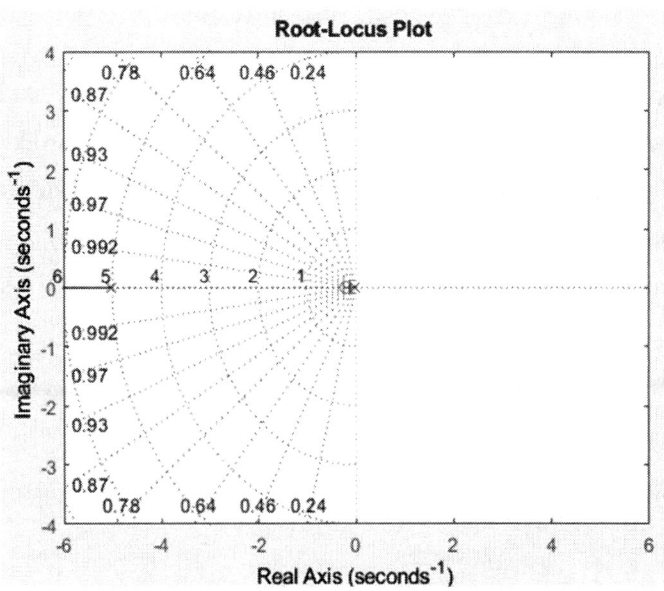


Figure 13-40: Compensated system root-locus plot

The step responses between the compensated system and the uncompensated system is shown in Figure 13-41. The unit-ramp response curves of designed systems is illustrated in Figure 13-42.

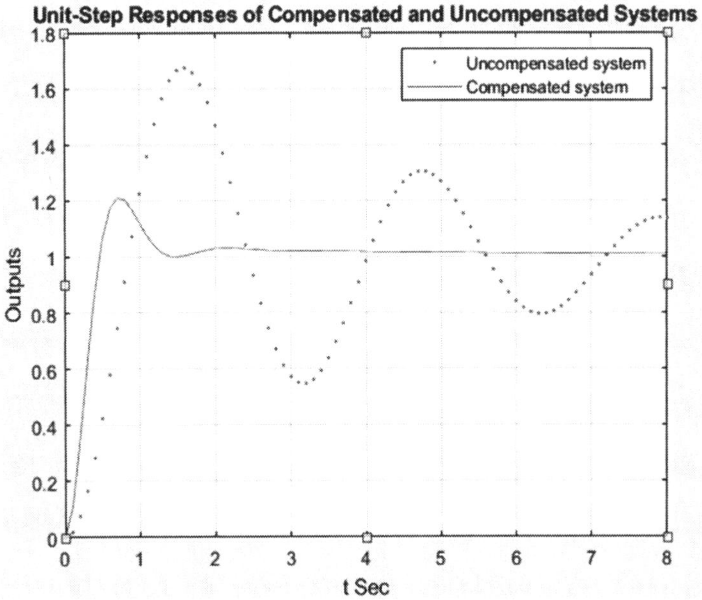


Figure 13-41: Unit-step responses of compensated and uncompensated systems

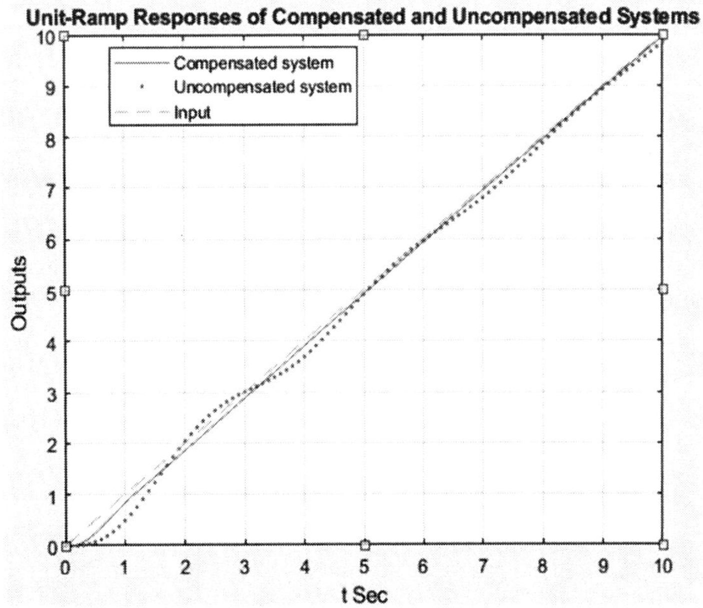


Figure 13-42: Unit-ramp responses of compensated and uncompensated systems

### 13.6.5 Application of Compensation Network in the Control Systems

In the nuclear power plant, typical controllers used for dynamic compensation of NCS control signals include:

- Lag Compensators
- Lead-Lag Compensators
- High Pass Filters
- Proportional Integral Differential Controllers.

#### (1) Lag Compensator

In the lag compensation network, when the input changes, the magnitude of the output is inversely proportional to the magnitude of the rate of change of the input. If the input remains stable, the magnitude of the output is the same as the input. In order to avoid sudden signal changes by eliminating high frequency noise. The lag compensator is used as steam flow input signal, supply flow input signal, and reactor output input signal. The formula of lag compensator is described as:

$$\text{Output} = \left(\frac{1}{(\tau s+1)}\right)\text{Input} \qquad \textbf{Eq. 13-40}$$

#### (2) Lead-Lag Compensator

The lead-lag compensator has a relatively high amplification effect on a rapidly changing input rather than on a slowly changing input. In the stable state, the output signal is the same as the input signal. The lead-lag compensator is used as steam generator level input signal, turbine runback request signal, and reactor output control system controller. The formula of lag compensator is described as:

$$\text{Output} = \left(\frac{\tau_2 s+1}{(\tau_1 s+1)}\right)\text{Input} \qquad \textbf{Eq. 13-41}$$

#### (3) High Pass Filter

The high bypass filter has a relatively high amplification effect on a rapidly changing input rather than on a slowly changing input. In the stable state, the output signal becomes "0". The high pass filter is used as Level compensation signal by steam flow rate and feed flow rate, Steam bypass control system Rapid open request signal, and Compensation signal due to deviation of turbine load index (TLI) and reactor output. The formula of lag compensator is described as:

$$\text{Output} = \left(\frac{\tau_2 s}{(\tau_1 s + 1)}\right) \text{Input} \qquad \textbf{Eq. 13-42}$$

### (4) Proportional Integral and Differential (PID) Controllers

The PID controller which is used to remove offset is commonly used for the process controller. Differential function is not used because it causes process instability due to sensitive reaction. The PI controller is used in the FWCS, SBCS, PPCS, and PLCS. The formula of lag compensator is described as:

$$\text{Output} = k\left(1 + \frac{1}{\tau_I s} + \tau_D s\right) \text{Input} \qquad \textbf{Eq. 13-43}$$

## 13.6.6 Compensation Network Response Simulation

The simulation are implemented through wolframalpha.com, and RRS compensation network is selected as a typical example to show unit step, ramp, and sinusoidal response plot for lag compensator, lead-lag compensator, and high pass filter.

### (1) Lead-Lag Compensator

- Case 1:

The system model is given by the transfer function as:

$$\frac{1 + 30s}{1 + 6s} \qquad \textbf{Eq. 13-44}$$

- Input Signal: Sin (10t)

It is obviously show that the unit step and ramp response plot are depicted in Figure 13-43 and Figure 13-44.

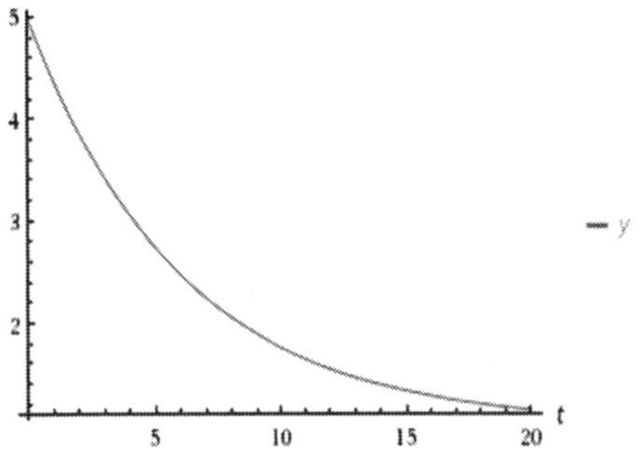


Figure 13-43: Unit step response plot

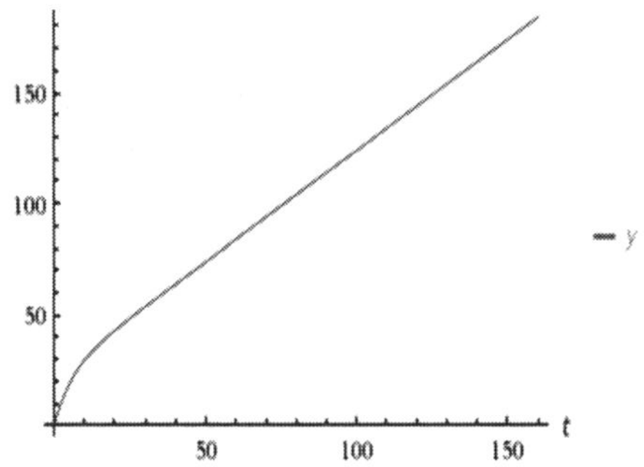


Figure 13-44: Unit ramp response plot

The sinusoidal response plot [sin (10t)] is shown in Figure 13-45.

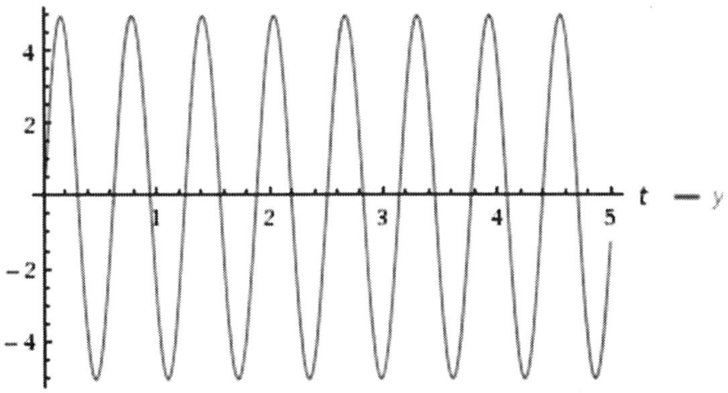


Figure 13-45: Sinusoidal response plot case 1

- Case 2:

The system model is changed as the transfer function:

$$\frac{1+6s}{1+30s}$$ **Eq. 13-45**

- Input Signal: Sin (10t)

As shown in Figure 13-46, Figure 13-47, it is straightforward to recognize the differences in unit step and ramp response, compared with case 1.

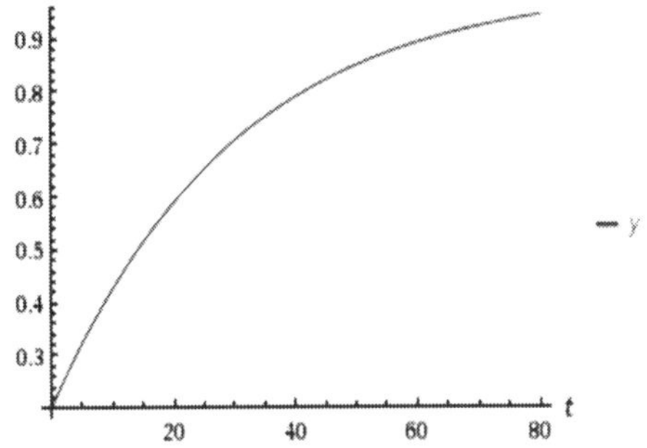


Figure 13-46 :Unit step response plot

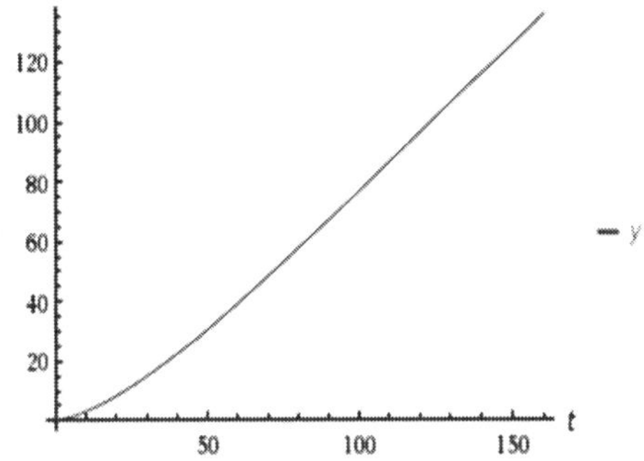


Figure 13-47: Unit ramp response plot

The sinusoidal response plot [sin (10t)] is shown in Figure 13-48.

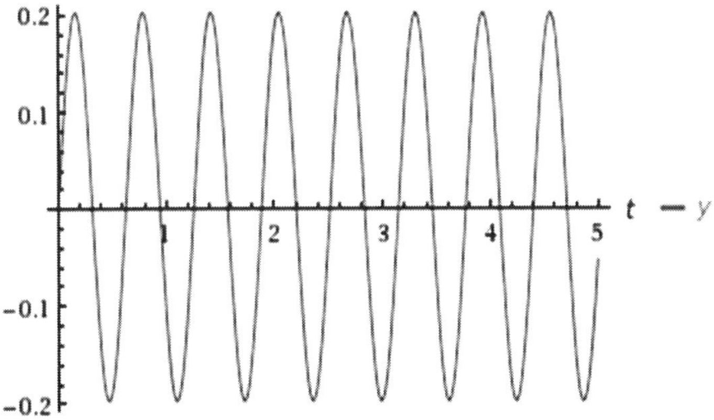


Figure 13-48: Sinusoidal response plot case 2

### (2) Lag compensator

The system model is illustrated as the transfer function:

$$\frac{1}{1+3s}$$ **Eq. 13-46**

And also the response to sin (10t), the unit step and ramp response are illustrated in Figure 13-49, Figure 13-50.

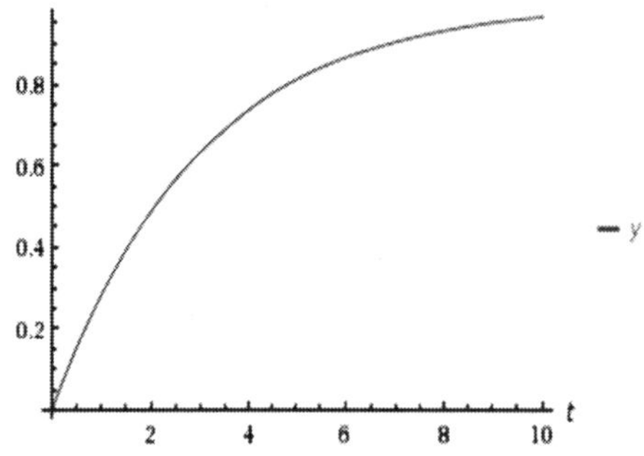


Figure 13-49: Unit step response plot

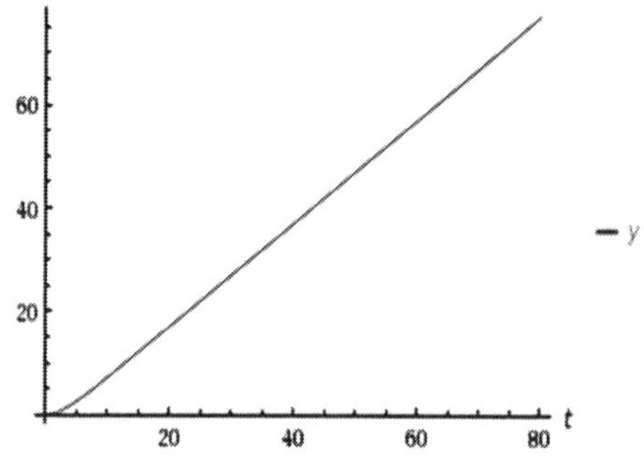


Figure 13-50: Unit ramp response plot

Sinusoidal response plot [sin (10t)] is shown in Figure 13-51.

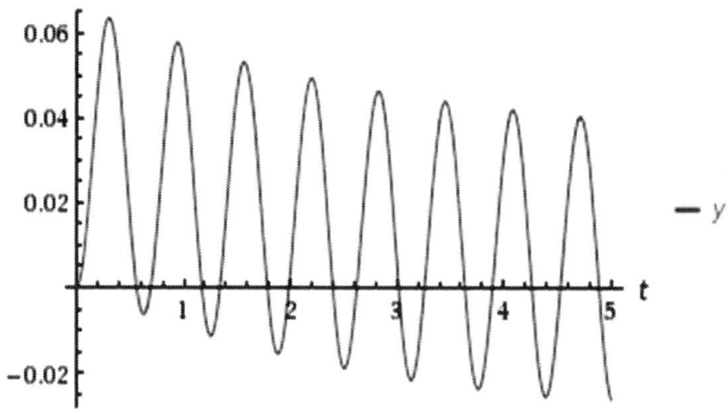


Figure 13-51: Sinusoidal response plot

## (3) High Pass Filter

The system model is illustrated as the transfer function:

$$\frac{12s}{1+12s} \quad \textbf{Eq. 13-47}$$

The unit step and ramp response are illustrated in Figure 13-52, Figure 13-53.

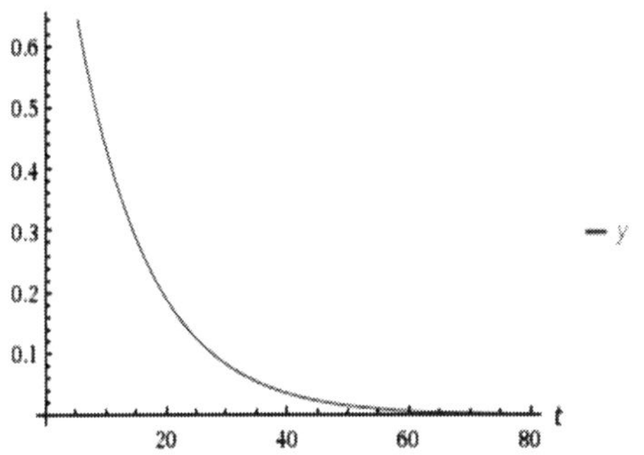


Figure 13-52: Unit step response plot

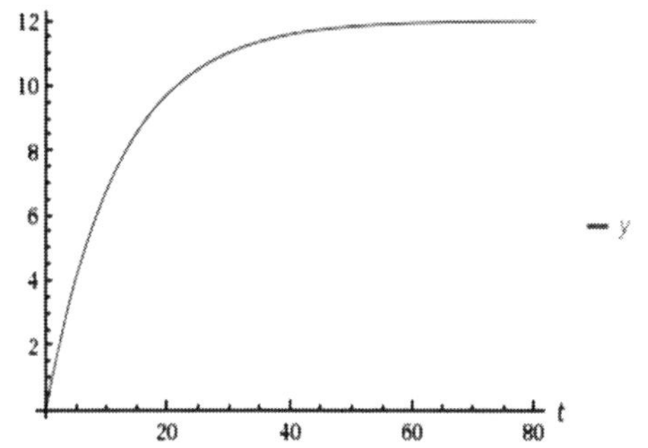


Figure 13-53: Unit ramp response plot

The Bode plot is shown in Figure 13-54. The Bode plot shows the magnitude (dB) and phase (degree) of the system.

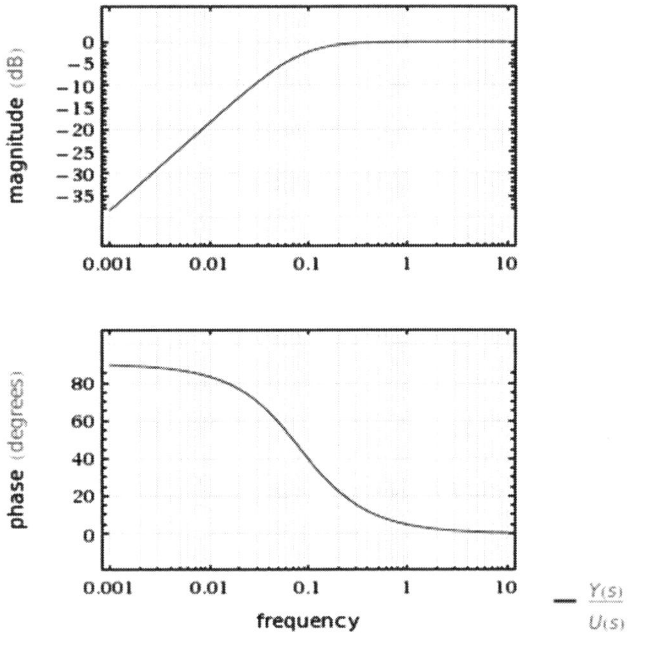


Figure 13-54: Bode plot

### 13.6.7 Control with Python on Anaconda-Jupyter IDE

Consider a real-life example of a Mass-Spring-Damper system belonging to a pneumatic valve shown in Figure 13-55.

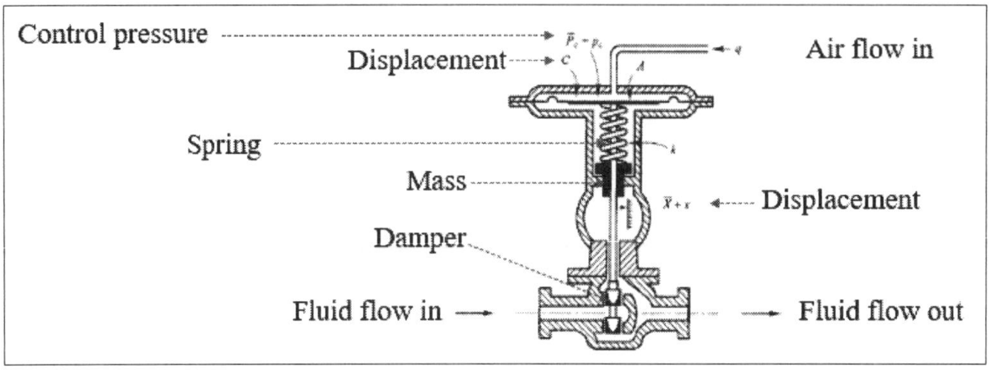


Figure 13-55: Mass-Spring-Damper system of Pneumatic valve

The transfer function of the above system is derived as indicated below:

i. If force applied on the spring $F(t)$, is considered as input and displacement $x(t)$ as output, then open loop transfer function $G(s)$ (based on Newton's Second law):

$$F(t) = k \times x(t) \qquad \textbf{Eq. 13-48}$$

$$F(s) = kX(s)$$

$$G(s) = \frac{X(s)}{F(s)} = \frac{1}{k}$$

ii. If force applied to damper $F(t)$, is considered as input and displacement $x(t)$ considered as output, then open loop transfer function $G(s)$ (based on Newton's Second law):

$$F(t) = k_d \times \frac{dx(t)}{dt} \qquad \textbf{Eq. 13-49}$$

$$F(s) = k_d \times sX(s)$$

$$G(s) = \frac{X(s)}{F(s)} = \frac{1}{K_d s}$$

iii. If force applied to the mass $F(t)$, is considered as input and displacement $x(t)$ considered as output, then open loop transfer function $G(s)$ (based on Newton's Second law):

$$F(t) = m \times \frac{d^2 t}{dt^2} \quad \textbf{Eq. 13-50}$$

$$F(s) = m \times s^2 X(s)$$

$$G(s) = \frac{X(s)}{F(s)} = \frac{1}{ms^2}$$

iv. Combining all equations above, the transfer function for the Mass-Spring-Damper system becomes:

$$G(s) = \frac{1}{ms^2 + k_d s + k} \quad \textbf{Eq. 13-51}$$

### (1) PID control of Mass-Spring-Damper system

The Mass-Spring-Damper system can be simulated using python on Jupyter as the Integrated Development Environment (IDE) with the correct libraries depicted in Figure 13-56 which include:

i. numpy

ii. matplotlib.pyplot

iii. matplotlib.patches

iv. control

```
import numpy as np
import matplotlib.pyplot as plt
import matplotlib.patches as mpatches
import control
from control import *
from control.matlab import *
from scipy import *
from scipy import linalg as la
from scipy.ndimage.filters import convolve
%matplotlib inline
```

Figure 13-56: Libraries imported on jupyter

To run a simulation we consider a Mass-Spring-Damper system with $m = 1$, $k_d = 15$, $k = 25$. The transfer function becomes:

$$G(s) = \frac{1}{s^2 + 15s + 25} \quad \textbf{Eq. 13-52}$$

To simulate this in python we write it in jupyter as in Figure 13-57:

$$sys = tf(1,[1,\ 15,\ 25])$$

```
sys = tf(1,[1,15,25])
sys
```

Figure 13-57: Mass-Spring-Damper transfer function on jupyter

To plot response of the system to a step input of 0.04 the following code is input to jupyter, this is depicted by Figure 13-58:

$$\begin{aligned}&[a,\ b] = step(sys)\\&plt.plot(b,a)\\&plt.plot(b,np.ones(b.shape)*0.04,'--')\\&plt.axis('tight')\\&plt.show()\end{aligned}$$

```
[a,b]=step(sys)
plt.plot(b,a)
plt.plot(b,np.ones(b.shape)*0.04,'--')
plt.axis('tight')
plt.show()
```

Figure 13-58: Jupyter codes to plot system response

The plot looks like shown in Figure 13-59.

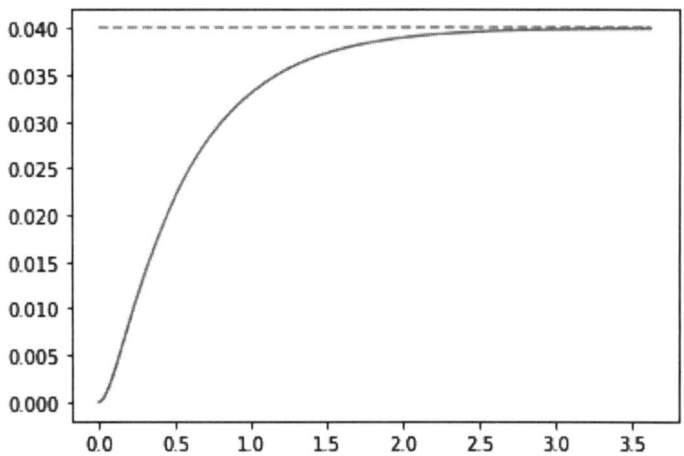


Figure 13-59: Plot of system response to step input

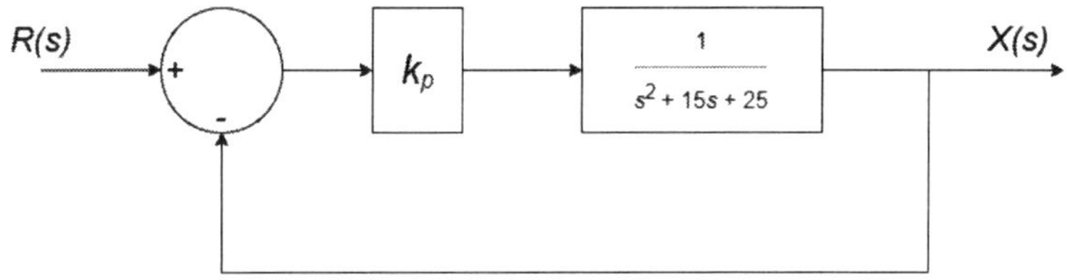


Figure 13-60: Control block diagram with proportional control

If we introduce a proportional controller to the system, the overall transfer function changes. The control block diagram for the system is as indicated in Figure 13-60.

The transfer function then becomes:

$$\frac{k_p}{s^2 + 15s + 25} \qquad \text{Eq. 13-53}$$

Let $k_p = 350$, then we simulate the proportional control on jupyter as:

$$P = tf(350,\ 1)$$
$$T_p = feedback(sys * P,\ 1)$$

To plot the system transfer function multiplied by proportional control, the following codes are run:

$$[c, d] = step(T\_p)$$
$$plt.plot(d,\ c)$$
$$plt.plot(d, np.ones(d.shape), '--')$$
$$plt.axis('tight')$$
$$plt.show()$$

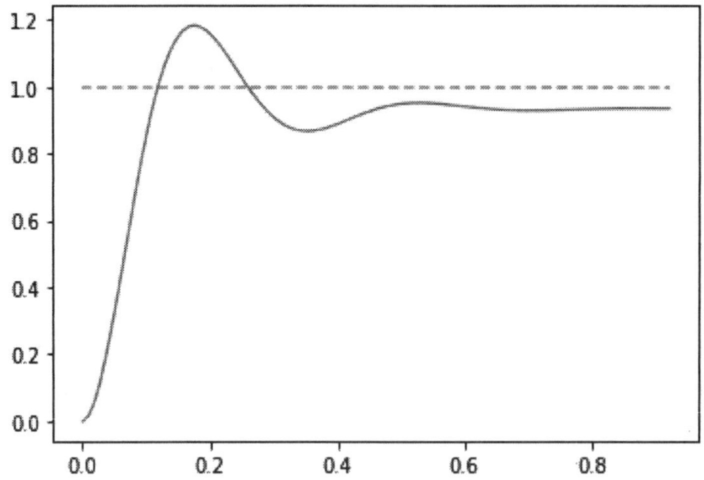


Figure 13-61: Plot of proportional control system response to step input

The output is as shown in Figure 13-61.

We vary the value of proportional gain and run simulations. Let $k_p$ , then new system transfer function is:

$$\frac{80}{s^2+15s+25} \qquad \textbf{Eq. 13-54}$$

To plot the system response, we input the following codes on jupyter:

$$[e,f] = step(T\_p)$$
$$plt.plot(f,e)$$
$$plt.plot(f,np.ones(f.shape),'--')$$
$$plt.axis('tight')$$
$$plt.show()$$

The plot is as shown in Figure 13-62.

From the proportional gain simulation, rise time is improved but an overshoot is generated and a steady state error is present too. For the next simulation we explore introducing a derivative term in the control system, the system control block diagram becomes as shown in Figure 13-63.

PART 3

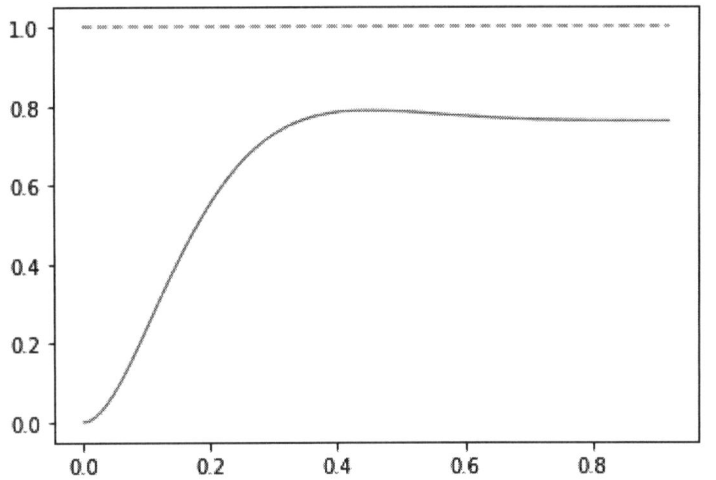


Figure 13-62: plot of system response with varied proportional gain

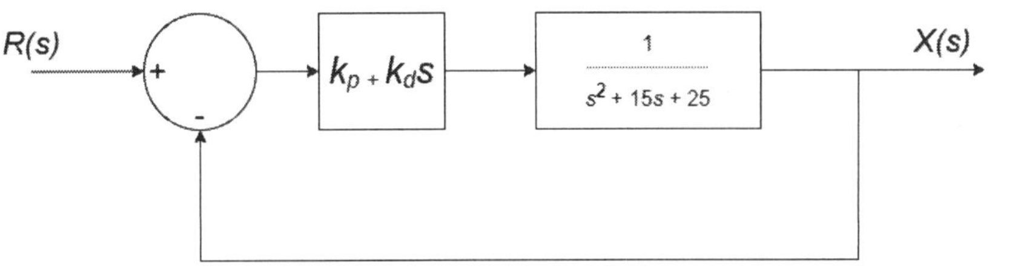


Figure 13-63: Control block diagram of a PD controller

Let $k_d = 10,\ k_p = 350$, then the overall system transfer function becomes:

$$T.F = \frac{10s + 350}{s^2 + 15s + 375} \qquad \textbf{Eq. 13-55}$$

To input the damper coefficient and spring constant on to jupyter we enter the following codes:

$$s = tf('s')$$
$$kp = 350$$
$$kd = 10$$
$$pd\_tf = tf(kp + kd * s)$$
$$pd\_tf$$

To code the overall transfer function on jupyter we define the new $T.F$ as:

$$T\_pd = feedback(sys*pd\_tf, 1)$$

The following codes are run on jupyter to plot system response to a step input:

```
[g,h] = step(T_pd)
plt.plot(h,g)
plt.plot(h,np.ones(h.shape),'--')
plt.axis('tight')
plt.show()
```

The plot for PD control simulated is depicted in Figure 13-64.

To eliminate the overshoot problem and increase the steady-state settling time, we simulate the system using a PI controller. The control block diagram of the new system is shown in Figure 13-65.

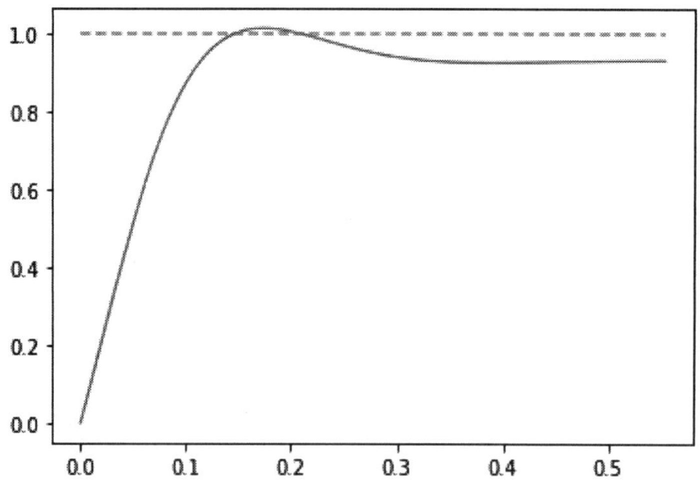


Figure 13-64: Plot of system response of a PD controlled system to a step input

PART 3

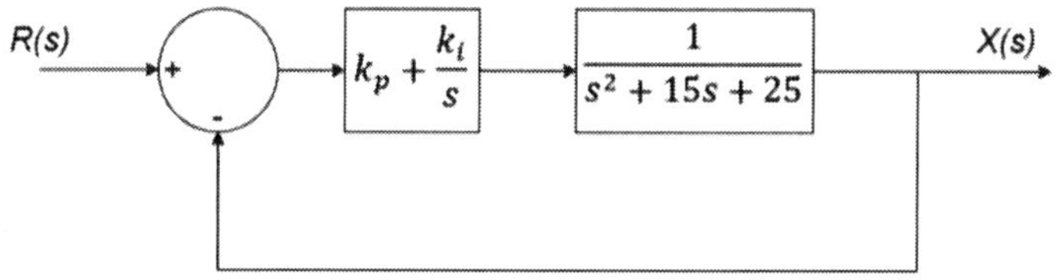


Figure 13-65: Control block diagram of a PI controlled system

Let $k_i = 70$, $k_p = 150$, then the overall system transfer function becomes:

$$PI = tf([150,70],[1,0])$$
$$T\_pi = feedback(sys * PI, 1) \qquad \textbf{Eq. 13-56}$$
$$T.F = \frac{150s + 70}{s^3 + 15s^2 + 175s + 70}$$

The following codes are run on jupyter to plot system response to a step input:

```
[g,h] = step(T_pi)
plt.plot(h,g)
plt.plot(h,np.ones(h.shape),'--')
plt.axis('tight')
plt.show()
```

The plot for PI control simulated is depicted in Figure 13-66.

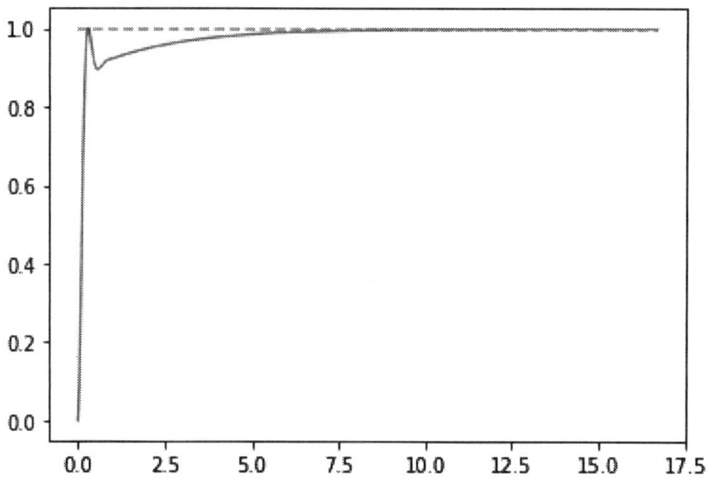


Figure 13-66: System response of a PI controlled system to a step input

### (2) Liquid level control using pneumatic valve

The pneumatic valve can be applied for liquid level control as depicted in the Figure 13-67. This control system can be simulated using python as follows.

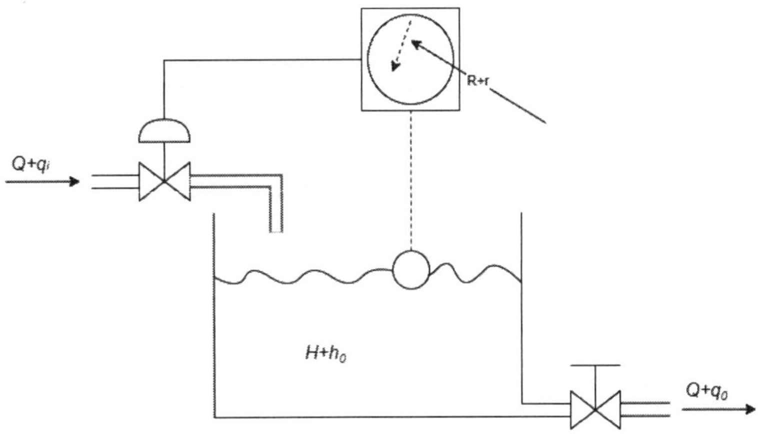


Figure 13-67: Liquid level control diagram

The control block diagram of the system with proportional control only is as shown in Figure 13-68.

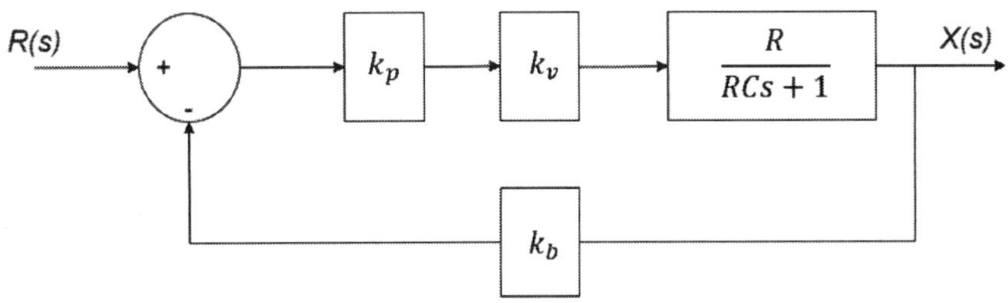


Figure 13-68: Control block diagram of a proportional control liquid level system

From this simplified block diagram (block reduction) we can deduce that:

$$X(s) = \left(\frac{1}{k_b}\right)R(s)$$

Where:

$$K = k_p k_v R k_b \quad \textbf{Eq. 13-57}$$

$$T = RC$$

$$Transfer\ Function = \frac{K}{Ts+1}$$

And for a unity feedback system, transfer function becomes:

$$\frac{H_1(s)}{X(s)} = \frac{K}{Ts+1+K} \quad \textbf{Eq. 13-58}$$

To simulate using python the following libraries are called on jupyter: The jupyter IDE equivalent is shown in Figure 13-69.

```
import numpy as np
import matplotlib.pyplot as plt
import matplotlib.patches as mpatches
import control
from control import *
from control.matlab import *
from scipy import *
from scipy import linalg as la
from scipy.ndimage.filters import convolve
%matplotlib inline
```

Figure 13-69: Libraries imported and called for liquid level control simulation

Using the information below, we simulate the system as:

i. Simulation one we have $k_p = 12$, $k_v = 3$, $R = 2$, $k_b = 1.5$ and $C = 6$
ii. Simulation two we have $k_p = 12$, $k_v = 3$, $R = 1$, $k_b = 1.5$ and $C = 3$
iii. Therefore, $K = 108$ and $T = 12$ from i, and $K = 54$. From ii, $T = 3$

The new transfer function becomes as equation below for case in (i)

This is shown in Figure 13-70.

```
e=108
f=12
sys = tf(e,[0,f,e+1])
sys
```

$$\frac{108}{12s + 109}$$

Figure 13-70: Liquid level control system transfer function

The plot of the response to a step input is as shown in Figure 13-71.

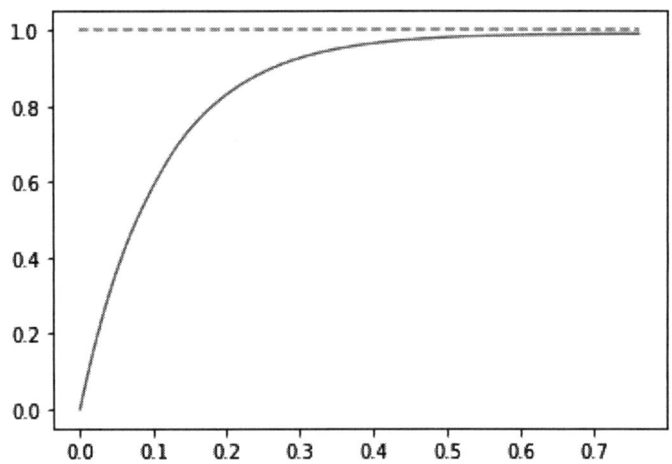


Figure 13-71: Plot of proportional control-liquid level control system response to a step input

The new transfer function becomes as equation below for case in (ii)

$$\frac{H_1(s)}{X(s)} = \frac{54}{3s + 55}$$

$$sys = tf(54,\ [0,\ 3,\ 20])$$

This is shown in Figure 13-72:

```
e=54
f=3
sys = tf(e,[0,f,e+1])
sys
```

$$\frac{54}{3s+55}$$

Figure 13-72: Liquid level control system transfer function with varied P-gain

The plot of the response to a step input is as shown in Figure 13-73:

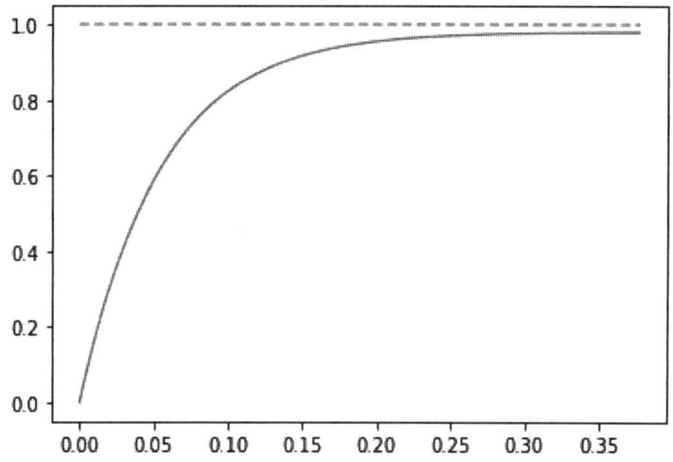


Figure 13-73: Plot of proportional control-liquid level control system response to a step input with varied P-gain

Both cases above when plotted show a steady-state error (offset) present, this is eliminated by introducing an integral controller in the system. Figure 13-74 shows the offset associated with proportional control.

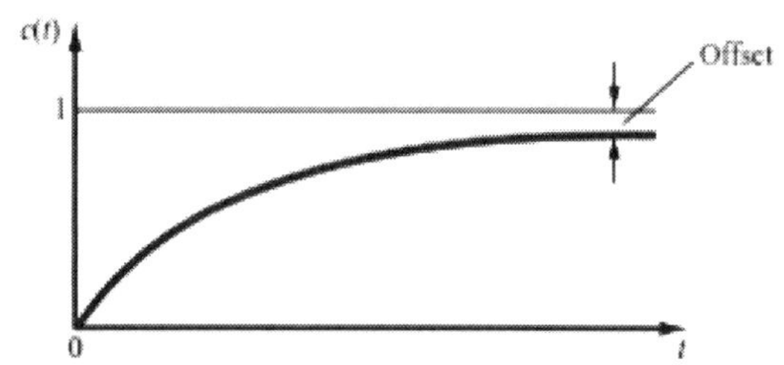


Figure 13-74: Steady-state error/Offset

The PI controlled system, while removing the offset, may lead to oscillatory response of slowly decreasing amplitude or even increasing amplitude, both of which are usually undesirable.

Figure 13-75 shows the control block diagram of the PI controlled system:

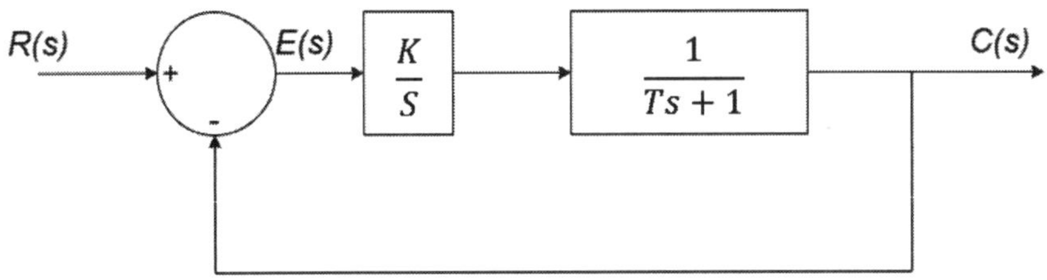


Figure 13-75: Control block diagram of a PI liquid control system

The overall transfer function of the system takes the form:

$$\frac{H(s)}{X(s)} = \frac{KR}{RCs^2 + s + KR}$$

Let $k_i = 50$, then from the scenario in (i) where; $\frac{H_1(s)}{X(s)} = \frac{108}{12s+109}$, the transfer function becomes:

$$\frac{H(s)}{X(s)} = \frac{108}{12s+109} \times \frac{2}{12s+1} = \frac{216s+100}{12s^2+217s+100}$$

This is coded as input on jupyter as shown in Figure 13-76:

```
PI = tf([108,50],[1,0])
PI
```

$$\frac{108s + 50}{s}$$

```
R = 2
C = 6
sys = tf(R,[0,R*C,1])
sys
```

$$\frac{2}{12s + 1}$$

```
T_pi = feedback(sys*PI,1)
T_pi
```

$$\frac{216s + 100}{12s^2 + 217s + 100}$$

Figure 13-76: Liquid level control system transfer function coding on jupyter

The plot is ploted using the codes:

$$[g,h] = step(T\_pi)$$
$$plt.plot(h,g)$$
$$plt.plot(h,np.ones(h.shape),'--')$$
$$plt.axis('tight')$$
$$plt.show()$$

And the plot is as shown in Figure 13-77.

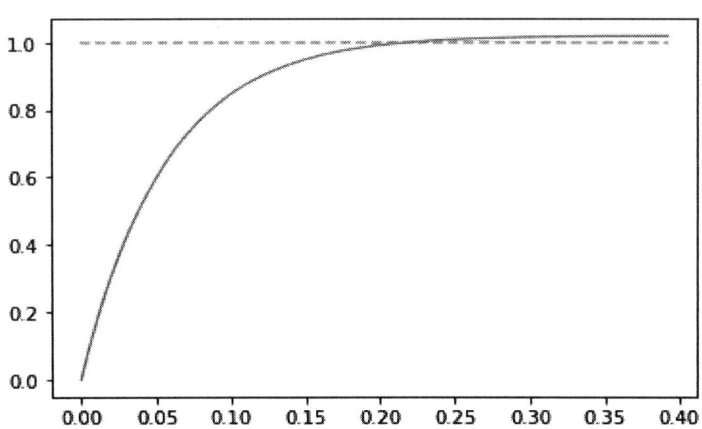


Figure 13-77: Plot of PI control liquid level system response to a step input

## 13.6.8 Control Action

The symbols for the different types of control logic are defined in Table 13-8.

Table 13-8: Control logic

| | |
|---|---|
| MANUAL INPUT | PM05 ○ HS-104 START (SC OR DC)<br>HS-104: CONTROL SWITCH TAG NO.<br>START: MANUAL CONTROL COMMAND.<br>(SC OR DC): TYPE OF CONTROL SWITCH (SC OR DC)<br>○: MOMENTARY CONTROL SWITCH.<br>●: MAINTAINED CONTROL SWITCH.<br>PM05: CONTROL PANEL NO. (THIS NUMBER IS NOT REQUIRED FOR SOFT CONTROL) |
| PROCESS INPUT | LX01 CNDSR A PRESS. HI / PS - 371<br>INPUT TO LOGIC EXISTS WHEN CONDENSER A HIGH PRESSURE SWITVH "PS-371" OPERATES<br>LX01: ORIGINATING POINT OF PROCESS INPUT (PS-371) |
| AND | 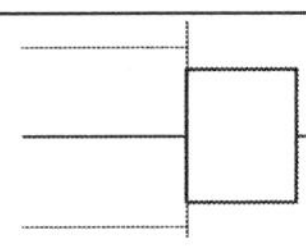 OUTPUT EXISTS ONLY WHEN ALL INPUTS ARE PRESENT. |
| COINCIDENCE LOGIC | 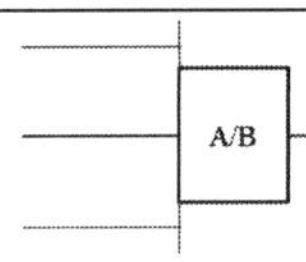 A/B — OUTPUT EXISTS ONLY WHEN AT LEAST A OUT OF B INPUTS ARE PRESENT. |
| OR | 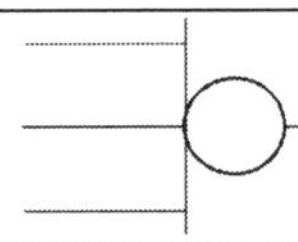 OUTPUT EXISTS ONLY WHEN ANY ONE OR MORE INPUTS ARE PRESENT. |
| NOT | 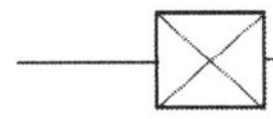 OUTPUT EXISTS ONLY WHEN INPUT IS NOT PRESENT. |
| ON DELAY | 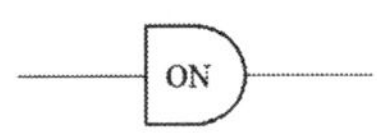 ON — OUTPUT EXISTS ONLY WHEN INPUT HAS BEEN CONTINUOUSLY PRESENT FOR A PRESET TIME AND REMAINS NOT PRESENT. |
| OFF DELAY (TIME MEMORY) | 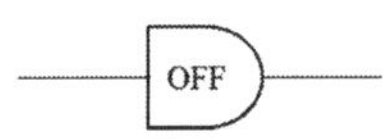 OFF — OUTPUT EXISTS WHEN INPUT IS PRESENT AND ALSO CONTINUES TO EXIST FOR A PRESET TIME AFTER THE INPUT IS NOT PRESENT. |
| S-R MEMORY | 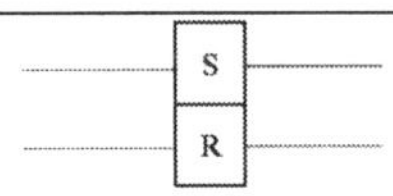 S / R — SET OUPUT EXISTS WHEN SET INPUT IS PRESENT AND ALSO CONTINUES TO EXIST UNTIL THE RESET INPUT IS PRESENT.<br>RESET OUTPUT EXISTS ONLY WHEN SET OUTPUT IS NOT PRESENT.<br>A CIRCLE AROUND Ⓢ INDICATES THAT SET OUTPUT EXISTS WHEN BOTH SET AND RESET INPUTS ARE PRESENT, OTHERWISE, RESET OUTPUT EXISTS WHEN BOTH SET AND RESET INPUTS ARE PRESENT. |
| LATCH MEMORY (OVERRIDE) | 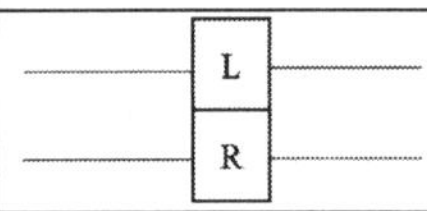 L / R — A DEVICE HAVING MEMORY AND ACTUATING SIGNAL BLOCK LOGIC FUNCTIONS AS INDICATED BELOW. |

| Function | Symbol | Description |
|---|---|---|
| | 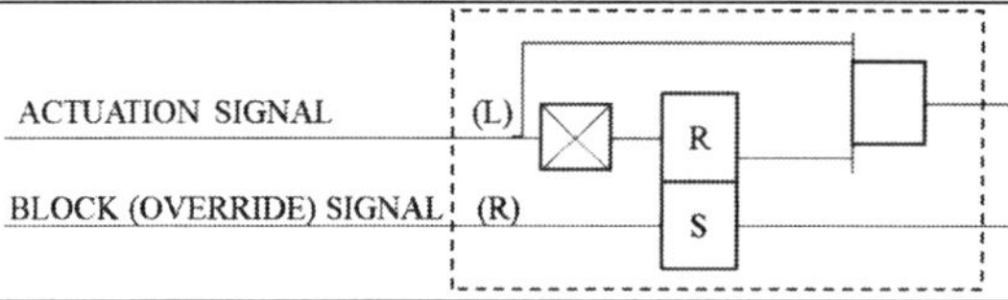  | OUTPUTS EXISTS WHEN LATCH INPUT (ACTUATION) SIGNAL IS PRESENT AND CONTINUES TO EXIST UNTIL THE RESET (BLOCK/OVERRIDE) SIGNAL IS PRESENT. THE LATCH INPUT IS CONTINUOUSLY BLOCKED EVEN AFTER THE RESET SIGNAL DISAPPEARS. THE BLOCK CAN BE REMOVED ONLY WHEN THE ACTUATION SIGNAL IS REACTUATED. |
| TOGGLE MEMORY | 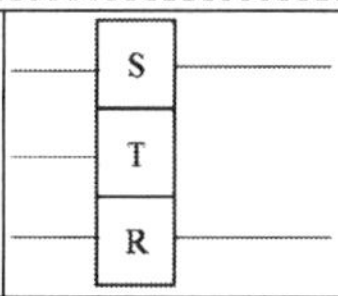  | OUTPUT CHANGES EACH TIME TOGGLE INPUT IS PRESENT.<br>SET AND RESET OPERATE THE SAME AS FOR THE S-R MEMORY GATE<br>THE TOGGLE INPUT IS NOT EFFECTIVE WHN SET OR RESET INPUTS ARE PRESENT. |
| SPECIAL FUNCTION | 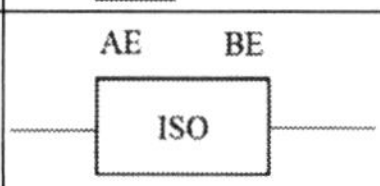  | ISOLATION-OUTPUT IS ISOLATED FROM INPUT WHENEVER ELECTRICAL CHANNEL CHANGES. WHEN THE CHANGE IS BETWEEN AE AND AB OR BE AND BB. THE CHANGE IS NOT SPECIFICALLY SHOWN ON AN INDIVIDUAL LOGIC. IN ALL OTHER CHANGE SUCH AS BETWEEN AE AND BE. IT IS SHOWN ON INDIVIDUAL LOGIC |
| | 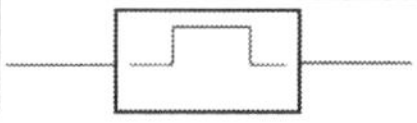 | PULSE – OUTPUT EXISTS FOR A PRESET TIME ONLY ONCE WHEN INPUT IS PRESENT. THE DEFAULT DURATION IS 1 SEC. UNLESS OTHERWISE NOTED. |
| | 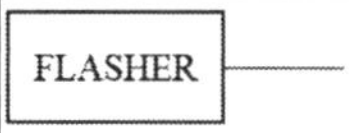  | FLASHER – FLASHER LOGIC SIGNAL GENERATOR |
| |   | POWER UP PULSE LOGIC UNIT |
| LIGHT (DISPLAY ON DC OR SC) | 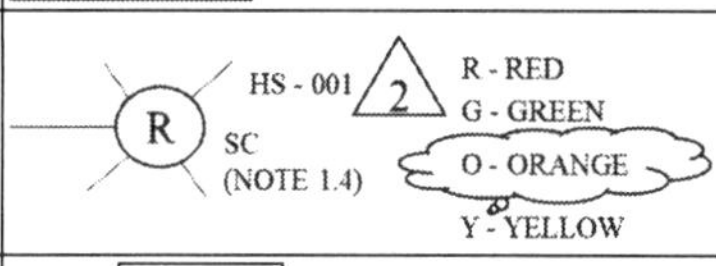  | - OPERATING, FLOWING OR INCREASING.<br>- NOT OPERATING, NOT FLOWING OR DECREASING.<br>- MANUAL, SYNCHRONIZE OR OPERATOR INFORMATION.<br>- TROUBLE/DISABLE |
| ALARM CONTROL (COMPARATOR) | 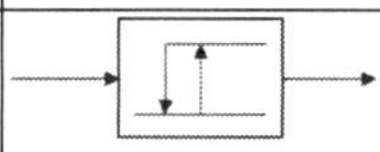 | OUTPUT SWITCHES ON/OFF PER ALARM SET-POINT.<br>HYSTERESIS (DEAD-BAND) APPLIED TO PROTECT SPURIOUS ON/OFF OUTPUT CHANGES. |
| X-Y CHARACTERIZER (Linear Piece-wise Function) | 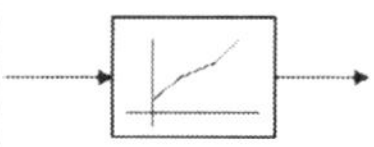 | OUTPUT CHANGES IN LINEAR FUNCTIONS BETWEEN THE GIVEN SET OF N (X,Y) POINTS. |
| TRANSFER GATE (switch function) | 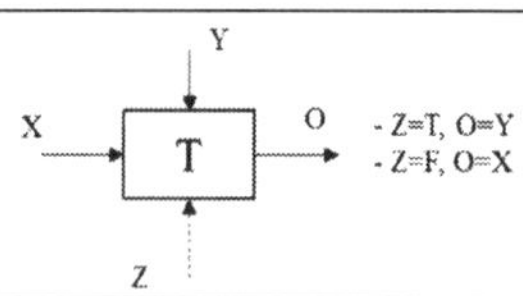  | |
| 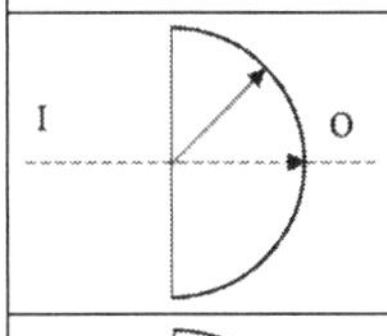  | | ADJUSTABLE ENERGIZING TIME DELAY<br>WHEN I = "0" THEN O = "0"<br>WHEN I = "1" THEN O = "1"<br>FOLLOWING A PRESET TIME DELAY |
| 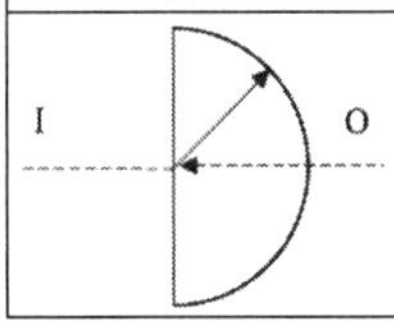  | | ADJUSTABLE DE-ENERGIZING TIME DELAY<br>WHEN I = "1" THEN O = "1"<br>WHEN I = "0" THEN O = "0"<br>FOLLOWING A PRESET TIME DELAY |

| | |
|---|---|
| 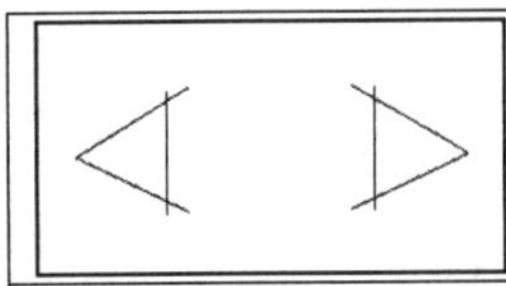 | LOW AND HIGH SIGNAL LIMITER |
| 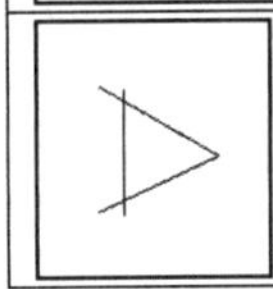 | HIGH SIGNAL LIMITER |

### 13.6.9 Flip-Flops

A flip-flop or latch is a circuit that has two stable states and may be used to store state information. The circuit can be made to change state by signals applied to one or more control inputs and will have one or two outputs. It is the basic storage element in sequential logic. Flip-flops and latches are fundamental building blocks of digital electronics systems used in computers, communications, and many other types of systems. In the scope of NPP, two types of flip flops: RS and LR flip flops are used in the systems.

#### RS Flip Flop

The RS Flip Flop is shown in Figure 13-78 and its truth table is defined in Table 13-9:

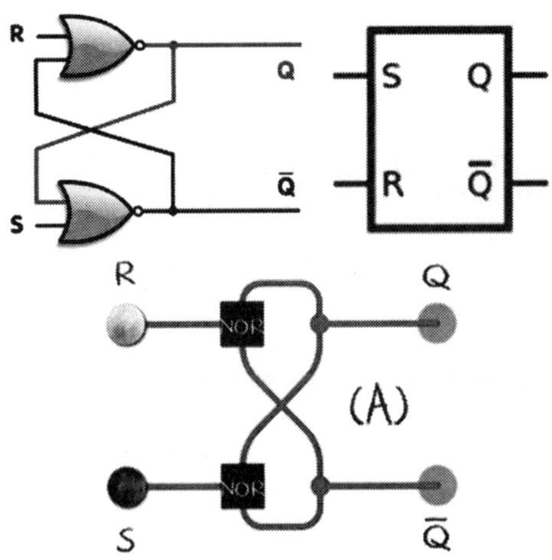


Figure 13-78: RS Flip Flop

Table 13-9: Truth table of RS Flip Flop

| S | R | Qnext | Action |
|---|---|---|---|
| 0 | 0 | Q | Hold |
| 0 | 1 | 0 | reset |
| 1 | 0 | 1 | set |
| 1 | 1 | X | Not Allowed |

## LR Flip Flop

The diagram of the LR Flip Flop is shown in Figure 13-79.

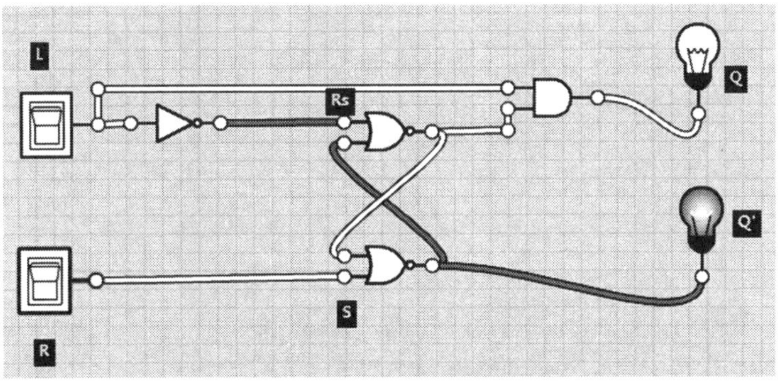


Figure 13-79: LR Flip Flop simulation

The simulations of LR Flip Flop are described in three cases.

- Case 1: If L = 0, R = 0, then Q = 0, Q' = 1. This input resets the flip-flop outputs

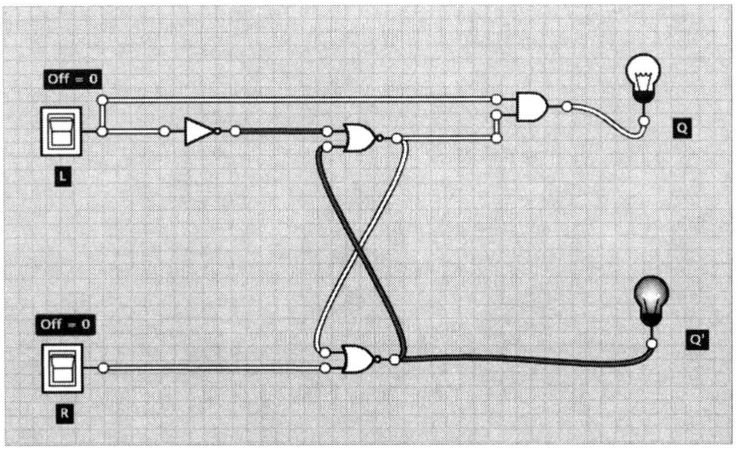

- Case 2: If L = 0, R = 1, then Q = 0, Q' = 0. This state is not allowed as Q and Q'.

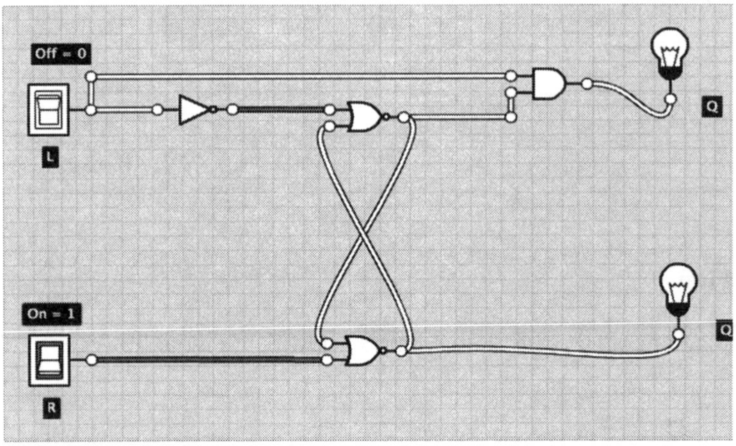


- Case 3: L =1, R = 0
  - If $Q_{prev}$ = 0

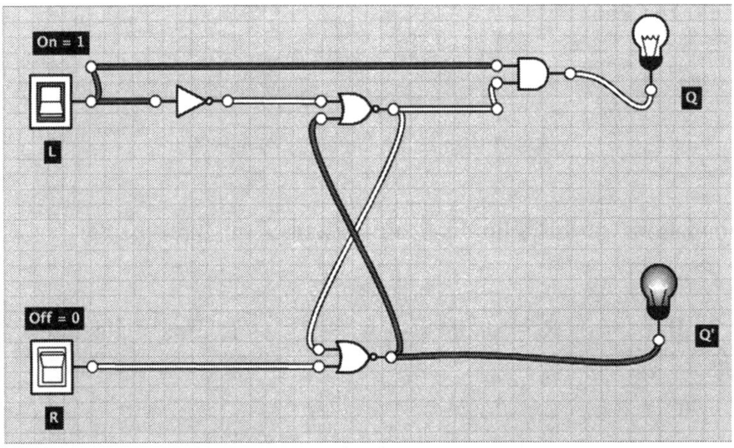

- If $Q_{prev} = 1$

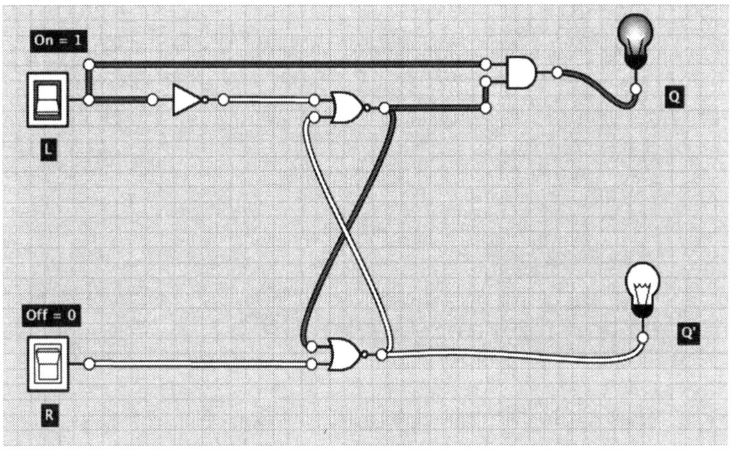


For this input state, the output is dependent on the previous value of Q ($Q_{prev}$).

Thus if $Q_{prev} = 0$, $Q_{next} = 0$.

Similarly, if $Q_{prev} = 1$, $Q_{next} = 1$.

Therefore, $Q_{next} = Q_{prev}$. The previous output value is held

.

- **Case 4**: If L = 1, R = 1, then Q = 1, Q' = 0. This input sets the flip-flop outputs Q to 1, and Q' to 0.

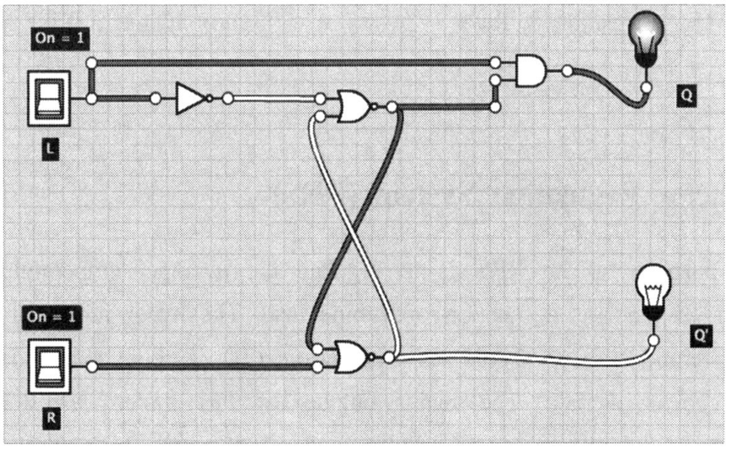

- Truth table in this situation is shown as below:

| L | R | Q | Q' | STATE |
|---|---|---|---|---|
| 0 | 0 | 0 | 1 | RESET |
| 0 | 1 | X | X | NOT ALLOWED |
| 1 | 0 | $Q_{prev}$ | $Q'_{prev}$ | HOLD |
| 1 | 1 | 1 | 0 | SET |

## 13.7 Power Control System (PCS)

In order to control the reactor power, the Power Control System (PCS), which consists of RRS, Digital Rod Control System (DRCS), and Reactor Power Cutback System (RPCS), is used in APR1400. Table 13-10 shows the controlled parameters and controlled components of the systems in the PCS.

Table 13-10: Controlled Parameters Components of the PCS Sub-Systems

| System | Controlled parameters | Controlled components |
|---|---|---|
| RRS | Reactor Coolant Temperature ($T_h$, $T_c$) | Regulating Bank (CEA Group) |
| RPCS | Step Reduction of Reactor Power & Turbine Power | Regulating Bank<br>Turbine Control Valve |
| DRCS | CEA Direction & Rate | Control Element Drive Mechanism |

### 13.7.1 Reactor Regulation System (RRS)

The purpose of the RRS is to regulate the average temperature ($T_{avg}$) to a set-point program as a function of turbine load. The major inputs of the RRS are hot leg temperature ($T_h$), cold leg temperature ($T_c$), Turbine load index (TLI), and reactor power signals. The major outputs of this system are CEA operation direction (withdraw, insert, and hold), and CEA speed (high low) as shown in Figure 13-80. As shown in Figure 13-58, there are the feedback and feed-forward loops in the RRS. These loops process the inputs to generate the outputs which are delivered to other systems, such as the DRCS, the SBCS, the PLCS, and the FWCS. The TLI is provided to a reference temperature (Tref) program which

PART 3

creates the expected average temperature. The hot leg and the cold leg temperature signals are averaged (Tavg) in the RRS. Hence, Tref signal is subtracted from the Tavg signal to give an error signal, named a temperature error signal. The TLI is subtracted from the reactor power to provide the compensation to the temperature error signal (Tavg-Tref) generated. The temperature error signal is dynamically compensated by a lead-lag network and the power error signal (-TLI) is compensated through a high-pass filter. The output of the lead-lag network and the output of the high-pass filter is summed to generate a signal compensated to the lag compensator. This signal is used to control the CEAs so as to determine whether the CEAs are to be moved at a high or low rate, and to a CEA direction program. The outputs of the rate and the direction programs are transmittal to the DRCS. The relationship between reactor coolant reference temperature and TLI is shown in Figure 13-81 [6].

The RRS also provides an Automatic Withdrawal Prohibit (AWP) signal to the DRCS for use when Automatic Sequential is selected to stop CEA withdrawal on a high deviation of reactor coolant average temperature from reactor coolant reference temperature. The RRS also provides an Automatic Motion Inhibit (AMI) signal to the DRCS for use when Automatic Sequential is selected to stop CEA motion on an input channel deviation.

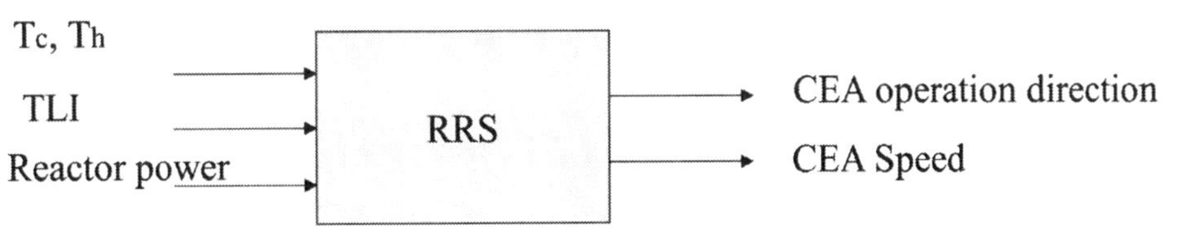


Figure 13-80: Inputs and outputs of RRS

## Design requirement

The RRS yields an automatic control for the reactor coolant average temperature and reactor power, in conjunction with the NSSS control system operation, for each of the following plant conditions during the plant operation from 15 percent and 100 percent power:

- Load rejection of any magnitude.
- Loss of one out of three operating feedwater pumps.

- ±10% step changes in NSSS load. The ultimate load after a step rise shall not surpass 100% of rated thermal power.
- ±5%/min changes in NSSS load. The ultimate load after a rise shall not surpass 100% of rated thermal power.
- Steady-state conditions.

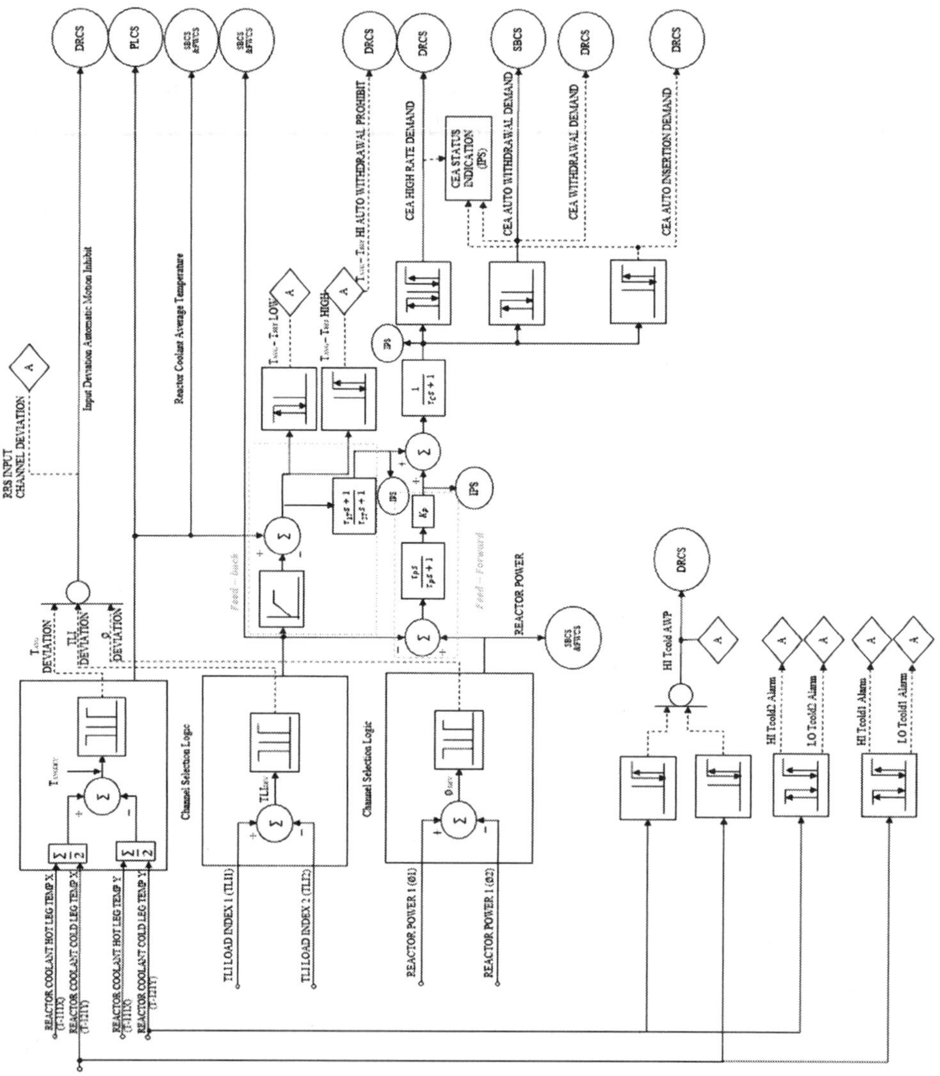

Figure 13-81 RRS Simplified Block Diagram

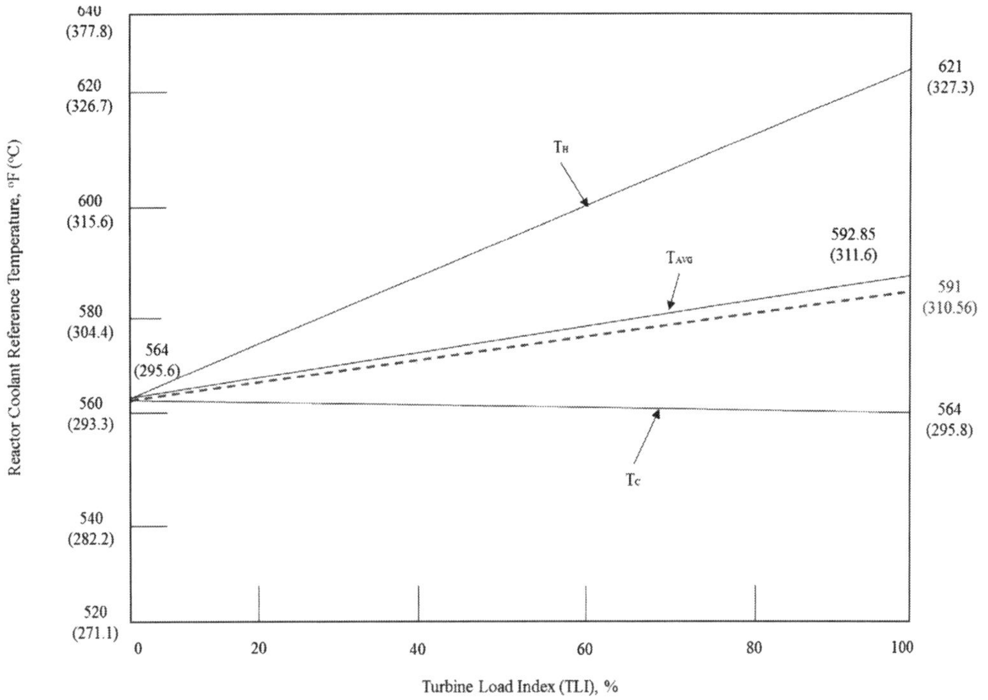


Figure 13-82: Reference temperature program

## 13.7.2 Reactor Power Cutback System (RPCS)

The RPCS is designed to accommodate certain types of imbalances through yielding a step reduction in reactor power. In order to obtain the step reduction, one or more preselected groups of full-strength regulating CEAs in the core is simultaneously dropped. The RPCS also supplies control signals to the turbine control system (TCS) to rebalance turbine and reactor power following the initial mitigation in the reactor power. The steam generator (SG) water level and pressure to their normal controlled values in association with the SBCS are restored when these control signals are provided. The SBCS reactor power cutback (RPC) demand, and loss of the feedwater pump (FWP) RPC demand are expressed as the major inputs of the RPCS. The major outputs of the RPCS include of CEA Bank Arm, CEA Bank Drop, CEA Motion Inhibit, Turbine (TB) Setback, TB Runback, TB load increase inhibit. Figure 13-83 depicts the major inputs and outputs. The RPCS simplified block diagram is shown in Figure 13-84. The SR flip flop which is used in this system processes all inputs to produce the outputs [6].

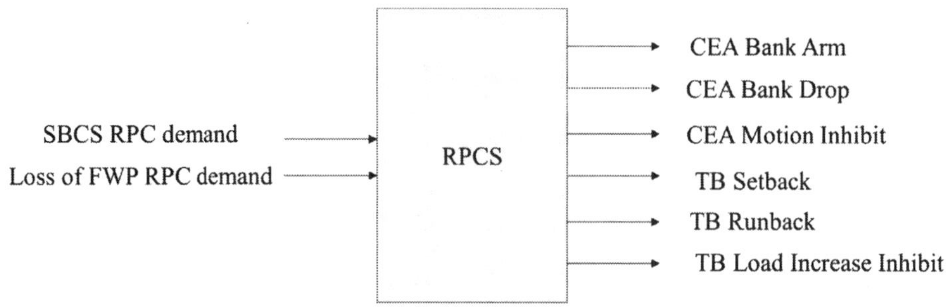


Figure 13-83: Inputs and outputs of RPCS

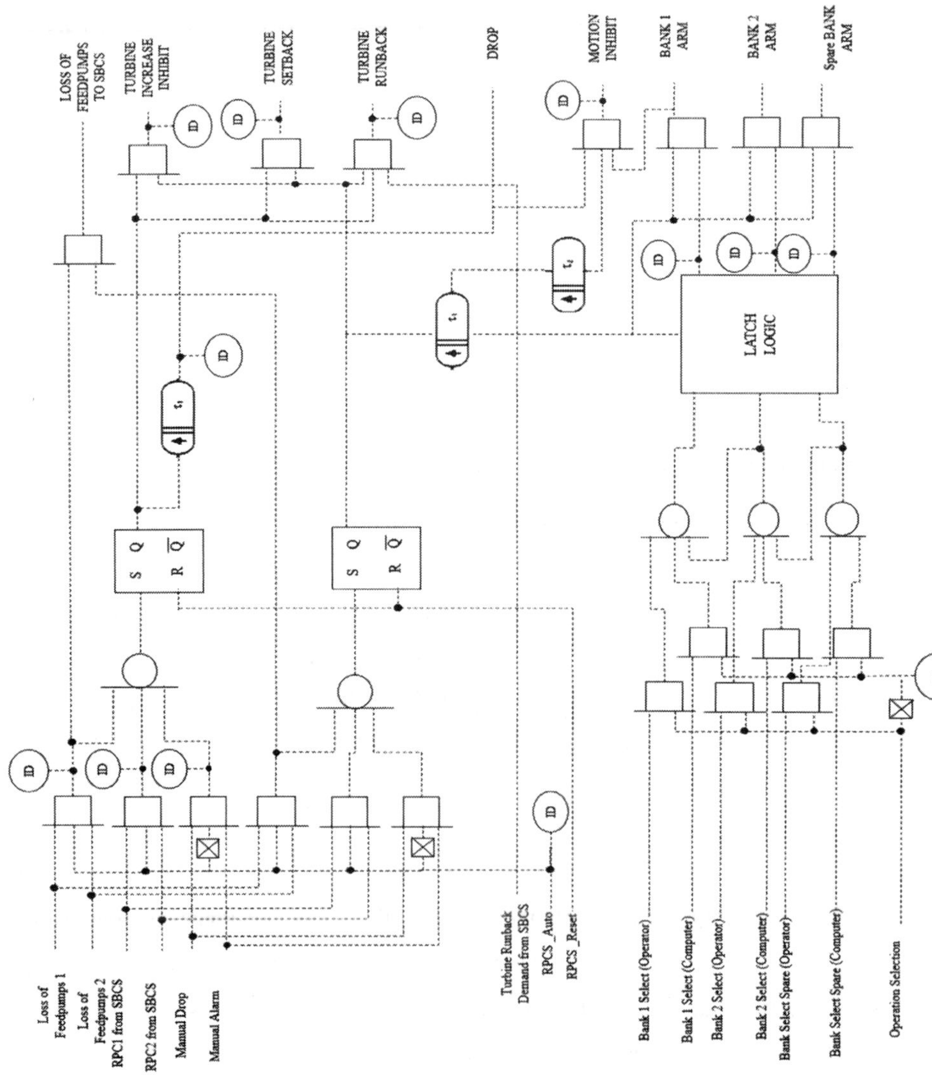


Figure 13-84 RPCS Simplified Block Diagram

## (1) Turbine Runback function

The RPCS provides a signal to the TCS to command a ramp reduction of turbine load at a controlled rate. The turbine runback signal is an output from the RPCS to the TCS if the TLI is above 55%, or if the turbine load surpasses reactor power. The function of turbine runback is to reduce load reference to less than 60%. RPCS operation causes reactor power to be set to turbine setback if it drops below 60%. The turbine runback signal command causes the TCS to decrease on the turbine admission valve area at the rate of 133%/minute, as shown Figure 13-85. Figure 13-86 shows the turbine runback area.

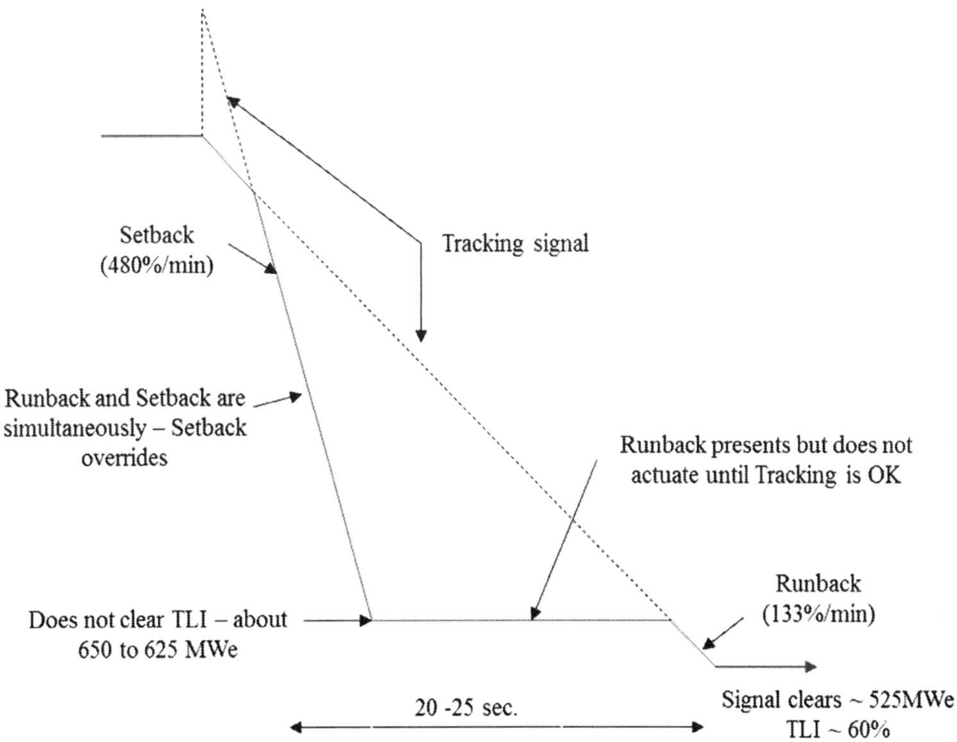


Figure 13-85: TBN output behavior

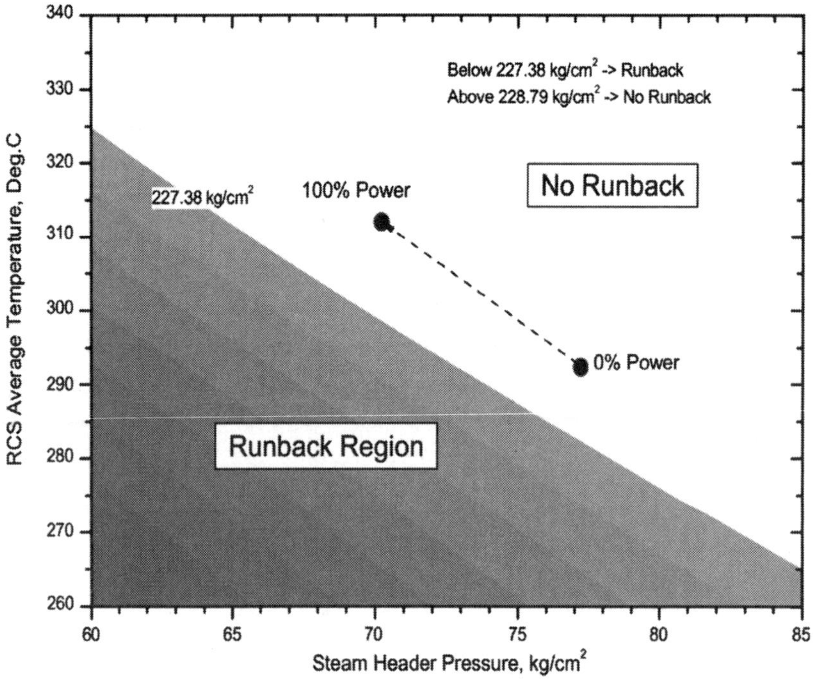


Figure 13-86: Turbine runback area

### (2) Design requirement

The RPCS mitigates the non-vital reactor trips by initiating a swift reduction of the reactor power when it senses of a considerable loss of load, an alleviated capability to sink heat from the core, or other conditions that requires a dramatic decrease in the core power so as to avoid a trip of the reactor.

The RPCS shall provide the NSSS with the likelihood to accommodate serve load rejection of any magnitude, including a turbine trip from full power and the loss of two main feedwater pumps by releasing preselected groups of CEAs which causes a swift reactor power mitigation. The function shall request 2 out of 2 coincidence logic provide by the RPCS.

The RPCS shall mitigate the reactor power to a state from 30% and 75% of rated power after a reactor power cutback based on the CEA selection function of the Nuclear Application Programs (NAPS) server of IPS, and include a means of mitigating turbine power to match reactor power after reactor power cutback.

The RPCS shall provide an operator interface to inform the operator of automatic RPCS actions, and to enable the operator to intervene and manually imitate a power reduction.

The selection of CEAs for RPCS shall be yielded by the NAPS server of the IPS during the computer CEA selection mode and available at any time during normal plant operation.

The RPCS shall be capable of initiating up to 3 banks simultaneously.

### 13.7.3 Digital Rod Control System (DRCS)

Figure 13-87 shows the Inputs and Outputs of the RPCS. The DRCS receives the automatic withdrawal prohibit (AWP), and the automatic motion inhibit (AMI) from the RRS and SBCS. The RRS also delivers the automatic withdrawal and insertion of the CEAs to the DRCS. The CEA withdrawal prohibit (CWP) is sent from the PPS to the DRCS. Additionally, the DRCS receives the signals, such the upper electrical limit (UEL), the lower electrical limit (LEL), the dropped CEA indications as from the Reed Switch Position Transmitters (RSPTs). The DRCS communicates with the RPCS through the CEA bank arm signals, the CEA bank drop signals, and the CEA motion inhibit signal. Further, the DRCS also receives the signals from the NAPS Server of the Information Processing System (IPS). Besides, the Soft Controls of the IPS sends several signals: the mode selection, the group selection, the individual CEA selection, the CEA withdrawal prohibit (CWP) bypass, the CEA insert/withdraw control, and CEA local control permission to the DRCS [6].

There are a number of outputs between the DRCS and the IPS. This system sends the CEA position information, the DRCS operating modes, the DRCS motion command, the sequential permissive, the DRCS operating modes, the constants, UGS status, LGS status, totalizer count, individual CEA selection, group selection, DRCS UEL/LEL, the automatic withdrawal prohibit (AWP), the automatic motion inhibit (AMI), the CEA withdrawal prohibit (CWP), the CEA motion inhibit (CMI), DRCS bus under-voltage, continuous gripper high voltage, continuous motion alarm, subgroup insertion status, subgroup withdrawal status, DRCS urgent alarm, DRCS non-urgent alarm to the IPS. In addition, the DRCS also transmits the signals to the TCS, the FWCS, the SBCS, the loose parts monitoring system (LPMS), the RRS, and the QIAS-N.

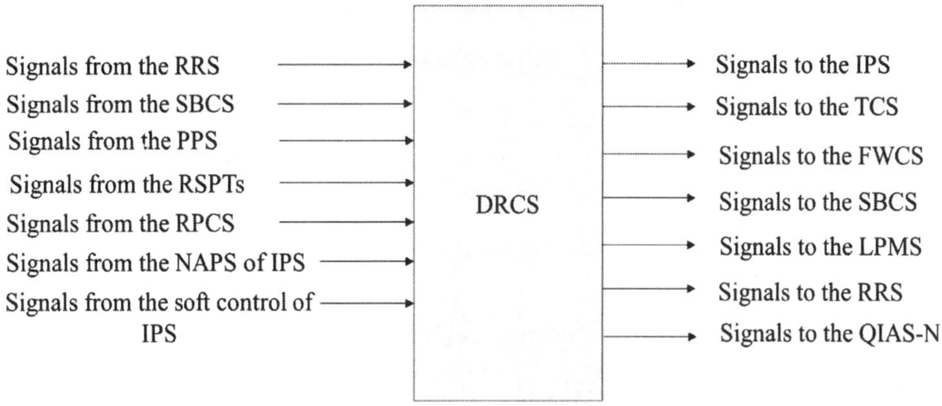


Figure 13-87: Inputs and Outputs of DRCS

### 13.7.4 System Interfaces

The interfaces of the PCS, such as the RRS, the RPCS, and the DRCS, and other systems as shown in Figure 13-88. The RRS receives Tavg (Tc, Th), TLI, and reactor power; and sends CEA withdrawal, and CEA Insertion to the DRCS. The major inputs of the SBCS are pressurized (PZR) pressure, reactor power, TLI, and steam flow. The outputs of this system: automatic withdrawal prohibit (AWP), and Automatic Motion Inhibit (AMI) sends the DRCS.

The P-CCS transmits loss of feedwater pump (LOFP) RPC demand to the RPCS, and the LOFP signal is sent to the SBCS in order to exhibit the loss of two main feedwater pumps from the RPCS. The RPCS receives the SBCS RPC demand signal from the SBCS to illustrate the need for the RPCS to accommodate a large turbine load rejection. The SBCS also delivers the turbine runback signal to the RPCS to show the need for the RPCS to control the runback of the main turbine when a reactor power cutback is initiated.

The RPCS also receives the three signal for dropping the CEA group: regulating group 5, regulating group 4, and spare from the nuclear application programs (NAPS) server of the IPS. Furthermore, when the reactor power is greater than 75%, the RPCS provides the turbine load increase inhibit, the turbine setback, and the turbine runback signals to the TCS. The RPCS also sends the CEA arm, CEA drop, and CEA Motion Inhibit (CMI) signals to the DRCS. The RPCS send the alarm signal to the IPS.

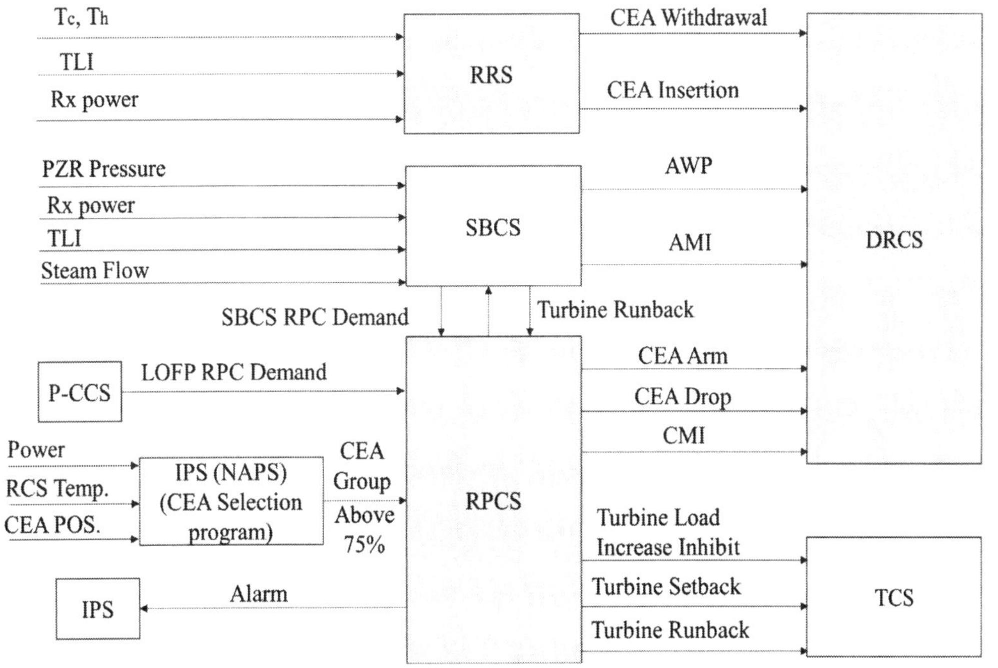


Figure 13-88: Schematic of Power Control System

## 13.8 NSSS Process Control System (NPCS)

There are several system such as the feedwater control system (FWCS), steam bypass control system (SBCS), pressurizer pressure control system (PPCS), pressurizer level control system (PLCS), and other miscellaneous NSSS control functions in the NPCS.

Table 13-11: NPCS Controlled Parameters and Components

| System | Controlled parameters | Controlled components |
|---|---|---|
| FWCS | Steam Generator Level | Feedwater Control Valve (downcomer and economizer)<br>Feedwater Pump Speed |
| SBCS | Steam generator pressure<br>Turbine Bypass Valves<br>Steam Header Pressure | Turbine Bypass Valve<br>Steam Header Pressure |
| PPCS | Pressurizer Pressure | Pressurizer Heater |
| PLCS | Pressurizer Level | Charging Control Valve<br>Letdown Isolation Valve |

### 13.8.1 Feedwater Control System (FWCS)

The function of the FWCS is to control the SG water level by regulating the feedwater flow rate. As shown in Figure 13-89, the major inputs of this system are Steam Flow, Feedwater flow, SG Water Level, Reactor Power, TAVG, SBCS Main Mod. Demand, TLI, Downcomer flow, Steam Header Press, FW Common Header Press, FW temperature. The major outputs are economizer/downcomer valve control output, main feedwater pump speed output.

The control methods, including feed-forward, and feed-back, of the FWCS are shown in Table 13-12. Figure 13-90 which is the functional block diagram of the FWCS also indicates the two controlled loops: feedback and feed forward. The main controller in the FWCS is the PI controller. The feedback part includes the PI controller, which receives the SG water level and reactor power through the lead/lag compensator. The feed forward part consist of the steam flow and feedwater flow signals, which are processed by the lag compensators and the high-pass filer. The main controller processes all inputs to generate the outputs, which are used to regulate the SG water level at different power levels: high power areas and low power areas. The additional functions of the FWCS consists of High Level Override (HLO), and Reactor Trip Override (RTO). Figure 13-67 also illustrates the post-trip refill control system, which is described in the next partition.

As shown in Figure 13-91, the feedback and the feed forward loops are displayed using the block diagram. The three signals, such as the SG level, the feedwater flow, and the steam flow, are transmittal to the PI controller to generate the flow demand signal. This flow demand signal goes to downcomer/economizer feedwater control valve programs and other FWCS, which is shown in Figure 13-90. The flow demand, which is generated through the PI controller, and the flow demand from other FWCS go to high select circuits where the higher flow demand will be selected. The selected flow demand is transmittal to the pump speed program. Figure 13-92 shows the control logic of the economizer valve, the downcomer valve, and the main feedwater pump [6].

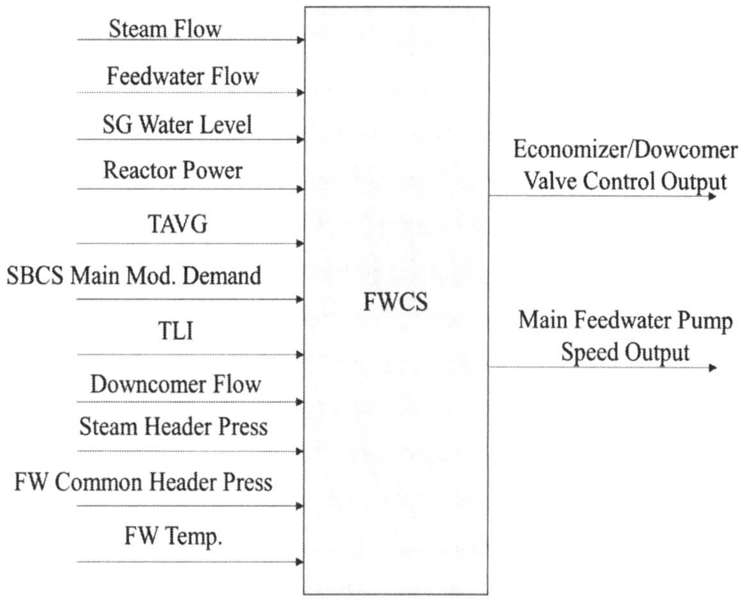


Figure 13-89: Inputs and outputs of FWCS

Table 13-12: Control method

| Division | Control method | Deviation calculation |
|---|---|---|
| Flow deviation | Feed-Forward | Steam flow rate - Feedwater flow rate |
| Water level deviation | Feed-back | Water level set-point value - Measurement level |

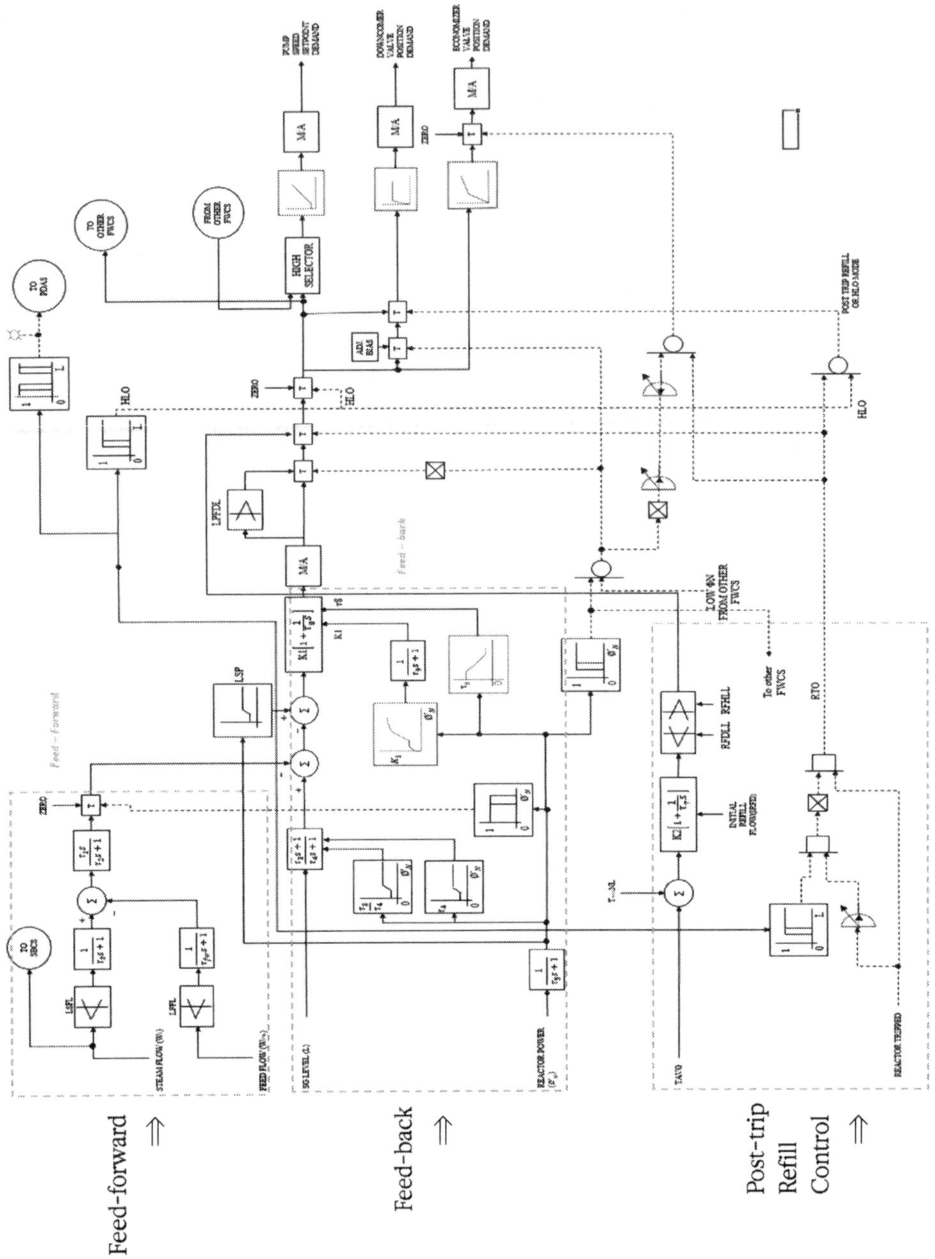


Figure 13-90 FWCS Functional Block Diagram

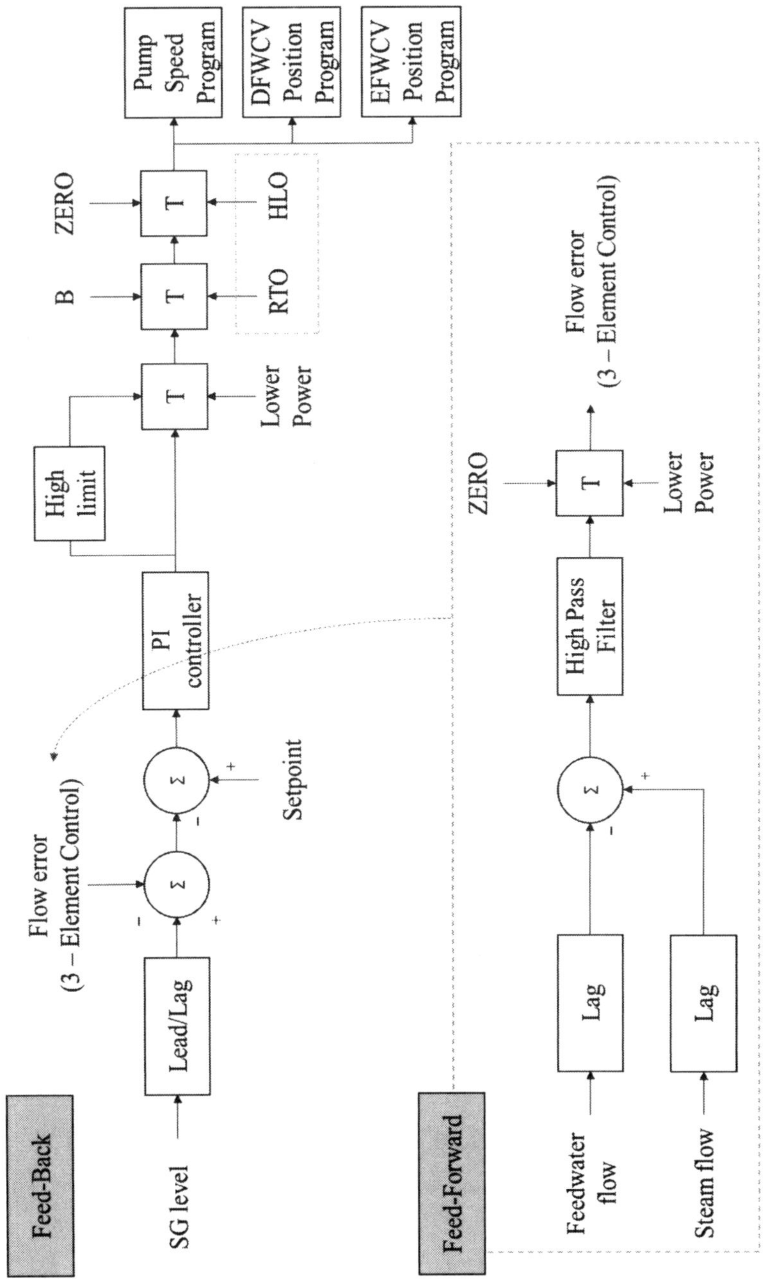


Figure 13-91 Feedback and Feed-forward Block Diagram of FWCS

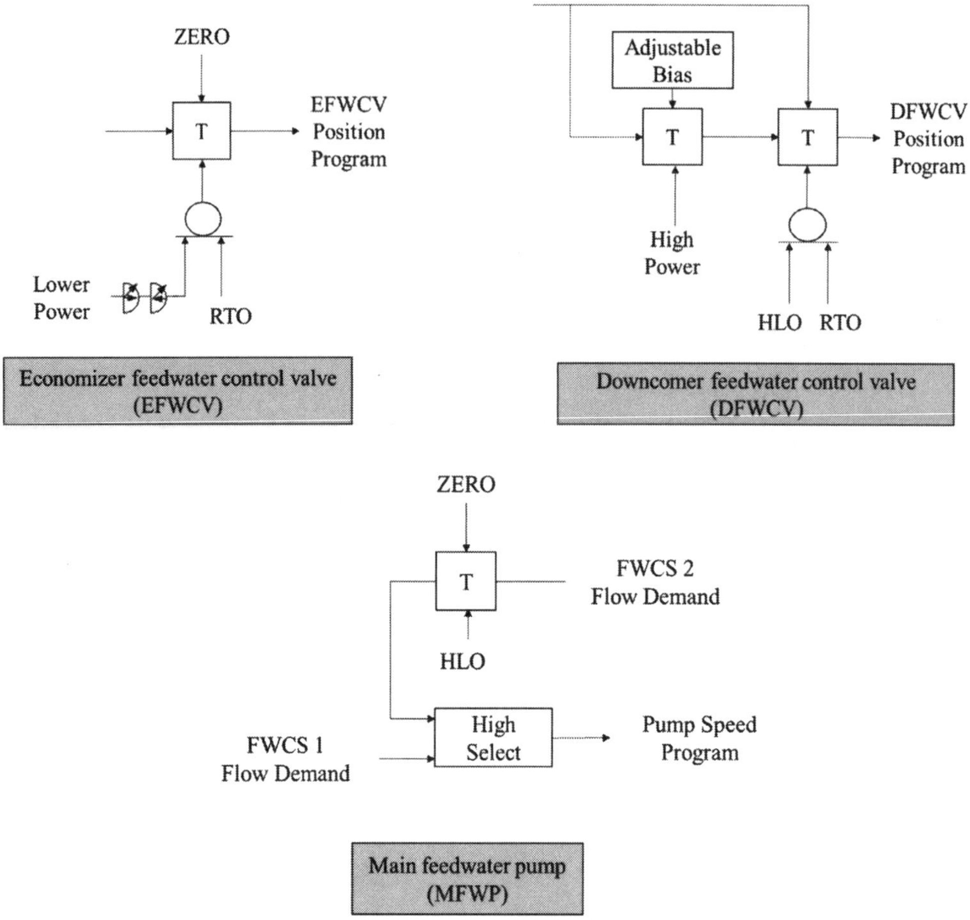


Figure 13-92: Output Control of FWCS

## (1) Post-Trip Refill Control System

One of the control function of the FWCS is post trip refill. After the reactor is tripped, the feedwater flow rate is controlled by employing the primary coolant loop average temperature ($T_{AVG}$) signal in order to prevent RCS overcooling. This is because that the SG shrinking due to sudden an increase in SG pressure in the early trip causing excess provision of the feedwater flow rate, and hence resulting in RCS overcooling. The $T_{AVG}$ is subtracted by the $T_{AVG,NL}$ set-point signal. Hence, this signal is processed by a PI controller, and the output of this controller is the flow demand signal after a reactor trip. During the steady-state, the error between $T_{AVG}$ and $T_{AVG,\ NL}$ is forced to be zero using the PI controller. Post-Trip Refill Control System is shown in Figure 13-93. The SG level response after reactor trip is shown in Figure 13-94. The level set-point and the post-trip exit condition are approximately 44%, and 40%, respectively. The high level override and the post trip enter condition are 20% and 80% of SG narrow range level, respectively.

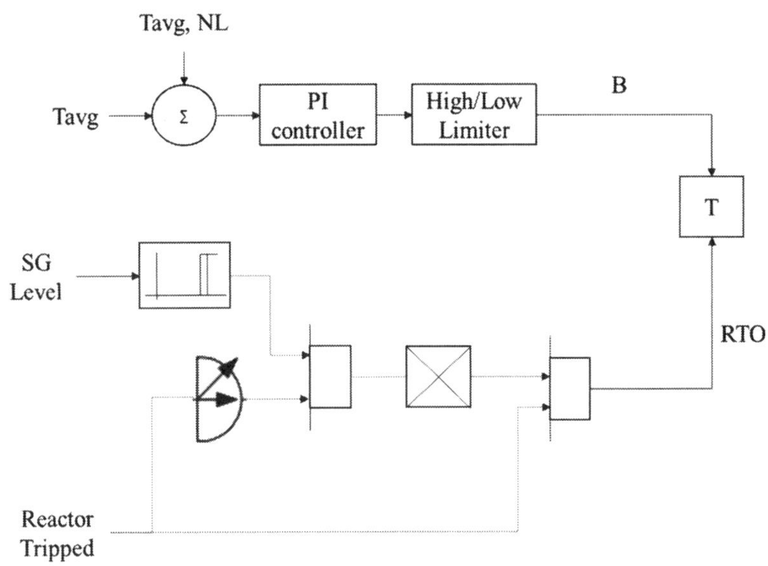


Figure 13-93: Post-Trip Refill Control System

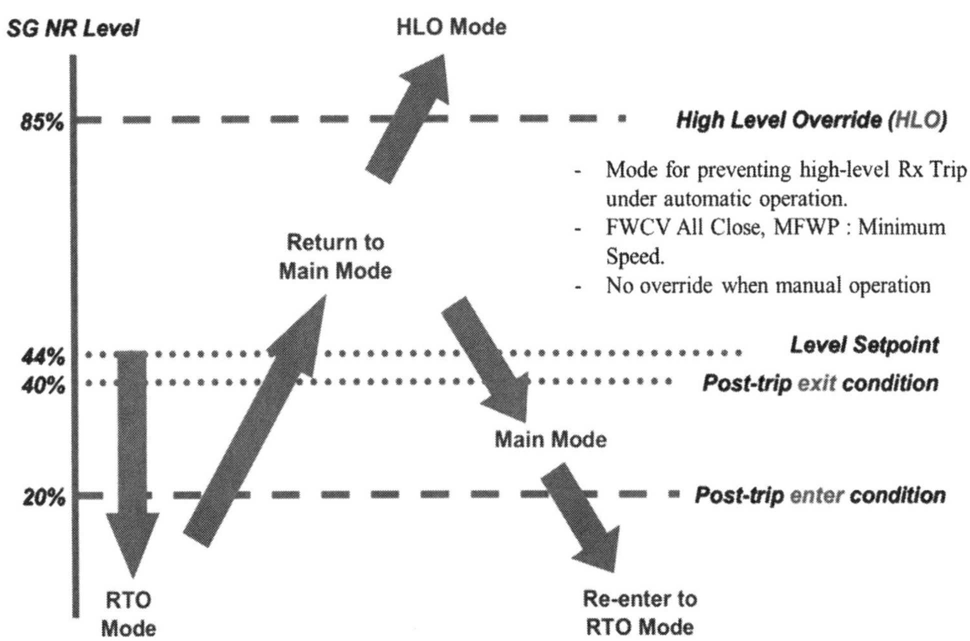


Figure 13-94: SG Level Response after Reactor Trip

Table 13-13 shows the feedwater operation according to output. There are many reactor power levels from the start-up to the normal operation of the FWCS. As shown in Table 13-13, each reactor power level has FWCS inputs, control valve and main feedwater pump, respectively. Figure 13-95 reveals clearly the diagram of feewater operation according to output. This diagram divides into three stage: manual operation (three elements), startup feedwater pump, and downcomer feedwater bypass valve. Figure 13-96 shows the operation of EFWCV and DFWCV. The EFWCV to each the SG is automatically adjusted using the FWCS for each SG at the reactor power levels above 20%. The DFWCV to each the SG is automatically adjusted using the FWCS for each SG from the start-up operation modes to the power operation mode.

Table 13-13: Feedwater operation according to output

| Reactor power | FWCS input | Control valve | Main feedwater pump |
|---|---|---|---|
| 0% - 5% | Manual operation | DFWBV (Downcomer Feedwater Bypass Valve) | Starting feed water pump |
| 5% - 15% | Steam generator level | DFWCV | One MFWP |
| 15% - 20% | Steam generator level, Weekly feeding, Main steam flow | | |
| 20% - 50% | | DFWCV + EFWCV | |
| 50% - 100% | | | Two MFWP |
| Reactor shutdown | | DFWCV | One MFWP |

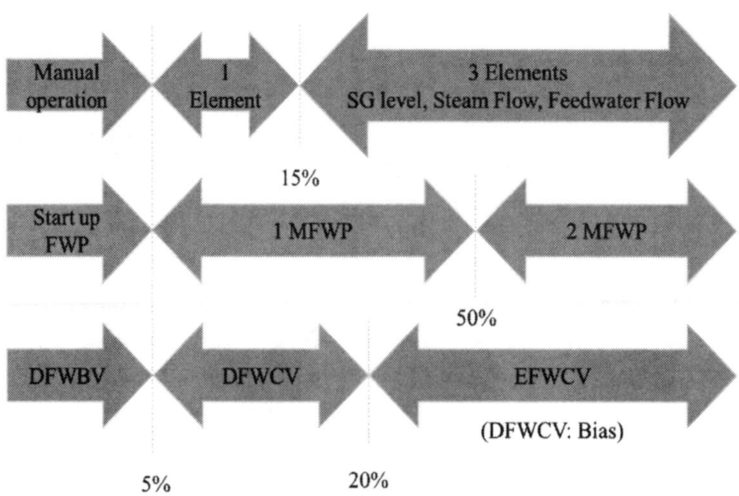


Figure 13-95: Diagram of Feedwater operation according to output

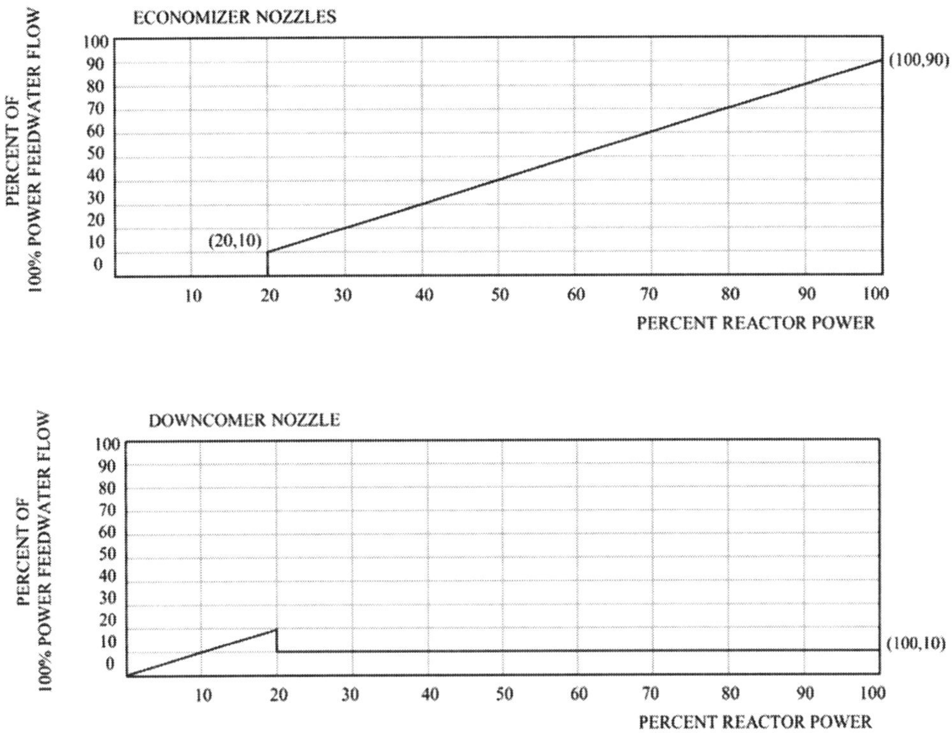


Figure 13-96: Operation of EFWCV and DFWCV

## (2) Design requirement

The SG level is controlled during the several modes of operation.

- The first mode is steady-state operation.
- The second mode consists of power ramping from 5% to 100% at a rate of 1% per minute between 5% and 15%, and a rate of 5% per minute between 15% and 100%.
- The third mode consists of power stepping from 5% to 100% at a rate of 1% between 5% and 15%, and a rate of 10% between 15% and 100%.
- The fourth state is caused by loss of one of three operating feedwater pumps.
- Load elimination of any magnitude is the final mode.

### 13.8.2 Steam Bypass Control System (SBCS)

The SBCS aims at dissipating excess energy in the NSSS by regulating the steam flow using turbine bypass valves. The steam flow, steam header pressure, and pressurizer pressure are the major inputs of the SBCS. There are several major outputs: modulation, quick open, RPC, AMI, and PRZ AWP as shown in Figure 13-97.

In the SBCS, there are two main functions, known as Quick Opening and Modulation. The control method and inputs of each function are shown in Table 13-14. The quick opening function is controller through the feed forward method, including the steam flow rate, and the pressurizer pressure signals as the inputs. The modulation function is controlled through the feedback method, including the steam flow (set-point), the steam header pressure, and the pressurizer pressure as the inputs.

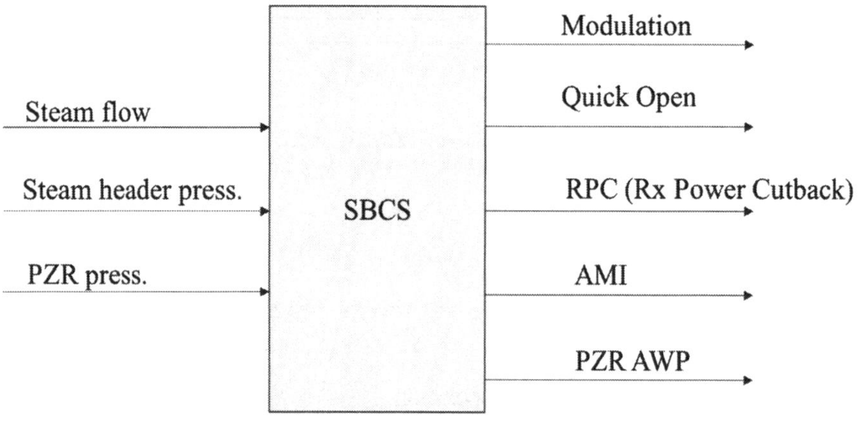


Figure 13-97: Inputs and outputs of SBCS

Table 13-14: SBCS Control method

| Function | Control method | Input |
|---|---|---|
| Quick Open | Feed-Forward | Steam flow rate<br>Pressurizer pressure |
| Modulation | Feed-back | Steam flow (Set-point)<br>Steam header pressure (Measurement)<br>Pressurizer pressure (Bias) |

The SBCS Functional Block Diagram is shown in Figure 13-75 in which the feed-forward and the feedback are pointed out. In the SBCS, there are the feed-forward and feed-back loops. The control loops process the inputs to generate the outputs which are used to operate eight turbine bypass valves. Figure 13-98 also shows the turbine runback and quick open block, the AMI station, the automatic permissive, and the turbine valves bypasses, which are debated clearly, in the continuous part.

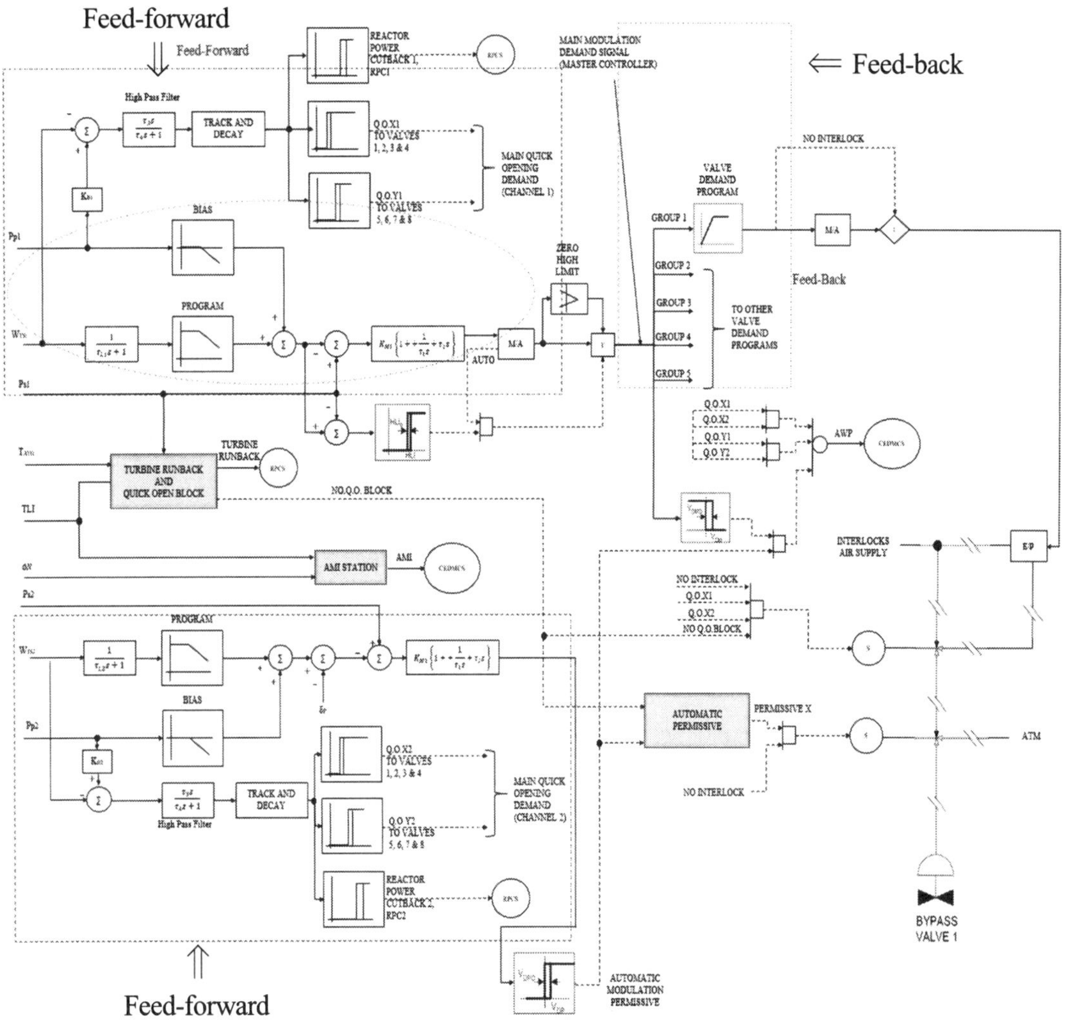


Figure 13-98: SBCS Functional Block Diagram

### (1) Quick opening Mode

As shown in Figure 13-99, when the reactor power trip or loss of main feedwater pump (LOMFP), the Quick Opening (Q.O) occurs if Tavg is greater than 578.4oF when loss of two out of three main feedpump status and the trip of reactor.

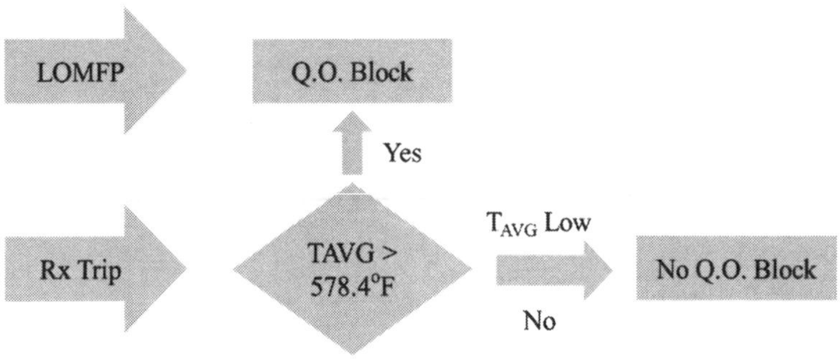


Figure 13-99: Quick Open Block

### (2) Modulation Mode

Valve Modulation is utilized to control relatively slow NSSS transients. The SBCS continuously monitors the main steam header pressure. When an excessive increase in above pressure is detected, steam from the main steam line is directly passed to the condenser through the turbine by pass valves.

For valve modulation, the SBCS have a comparison between an actual steam header pressure and a steam pressure header set-point produced as a function of NSSS power, with steam flow used as a power index. At the same time, the pressurizer pressure bias program is produced from the pressurizer pressure. The outcome of error signal is processed by a PID controller and the output transmits through a master controller to actuate valve.

Two steam header pressure set points are generated in an identical and independent fashion. The first steam header pressure is selected for use in the Main Modulation Control Channel, which produces the valve control signals. The second one is used for the Permissive Modulation Control Channel, which generates the valve permissive signals.

The Main (Permissive) Steam Header Pressure Set Points $SP_M$ ($SP_P$) are identically set points representing the pressure which the SBCS will maintain whenever bypass action is demanded. Refer to Figure 13-100 to see the Main steam header pressure set-point. Figure 13-101 shows the SBCS Modulation Set-point Program. Figure 13-102 illustrates the computation of pressure set-point [6].

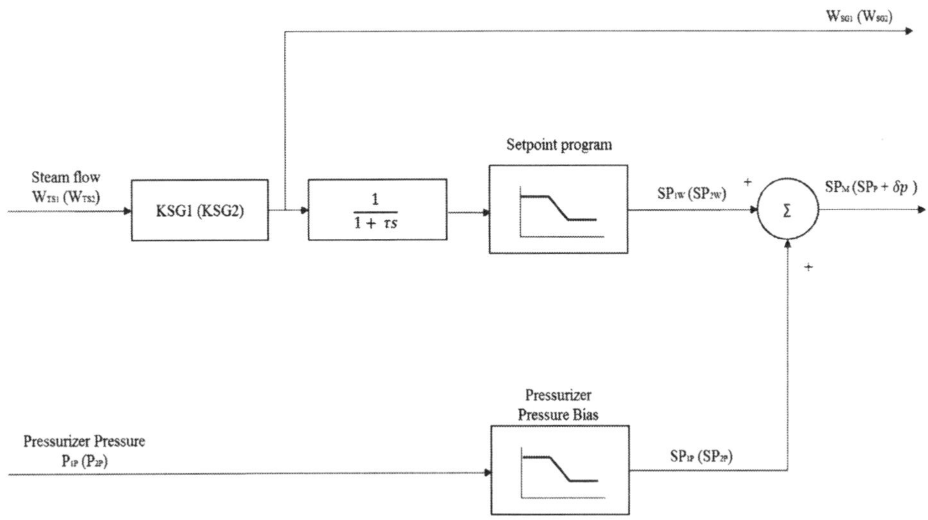


Figure 13-100: Main steam header pressure set-point

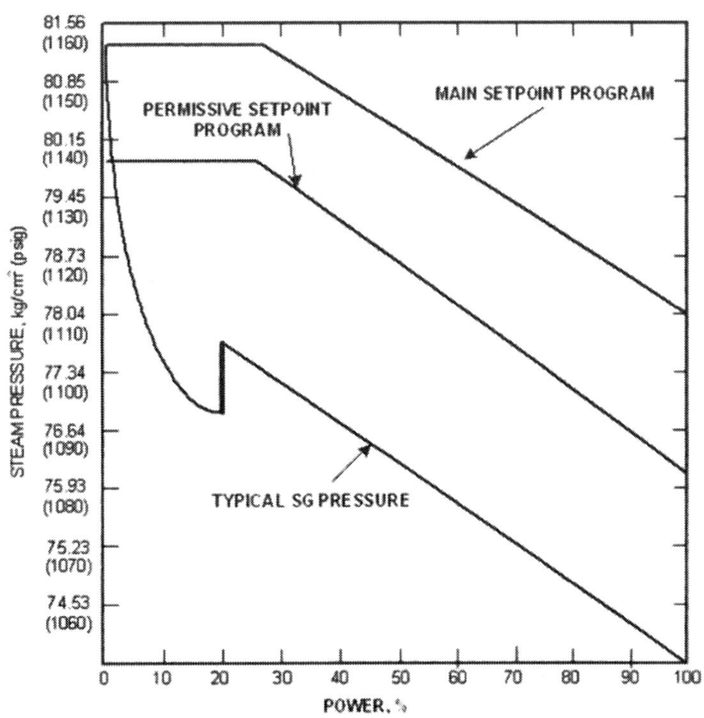


Figure 13-101: SBCS Modulation Set-point Program

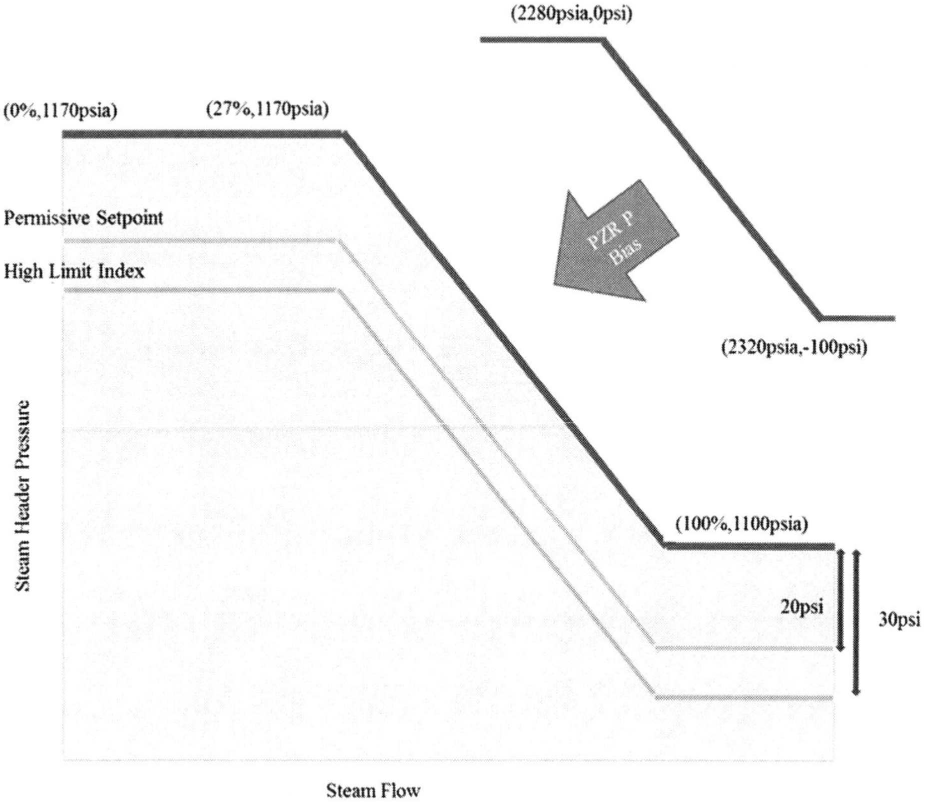


Figure 13-102: Calculation of pressure set-point

### (3) Turbine Bypass Valve Opening Program

There are eight (8) turbine bypass valves in the SBCS which discharge steam to the condenser by bypassing the main turbine. The turbine bypass valves will not open if the condenser vacuum is not maintained, and shall be closed, if open during plant operation. All turbine bypass valves shall open in sequence and shall be exhausted to the main condenser. The turbine bypass valve opening program is shown in Figure 13-103. In the modulation mode, the SBCS opens the turbine bypass valves sequentially by group. There are five groups: group I is valve 1, group II is valve 2, group III consists of valves 3 and 4, group IV consists of valves 5 and 6, and group V includes valves 7 and 8. In the quick opening mode, the SBCS results in quick opening of the turbine bypass valves in two groups, named group X and group Y. These groups shall be opened sequentially as required based on the load rejection magnitude. Group X includes valves 1, 2, 3, and 4. Group Y includes valves 5, 6, 7 and 8 [6].

PART 3

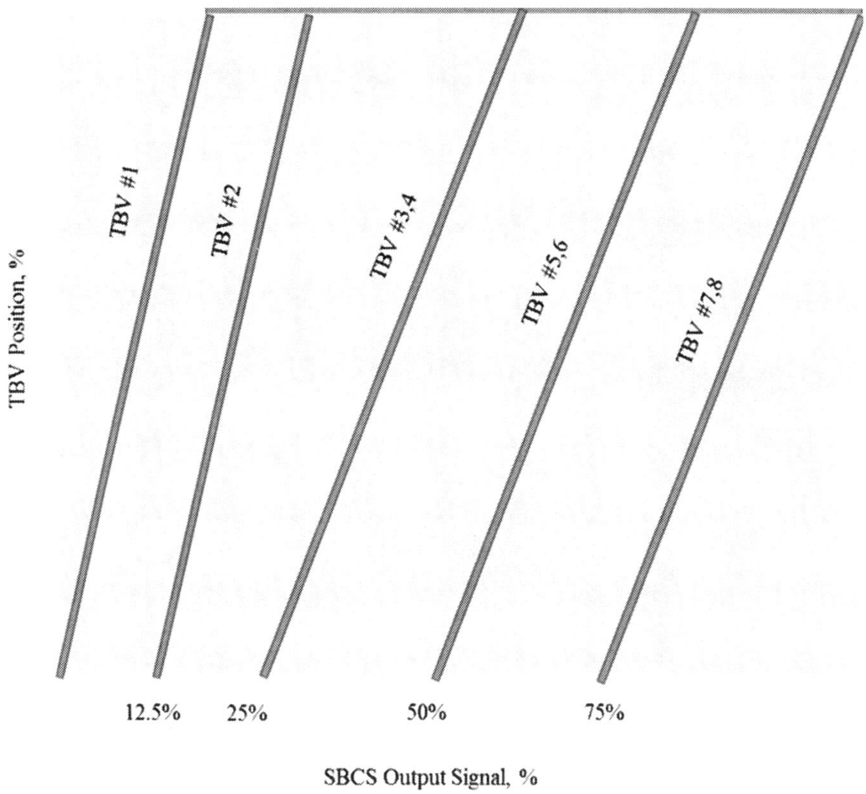


Figure 13-103: Turbine bypass valve for modulation mode

## (4) CEA Automatic Operation Prohibition (AMI)

The SBCS consists of the Automatic Motion Inhibit (AMI) feature to provide additional plant operating flexibility and simplify the operator's task, eliminating in some instances the urgency for intervention of the operator. This feature enables the operator to choose a reactor power level (AMI set-point) below which the operator desires that automatic reactor power control to be blocked if turbine power is below another level (AMI Permissive set-point). However, the AMI set-point is automatically limited to a maximum value corresponding to the steady state heat dissipating capacity of the SBCS/turbine combination, then generating the AMI threshold, which is the reactor power level below which the AMI will happen. However, regardless of any values, the AMI action always occurs below about 15% reactor power level. This level is the lower boundary of the range of the RRS automatic operation. The operator can take an immediate action whenever the turbine power goes down to below 15% [6].

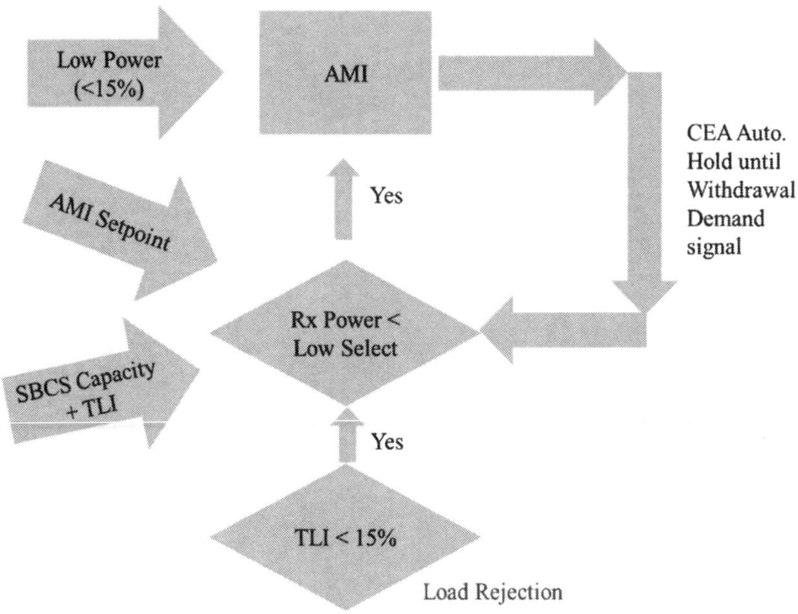


Figure 13-104: Diagram of CEA Automatic operation prohibition (AMI)

As shown in Figure 13-104, the AMI feature aims at automatically maintaining a high reactor power after load rejections, such as turbine trips and load rejections to house load which is resulted in by a temporary fault, and allowing a quick reloading of the unit after aforementioned fault is corrected. The RRS will be prevented from inserting or withdrawing CEAs at the oneset of the load rejection if a load rejection below the AMI Permissive set-point happens and if the initial reactor power level is below the AMI threshold. Otherwise the CEAs are driven in by the RRS until the reactor power is mitigated to the AMI threshold level, at which point further automatic CEA motion is inhibited.

### (5) CEA Automatic Withdrawal (AWP)

The SBCS produces an AWP signal whenever SBCS demand for opening the turbine bypass valves exists. As shown in Figure 13-105, the AWP signal is generated when the main controller output signal and the permissive controller output signal are presented or the presentation of the main quick opening signal and the permissive quick opening of channel X, or the illustration of the main quick opening signal and the permissive quick opening of channel Y.

PART 3

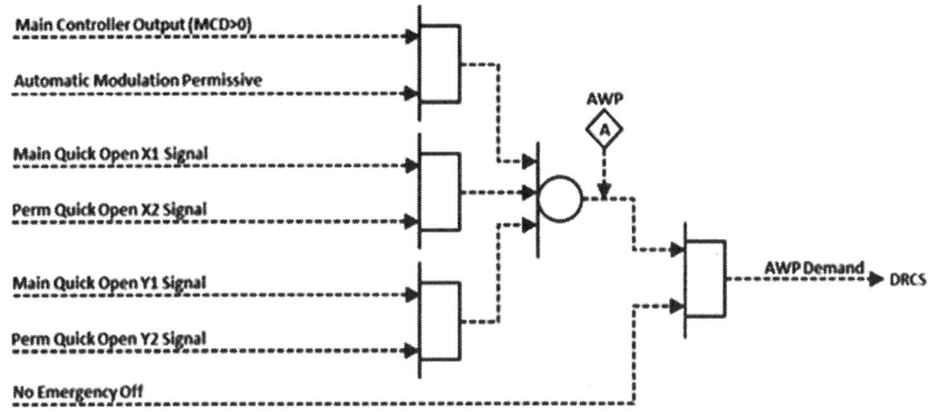


Figure 13-105: Diagram of CEA Automatic withdrawal (AWP)

## (6) Enable Signal (Automatic)

As shown Figure 13-106, the automatic signal is enabled when the quick opening signal and the signal of no quick opening block or the automatic modulation permissive is presented.

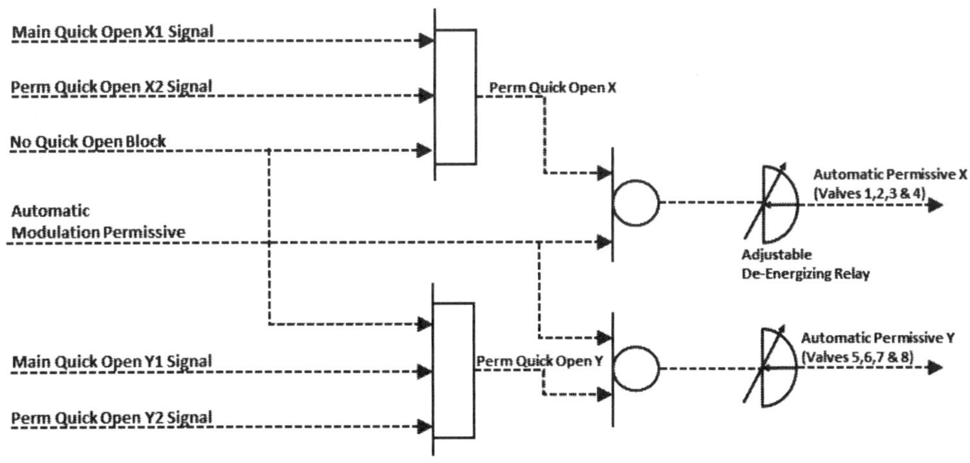


Figure 13-106: Diagram of enable signal (Automatic)

## (7) Design requirement

The SBCS shall be designed to have the following load change capabilities:

- A single thermal power step change of 10 (±10) % of rated output from 15% to 100% of rated output for 90% of core life.
- Ramp load changes up to five (±5) % per minute from 15% to 100% of rated output for 90% of core life.
- Accommodating, at any load, without reactor trip and without opening safety valves, a complete loss of load defined as either:
- An instantaneous load rejection imposed on the NSSS of up to 50% of related capacity, or
- An automatic turbine generator runback to house load following a loss of net load, provided the Balance of Plant has 55% condenser stem dump capability.
- A 100-50-100% daily load cycle with power sustained initially at 100% rated output for 10-16 hours, ramping down to 50% over two hours, operating for four hours to 10 hours at 50% rated output and hence ramping up to 100% over the next two hours.

### 13.8.3 Pressurizer Pressure Control System (PPCS)

The PPCS gives a mechanism for the automatic and manual control of the pressurizer pressure through controlling heater power and spray flow. As illustrated in Figure 13-107, the major inputs are PZR pressure, and process representative value (PRV); and the major outputs are Spray valve control output, PZR high pressure heater cutoff, and PZR low pressure (backup heater on). Refer Figure 13-108 to see the PPCS Functional Block Diagram. In the PPCS, the proportional and integral(PI) controller is used to modulate the proportional heaters and the spray valve controller controls the spray valves. The PZR level and pressure interlock signal control the proportional heaters and backup heaters. There are a lower level interlock and a high level interlock in the control system [6].

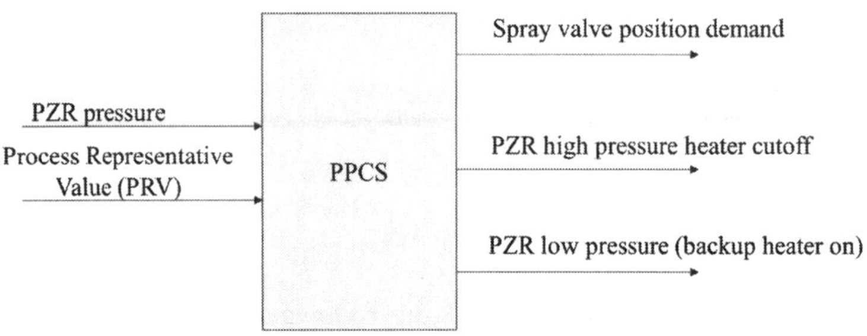


Figure 13-107: Inputs and outputs of PPCS

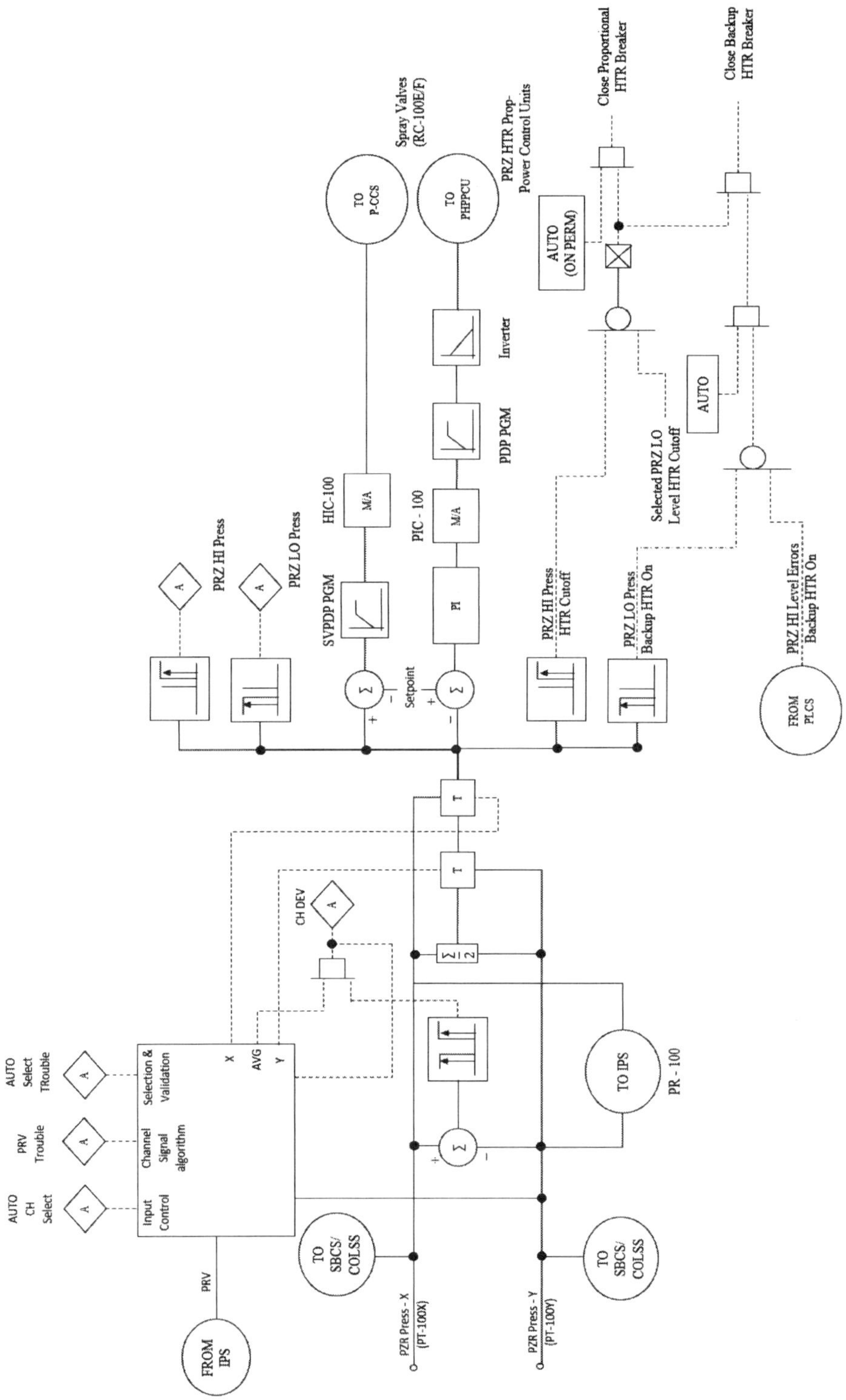


Figure 13-108: PPCS: Functional Block Diagram

## Pressurizer Pressure Control System Set Point

In the PPCS, there are two parallel spray valves which are available to mitigate the PZR pressure. There are also eight (8) banks of independently powered immersion heaters which are available to add energy to the PZR in order to increase the PZR pressure. As shown in Figure 13-109, the pressure set-point is normally set at 2250 psia, and there is a list of sequence actions which are performed when a continuous pressure increase above the normal pressure set-point and A continuous pressure decrease below 2250 psia.

If the pressure in the pressurizer decreases from 2250 psia to 2225 psia, all the banks of backup heaters are turned off. If the pressure continues to decrease of 2200 psia, all banks of backup heaters are turned on.

If the pressure in the pressurizer increases from 2250 psia to 2275 psia, both spray valves are fully closed. If the pressure continues to increase of 2300 psia, both spray valves are fully opened.

The proportional heater "ON" and "OFF" is ranging from 2225 psia to 2275 psia.

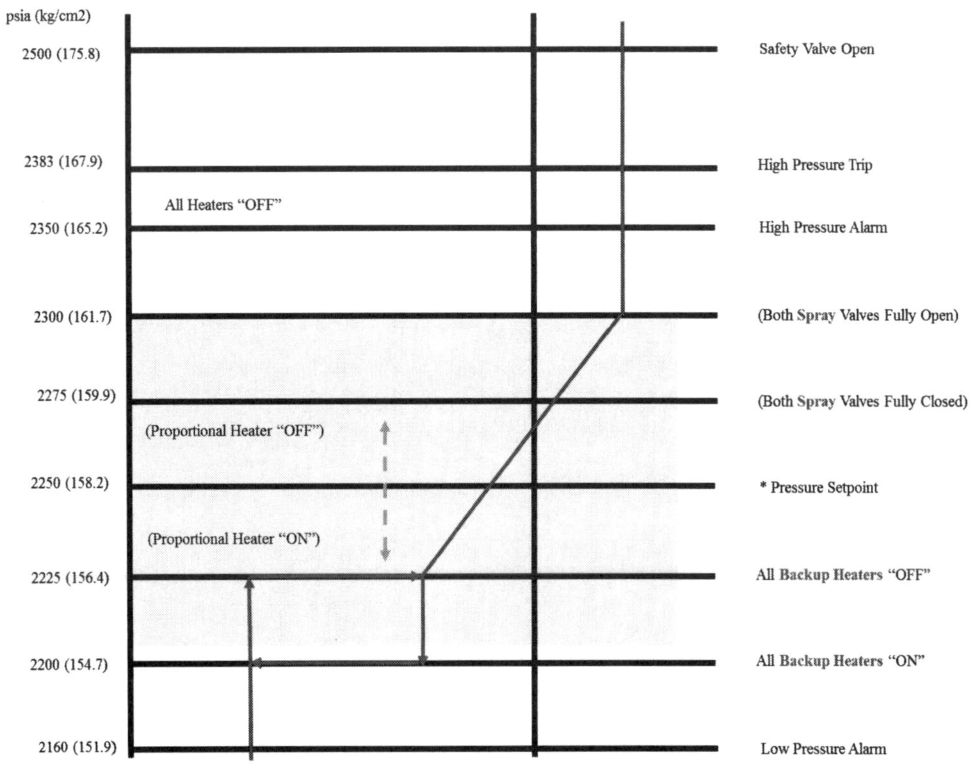


Figure 13-109: Pressurizer Pressure Control System Set Point

### 13.8.4 Pressurizer Level Control System (PLCS)

The PLCS provides the automatic and manual control of the PZR level in order to accommodate applicable performance of related design bases events (PRDBEs) association with exact operation of other control systems through control the charming and letdown flow rate. As shown in Figure 13-110, the major inputs are PZR level, TAVG and process representative value (PRV), and the major outputs are charging valve position demand, PZR high level error (backup heater on), and PZR low level heater cut off. Figure 13-111 shows the PLCS Simplified Logic Diagram. The control system has a fine control and coarse control. During the normal operation, the level is programmed as a Tavg function. The Tavg flows through a level set-point program and the set-point program signal is compared to the actual level. The level error signal is employed to control the charging flow control valve through the PI controller. The level error program controls letdown orifice isolation valves and backup heaters [6].

As shown in Figure 13-111, the Auxiliary Process Cabinet-Safety (APC-S) sends two analog PRZ level signals to the PLCS, these signals are utilized to supply the alarm and control functions. The RRS delivers a signal analog RCS average temperature signal (Tavg) to the PLCS, the IPS also sends a PRV signal to the PLCS. In the PLCS, the demand signal is produced through the PI charging valve controller. In order to control the PRZ water level to the programmed level and zero-out the PZR water level error, the demand signal modulates the charging flow control valves through the fine control of the charging flow. The letdown flow maintains a minimum whenever the level error signal is less than the set-point through the demand generated. The letdown flow path is formed through the first letdown orifice which limits the letdown flow at minimum required flow rate. Moreover, the demand signal is produced to yield a maximum letdown flow whenever the level error signal is greater than the set-point. All isolation valves are opened and the letdown flow path is maintained at the maximum required flow rate. The generation of the demand signal is to preclude a pressure decrease and mitigate thermal stresses on the PZR inner structures during insurge of relatively cold coolant into PZR whenever the level error signal is greater than the set-point. This signal is transmitted to the PPCS. The demand signal for each level is generated to preclude heater burnout, if all heaters are in danger of becoming uncovered and the level signal is less than the set-point. This signal is sent to the PPCS and de-energizes all heaters.

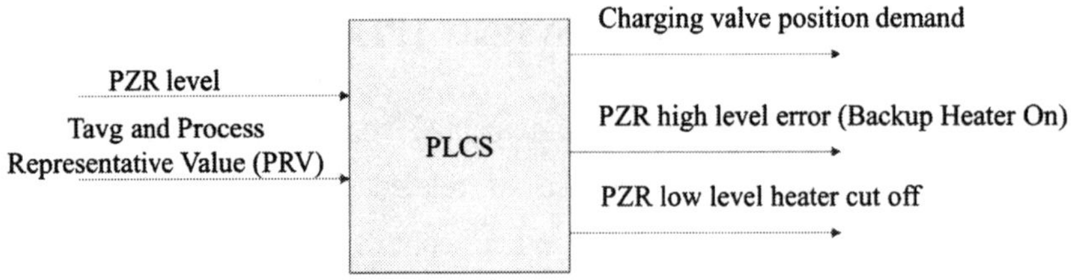


Figure 13-110: Inputs and outputs of PLCS

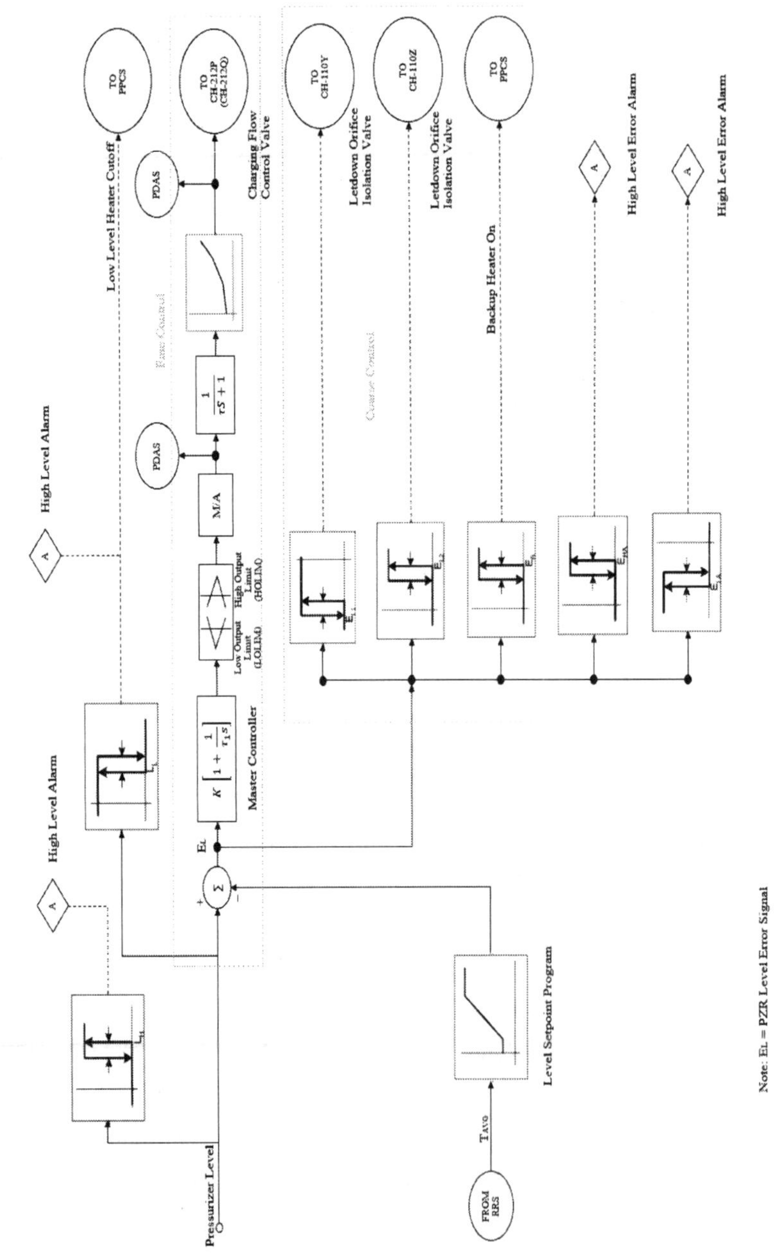


Figure 13-111: PLCS Simplified Logic Diagram

### (1) Pressurizer Level Set-point Program

The pressurizer level set-point program is either manually set or programmed as a function of the RCS $T_{avg}$, which is provided from the RRS. The pressurized level is proportionally programmed as a function of the RCS average temperature with specific upper and lower limits. Figure 13-112: shows the Pressurizer Level Set-point Program.

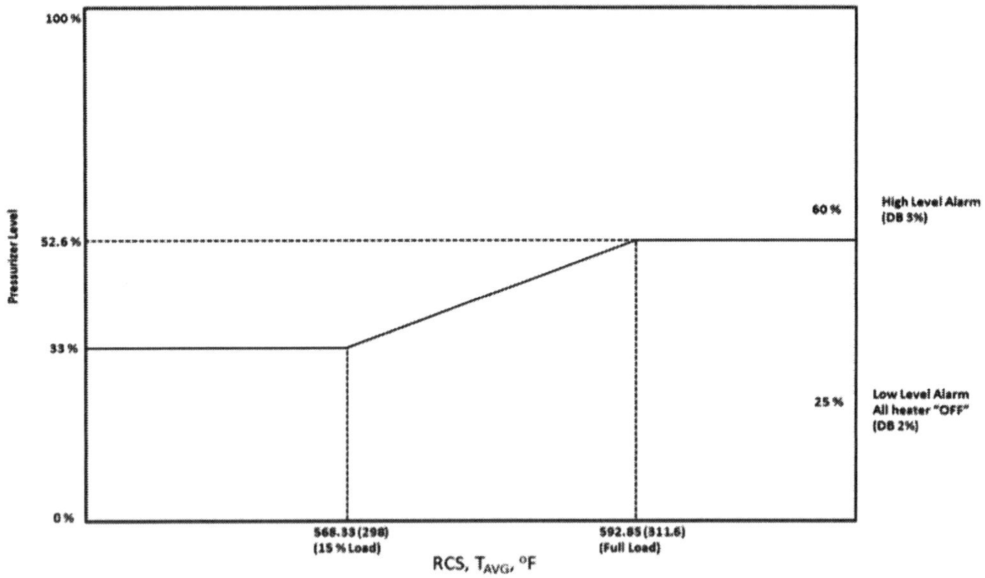


Figure 13-112: Pressurizer Level Set-point Program

### (2) Charging control valves and Letdown Orifice Isolation Valves Control

As shown in Figure 13-113, if the PRZ level error reaches the high level error set-point after the charging flow control valve reaches its low limit, the PLCS opens the letdown orifice isolation valve #3 (CH-100Z) for maximum letdown flow.

As the high-level error clears, the letdown orifice isolation valve #3 (CH-100Z) closes, and then the charging flow control valve modulates.

If the level error reaches the low-level error set-point, the PLCS closes the letdown orifice isolation valve #2 (CH-110Y) for minimum letdown flow.

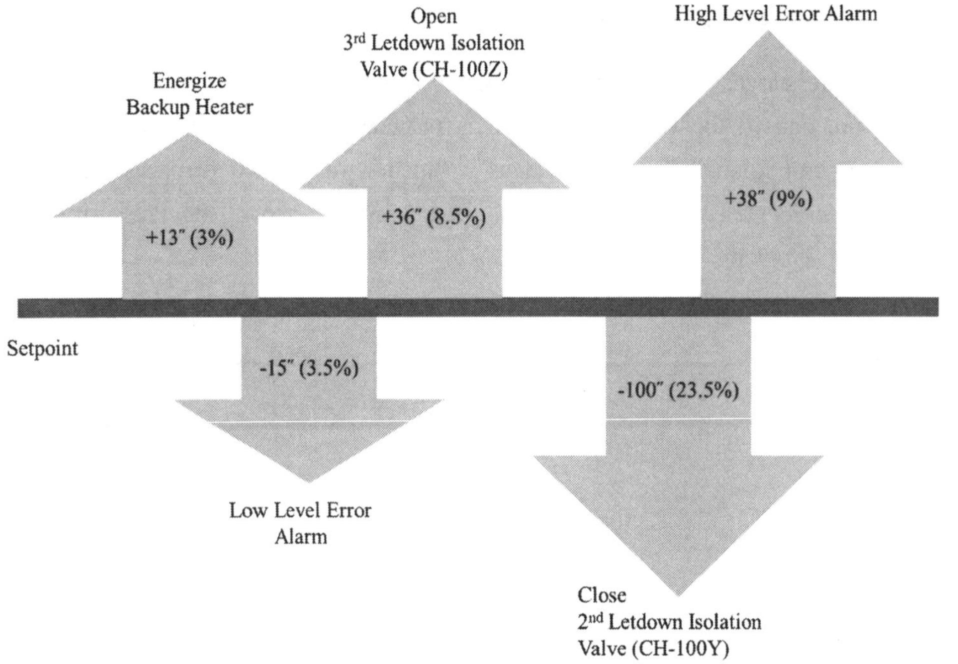


Figure 13-113: PLCS Level Error Set-point

### (3) Design requirement of both PPCS and PLCS

These systems provides the capabilities to control the following Performance Related Design Bases Events (PRDBEs):

- ±10% step change between 15% and 100% power
- ±1% step change between 5% and 15% power
- ±5%/minute ramp change between 15% and 100% power
- ±1%/minute ramp change between 5% and 15% power
- A loss of load or turbine trip
- A loss of a main feedwater pump
- Feedwater valve transfer
- 100-50-100% load cycle

PART 3

## 13.8.5 System Interface

There are many control systems in the non-safety control system. These control systems make the NPPs operate efficiently under transient conditions as well as under steady state conditions. The FWCS directly interfaces with RRS, SBCS, and DRCS. The RRS receives the temperature cold leg, temperature hot leg, turbine load index (TLI), and reactor power to supply for the FWCS and the SBCS. The RRS also interfaces with the DRCS by sending the following signals, RRS AWP, CEA Withdrawal, CEA insertion, Low rate/High rate RRS AMI. Based on the inputs to the RRS, and other inputs such as steam flow, pressurized pressure (PRZ Press.), and steam header pressure (STM HDR Press.), the SBCS sends the total steam flow and SBCS modulation demand to the FWCS. The reactor trip signal is received from DRCS and transmitted between the FWCS and the SBCS. Figure 13-114 depicts the interface between the control systems and FWCS.

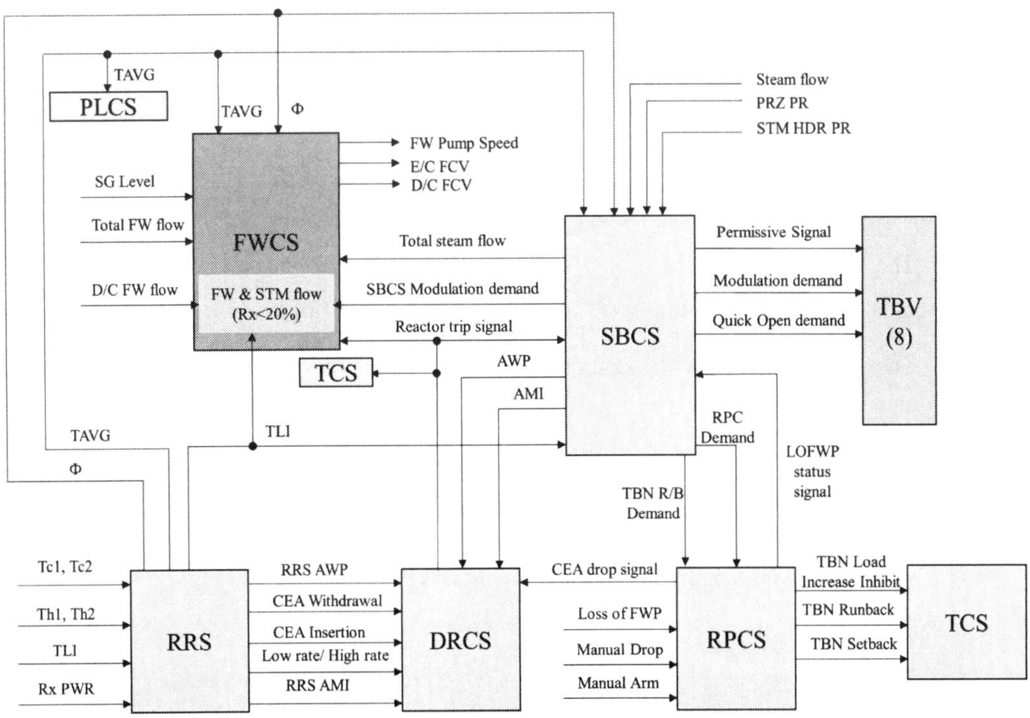


Figure 13-114: System Architect of NPCS

## 13.9 Control System CCF Analysis

According to [7], the following Failure Types due to a shared signal failure and control system common cause failures (CSCCFs) are evaluated to confirm that the DCD Chapter 15 analysis acceptance criteria are met. There are four types of failures as follows:

1. Failure Type 1: multiple function failures due to a single failure of a shared signal.
2. Failure Type 2: multiple failures of a single control group due to a CSCCF.
3. Failure Type 3: multiple failures of more than one control group due to a CSCCF.
4. Failure Type 4: multiple failures of Information Flat Panel Display (IFPD) control commands due to a CSCCF

All four-failure types are evaluated to meet the AOO (type 1 and 2) and PA (type 3 and 4) acceptance criteria of the DCD Chapter 15.

### 13.9.1 Applicable Codes and Regulations

The 10 CFR 50.55a (h), "Protection and Safety Systems" and the IEEE Standard 603 provide the applicable codes and regulations.

The 10 CFR 50.55a (h) complies with the IEEE Std. 603.

### 13.9.2 Control System Design Features to Prevent CCF

This subsection shows the design features of the non-safety control systems and safety systems that:

(1) Prevent failures that could otherwise lead to a CCF,

(2) Mitigate the adverse effect of CCFs or

(3) Enable coping with CCFs

### (1) Boundary of Credible Failure

The control systems are controlled by a Distributed Control System (DCS)-based common platform that has been demonstrated by operating experiences in the nuclear industry. The DCS implements a variety of functions, such as operator interface, component level control, automatic process control, high-level group control, and data processing for normal operation. So as to accomplish the high reliability, to minimize, and prevent the failure of a single component from causing a spurious plant trip, the DCS is designed with a redundant and fault-tolerant architecture. It is due to the fact that the non-safety control systems are software-based systems that are susceptible to a software defect, the design features and evaluation are necessary to yield Control System CCF(CSCCF) [7].

### (2) Credible Failure Types of Control System CCF

Credible failures of the CSCCF are initial events caused by the control system failure that may impact the critical safety functions. Because the major control function is assigned to a separate control group which includes at least one controller, the four credible failure types are their initial events and frequencies are given in Table 13-15, and Table 13-16 respectively.

Table 13-15: Types of Credible Failure

| Failure Type | Evaluation Criteria |
|---|---|
| Failure Type 1 : Multiple function failures due to a single failure of a shared signal | To be bounded by DCD Chapter 15 AOO acceptance criteria |
| Failure Type 2 : Multiple failures of a single control group due to CSCCF | |
| Failure Type 3 : Multiple failures of more than one control group due to CSCCF | To be bounded by DCD Chapter 15 PA acceptance criteria |
| Failure Type 4 : Multiple failures of IFPD control commands due to CSCCF | |

Table 13-16: Initial Events and Frequencies [1]

| Event | Frequency of Event |
|---|---|
| Increase in heat removal by the secondary system | — |
| Decrease in feedwater temperature | AOO |
| Increase in feedwater flow | AOO |
| Increase in steam flow | AOO |
| Inadvertent opening of a steam generator relief or safety valve | AOO |
| Steam system piping failures inside and outside the containment | PA |
| Decrease in heat removal by the secondary system | — |
| Loss of external load | AOO |
| Turbine trip | AOO |
| Loss of condenser vacuum | AOO |
| Closure of main steam isolation valve | AOO |
| Steam pressure regulator failure | N/A |
| Loss of nonemergency AC power to the station auxiliaries | AOO |
| Loss of normal feedwater flow | AOO |
| Feedwater system pipe break inside and outside the containment | PA |
| Decrease in reactor coolant system flow rate | — |
| Loss of forced reactor coolant flow | AOO |
| Flow controller malfunctions | N/A |
| Reactor coolant pump rotor seizure | PA |
| Reactor coolant pump shaft break | PA |
| Reactivity and power distribution anomalies | — |
| Uncontrolled control rod assembly withdrawal from subcritical or low-power startup conditions | AOO |
| Uncontrolled control element assembly withdrawal at power | AOO |
| Control element assembly misoperation | AOO |
| Startup of an inactive reactor coolant pump | AOO |
| Flow controller malfunction causing an increase in BWR core flow rate | N/A |
| Inadvertent decrease in boron concentration in the reactor coolant system | AOO |
| Inadvertent loading and operation of a fuel assembly in an improper position | AOO |
| Spectrum of control element assembly ejection accidents | PA |
| Increase in reactor coolant inventory | — |
| Inadvertent operation of the emergency core cooling system that increases the reactor coolant inventory | AOO |

| Event | Frequency of Event |
|---|---|
| Chemical and volume control system malfunction that increases the reactor coolant inventory | AOO |
| Decrease in reactor coolant inventory | — |
| Inadvertent opening of a PWR pressurizer pressure relief valve | PA |
| Failure of small lines carrying primary coolant outside the containment | AOO |
| Steam generator tube failure | PA |
| Radiological consequences of main steam line failure outside the containment (BWR) | N/A |
| Loss-of-coolant accidents resulting from spectrum of postulated piping breaks within the RCPB | PA |
| Radioactive material release from a subsystem or component | — |
| Radioactive gas waste system leak or failure | PA |
| Radioactive liquid waste system leak or failure | N/A |
| Postulated radioactive releases due to liquid-containing tank failures | PA |
| Fuel handling accident | PA |
| Spent fuel cask drop accident | PA |
| Anticipated transient without scram | N/A |

### (3) Control System Design Features

In order to alleviate the probability of CCF, the control system is designed with the following design features:

- The control function is assigned to a separate control group which contains at least one separate controller to constrain the failure in the control group (segmentation). Refer to Table 13-17.
- A reduplicated controller so as to raise the availability of the control system
- Interlock/permissive functions by an independent control group or safety system to constrain the failure impacts (e.g., control element assembly (CEA) withdrawal interlock signals, turbine bypass valve (TBV) permissive signals). Refer to Table 13-18.
- Control signal validation to constrain a single input failure of reduplicated channel inputs (i.e., large deviation of reduplicated inputs).
- Reduplicated analog input modules with auto signal selection algorithm to constrain the failure impact of a single module (i.e., out of range).
- Hardwired signal interface of shared signals between the control groups within PCS and NPCS

- Diagnostic and alarming functions
- In order to defend against a design basis event (DBE) because of single random hardware failure, it is necessary to design features of the information flat panel display (IFPD). Refer to Table 13-19 for the Key Available Information for Operators for DBEs with a CCF in Digital Safety I&C Systems.

Table 13-17: Controller Grouping in the NSSS Control System [1]

| Controller Group | Control System | Postulated Event Due to a Single Failure in the Corresponding Controller Group (1) | | | | | |
|---|---|---|---|---|---|---|---|
| | | Excessive or Deficient Feedwater Flow | Full Open of Any One TBV | Excessive Charging Flow or Deficient PZR Spray | Uncontrolled CEA Withdrawal | Inadvertent Deboration | Refer to |
| Group 1 Controller | SG 1 Feedwater Control (FWCS 1) | × | | | | | Chapter15.1.2, DCD, Tier 2 |
| Group 2 Controller | SG 2 Feedwater Control (FWCS 2) | × | | | | | Chapter15.1.2, DCD, Tier 2 |
| Group 3 Controller | PZR Pressure Control (PPCS) | | | × (Deficient PZR Spray) | | | Chapter15.5.2, DCD, Tier 2 |
| Group 4 Controller | PZR Level Control (PLCS) | | | × (Excessive Charging) | | | Chapter15.5.2, DCD, Tier 2 |
| Group 5 Controller | Turbine Bypass Control (SBCS Main) | | ×(2) | | | | Chapter15.1.4, DCD, Tier 2 |
| Group 6 Controller | Turbine Bypass Control (SBCS Permissive) | | × | | | | Chapter15.1.4, DCD, Tier 2 |
| Group 7 Controller | Reactor Makeup Control (CVCS) | | | | | × | Chapter15.4.6, DCD, Tier 2 |
| Group 8 Controller | Control Rod Control (RRS) | | | | × | | Chapter15.4.1, 15.4.2, DCD, Tier 2 |
| Group 9 Controller | Control Rod Control (DRCS) | | | | ×(2) | | Chapter15.4.1, 15.4.2, DCD, Tier 2 |

(1) This table illustrates that one controller failure does not cause credible failures in other controller groups.

(2) An interlock is provided (for this control system) in a separate controller group or safety systems, to limit the effect of this single controller failure.

Table 13-18: Control Limit and Interlocks on Digital Rod Control System [1]

| Conditions of Interlocks | Functions | Signal Path |
|---|---|---|
| Upper electrical limit (UEL) and lower electrical limit (LEL) signals from reed switch position transmitter (RSPT). | Interlock: Blocks control rod withdrawal or insertion on automatic, manual group and manual individual DRCS control modes. | RSPT → DRCS (UEL, LEL) |
| Automatic withdrawal prohibit (AWP) signals from RRS and SBCS when Tavg is much higher than Tref, Tcold is high, or any opening demand of TBVs is generated in accordance with excessive energy in the NSSS. | Interlock: Blocks control rod withdrawal on automatic DRCS control mode. | RRS → DRCS (AWP)<br>SBCS → DRCS(AWP) |
| Upper group stop (UGS) and lower group stop (LGS) function in the DRCS | Control Limit: Blocks control rod withdrawal or insertion on automatic and manual group DRCS control modes. | DRCS (UGS, LGS) |
| CEA withdrawal prohibit (CWP) signal from PPS. | Interlock: Blocks control rod withdrawal on automatic, manual group and manual individual DRCS controlmodes. | PPS → DRCS (CWP) |

Table 13-19: Key Available Information for Operators for DBEs with a CCF in Digital Safety I&C Systems [8]

| Event | Alarms and Indications Available | Available Systems in Non-Safety Platforms | Available DMA Switches |
|---|---|---|---|
| 1. Increase in Feedwater flow | * Reactivity control<br>-Alarms<br>High core power rate<br>DNBR<br>-Indication<br>Core power<br>RCS inlet temperature<br>Control rod bottom contact | IPS<br>IPS, LDP<br>IPS, LDP, DIS<br>IPS, LDP,<br>DIS(WR3))<br>IPS | Safety injection |
| | * RCS heat removal<br>-Alarm<br>High core power rate<br>High SG level<br>-Indication<br>Core power<br>SG level<br>SG pressure<br>Feedwater flow<br>RCS flow<br>RCS outlet temperature<br>RCS inlet temperature | IPS<br>IPS, LDP<br><br>IPS, LDP, DIS<br>IPS,LDP, DIS<br>LDP, DIS<br>IPS, LDP, DIS<br>IPS, LDP<br>IPS, LDP, DIS(WR)<br>IPS, LDP, DIS(WR) | Auxiliary feedwater |
| | * RCS pressure control<br>-Alarm<br>Low PZR pressure<br>-Indication<br>PZR pressure | IPS, LDP<br><br>IPS(NR5)), LDP(NR), DIS(WR) | |

| 2. Main Steam Line Break Outside Containment | * Reactivity control<br>-Alarms<br>High core power rate<br>DNBR<br>-Indication<br>Core power<br>RCS inlet temperature<br>Control rod bottom contact | IPS<br>IPS, LDP<br>IPS, LDP, DIS<br>IPS, LDP, DIS(WR)<br>IPS | Safety injection |
|---|---|---|---|
| | * RCS heat removal<br>-Alarm<br>High core power rate<br>-Indication<br>Core power<br>SG level<br>SG pressure<br>Feedwater flow<br>RCS flow<br>RCS outlet temperature<br>RCS inlet temperature | IPS<br><br>IPS, LDP, DIS<br>IPS, LDP, DIS<br>LDP, DIS<br>IPS, LDP, DIS<br>IPS, LDP<br>IPS, LDP, DIS(WR)<br>IPS, LDP, DIS(WR) | Auxiliary feedwater |
| | * RCS pressure control<br>-Alarm<br>Low PZR pressure<br>-Indication<br>PZR pressure | IPS, LDP<br><br>IPS(NR), LDP(NR), DIS(WR) | |
| 3. Total loss of Flow | * Reactivity control<br>-Alarm<br>DNBR<br>-Indication<br>Core power<br>RCS inlet temperature<br>Control rod bottom contact | IPS, LDP<br><br>IPS, LDP, DIS<br>IPS, LDP, DIS(WR)<br>IPS | Safety injection |
| | * RCS heat removal<br>-Alarm<br>High RCS outlet temperature<br>-Indication<br>Core power<br>SG level<br>Feedwater flow<br>RCS flow<br>RCS outlet temperature<br>RCS inlet temperature<br>RCP status | IPS, LDP<br><br>IPS, LDP, DIS<br>IPS, LDP, DIS<br>IPS, LDP, DIS<br>IPS, LDP<br>IPS, LDP, DIS(WR)<br>IPS, LDP, DIS(WR)<br>IPS | Auxiliary feedwater |
| | * RCS pressure control<br>-Alarm<br>High PZR pressure<br>-Indication<br>PZR pressure | IPS, LDP<br><br>IPS(NR), LDP(NR), DIS(WR) | |

| | | | |
|---|---|---|---|
| 4a. RCP Shaft Seizure | * Reactivity control<br>-Alarm<br>DNBR<br>-Indication<br>Core power<br>RCS inlet temperature<br>Control rod bottom contact | IPS, LDP<br><br>IPS, LDP, DIS<br>IPS, LDP, DIS(WR)<br>IPS | Safety injection |
| | * RCS heat removal<br>-Alarm<br>High RCS outlet temperature<br>-Indication<br>Core power<br>SG level<br>Feedwater flow<br>RCS flow<br>RCS outlet temperature<br>RCS inlet temperature<br>RCP status | IPS, LDP<br><br>IPS, LDP, DIS<br>IPS, LDP, DIS<br>IPS, LDP, DIS<br>IPS, LDP<br>IPS, LDP, DIS(WR)<br>IPS, LDP, DIS(WR)<br>IPS | Auxiliary feedwater |
| | * RCS pressure control<br>-Alarm<br>High PZR pressure<br>-Indication<br>PZR pressure | IPS, LDP<br><br>IPS(NR), LDP(NR), DIS(WR) | - |
| 4b. RCP Shaft Break | Same as the RCP Shaft Seizure | | |
| 5a. CEA Ejection without Primary System Rupture | * Reactivity control<br>-Alarms<br>High core power rate<br>DNBR<br>-Indication<br>Core power<br>RCS inlet temperature<br>Control rod bottom contact | IPS<br>IPS, LDP<br><br>IPS, LDP, DIS<br>IPS, LDP, DIS(WR)<br>IPS | Safety injection |
| | * RCS heat removal<br>-Alarm<br>High RCS outlet temperature<br>-Indication<br>Core power<br>SG level<br>Feedwater flow<br>RCS flow<br>RCS outlet temperature<br>RCS inlet temperature | IPS, LDP<br><br>IPS, LDP, DIS<br>IPS, LDP, DIS<br>IPS, LDP, DIS<br>IPS, LDP<br>IPS, LDP, DIS(WR)<br>IPS, LDP, DIS(WR) | |
| | * RCS pressure control<br>-Alarm<br>High PZR pressure<br>-Indication<br>PZR pressure | IPS, LDP<br><br>IPS(NR), LDP(NR), DIS(WR) | |
| 5b. CEA Ejection With Primary System Rupture | Same as the Loss of Coolant Accident | | |

| | | | |
|---|---|---|---|
| 6. Steam Generator Tube Rupture | * Reactivity control<br>-Indication<br>Core power<br>RCS inlet temperature<br>Control rod bottom contact | IPS, LDP, DIS<br>IPS, LDP, DIS(WR)<br>IPS | Safety injection |
| | * RCS heat removal<br>-Alarm<br>Low PZR level<br>High SG level<br>-Indication<br>Core power<br>SG level<br>SG pressure<br>PZR level<br>Feedwater flow<br>RCS flow<br>RCS outlet temperature<br>RCS inlet temperature<br>Subcooling margin | IPS, LDP<br>IPS, LDP<br>IPS, LDP, DIS<br>IPS, LDP, DIS<br>LDP, DIS<br>IPS, LDP, DIS<br>IPS, LDP, DIS<br>IPS, LDP<br>IPS, LDP, DIS(WR)<br>IPS, LDP, DIS(WR)<br>LDP, DIS | Auxiliary feedwater |
| | * RCS pressure control<br>-Alarm<br>Low PZR pressure<br>High SG level<br>-Indication<br>PZR pressure<br>SG level | IPS, LDP<br>IPS, LDP<br>IPS(NR), LDP(NR), DIS(WR)<br>IPS, LDP, DIS | |
| | * Radioactivity Control<br>-Alarms<br>Main steam line(N-16) Radiation<br>-Indication<br>Main steam line(N-16) Radiation | IPS<br>IPS | Main steam isolation<br>Containment isolation |
| 7. Loss of Coolant Accident | * Reactivity control<br>-Alarm<br>DNBR<br>-Indication<br>Core power<br>RCS inlet temperature<br>Control rod bottom contact | IPS, LDP<br>IPS, LDP, DIS<br>IPS, LDP, DIS(WR)<br>IPS | Safety injection |
| | * RCS heat removal<br>-Alarm<br>Low PZR level<br>-Indication<br>Core power<br>SG level<br>SG pressure<br>PZR level<br>Feedwater flow<br>RCS flow<br>RCS outlet temperature<br>RCS inlet temperature<br>Subcooling margin | IPS, LDP<br>IPS, LDP, DIS<br>IPS, LDP, DIS<br>LDP, DIS<br>IPS, LDP, DIS<br>IPS, LDP, DIS<br>IPS, LDP<br>IPS, LDP, DIS(WR)<br>IPS, LDP, DIS(WR)<br>DIS | Auxiliary feedwater |

| | | | |
|---|---|---|---|
| 7. Loss of Coolant Accident | * RCS pressure control<br>-Alarm<br>Low PZR pressure<br>-Indication<br>PZR pressure | IPS, LDP<br><br>IPS(NR), LDP(NR), DIS(WR) | - |
| | * Containment pressure control<br>-Alarms<br>High containment pressure<br>-Indication<br>Containment pressure<br>Containment temperature | IPS, LDP<br><br>IPS, LDP, DIS<br>DIS | Containment spray<br>Containment isolation |
| 8. Steam Line Break Inside Containment | * Reactivity control<br>-Alarm<br>High core power rate<br>-Indication<br>Core power<br>RCS inlet temperature<br>Control rod bottom contact | IPS<br><br>IPS, LDP, DIS<br>IPS, LDP, DIS(WR)<br>IPS | Safety injection |
| | * RCS heat removal<br>-Alarm<br>Low PZR level<br>-Indication<br>Core power<br>SG level<br>SG pressure<br>PZR level<br>Feedwater flow<br>RCS flow<br>RCS outlet temperature<br>RCS inlet temperature<br>Subcooling margin | IPS, LDP<br><br>IPS, LDP, DIS<br>IPS, LDP, DIS<br>LDP, DIS<br>IPS, LDP, DIS<br>IPS, LDP, DIS<br>IPS, LDP<br>IPS, LDP, DIS(WR)<br>IPS, LDP, DIS(WR)<br>DIS | Auxiliary feedwater |
| | * RCS pressure control<br>-Alarm<br>Low PZR pressure<br>-Indication<br>PZR pressure | IPS, LDP<br><br>IPS(NR), LDP(NR), DIS(WR) | - |
| | * Containment pressure control<br>-Alarms<br>High containment pressure<br>-Indication<br>Containment pressure<br>Containment temperature | IPS, LDP<br><br>IPS, LDP, DIS<br>DIS | Containment spray<br>Containment isolation |

### (4) Evaluation of Control System Software Common Cause Failure

Within the scope of this book, failure type 3 defined as multiple failures of more than one control group due to a CSCCF, is selected to evaluate the Control System Software Common Cause Failure. In terms of the failure type 3, two worst combinations of failures in control groups, i.e. fuel cladding and primary system integrity are quantitatively evaluated employing a realistic methodology with the RELAP5/MOD3.3 computer code. These combinations include the normal operation of several control groups if the normal operation causes the worst event results [9].

The realistic analysis method is used to analyze the failure type 3 owing to the design features such as appropriate control system segmentation and the interlock/ permissive signals from the separate control group or safety system which reduce the likelihood of failure types 3. The RELAP5/MOD3.3 computer code is utilized for the analysis of the thermal-hydraulic behavior during failure type 3. The nodalization for modeling of the APR1400 consists of 305 hydraulic components and 407 junctions. NSSS and non-safety control system related to evaluation of failure type 3 event are modeled.

### (5) Initial Conditions

Table 13-20 illustrates the initial conditions which are employed to analyze the NSSS and the core thermal hydraulic response to failure type 3. The full power plant conditions are normally considered to the analyses.

Table 13-20: Initial Conditions for Failure Type 3

| Parameters | Design Value |
|---|---|
| Core Thermal Power, MWt | 3,983.00 |
| Core Inlet Temperature, oC (oF) | 290.56 (555) |
| Pressurizer Pressure, kg/cm2A (psia) | 158.191 (2,250) |
| RCS Flow, 106 kg/hr (lbm/hr) | 75.57 (166.6) |
| Pressurizer Water Volume, m3 (ft3) | 33.16 (1,171.17) |
| SG Wide Range Level, % | 76.83 |

### (6) Major Assumptions

The worst combination of failure status in each control system has been created for the fuel cladding integrity and the primary system integrity, as shown in Table 13-22.

### (7) Fuel Cladding Integrity (Case 1)

The reactor coolant system (RCS) cools down because of the multiple failures of the high pressure (HP) feedwater (FW) heater and the deaerator heater, the SBCS and the FWCS. This cooldown causes an increase of the core power owing to moderator density feedback effect. The increased core power threatens fuel cladding integrity. The moderator temperature coefficient (MTC), fuel temperature coefficient (FTC), and radial peaking factor are the three significant parameters with the biggest impact on the minimum departure from nucleate boiling ratio (DNBR).

The sequence of events for the case 1 given in Table 13-21.

Table 13-21: Major Sequences of the Fuel Cladding Integrity

| Time (sec) | Event of Fuel Cladding Integrity |
|---|---|
| 0.0 | Failure Type 3 Event occurs |
| 6.0 | VOPT trip occurs |
| 7.0 | Power Reaches the Maximum Value, % |
| 7.5 | DNBR Reaches the Minimum Value |
| 21.0 | MSIS is generated by LSGP, kg/cm2A (psia) |
| 780.0 | POSRV opens first, kg/cm2A (psia) |
| 1,400.0 | MSSV bank 1 opens, kg/cm2A (psia) |
| 1,800.0 | Operators initiate plant cooldown |

As shown in Figure 13-115, the core power went up owing to inadvertent opening of all TBVs and the maximum FW flow. After that, the reactor is tripped by the variable overpower trip (VOPT) of the PPS at approximately 6.0 seconds and the maximum power is reached a peak of 116% of the initial power.

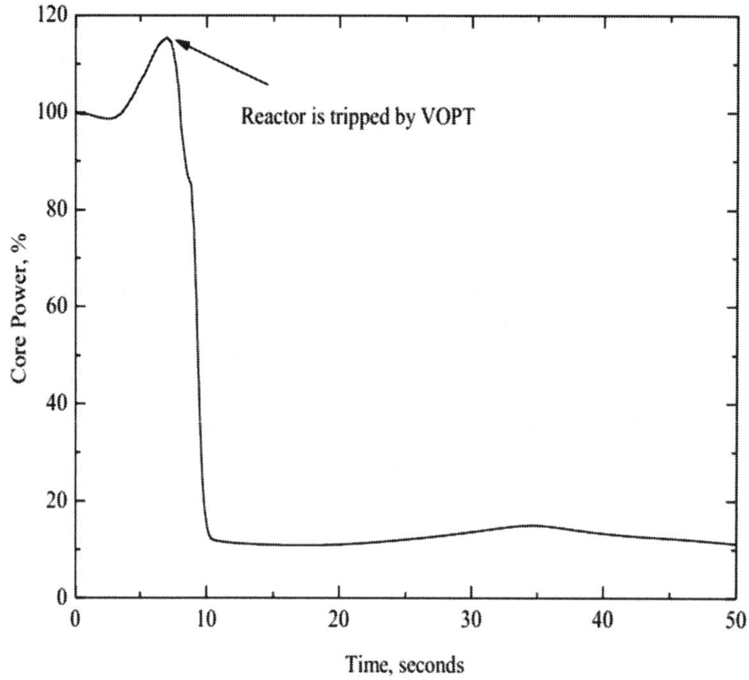


Figure 13-115: Core Power at the case of the Fuel Cladding Integrity

Figure 13-116 shows the System Pressure at the case of the Fuel Cladding Integrity which is went down in the early period of the event because of RCS overcooling by inadvertent opening of all TBVs and the maximum FW. After the trip of reactor, the RCS pressure rises owing to the discontinuation of primary feedwater to both steam generators. Additionally, because of closing the MSIVs, this leads to the increase of the SG pressure. Then, the repeating of automatic opening and closing pilot operated safety relief valves (POSRVs) and main steam safety valves (MSSVs) causes oscillations of the system pressure between set-point of opening and closing pressure of POSRVs and MSSVs.

Throughout Figure 13-117, the minimum outcome of the DNBR is around 2.4 at 7.5 seconds and the DNB-SAFDL is not violated. After the event initiation of 30 minutes, through the appropriate emergency procedures, the operator should initiate plant cool down in order to bring the plant into a safe shutdown condition.

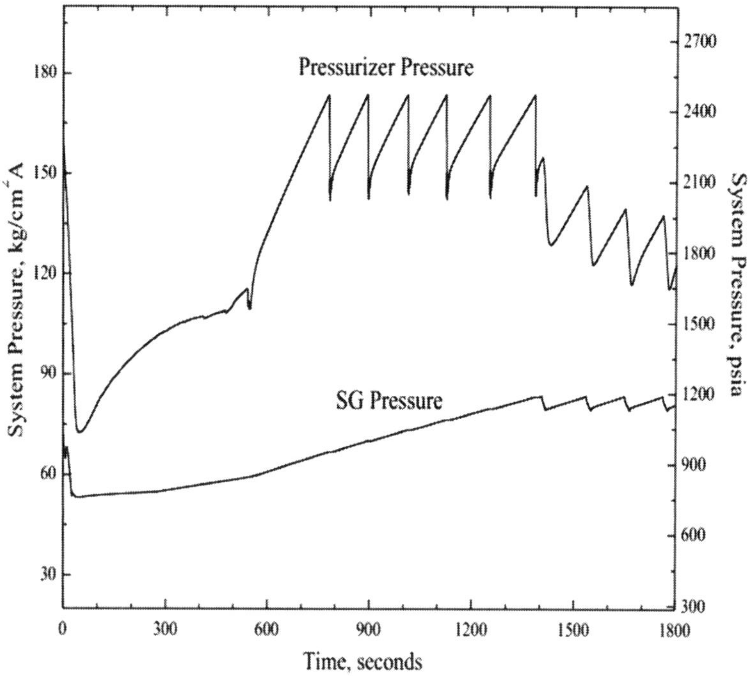


Figure 13-116: System Pressure at the case of the Fuel Cladding Integrity

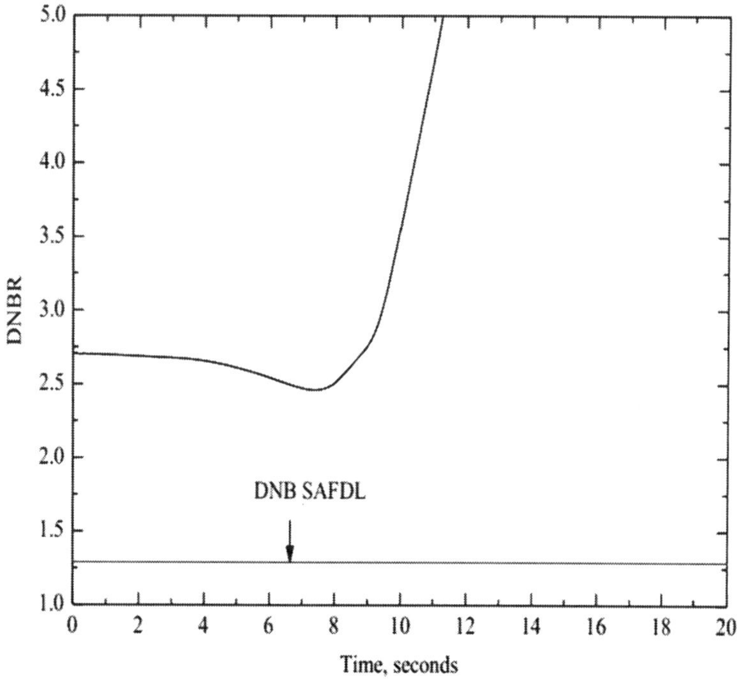


Figure 13-117: DNBR in the case of the Fuel Cladding Integrity

### (8) Primary System Integrity (Case 2)

Heat removal from the primary system to the secondary system is performed through the FWCS and the turbine control system (TCS). The loss of heat removal in the secondary system primarily leads to overpressure of the primary system.

Table 13-22: Assumptions for Failure Type 3

| Control Group | Fuel Cladding Integrity | Primary System Integrity |
|---|---|---|
| SBCS | Inadvertent Opening of all (8) Turbine Bypass Valves (TBVs) | No Opening of TBVs |
| FWCS 1 | Maximum FW flow during Power Operation, Inadvertent Opening of All FW Control Valves, Inadvertent Speed Increase of MFWPs | Termination of Feedwater Supply, Inadvertent Close of All FW Control Valves, Stop of 3 MFWPs |
| FWCS 2 | | |
| MFWP | | |
| PPCS | All Heaters Off and All Spray Valves Open | All Heaters On and All Spray Valves Close |
| PLCS | Minimum Charging and Maximum Letdown | Maximum Charging and Minimum Letdown |
| CVCS | Minimum Charging and Maximum Letdown | Maximum Charging and Minimum Letdown |
| RRS/RPCS | Inadvertent CEA Withdrawal Demand | Inadvertent CEA Withdrawal Demand |
| DRCS | Inadvertent CEAs Withdrawal with High Speed | Inadvertent CEAs Withdrawal with High Speed |
| RCP 1,2,3,4 | Normal Operation | Normal Operation |
| Condenser Vacuum | Normal Operation | Condenser Vacuum Pump Trip |
| LP FW Heater | Loss of Condensate Heating of LP FW Heater | Normal Operation (Minor Effect) |
| Non-1E Power | Normal Operation | Normal Operation |
| TCS | Turbine Control Valve Wide Open | Turbine Trip due to Inadvertent LOCV Signal |
| HP FW Heater | Loss of HP FW Heater | Close of Inlet and Outlet Valve of HP FW Heater |
| Miscellaneous BOP | Loss of Condensate Heating of Deaerator | Normal Operation (Minor Effect) |

The sequence of events for the Primary System Integrity is shown in Table 13-23.

Table 13-23: Major Sequences of the Primary System Integrity

| Time (sec) | Event of the Primary System Integrity |
|---|---|
| 0.0 | Failure Type 3 occurs |
| 5.1 | Pressurizer pressure reaches HPP trip set-point, kg/cm2A (psia) |
| 5.6 | MSSV bank 1 opens, kg/cm2A (psia) |
| 7.4 | POSRV opens, kg/cm2A (psia) |
| 7.7 | RCP discharge pressure reaches maximum value, kg/cm2A (psia) |
| 7.9 | MSSV bank 2 opens, kg/cm2A (psia) |
| 1,800.0 | Operators initiate plant cooldown |

A loss of condenser vacuum (LOCV) may occur due to the failure of the circulating water system to supply cooling water, failure of the main condenser evacuation system to remove non-condensable gases, or excessive inleakage of air. As the secondary heat removal drops sharply by the LOCV, the primary and secondary pressure rises considerably. Furthermore, the raised core power by the inadvertent control element assembly (CEA) withdrawal raises the pressure of the primary system. The complete steam flow reduction and termination of the feedwater flow cause a reactor trip at 6.0 seconds on high pressurizer pressure due to reduced RCS cooling at 5.10 seconds as shown in Figure 13-118.

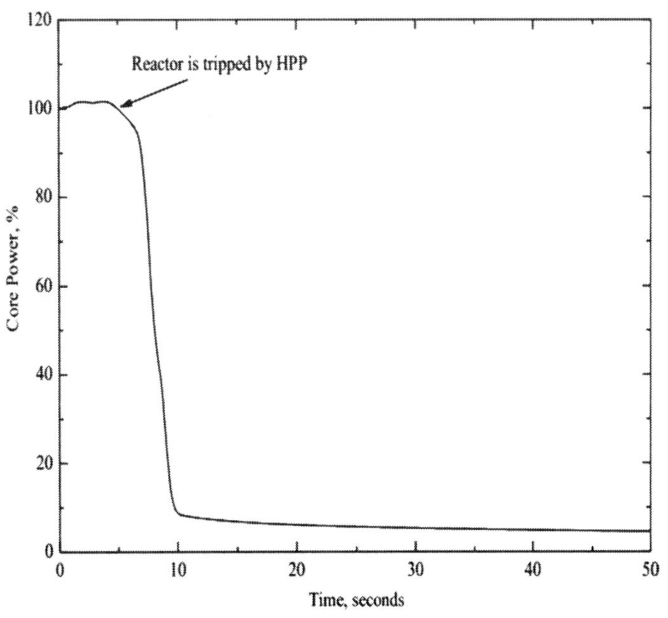


Figure 13-118: Core Power at the case of the Primary System Integrity

After reactor trip due to the failuer of all the pressurizer heaters, the RCS pressure decreases and stabilizes. Ultimately, the POSRVs and the MSSVs repeat open and close automatically to constrain the system pressure which oscillates between set-point of opening and closing pressure of POSRVs and MSSVs, as illustrated in Figure 13-119.

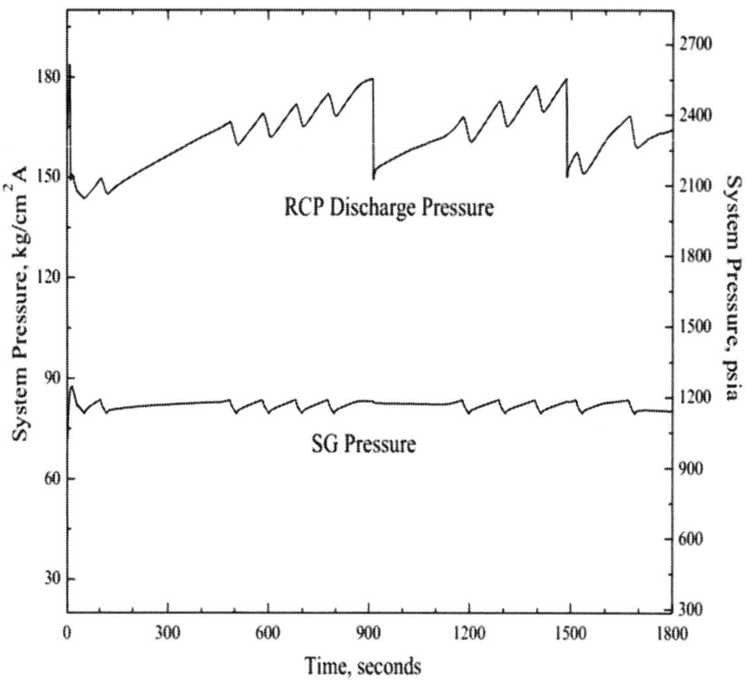


Figure 13-119: System Pressure at the case of the Primary System Integrity

As a result, the primary and secondary pressure are sustained below the acceptable design limits. After the event initiation of 30 minutes, the operator uses the appropriate emergency procedures to initiate plant cool down to bring the plant into a safe shutdown condition.

## PROBLEMS

1. Which are the systems in the NSSS? What are the major functions, inputs, and outputs of each system?
2. Explain what is meant by PRDBE and SRDBE.
3. Using Matlab, implement the simulations of the lead, lag, and lead-lag compensation networks.
4. Using Matlab, implement the simulations of the root locus plot based on the above-mentioned compensators.
5. Desribe the feed-forward and feedback loops of the FWCS.
6. Descrbe the compensation networks in the RRS.
7. Describe the applicable codes and regulations for control CCF analysis.

## FURTHER READING


- USNRC, "APR1400 Design Control Document: Tier 2," [Online]
- KHNP, "APR1400 Control Systems," Training Textbook
- KHNP & KEPCO, "Control System CCF Analysis," APR1400-Z-J-NR-14012-NP, rev. 0. 2014

Chapter 14

# A BUMPLESS CONTROL TRANSFER AND AN ANTI-WINDUP SCHEME FOR THE FEEDWATER CONTROL SYSTEM USING MIGO METHOD


## SUMMARY

Failure to control the steam generator (SG) water levels causes unexpected shutdowns in the nuclear power plants. This impacts maintenance, operational life, and safety critically. Therefore, it is vital to stabilize the SG water level at different power levels, particularly at low powers, by improving the performance of the proportional integral (PI) controller in the feedwater control system (FWCS). The PI controller designed using Ms-Constrained Integral Gain Optimization (MIGO) mitigated this problem significantly. In this chapter, in order to design this PI controller for the FWCS, the SG model was developed as a process model with a system identification algorithm in MATLAB. The algorithm processed data sets obtained from the thermal-hydraulic model by the Safety and Performance Analysis Code (SPACE). The MIGO and the Ziegler-Nichols methods were each used to produce a design for the PI controller, and the controllers were compared. Simulations were run in MATLAB for resulting designs. The PI controller designed by the MIGO is more stable than the controller using the Ziegler-Nichols approach is. However, so as to prove the robustness of the MIGO PI controller, this controller was again compared with the PI controller which was designed by genetic algorithm (GA) approach. Finally, the MIGO PI was applied to an overall schematic block diagram of the FWCS. The simulations of the diagram in MATLAB Simulink demonstrate the MIGO PI controller offers superior control and enhanced robustness in the FWCS. Additionally, this chapter also illustrates the anti-windup scheme for the MIGO PI controller through a back-calculation method as well as show the bumpless control transfer technique by using track mode of the MIGO PI controller. This chapter presents the work of Cong Tranh and Jung Jae Cheon [10]

The main content of this chapter is shown in Figure 14-1.

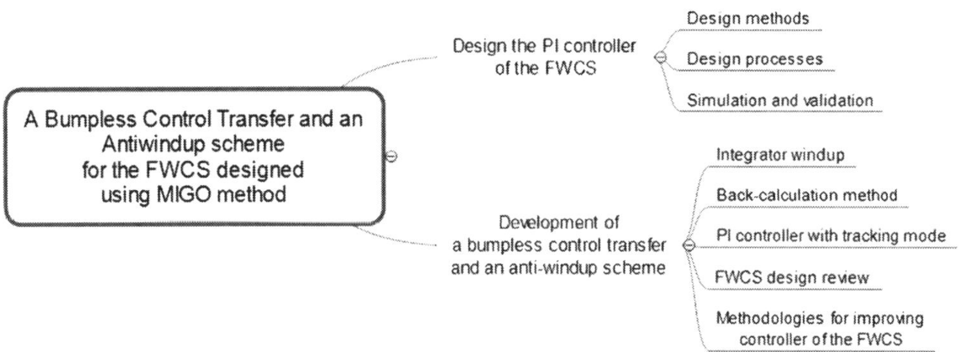


Figure 14-1: Mind Map for Chapter 14

## OBJECTIVES

- Design a novel PI control to regulate SG water level at different power levels (100%, 80%, 30%, and 5%) using the MIGO method
- Develop a bumpless control transfer and anti-windup scheme for the FWCS

# 14.1 Design of the PI controller of the FWCS

## 14.1.1 Design Methods

The controller was designed using two methods, for the purpose of comparative analysis:

1. Ziegler Nichols Method
2. Ms-Constrained Integral Gain Optimization (MIGO) Method

### (1) Ziegler Nichols Method

Two classic methods: the step response method and the frequency response method are presented by Ziegler Nichols so as to develop the values of the PID controller. In this research, we mention the step response method which may be considered as the conventional approach depending on modeling and control. The closed-loop controller of a plant is illustrated in Figure 14-2, where U(s) is the control signal, and C(s) is the output of a system. In this approach, we accomplished experimentally the response of the plant to a unit-step input, as illustrated in Figure 14-3. The unit-step response curve can look S-shaped, as depicted in Figure 14-4, if the plant does not involve integrator(s) and dominant complex-conjugate poles. This approach applies if the response to a step input illustrates an S-shaped curve. Based on either experiment or a dynamic simulation of the plant, the step response curves can be created [3].

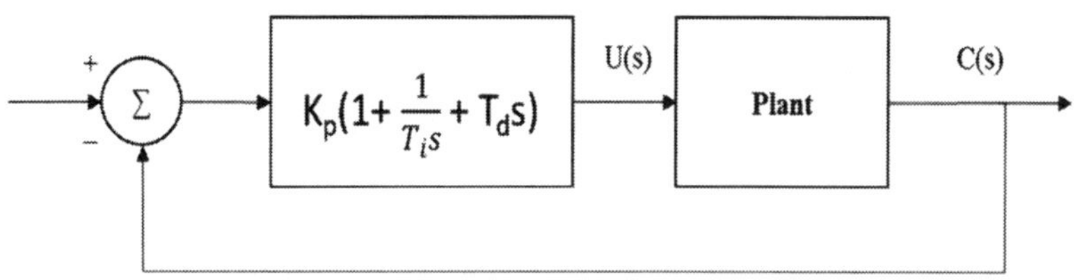


Figure 14-2: Closed-loop controller of a plant

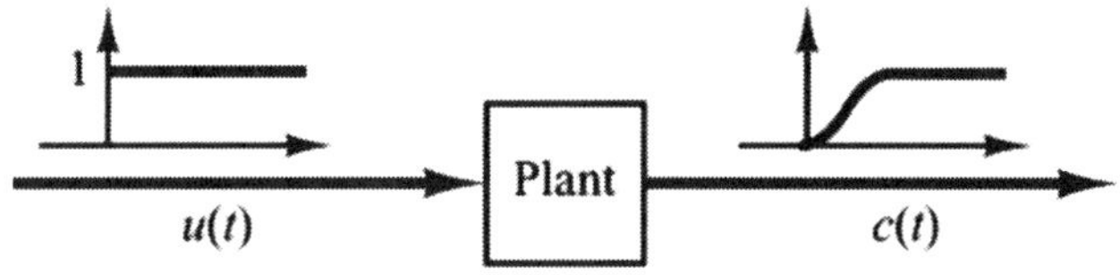


Figure 14-3: Plant step response

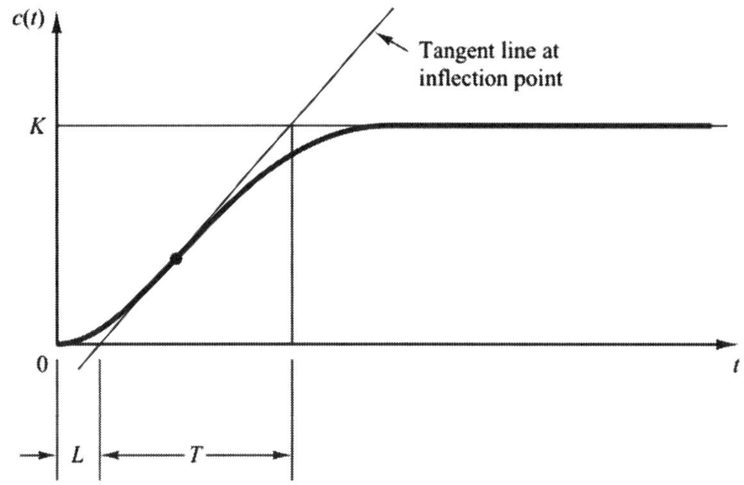


Figure 14-4: S-shaped response shape

The S-shaped curve is specified by two parameters: delay time L and time constant T. A tangent line is drawn at the inflection point of the S-shaped curve. The intersections between the tangent and coordinate axes (time axis and line c(t)=K) give L and T parameters, as shown in Table 14-1. The transfer function C(s)/U(s) is then approximated by a first-order system as follows:

$$\frac{C(s)}{U(s)} = \frac{Ke^{-Ls}}{Ts+1} \qquad \text{Eq. 14-1}$$

Table 14-1: Ziegler-Nichols approach using plant step response

| Type of Controller | $K_p$ | $T_i$ | $T_d$ |
|---|---|---|---|
| P | $\frac{T}{L}$ | $\infty$ | 0 |
| PI | $0.9\frac{T}{L}$ | $\frac{T}{0.3}$ | 0 |
| PID | $1.2\frac{T}{L}$ | 2L | 0.5L |

Based on Table 14-1, the PID controller parameters were tuned by the first method of Ziegler–Nichols rules as follow:

$$G_c(s) = K_p(1 + \frac{1}{T_i} + T_d s)$$ **Eq. 14-2**

$$= 1.2\frac{T}{L}(1 + \frac{1}{2Ls} + 0.5Ls)$$

$$= 0.6T\frac{(s + \frac{1}{L})^2}{s}$$

### (2) Ms-Constrained Integral Gain Optimization Method

It is acknowledged that simple approaches for tuning the PI controller were initiated by Ziegler and Nichols in [11]. The Ziegler-Nichols method is influential on the design and utilization of a PID control. However, there are two notable drawbacks to this method. First, the rules use inadequate process dynamics information. Second, the Ziegler and Nichols design method, i.e. quarter amplitude damping, produces closed-loop systems that perform poorly in terms of robustness and damping [12].

MIGO was established as an alternative method to tune the PI controller parameters based on a robust loop shaping [13], and it is effective in designing PI controllers. There are various versions of PI controller designed according to the MIGO method in use as of publication date of this research. In this paper, the PI controller uses the following equation:

$$u(t) = k(by_{sp}(t) - y(t)) + k_i\int_0^t (y_{sp}(\tau) - y(\tau))d\tau$$ **Eq. 14-3**

Where u(t) is the control signal, ysp is the set-point, y is the process output, and k, ki, and b are controller parameters.

When there is substantial measurement noise, the typical filter transfer function Gf(s) is the first order with time constant Tf.

$$G_f(s) = \frac{1}{1 + sT_f}$$ **Eq. 14-4**

Then, there are three or possibly four parameters, such as k, ki, b, and Tf in the PI controller.

The illustrated controller by Eq. 14-3 is considered to have two degrees of freedom when b ≠ 1. The benefits of this algorithm with two degrees of freedom have been expressed in [14], and in [15], where the functions of such a structure are debated.

There are several requirements that should be taken into account in the controller design, such as load disturbance response, robustness with respect to model uncertainties, set point response, and measurement noise [16]. Achieving good rejection of the load disturbances is the major purpose of the process control. Nonetheless, in [17], the robustness and the measured noise must also be considered. The requirements of the set point response can be separately tackled by utilizing the controller architecture having two degrees of freedom. The set point response for the PI controller was a primary focus of research in [18]. In most applications, a tuning parameter is used to adjust the balance aggressiveness versus robustness. In this method, the tuning parameter is identified to indicate the sensitivity to the model uncertainties.

### (3) Load Disturbance Attenuation

The most popular performance criterion is minimizing the integrated absolute error (IAE) at the load disturbances, and the IAE has the form as follows:

$$IAE = \int_0^\infty |e(t)|\, dt \qquad \textbf{Eq. 14-5}$$

However, such a criterion is difficult to tackle analytically and the integrated error (IE) criterion is utilized to instead. As the value of IE is directly involved in the integral gain ki of the PI controller, and it allows for straightforward and adequate design algorithms. In [19], it is depicted IE = 1/ki, and hence, the IE is defined by:

$$IE = \int_0^\infty e(t)dt = \frac{1}{k_i} \qquad \textbf{Eq. 14-6}$$

When the error is positive, the IE and IAE criteria are same. Likewise, when there is a well-damped system, the two criteria will be close which will be assured by the sensitivity constraints.

### (4) Robustness

Stability and robustness to process uncertainties is expressed by the sensitivity function S(s) and the complementary sensitivity function T(s). Notably, the loop transfer function is defined by L(s) = G(s)Gc(s), where G(s) is the process transfer function, Gc(s) is the controller transfer function. The sensitivity functions have the property S(s) + T(s) =1.

The maximum sensitivity is

$$S(s)=\frac{1}{1+G(s)G_c(s)}=\frac{1}{1+L(s)} \qquad \textbf{Eq. 14-7}$$

After that, the maximum sensitivity is defined by

$$M_s=\max\left|\frac{1}{1+L(iw)}\right|=\max|S(iw)| \qquad \textbf{Eq. 14-8}$$

We can determine the quantity Ms by simply inversing of the shortest distance from the Nyquist Curve of the loop of the transfer function to the critical point -1. The values of reasonable Ms are in range 1.2 to 2.

The complementary function is defined as follow,

$$T(s)=1-S(s)=\frac{G(s)G_c(s)}{1+G(s)G_c(s)}=\frac{L(s)}{1+L(s)} \qquad \textbf{Eq. 14-9}$$

The sensitivity can also be shown by the largest value of the complementary function,

$$M_p=\max\left|\frac{L(iw)}{1+L(iw)}\right|=\max|T(iw)| \qquad \textbf{Eq. 14-10}$$

Typical values of Mp belong to the range from 1.0 to 1.5.

### (5) Set Point Response

The transfer function from the set point to process output is:

$$\frac{Y(s)}{Y_{sp}(s)} = \frac{k_i + bks}{k_i + ks}\frac{L(s)}{1+L(s)} = G_{sp}(s) \qquad \textbf{Eq. 14-11}$$

The resonance peak of the transfer function is given by:

$$M_{sp} = \max \mid G_{sp}(iw) \mid \qquad \textbf{Eq. 14-12}$$

As a result, the b-value is determined based on Eq. 14-12, and it can be noted that Msp < Mp when b<1.

### (6) Measurement Noise and Filtering

The measurement noise can cause undesirable control actions and degrade performance. The PI algorithm typically has a high-frequency gain k. The influences of measurement noise is significantly alleviated by filtering the signal, as mentioned in Eq. 14-4. The requirements could be interpreted as relevant to the magnitude of high-frequency gain of the PI controller. However, these requirements are optional and only employed in select circumstances.

### (7) Design Method

The design method is utilized to give the parameter of the PI controller which maximizes integral gain subject to a robustness constraint. The robustness is specified related to the maximum value of the sensitivity function (Ms). This approach may be explained as a minimization of the integrated error at step load disturbances, which are subject to limitations on the maximum sensitivity. Consequently, the controller is capable of easily mitigating the load disturbances, while also being robust. As a result, the Ms value is a good design parameter, with a rational range from 1.2 to 2.0. This is because the lower values allow for higher levels of robustness, but poorer elimination of the load disturbances. In the simple rule, the selected Ms value is chosen as 1.4 as it yields an acceptable balance between performance and robustness. The closed-loop responses rarely have oscillations with this value [12].

### 14.1.2 Design Processes

#### (1) Model of Steam Generator Thermal-Hydraulic Using SPACE Code

The model of the APR 1400 SG thermal-hydraulic is built utilizing the Safety and Performance Analysis Code (SPACE) as illustrated in Figure 14-5. SPACE develops thermal-hydraulics system analysis code by adopting advanced physical modeling of two-phase flows, primarily using two-fluid three-field models which encompass gas, continuous liquid, and droplet fields. SPACE also has the capability to simulate 3D effects by the use of structured and/or non-structured meshes. A key feature in the SPACE is a hydrodynamic model package, which includes a hydraulic solver, constitutive models, special process models and component models. The two-fluid, three-field governing equations are used as the basic foundation for the hydraulic solver. The finite volume method is applied to discretize each field equation to the unique SPACE mesh system. This mesh system consists of several three dimensional mesh systems that can be either structured or non-structured, as well as one-dimensional pipe meshes.

The first-order Taylor expansion technique is used to linearize non-primitive variables appearing in the temporal terms. A semi-implicit scheme is selected as the basic numerical time advancement scheme. The constitutive model has some correlations such as friction and heat transfer which required governing equations. The special process and the system component models restrict or reform the solution of the basic governing equations and contribute to the accuracy of the system simulations of the NPPs [20].

In this study, a downcomer region, a tube bundle region (an evaporator/ riser/economizer), and a steam dome region are developed in the SPACE model.

PART 3

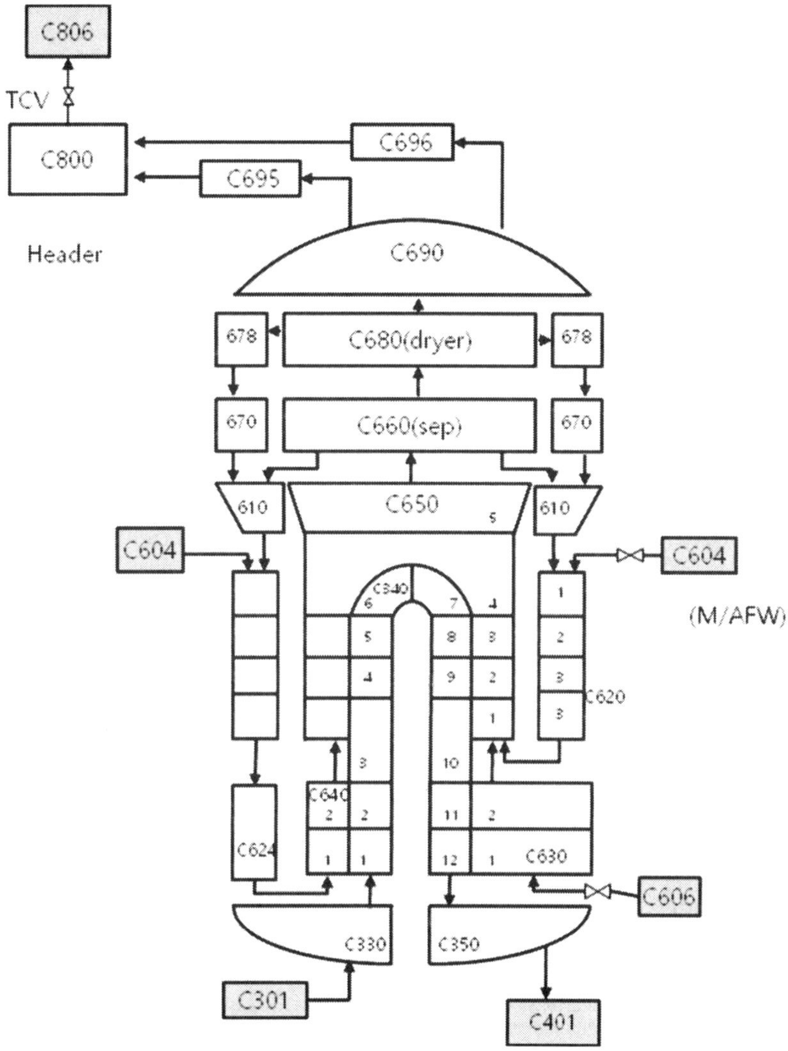


Figure 14-5: SPACE nodalization for the APR 1400 steam generator

## (2) Primaries of the Steam Generator

The APR 1400 SG nodalization for one of the two SG primaries is depicted in Figure 14-5. Flow from the hot leg (C301) goes to the SG inlet plenum (C330). The SG tube primaries are grouped together in a 12 cell pipe component. The C340-1 through to the C340-6 pipes represent the hot side U-tubes of the SG. The cold side U-tubes of the SG are represented by the C340-7 to the C340-12 pipes. The branch C350 represents the SG outlet plenum [21].

### (3) Secondary of the Steam Generator

On the secondary side, the main feedwater is transmitted through both downcomer and economizer feedwater lines into the SG downcomer and economizer annulus through their respective control valves. The downcomer feedwater (C604) flows through the downcomer feedwater control valve and enters into the SG downcomer annulus at C620-1 where it combines with the recirculation flow from the separator (C660). The combined flow goes down via the downcomer annulus (C620, C624) before entering the evaporator region (C640, C650).

The SG annulus at C630-1 receives the economizer feedwater (C606) through the economizer feedwater control valve. It then combines with the downcomer flow from C620. The mixed flow moves up and infiltrates the evaporator zone (C650).

The two-phase mixture which escapes the evaporator region exudes through the SG's mid region (C650) and enters the separator (C660).

The separators have a span for elevation. There are two differentiated regions in the U-tube type steam generator that run over this span. The first region comprises of the volume inside of the separator cylinders. The second region encompasses volume between the outermost steam generator cylinders and the inside of the SG shell. These two areas are designated in the model as cell C660 and cell C670. The separator-area nodalization permits the depiction of separate flows of fluid characteristics inside and outside of the separator cylinders.

The first region is one of high resistance to flow, because internal walls force the flow to swirl, and then separate steam and water. During the secondary – side transients, the behavior of the flow within the two regions becomes significantly and increasingly different. Therefore, it is critical that the two regions are modeled separately in this scenario, so that the simulation of the liquid flow accurately accounts for movement through the separators, around the separators, and out of the steam nozzle [21]. The model uses heat structures to represent the SG tubes, the inner and outer shell and steam head, as well as the internals of the separator and the steam dome regions.

### (4) Identification of Steam Generator Model for Control

Developing a Steam Generator Model requires that the fundamental understanding of the design of the plant be matched with the observed input-output data from system tests run during the plant operation. The model should have the ability to reliably represent and predict the behaviors and characteristics of the plant. The process of system identification algorithm requires four things: measurement of the input and output signals from the system in terms of time domain or frequency domain, choice of a model structure, application of a approach to develop the values for the adjustable parameters of the model being considered, and assessment of the estimated model to see if it is adequate for the application's needs. Figure 14-6 illustrates the system identification loop [22].

The accuracy of the estimates generated by a particular model will depend on its purpose. In some circumstances, the models need to accurately identify the forms of the linear state-space model, while other simplified models will capture the foundational dynamic system behaviors, such as time constants, static gains, and delay times. In order to design the PI feedwater controller to control the APR 1400 SG, the SG model must be developed as the process model. The thermal–hydraulic model relied on physical considerations is made by the SPACE as shown in Figure 14-5. However, this model is a complicated, high order model, and cannot be used directly for designing the controller. In this case, a simplified transfer functions model of the SG is produced by the system identification procedure, based on the input-output measurements.

There are four parameters that affect SG level changes: steam flow rate, feedwater flow rate, feedwater temperature, and reactor power. The first two parameters directly change the water mass in the SG. The last two impact the thermodynamic condition of the SG, determining the changes in the SG water level. Therefore, changes in the feedwater flow rate and the steam flow rate have the greatest impact on the SG secondary water mass. The feedwater temperature and the reactor power indirectly influence the thermal-hydraulic conditions of the SG, so their effect on the SG water level is much weaker [21]. In this study, the feedwater flow rate is a major parameter in defining the mathematical SG model for the PI feedwater controller. This is mainly because the feedwater flow is the critical control parameter in regulating the SG water mass and level. There

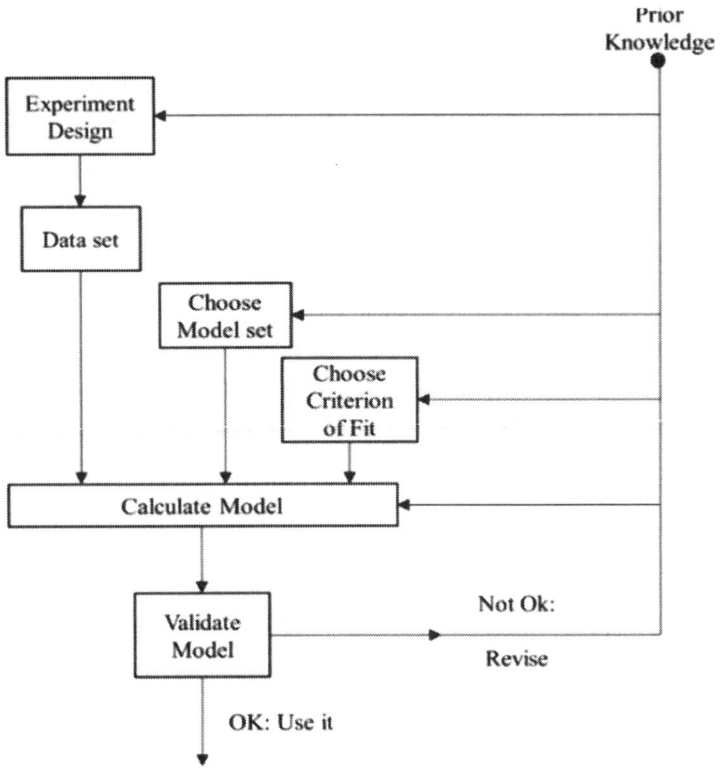


Figure 14-6: System identification loop

are two main operating power regions in the SG. The low power operating region is below 20% power level, while the high power operating area is above 20% power level. Finally, to acquire the detailed data sets from the SG thermal-hydraulic model, 100% power, 80% power and 30% power are selected as the high power regions, and 5% power is chosen as the low power region.

### (5) Steam Generator Model

The Simple Process Model [23] is employed to create a simple continuous-time process that describes the SG model for control in terms of one or more of the following elements: static gain, one or more time constants, process zero, a possible delay, or possible enforced integration. The measured data sets obtained from the simulation of the SPACE SG model in the preceding section are applied to determine the aforementioned elements at different power levels. The most prevalent identification approach in the industrial practice is to generate a step response experiment based on the data, and then adjust two or three parameters to the measured response. Generally, the simple two or three parameter models commonly employed to tune PI or PID controllers rely on such parameters. The chief principles for estimating such parameters and how estimated models fit into a general framework are reviewed in [22], and the research of the Simple Process Models is introduced in [23].

The most popularly utilized Simple Process Model is:

$$G(s) = \frac{K}{1 + sT_{p1}} e^{-sT_d} \qquad \textbf{Eq. 14-13}$$

A model without delay (Td = 0) which is one of variants of this model is:

$$G(s) = \frac{K}{1 + sT_{p1}} \qquad \textbf{Eq. 14-14}$$

An Enforced Integration (Self-Regulating Process) model is:

$$G(s) = \frac{K}{s(1 + sT_{p1})} e^{-sT_d} \qquad \textbf{Eq. 14-15}$$

Additionally, a Two Real Poles model with or without a Zero:

$$G(s) = \frac{K(1 + sT_z)}{(1 + sT_{p1})(1 + sT_{p2})} e^{-sT_d} \qquad \textbf{Eq. 14-16}$$

Another possible model which allows resonant poles is an Under-Damped Model:

$$G(s) = \frac{K(1 + sT_z)}{1 + 2\xi s T_r + (sT_r)^2} e^{-sT_d} \qquad \textbf{Eq. 14-17}$$

In Eq. 14-13 to Eq. 14-17, Tp1, Tp2, Tr, and Tz are time constant parameters, where Td is a time delay, K is a constant parameter, and ξ is the associated damping constant in the model. The process model's estimations utilize the measured input-output data sets, and are implemented in MATLAB [24]. The identified transfer function from feedwater flowrate to steam generator water level is as follows:

$$G(s) = \frac{K_f(1+sT_{zf})}{s(1+2s\xi_f T_{rf}+(sT_{rf})^2)}e^{-sT_{df}} \quad \textbf{Eq. 14-18}$$

Where Kf is a feedwater flowrate constant parameter, Trf and Tzf are time constant parameters of the feedwater flowrate, ξf is the associated damping constant of the feedwater flowrate, and Tdf is a time delay of the feedwater flowrate. Nevertheless, based upon the identification process, the transfer function from feedwater flowrate to steam generator water level at 80% reactor power does not have a time delay.

In Eq. 14-18 the parameters at each power level are illustrated in Table 14-2.

Table 14-2: Parameters of transfer function at each power level

| Parameter | 5% | 30% | 80% | 100% |
|---|---|---|---|---|
| Kf | 0.0029 | 0.0024 | 0.0023 | 0.0028 |
| Trf | 0.455 | 1.256 | 0.091 | 0.236 |
| f | 1.461 | 0.637 | 6.764 | 7.369 |
| Tdf | 2.114 | 0.546 | 0 | 3.0 |
| Tzf | 1.953 | 0.148 | 0.330 | 3.389 |

## (6) Simplified APR 1400 Feedwater Control System

Figure 14-7 reveals the simplified control block diagram of the APR 1400 FWCS [21]. The FWCS is designed to automatically control the steam generator water level by regulating the feedwater flow rate from the startup mode to full power operation. The required feedwater flow is calculated by processing the summed error signals, i.e. the compensated level error and selected flow errors, through the PI controller. The compensated level error is measured by the difference between a compensated SG downcomer water level signal and the reference level (constant set point).

When operating at low power levels (below 20%), the compensated level error signal is combined with the SG level error signal. The economizer feedwater control valve position demand signal is reset to zero causing the economizer feedwater control valve to close, and the pump speed set point demand signal to drop to its minimum value. Thus, the feedwater flow rate is regulated by the downcomer feedwater control valve.

When operating at high power levels (above 20%), the compensated level error signal is combined with the difference between the total feedwater flow signal and the total steam flow signal. The flow demand to the downcomer feedwater control valve program is set to an adjustable bias set point which positions the downcomer valve to pass approximately 10% of the full-power feedwater flow. The economizer feedwater control valve and main feedwater pump speed determine the feedwater flow rate. Figure 14-8 shows the feedwater flow split.

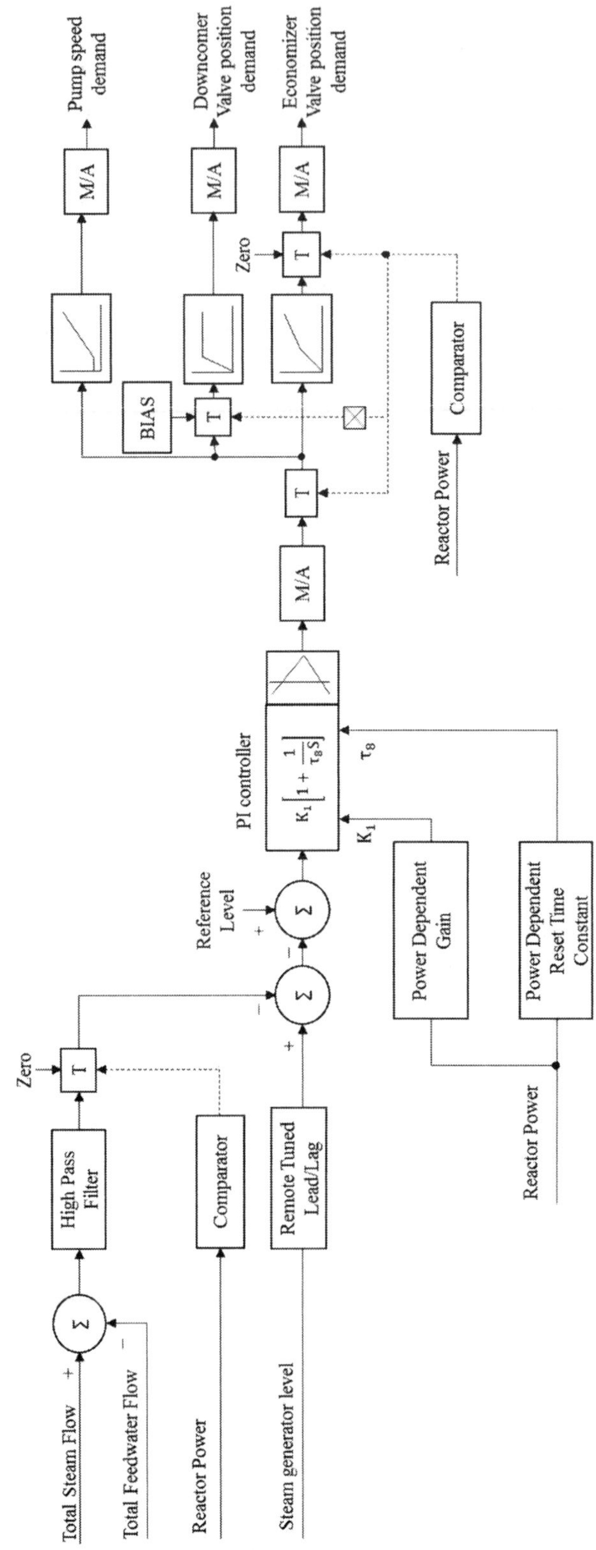


Figure 14-7: Simplified block diagram of the APR 1400 FWCS

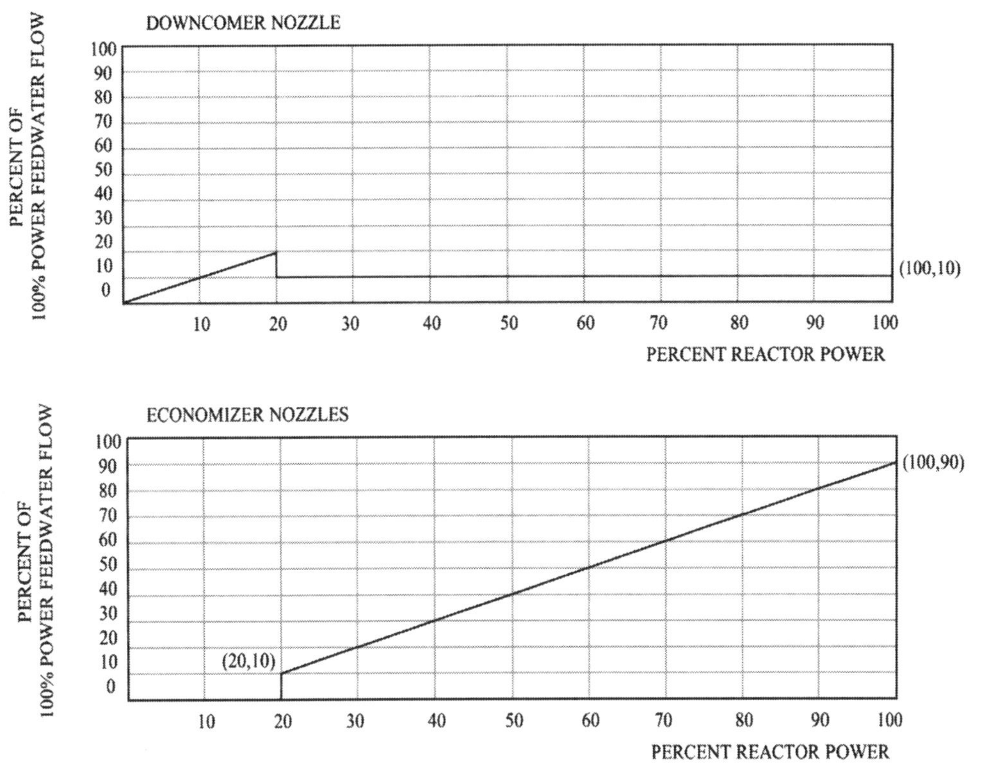


Figure 14-8: Feedwater flow split

## (7) Tuning PI Controller

- **MIGO PI and Ziegler-Nichols PI Controllers**

The feedwater PI controller design is completed with the Control System Designer app in MATLAB®. The feedback control architecture is chosen as the single feedback loop. The system compromises controller (Gc), plant (G) in forward path, sensor dynamics (H) in feedback path, and prefilter F [25]. Figure 14-9 depicts the single-loop feedback configuration. Table 14-3 illustrates the obtained PI controller of the MIGO method and the Ziegler-Nichols method.

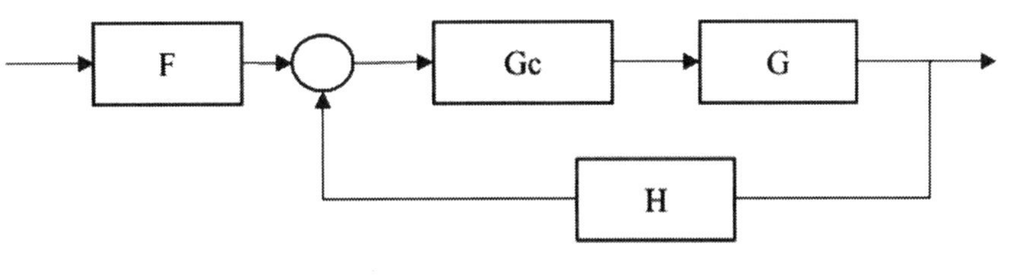


Figure 14-9: Feedback controller architecture

The comparisons of the overshoot percentages of step responses using the Ziegler-Nichols closed-loop tuning approach and the MIGO closed-loop tuning approach at different power levels (100%, 80%, and 30% powers) are given in Table 14-4. It can be obvious that the overshoot percentages of the closed-loop system employing the MIGO approach are substantially lower than the ones of the closed-loop system employing the Ziegler-Nichols approach. Notably, there is a gap in the data set at 5% power for the Ziegler-Nichols PI controller. This was due to the fact the Ziegler-Nichols method cannot produce a stable feedback loop at 5% power. The overshoot percentage of the MIGO closed-loop approach is reasonable, approximately 14.9%.

Table 14-3: PI controller of MIGO and Ziegler -Nichols methods

| PI controller | MIGO | Ziegler -Nichols |
|---|---|---|
| 5% | $\frac{80.284(s+0.05004)}{s}$ | |
| 30% | $\frac{72.198(s+0.03735)}{s}$ | $\frac{185.65(s+0.1668)}{s}$ |
| 80% | $\frac{98.45(s+0.04786)}{s}$ | $\frac{253.16(s+0.2138)}{s}$ |
| 100% | $\frac{40.472(s+0.02422)}{s}$ | $\frac{104.07(s+0.1082)}{s}$ |

Table 14-4: Comparison of the overshoot percentages between the MIGO method and the Ziegler-Nichols method

| Power levels | Overshoot of MIGO method (%) | Overshoot of Ziegler and Nichols method (%) |
|---|---|---|
| 30% | 16.2 | 103 |
| 80% | 17.6 | 62.8 |
| 100% | 16.7 | 94.6 |

These results strongly suggest that the MIGO method is more robust than the Ziegler-Nichols method. Figure 14-10 depicts the step response of the closed-loop system for the PI controller based on the MIGO method at 5% reactor power. Figure 14-11, Figure 14-12, and Figure 14-13 show the comparison step responses of the closed-loop system for the PI controllers of the two mentioned methods at 30%, 80%, and 100% power, respectively. From the simulation, we can observe that the MIGO PI performs the improved performance than the PI designed by Ziegler-Nichols. Through the simulated results, the PI controller which is designed by MIGO method is selected as the novel PI controller of the FWCS.

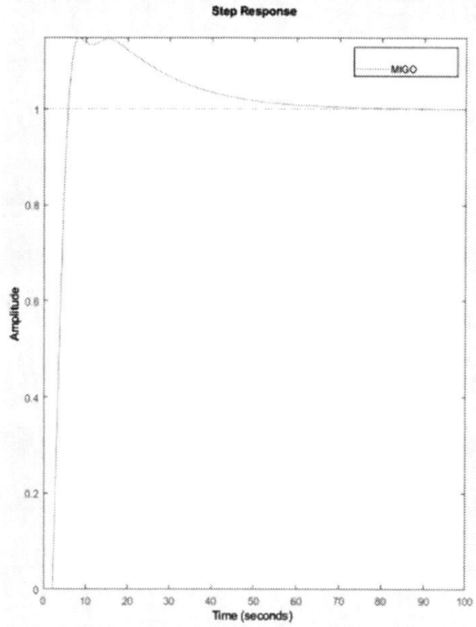


Figure 14-10: Step response for PI controller using MIGO method at 5% power

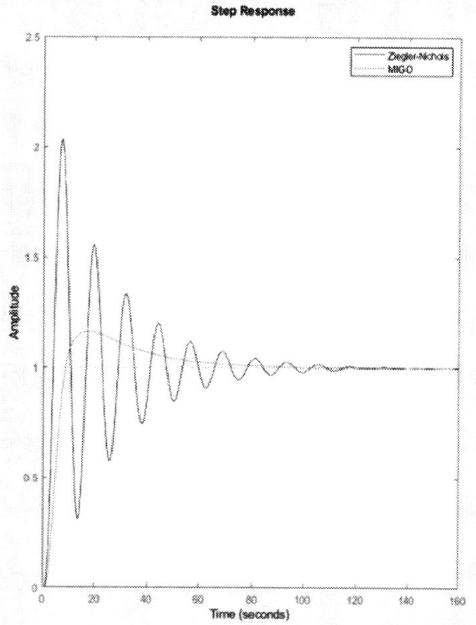


Figure 14-11: Step responses for PI controllers using MIGO and Ziegler-Nichols methods at 30% power

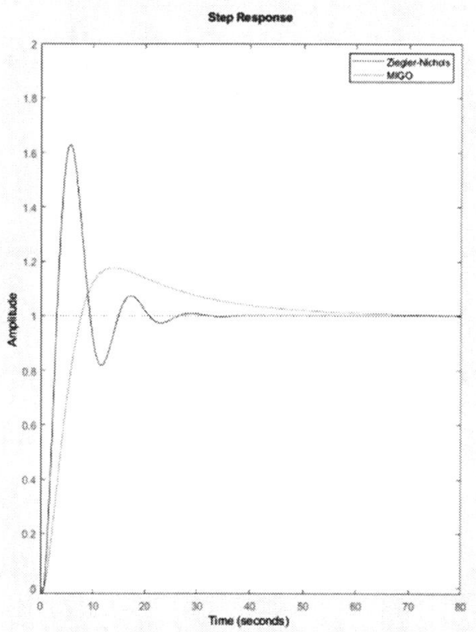


Figure 14-12: Step responses for PI controller using MIGO and Ziegler-Nichols methods at 80% power

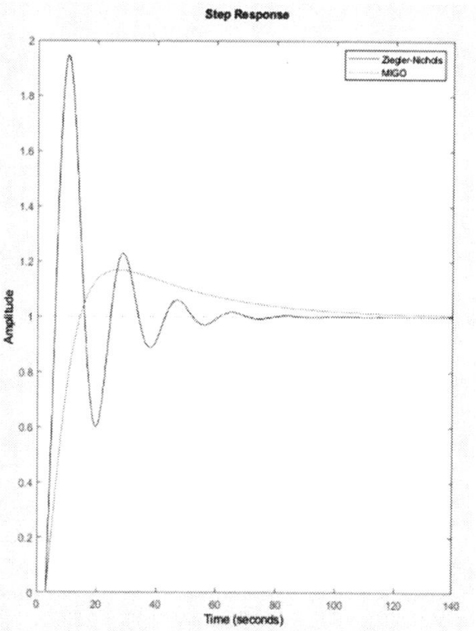


Figure 14-13: Step responses of PI controller using MIGO and Ziegler-Nichols methods at 100% power

### (8) GA PI Controller

There are a variety of operating reactor power levels between start-up mode and the full-operation mode in the APR 1400 FWCS. Because collecting the data set by SPACE is very difficult and time-consuming, so the quality of the obtained data is not so good. In this section, we just choose a case study of 100% reactor power level as a typical circumstance to design the PI algorithm using the GA method. With the help of the optimization app in MATLAB, the PI controller designed by GA of the FWCS was obtained. The objective function value is 0.5 when the value of Kp is 50.5075 and the value of Ki is 1.2566 in Eq. 14-19. Based on the GA approach, we may accomplish the stable step response of the closed-loop system using GA PI controller. Through Simulink/Matlab, the block diagram of the feedback control system based on GA PI controller was developed and the step response is shown in Figure 14-14.

$$PI = 50.5075 + \frac{1.2566}{s} \qquad \textbf{Eq. 14-19}$$

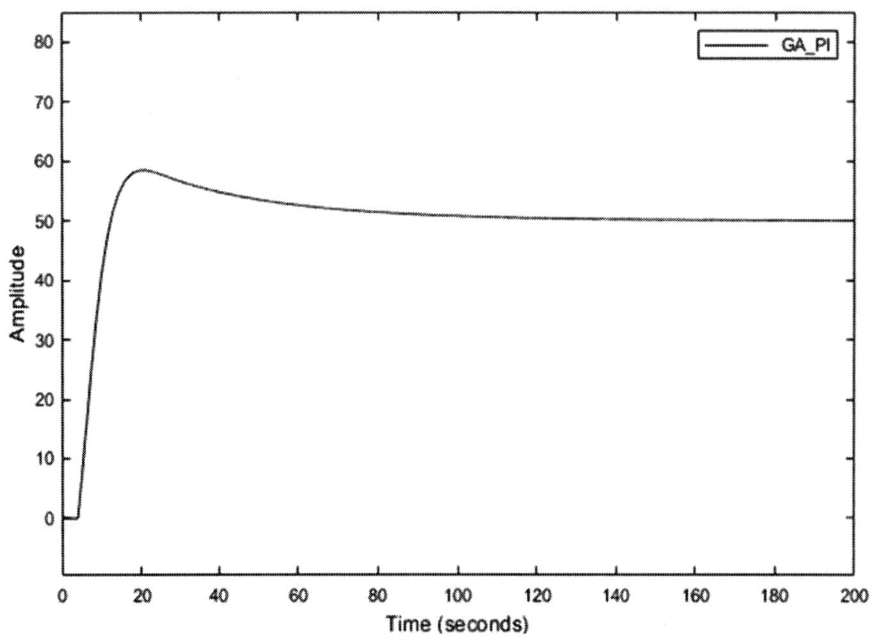


Figure 14-14: Step response of GA-PI technique

### 14.1.3 Simulation and Validation

There are two methods: MIGO, and GA which are able to design the PI controller of the FWCS. Figure 14-15 describes the block diagram of an overall closed-loop control system which is simulated by Simulink. The overall closed-loop system also expresses the valve and pump model as shown in Figure 14-15. This assumes that the feedwater control valve and pump model has the transfer function which is considered as a first order system with a time constant of 1 [21]. Additionally, the simulation of the feedback system with the existing PI controller employs the same SG model and valve and pump model as illustrated in Figure 14-15.

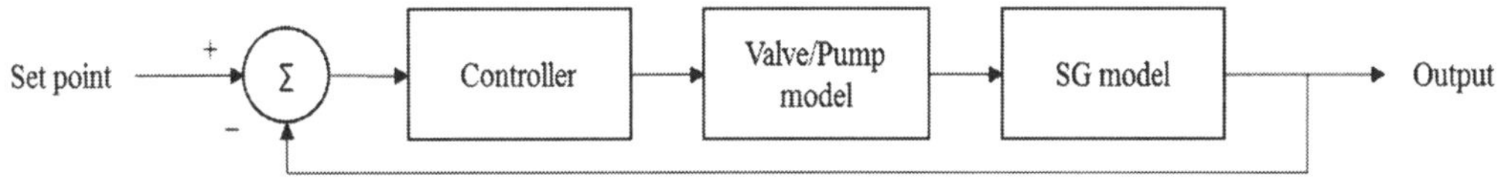


Figure 14-15: Schematic diagram of an overall control system

At 100% reactor power level, when both methods were applied in the overall closed-loop system, they generate the stable step response of the closed-loop system. It means that two controllers: MIGO PI, and GA PI work effectively and have better performance than the existing PI controller. Figure 14-16 shows overall closed-loop system of GA_PI, MIGO_PI and current controllers at 100% power. Figure 14-17 indicates the performance of three feedwater controllers: MIGO PI, GA PI, and existing PI. However, the performance of the MIGO PI controller works more effective than the GA PI performance. Figure 14-18 shows overall closed-loop system of GA_PI and MIGO_PI controllers at 100% power. Figure 14-19 depicts the step response of the closed-loop system between MIGO and GA controllers.

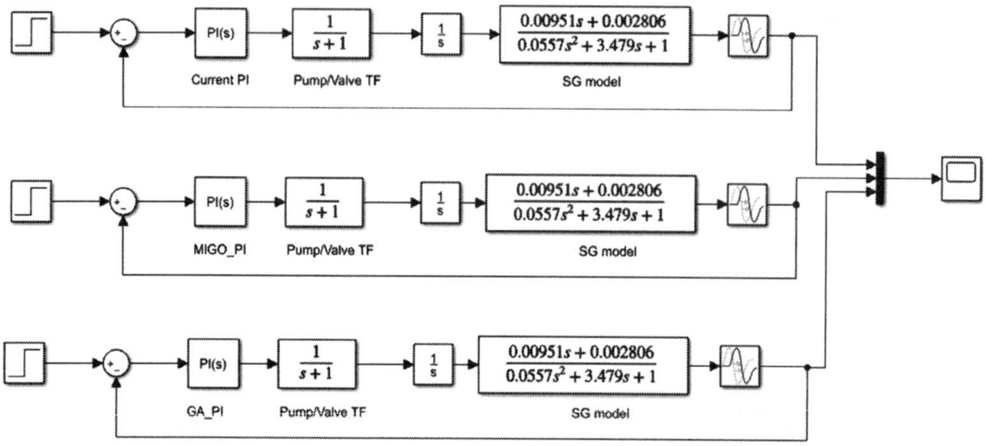


Figure 14-16: Overall closed-loop system of GA_PI, MIGO_PI and current controllers at 100% power

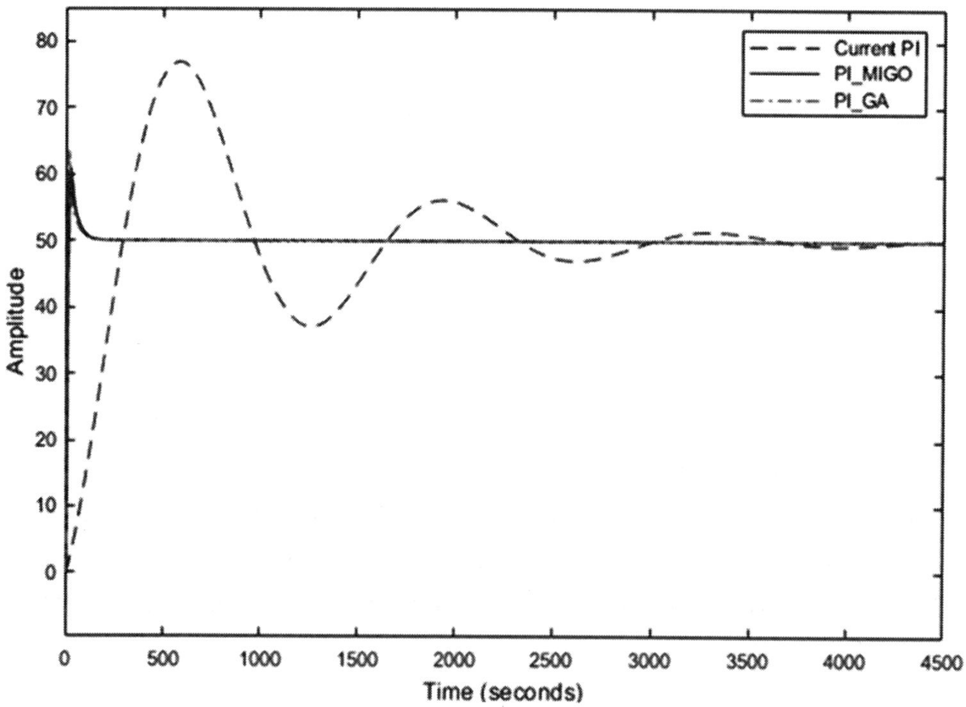


Figure 14-17: Step responses of closed-loop system at 100% power between MIGO PI, GA PI, and existing PI controllers

P A R T 3

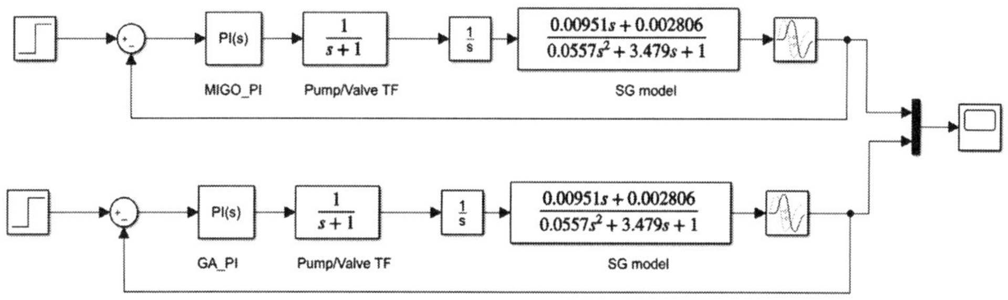


Figure 14-18: Overall closed-loop system of GA_PI and MIGO_PI controllers at 100% power

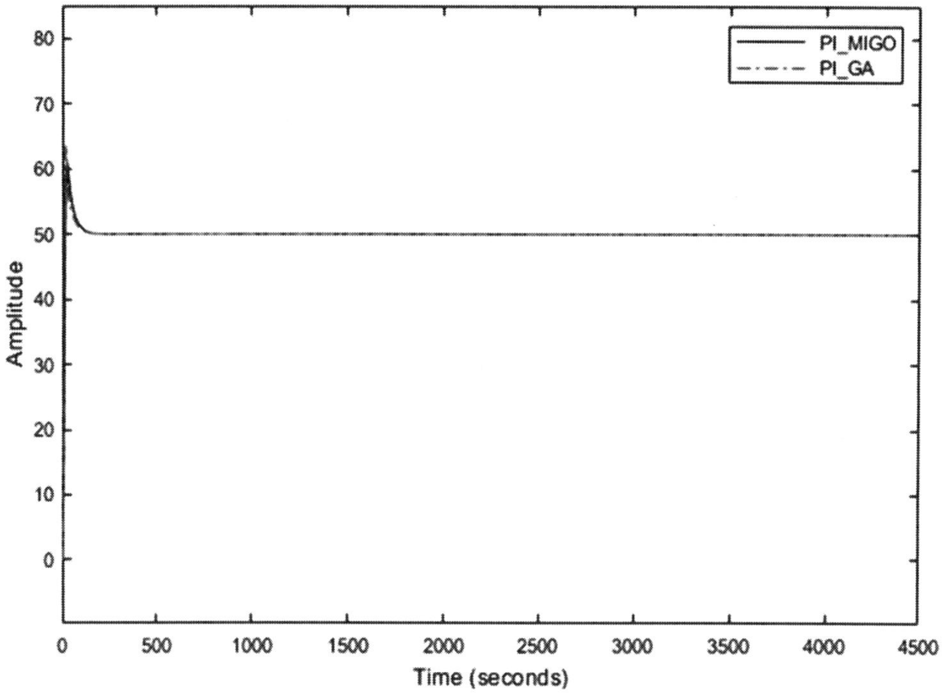


Figure 14-19: Step response of the closed-loop system at 100% reactor power level between MIGO and GA controllers

Consequently, we made a final decision to choose the MIGO approach as the main method for designing the feedwater PI controller. The simulations of the developed PI feedwater controller using the MIGO approach are performed to reaffirm the performance and validate the robustness at the above-mentioned power levels. The overall closed-loop system of MIGO_PI controller at 5%, 30%, 80%, 100% powers are shown in Figure 14-20, Figure 14-24, Figure 14-28, and Figure 14-32, respectively. The step response of the overall closed-loop systems using the MIGO PI at 5%, 30%, 80%, and 100% powers are very stable as shown in Figure 14-21, Figure 14-25, Figure 14-29, and Figure 14-33 respectively. The overall closed-loop system of MIGO_PI and exist controllers at 5%, 30%, 80%, 100% powers are illustrated in Figure 14-22, Figure 14-26, Figure 14-30, and Figure 14-34, respectively. Figure 14-23, Figure 14-27, Figure 14-31, and Figure 14-35 illustrate the comparisons of the step responses of the overall closed-loop systems between the two controllers, namely existing PI and MIGO PI at 5%, 30%, 80% and 100% powers, respectively. Through the validation of the simulation results at four different power levels, it is proved that the MIGO PI controller has an effective performance and robustness when comparing to the existing PI controller. Particularly, the performance of the system at the lower power region, such as 5% power, is significantly improved when using the MIGO PI controller.

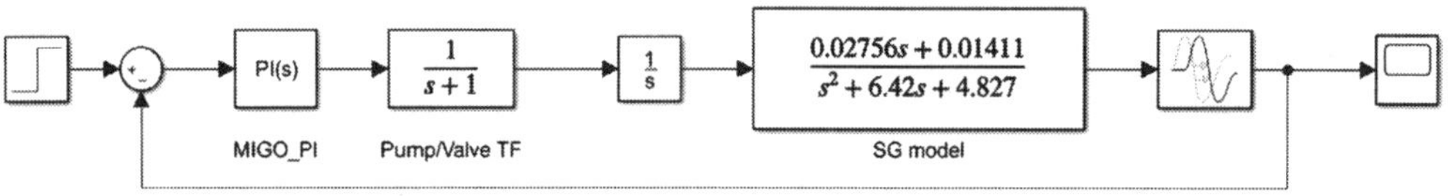


Figure 14-20: Overall closed-loop system of MIGO_PI controller at 5% power

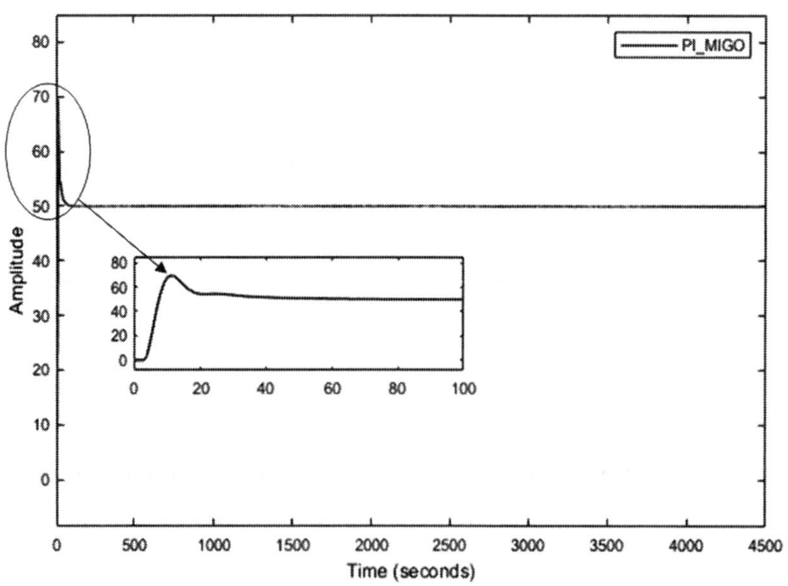


Figure 14-21: Step response of the overall closed-loop system at 5% power using MIGO PI controller

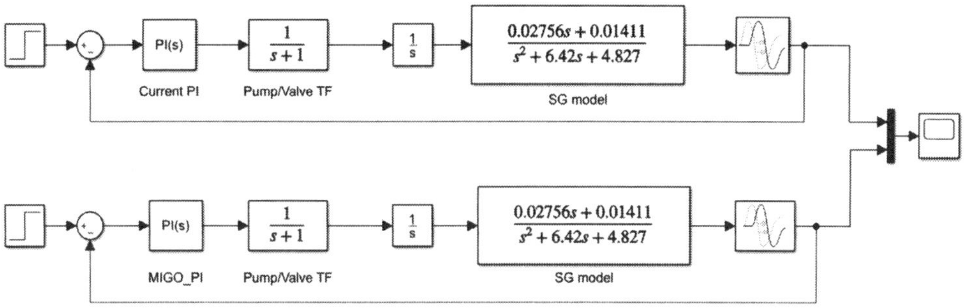


Figure 14-22: Overall closed-loop system of MIGO_PI and current controllers at 5% power

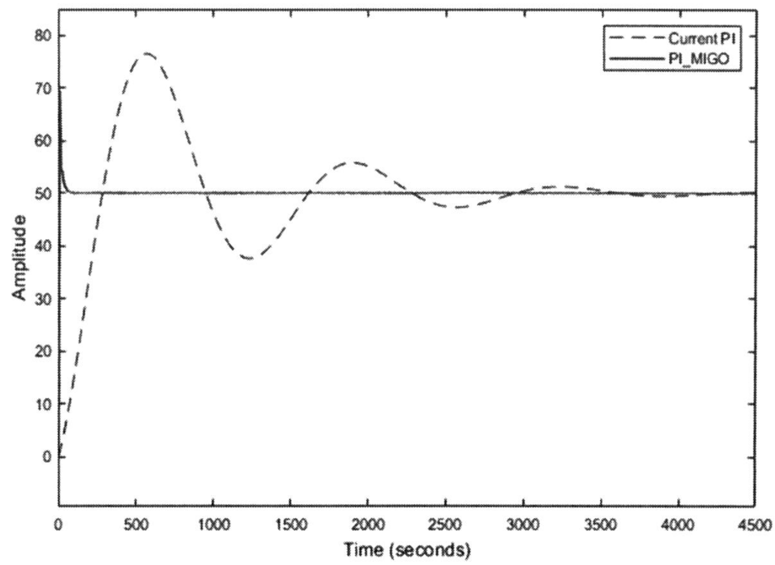


Figure 14-23: Step responses of the overall closed-loop systems at 5% power using existing PI and MIGO PI controllers

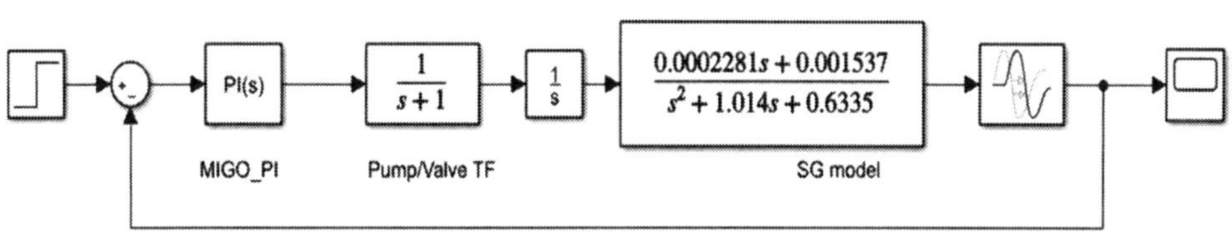


Figure 14-24: Overall closed-loop system of MIGO_PI controller at 30% power

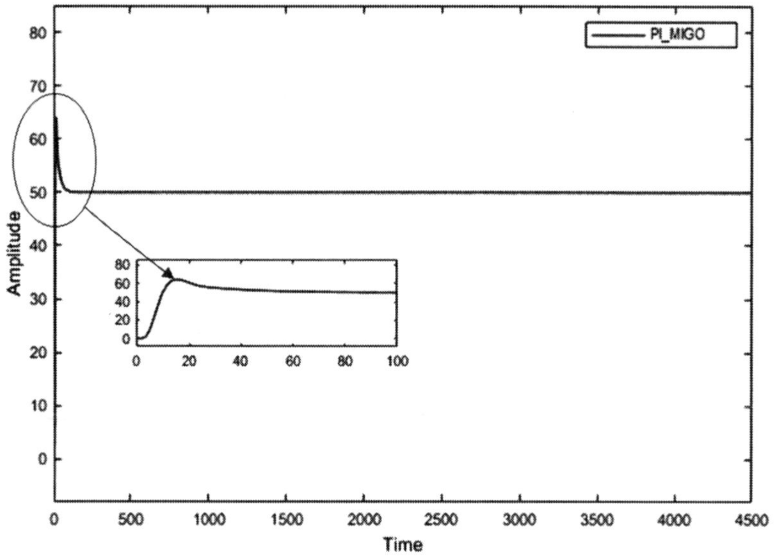


Figure 14-25: Step response of the overall closed-loop system at 30% power using MIGO PI controller

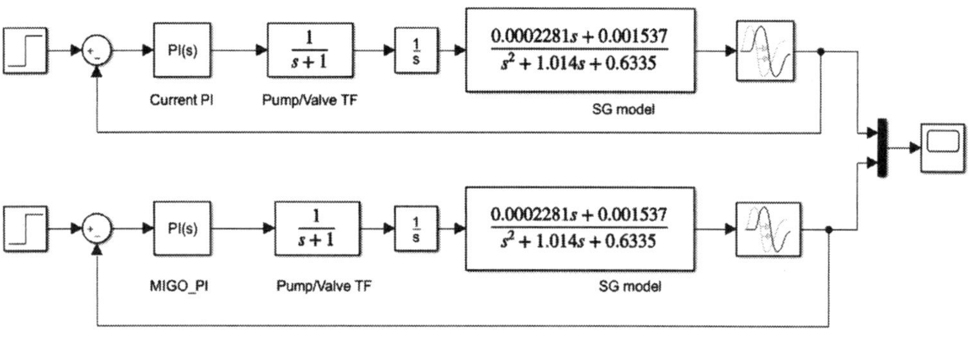


Figure 14-26: Overall closed-loop system of MIGO_PI and current controllers at 30% power

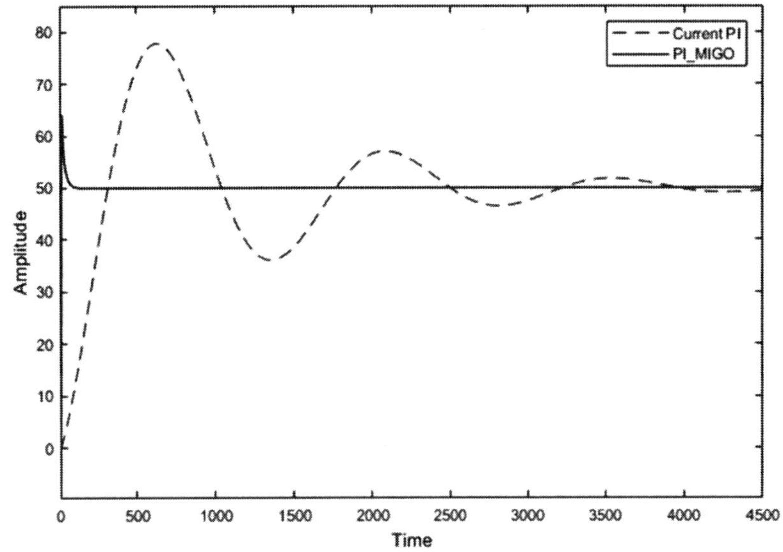


Figure 14-27: Step responses of the overall closed-loop systems at 30% power using existing PI and MIGO PI controllers

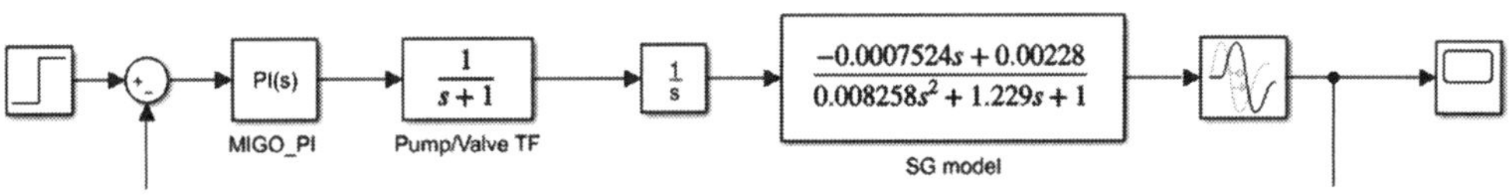


Figure 14-28: Overall closed-loop system of MIGO_PI controller at 80% power

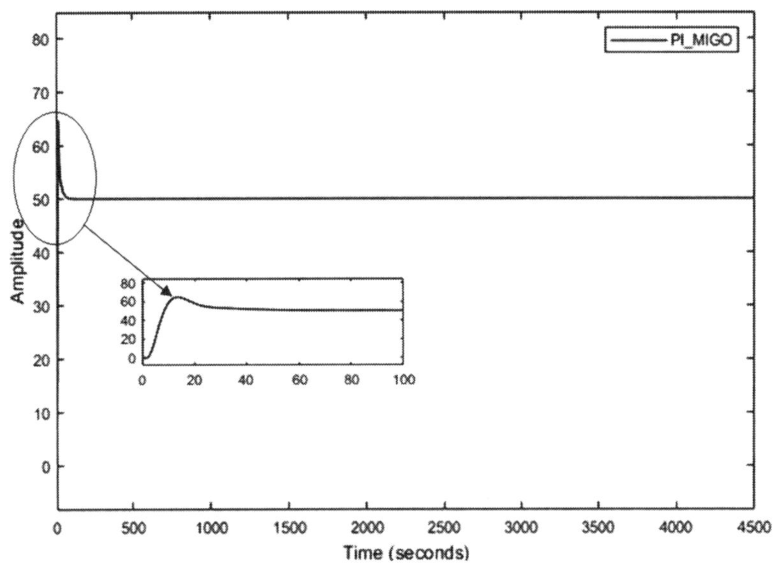


Figure 14-29: Step response of the overall closed-loop system at 80% power using MIGO PI controller

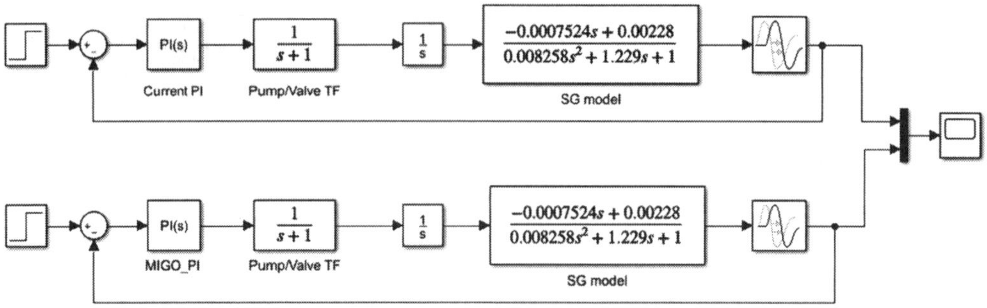


Figure 14-30: Overall closed-loop system of MIGO_PI and current controllers at 80% power

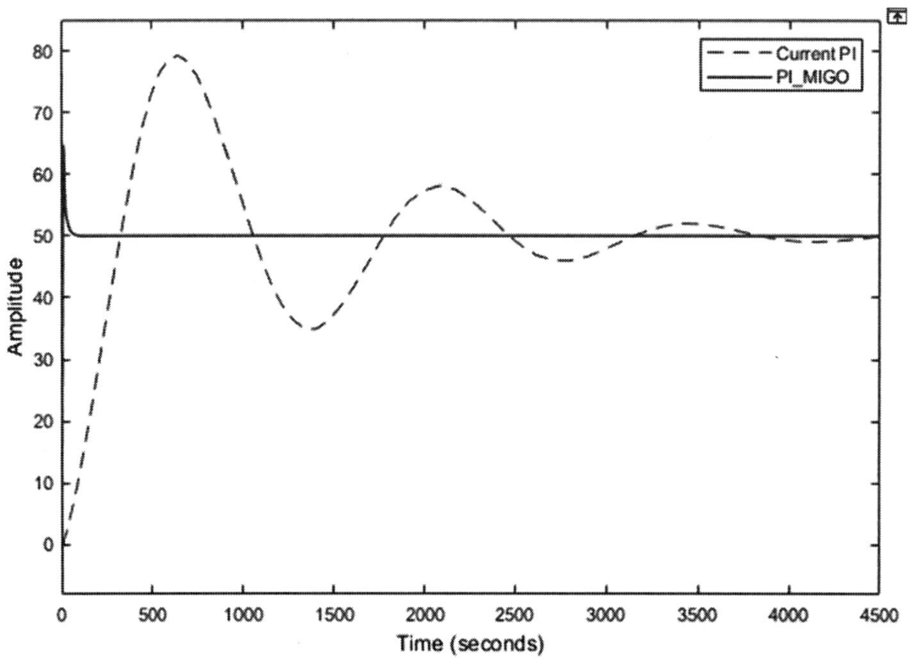


Figure 14-31: Step responses of the overall closed-loop systems at 80% power using existing PI and MIGO PI controllers

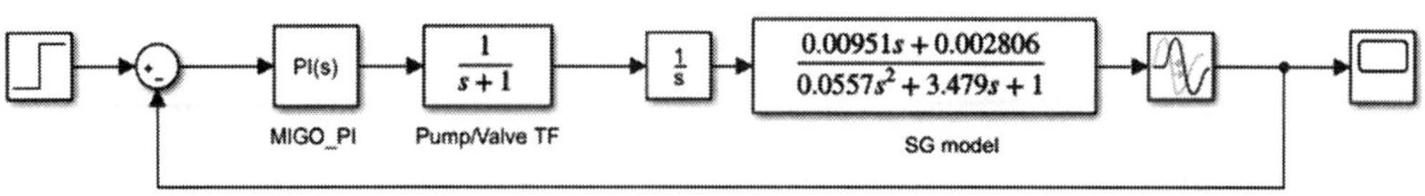


Figure 14-32: Overall closed-loop system of MIGO_PI controller at 100% power

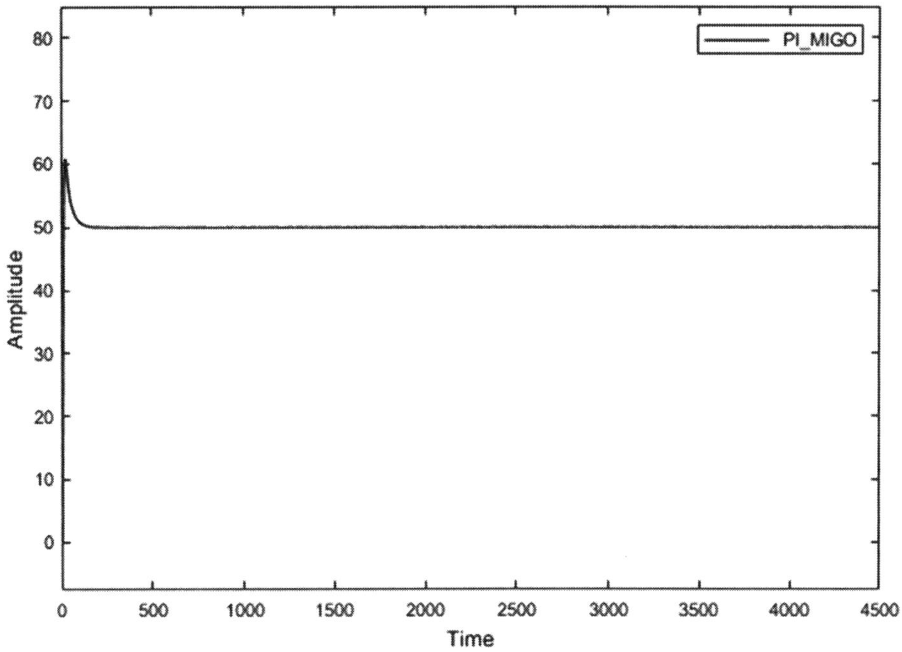


Figure 14-33: Step response of the overall closed-loop system at 100% power using MIGO PI controller

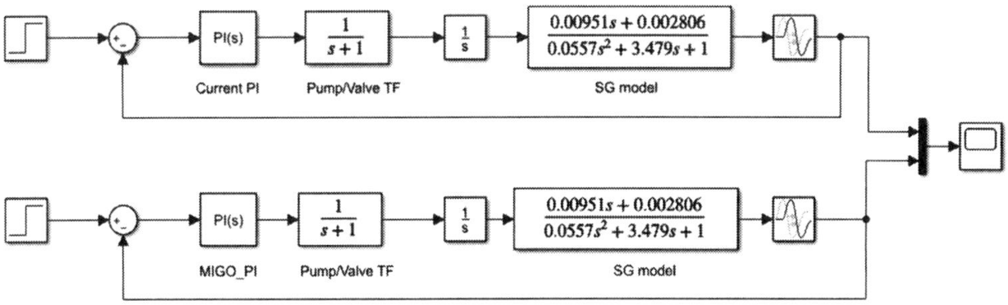


Figure 14-34: Overall closed-loop system of MIGO_PI and current controllers at 100% power

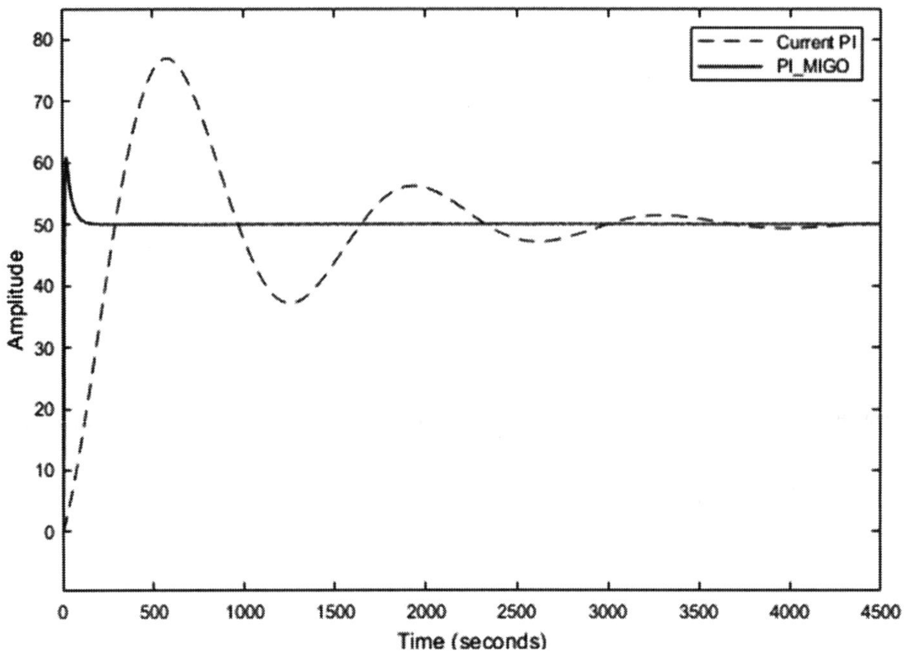


Figure 14-35: Step responses of the overall closed-loop systems at 100% power using existing PI and MIGO PI controllers

## 14.2 Development of a Bumpless Control Transfer and an Anti-Windup Scheme

### 14.2.1 Integrator Windup

Due to the fact that a control system has a wide range of operating conditions, the control variable may reach the actuator limits. In addition, switching between manual mode and automatic mode causes that replacement change of the controller input may have the same effect. The result is that the feedback loop is broken, and the closed-loop system becomes an open-loop system as the actuator remains at its limits regardless of the process output. When the controller uses the integral action and if the algorithm is not exactly designed, the error might continue to be integrated. In other words, the integral term could become very large, or, it winds up. Thus, the result is that the error has an opposite sign for a long time period before the system operates normally. Therefore, any controller

PART 3

which has the integral function will give large transients if the actuator saturates or plant input substitution happens. The controller does not control the plant when wind-up occurs, and the states of the controller are inadequately updated [19].

### 14.2.2 Back-Calculation Method

The block diagram of the PI controller with anti-windup is illustrated in Figure 14-36. When the output of the system saturates, the integral term in the algorithm is recalculated so that the new value gives an output at the saturation limit. This does not reset the integrator instantaneously but dynamically with a time constant Tt.

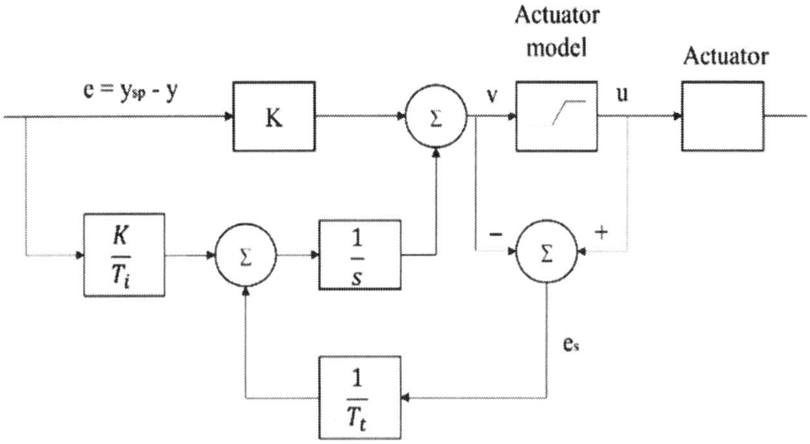


Figure 14-36: PI controller with anti-windup

As shown in Figure 14-36, an extra feedback path is added to the system. The extra feedback is produced by measuring the difference between the controller output (v) and the actuator output (u). An error signal (es) which is formed from this difference is an input to the integrator through the gain 1/Tt. The signal is reduced zero when there is no saturation. Thus, when the actuator does not saturate, it will not have any impact on the normal operation. However, if there is saturation, the signal (es) is not equal to zero. The normal feedback path around the process is broken as the process input remains constant. However, there is the back-calculation around the integrator, and hence, the integrator output is driven towards a value so that the integrator input is zero.

The integrator input is as follow:

$$\frac{1}{T_t}e_s + \frac{K}{T_i}e \qquad \text{Eq. 14-20}$$

where e is the error controller. Thus,

$$e_s = -\frac{KT_t}{T_i}e \qquad \text{Eq. 14-21}$$

in the steady-state. Because of $e_s$ = u − v,

$$v = u_{\lim} + \frac{KT_t}{T_i}e \qquad \text{Eq. 14-22}$$

where $u_{\lim}$ is the saturating value. It is true that e and ulim signals have the same sign, so the magnitude of v is always bigger than the magnitude of u. This avoids the wind-up phenomenon of integrator. The parameter $T_t$ is expressed as the tracking time constant, and this parameter determines how quickly the reset of integral term occurs [19].

### 14.2.3 PI Controller with a Tracking Mode

The PI structure with the back-calculation consists of two modes: the normal control mode and a tracking mode. When the tracking signal (ω) is equal to the controller output (v), tracking is automatically inhibited. It is not essential to have a logical signal for mode switching, although the PI structure can be operated in both modes. Figure 14-37 represents the controller with a tracking mode. The module has two inputs: the set-point (SP), and a tracking signal (TR) [19].

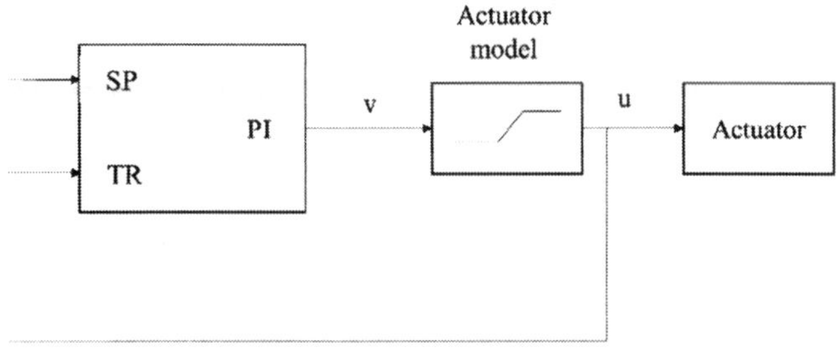


Figure 14-37: Controller with a tracking mode

### 14.2.4 FWCS Design Review

The FWCS plays an important role in the non-safety control systems of the nuclear power plants. The purpose of this system is to automatically control the steam generator downcomer water level by regulating the feedwater flow from start-up mode until full power operation [1].

#### (1) Manual and Automatic Control of Feedwater Control System

There are two operating modes: manual mode and automatic mode in the master controllerof the FWCS. The master control logic is shown in Figure 14-38.

The compensated error to the master controller (CEMC) is calculated by the selected flow error (WSE) signal and the compensated level error (CLE). Where the selected error is the difference between steam flow signal and the total feedwater flow signal at the high power levels, or the difference between the downcomer feedwater flow and calculated steam flow at the low power levels. The compensated level error is defined as the difference between a compensated SG downcomer water level signal and the set-point. The CEMC signal is dynamically compensated by the PI algorithm. The output of this algorithm is known as the compensated summer error (CSE) signal. This algorithm has the steady-state characteristic when the CEMC signal is zero, the PI output signal is constant. A non-zero CEMC signal is integrated causing the output signal to move forward to its maximum or minimum value. Hence, any CEMC is forced to become zero during steady-state. The $K_1$ parameter is the gain and the $\tau 8$ is the time constant of the PI controller. These parameters vary as a function of reactor power [6].

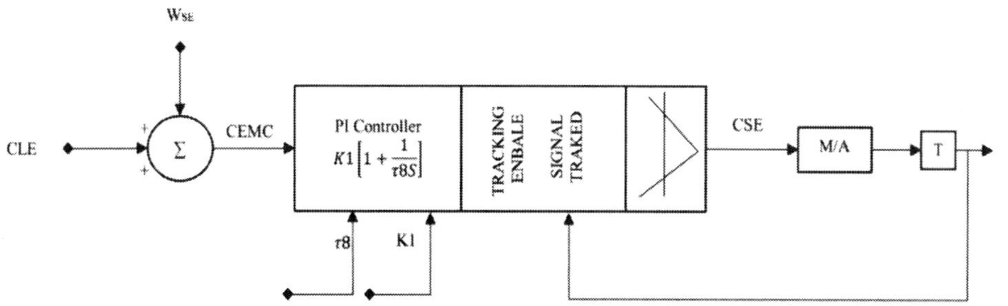


Figure 14-38: Master Control Logic

### (2) Transfer Function of Steam Generator and PI Controller

In the nuclear power plants, there are several reactor power levels from start-up mode to power operation mode. In this study, we choose 100% reactor power to prove the anti-windup based on the back-calculation, and to show the bumpless control transfer between the manual phase and the automatic phase. In order to accomplish the steam generator (SG) model, there are several implementing steps. The first step is that based on the Safety and Performance Analysis Code (SPACE) for Nuclear Power Plants, the input/output data set is acquired through model experiments. The second step is to process the measured data set by system identification toolbox in MATLAB. Ultimately, we obtained the SG model as an under-damped model relied upon the fundamental principles. The PI controller is tuned through the robust loop shaping using Ms Integral Gain Optimization (MIGO) method [26]. The new PI controller offers improved performance rather than the existing PI controller of the FWCS.

We obtained the following transfer function of SG model:

$$G(s) = \frac{0.00951s + 0.002806}{s(0.0557s^2 + 3.479s + 1)} e^{-3s} \qquad \text{Eq. 14-23}$$

The equation fo the PI controller is given as:

$$PI = 40.472 + 0.98\frac{1}{s} \qquad \text{Eq. 14-24}$$

## 14.2.5 Methodologies for improving Controller of FWCS

### (1) Anti-Windup Control

First, we performed the simulation without using anti-windup scheme. We examined the impact of saturation on the closed-loop when the saturation model is not considered by the PI structure block. Figure 14-39 shows the block diagram of the feedwater controller with saturation is simulated in MATLAB. Figure 14-40 indicates set-point and measured output without anti-windup. Figure 14-41 depicts controller output and saturated input without anti-windup.

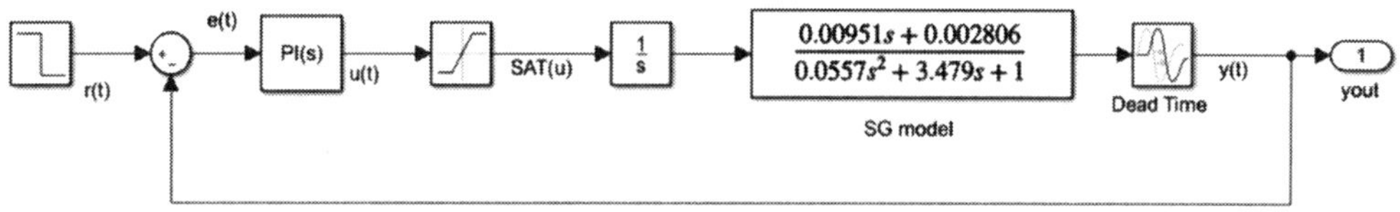


Figure 14-39: Block diagram of the feedwater controller with saturation

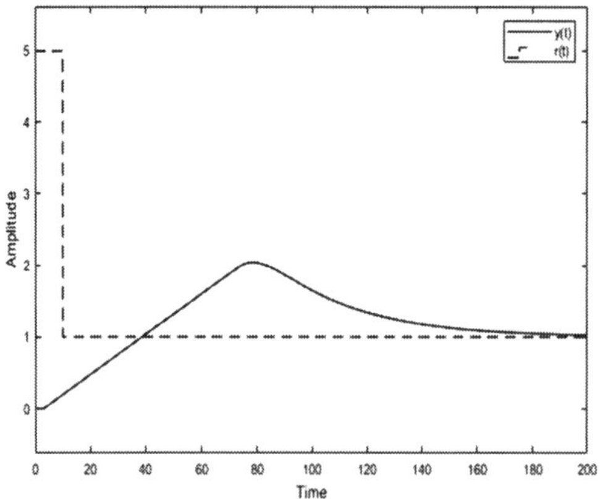


Figure 14-40: Set-point and measured output with no anti-windup

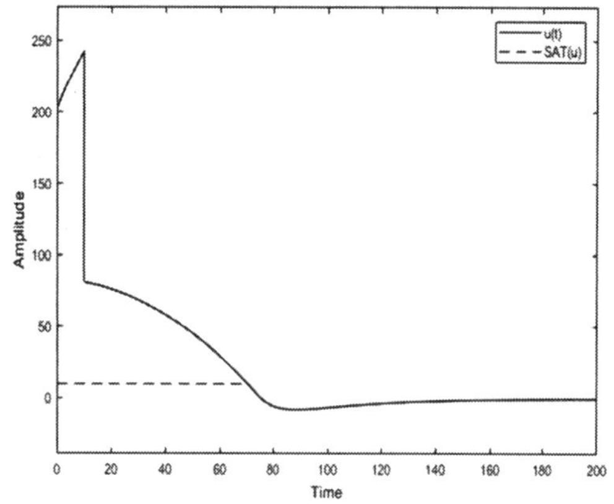


Figure 14-41. Controller output and saturated input with no anti-windup

Figure 14-40 and Figure 14-41 depict two of the emergent problems when controlling the FWCS with input saturation. First, when the set-point value is 5, the PI control signal is instability and takes a long time before coming inside the range of the saturation. Consequently, the PI algorithm is operating in a nonlinear region. When the control signal is increased, this action does not impact the system output. Secondly, once the set-point value becomes 0, there is a substantial delay before the PI controller output returns to within the saturation range.

Secondly, configuring the PI block for anti-windup based on back-calculation approach, the PI block has an internal tracking loop in order to discharge the saturated integrator output [24]. Figure 14-42 illustrates set-point and measured output with back-calculation. Figure 14-43 shows controller output and saturated input with back-calculation.

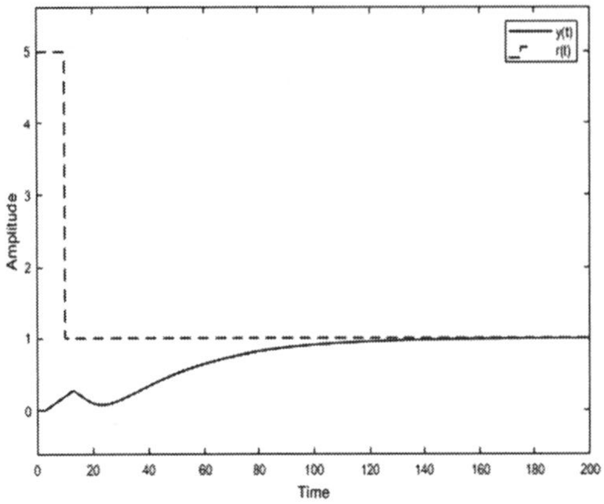


Figure 14-42: Set-point and measured output with back-calculation

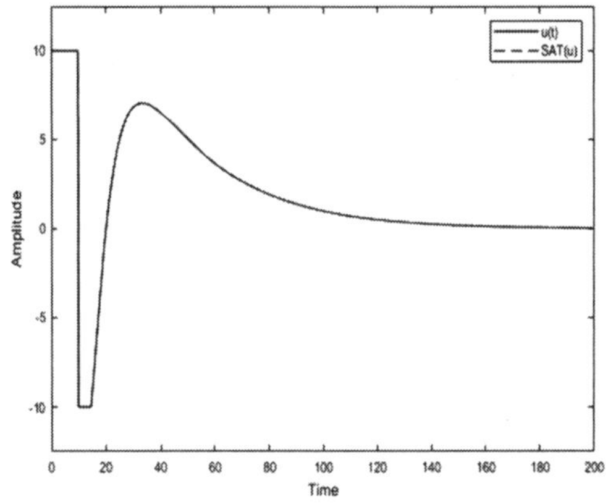


Figure 14-43: Controller output and saturated input with back-calculation

Figure 14-43 illustrates that the PI controller output u(t) and the saturated input SAT(u) coincide with each other as Limit output is enabled. The plant measured output y(t), with and without anti-windup, is shown in Figure 14-44 in order to better visualize the effect of the anti-windup configuration.

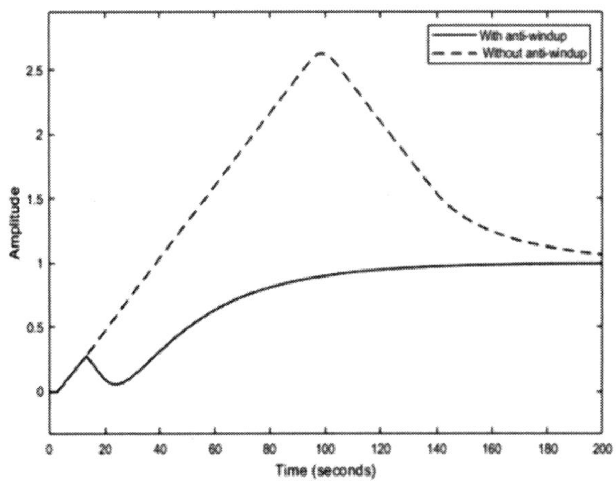


Figure 14-44: Measured output with and without using the back-calculation method

## (2) Bumpless Control Transfer

As the FWCS is a dynamic system, it is pivotal to ensure the state of the system is correct when switching the feedwater control between the automatic mode and the manual mode. When the FWCS is in the manual mode, the control algorithm generates a control signal which may not be same the manually produced control signal. It is essential to ensure that the two outputs concur at the time of switching. Figure 14-45 shows a Simulink model of PI control without bumpless transfer. Figure 14-46 illustrates the set-point and measured output with no bumpless transfer. Figure 14-47 shows the control signal switching between PI signal and manual signal with bump.

Figure 14-46 and Figure 14-47 describe the performance in control transfer with bumps that do not enable tracking mode.

As shown in Figure 14-46 and Figure 14-47, the plant is under manual control can cause unexpected large transients upon switching.

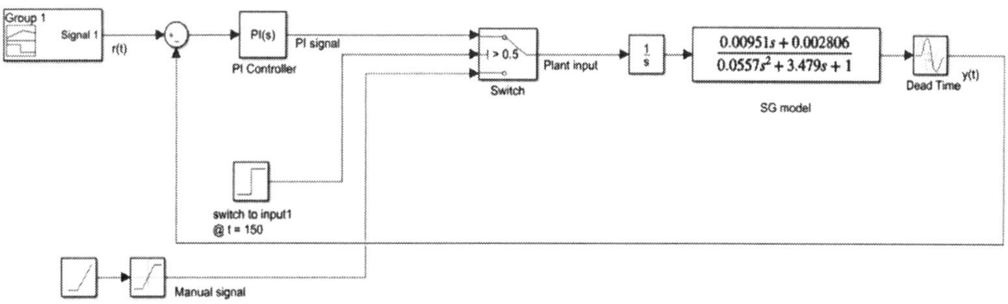


Figure 14-45: Control transfer with Bumps

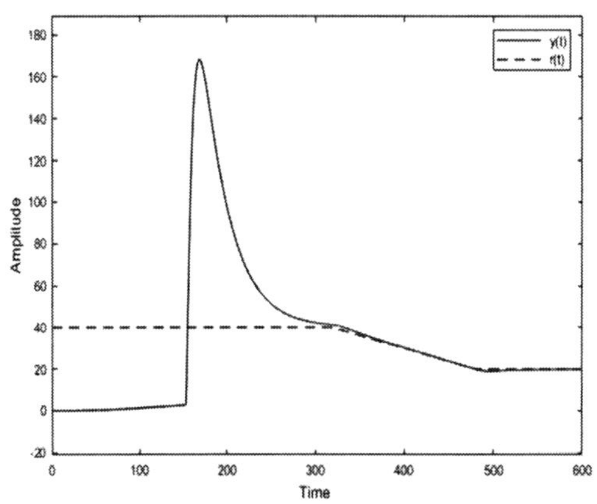


Figure 14-46: Set-point and measured output with bump

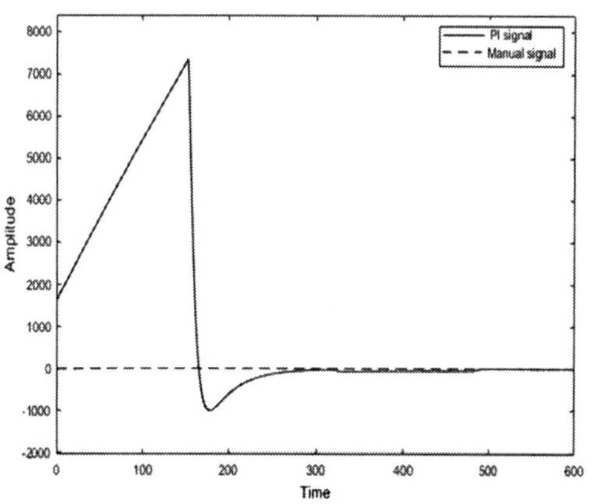


Figure 14-47: Control signal switching between PI signal and manual signal with bump

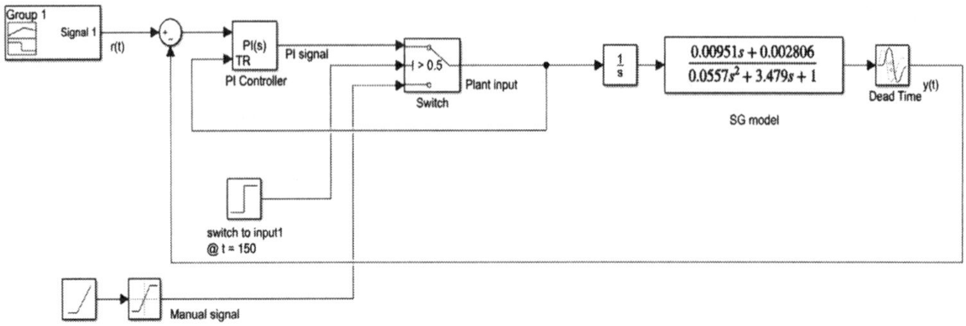


Figure 14-48: Control transfer with Bumpless

Next, a Simulink model of PI control with bumpless transfer was developed and is presented in Figure 14-48.

In this work we assumed that the control process in an open-loop manner by feeding the plant input with a saturating ramp signal to drive the output of the plant deliberately to a desirable steady-state value of 40. A control transfer is scheduled to occur at t = 150s. This transition between open-loop control and closed-loop control relates two control operational phases. In the manual phase, from start-up to t = 150s, where a saturated ramp signal is input to the plant input. In the automatic phase, the controller of the FWCS begins to operate the plant at t = 150s, and must take control of the process without causing bumps at the plant input [24].

In order to provide smooth control transition, the PI block in the MATLAB Simulink has two operational modes: a tracking mode and a control mode. In the control mode, the PI controller block operates normally. However, in the tracking mode, the block has an extra input that enables the PI block to have an adjustment its internal state by changing its integrator output so that the block output tracks an assigned signal feeding this extra input port. Therefore, to accomplish bumpless control transfer, the PI algorithm block is in the tracking mode when the plant is in the manual control phase (open-loop control), and in control mode when the plant is in the automatic control phase (closed-loop control).

As shown in Figure 14-46, there is a large deviation between the set-point and measured output when absence of an appropriate bumpless control transfer technique. However, when utilizing the tracking mode of the PI controller of the

FWCS, the measured output tracks the set-point smoothly at the switching time (t = 150) as illustrated in Figure 14-49. The control signal switching between PI signal and manual signal with bumpless transfer is shown in Figure 14-50. The plant input with bumpless is described in Figure 14-51. Compared with the control signal switching between PI signal and manual signal without enabling the tracking mode as shown in Figure 14-47, the control transfer happens in smooth bumpless fashion as shown in Figure 14-50. Additionally, Figure 14-50, and Figure 14-51 indicate that in the circumstance of switching control transfer, there are no step changes in the plant input when applying the bumpless strategy. Consequently, the smooth transition in the control transfer of the feedwater PI controller is successfully accomplished.

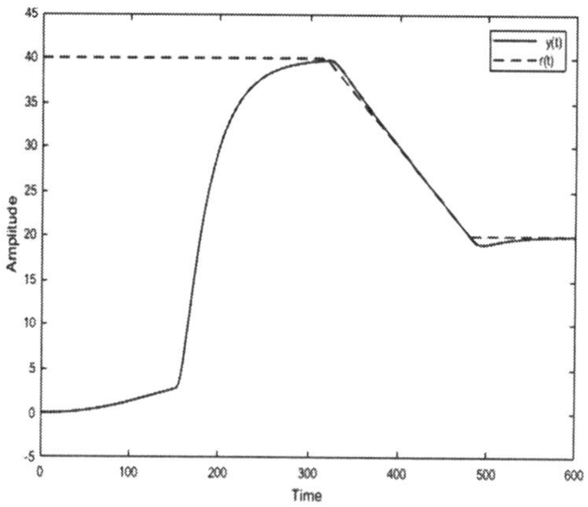


Figure 14-49: Set-point and measured output with bumpless

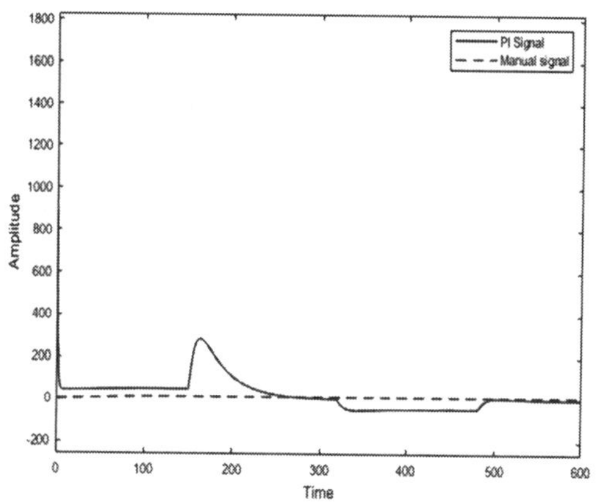


Figure 14-50: Control signal switching with bumpless

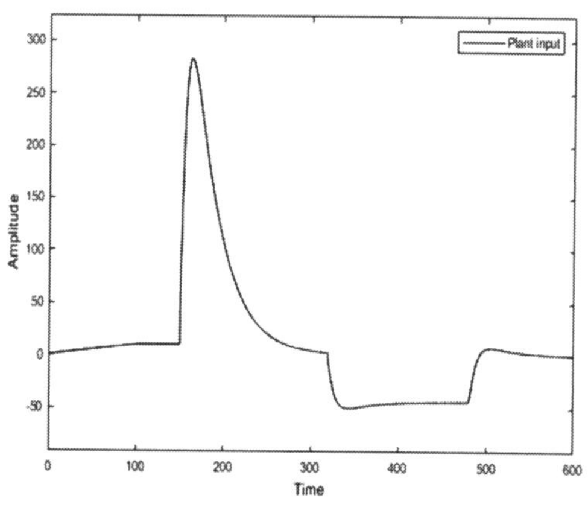


Figure 14-51: Plant input with bumpless

## PROBLEMS

1. What is the difference between one control element and three control elements in the FWCS?
2. Describe the MIGO method.
3. Describe the FWCS function based on the bock diagram.
4. Explain the master control logic of the FWCS.
5. Describe the identification for control method.
6. Using Matlab or Python tools to implement examples related to this method in order to understand the method of identification for control.
7. Describe of the the feedback controller architecture.
8. Description of the schematic diagram of an overall control system
9. What is the Simple Process Model?
10. What is the identified transfer function from feedwater flowrate to steam generator water level? Describe of its parameter for each reactor power levels.
11. How many parameters affect the SG level change?
12. What is the windup phenomenon?
13. What is bumpless control transfer?
14. Describe the back calculation method for the PID controller.
15. What is the tracking mode for PID controller?
16. Explain the Swelling and Shrinking phenomena of the SG.

## FURTHER READING


- K.J. Aström and T. Hägglund, "Advanced PID controller," ISA. 2006.
- L. Ljung, "System Identification: Theory for the User," Prentice Hall, 1999.
- L. Ljung, "Identification for control: simple process models," Proceedings of the 41st IEEE Conference on Decision and Control, pp. 4652–4657. 2002.
- J. J. Sohn and P. H. Seong, "A Steam Generator Model Identification and Robust H-infinity Controller Design with nu-gap Metric for a Feedwater Control System," Ann. Nucl. Energy, v.37, no.2, pp.180 – 195, 2010.

- T. Tran, J. C. Jung, "Controller Tuning Constants for the APR1400 Control System Based on Ms-constrained Integral Gain Optimization Method," Ann. Nucl. Energy, 127, 39–52, 2019.
- T. Tran, J. C. Jung, "Development of Anti-windup PI Control and Bumpless Control Transfer Methodology for Feedwater Control System," Ann. Nucl. Energy 131, 233–241. 2019.
- T. Tran, J. C. Jung, "A systematic Engineering Approach to design the controller of the advanced power reactor 1400 Feedwater control system using Genetic Algorithm," Journal of KOSSE, 14, 58-66. 2018

Chapter 15

# IMPROVED RRS LOGICAL ARCHITECTURE USING GENETIC ALGORITHM


## SUMMARY

An improved RRS logic architecture, which is combined with the Genetic Algorithm (GA), is implemented in this work to provide an optimal solution to a given system. The current system works desirably and has been contributed to the safe and stable Nuclear Power Plant (NPP) operation. However, during the ascent and decent section of the reactor power, the RRS output reveals a relatively high steady state error and also carries a considerable level of overshoot. In an attempt to consolidate conservatism and minimize the error, this work proposes applying GA to RRS and suggests reconfiguring the system. Prior to the use of GA, reverse-engineering is implemented to build an executable RRS model using a full scope APR1400 NPP simulator named Nuclear Plant Analyzer (NPA) and Simulink. RRS-related inputs and outputs are retrieved by NPA and the model is verified by comparing its outputs and the retrieved outputs. Re-engineering is followed to devise an improved RRS and concerns adding a compensator to the model. The key parameters of the compensator are explored by a GA module built in Labview. The improved RRS model complete with the compensator generates outputs that have reduced steady state errors and diminished overshoot levels. Once deployed in the APR1400 NPP, the model would contribute to more stabilizing the reactor regulation with its improved features. This chapter presents the work of Hyo-Sub Shim and Jung Jae Cheon [27]

The main content of this chapter is shown in Figure 15-1.

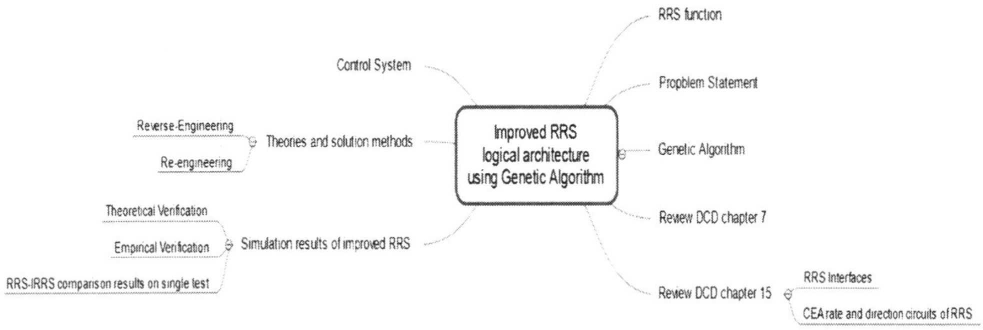


Figure 15-1: Mind Map for Chapter 15

## OBJECTIVES

- Improvement of the RRS architecture using a Genetic Algorithm approach, to design the additional lead-lag compensator

# 15.1 RRS Function

To control the reactor power, PCS (Power Control System) which consists of RRS, DRCS (Digitalized Reactor Power Control System) and RPCS (Reactor Power Cutback System) has been used in APR1400 NPP. The RRS automatically adjusts reactor power and reactor coolant temperature to follow the turbine load transients within established limits. Providing the limits and interlocks prevents abnormal power and temperature conditions that could result from excessive control rod withdrawal initiated by a control system malfunction or by an operator [1].

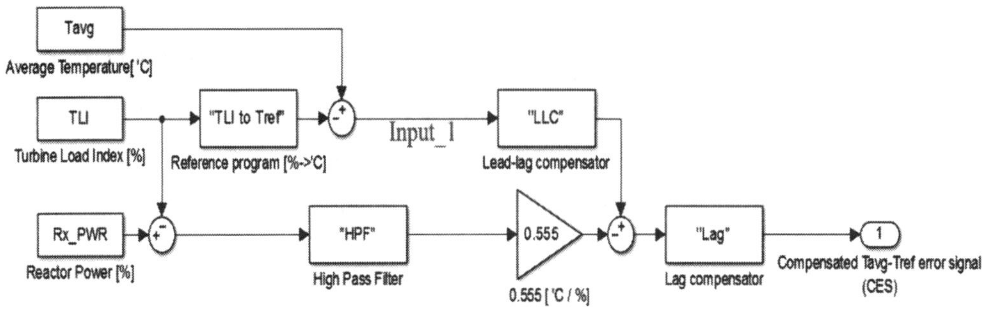


Figure 15-2: RRS Schematic Diagram

In Figure 15-2, Tavg signal is the average temperature of the hot leg and cold leg which are the reactor coolant inlet and the outlet respectively. The Tref signal is then subtracted from the Tavg signal to provide a temperature error signal. Power range neutron flux (Rx-PWR) is subtracted from the turbine load index to provide compensation to the generated error signal. The compensated error signal and the uncompensated Tavg-Tref are referred to as CES and input_1 respectively. CES is fed to a CEA (Control Element Assembly) rate program to determine whether the CEAs are to be adjusted at a high or low rate, and to a CEA motion demand program that determines if the CEAs are to be withdrawn, inserted, or held. The outputs of the rate and motion demand programs are used by the DRCS. If the temperature error signal is high, the RRS provides an AWP (automatic withdrawal prohibit) signal to the DRCS. The withdrawal of CEAs causes the Tavg to increase and prohibiting a withdrawal prevents an increase in the error signal [1].

As shown in Figure 15-3, the RRS schematic diagram at 50% reactor power level is simulated in matlab. The simulation of RRS at 100% reactor power level is shown in Figure 15-4.

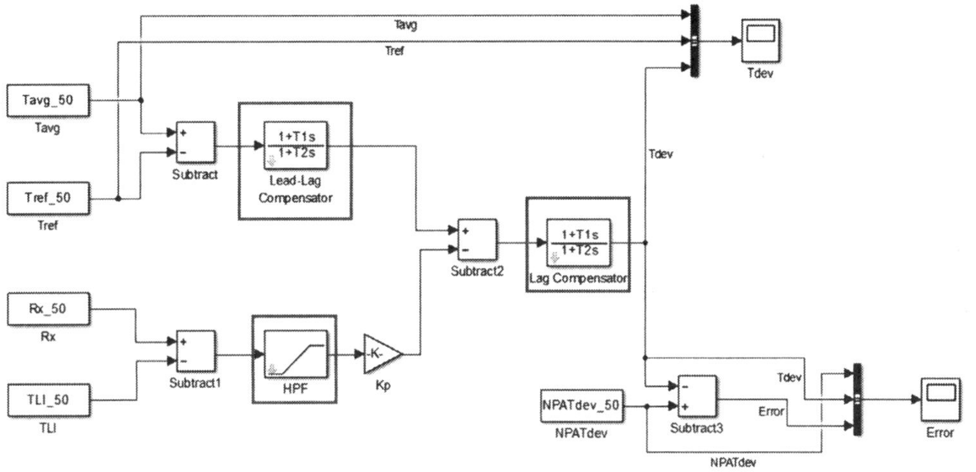


Figure 15-3: Simulation of RRS at 50% reactor power level_case1

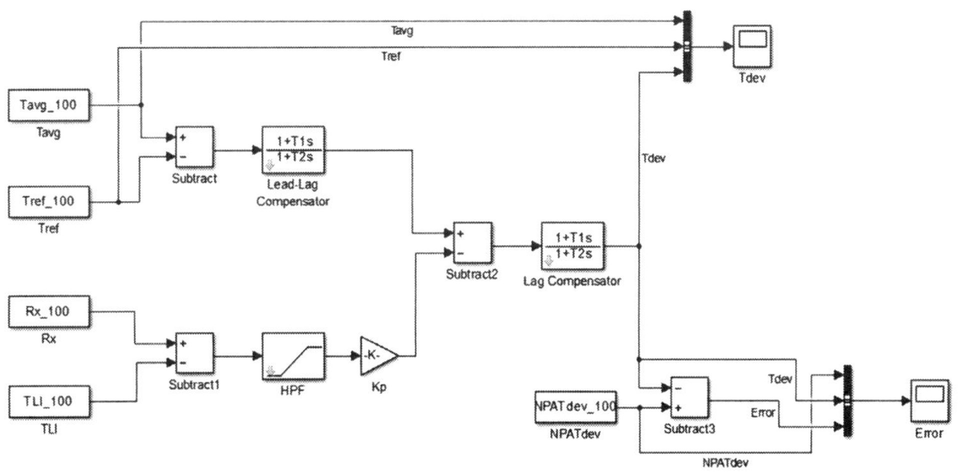


Figure 15-4: Simulation of RRS at 100% reactor power level

## 15.2 Problem Statement

A comprehensive investigation into RRS was performed by exploiting the APR 1400 simulator to measure the stability and accuracy of RRS. The APR1400 simulator provides a multitude of scenarios and among them, the investigation encompassed four scenarios: 25%, 75%, 100% power levels and one transient condition at the 75% power level. First, the CES at low power (25% power level) was monitored and the initial CES commenced from -0.664°C despite input_1 at the same time being -0.038°C as visualized in Figure 15-5. The selection of 25% power level was ascribed to the fact that the total energy at low power is greater than during the subcritical case.

Afterwards, RRS ramp response at 75% was checked by adjusting upwards and downwards the turbine load to observe a RRS deviation, given that the lower the deviation of a system is, the better accurate it is [1]. There are four predefined power level scenarios of the simulator: 25%, 50%, 75%, and 100%. Except 25%, the input_1- CES deviation at other power levels were identically observed in Figure 15-5, Figure 15-7, Figure 15-8 and it was attributed to the fact that NPP operations at 50, 75 and 100% power levels are normal and stable. Among the levels, 75% was considered desirable to measure the ramp response as the level has 25% margin to the maximum normal operating limit of 100% and 25% margin to the middle power level of 50%.

With the measurement lasting 600 seconds, the input_1-CES deviation was monitored. The average deviation of 0.17923°C was calculated based on Figure 15-6. Afterwards, the 100% power level was run and its maximum overshoot was gauged at of 47.104°C as shown in Figure 15-7. As the last investigation, a sudden load change such as a load rejection was simulated by manually tripping two Feed Water Pumps (FWPs) in sequence. This simulation was devised to generate a sever power level drop and demonstrated in Figure 15-8 the average deviation of 0.01756°C.

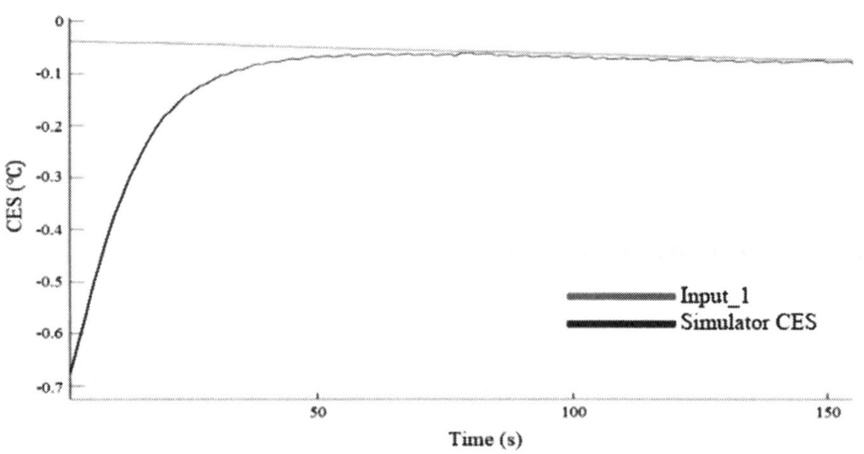


Figure 15-5: Initial output at 25% power level

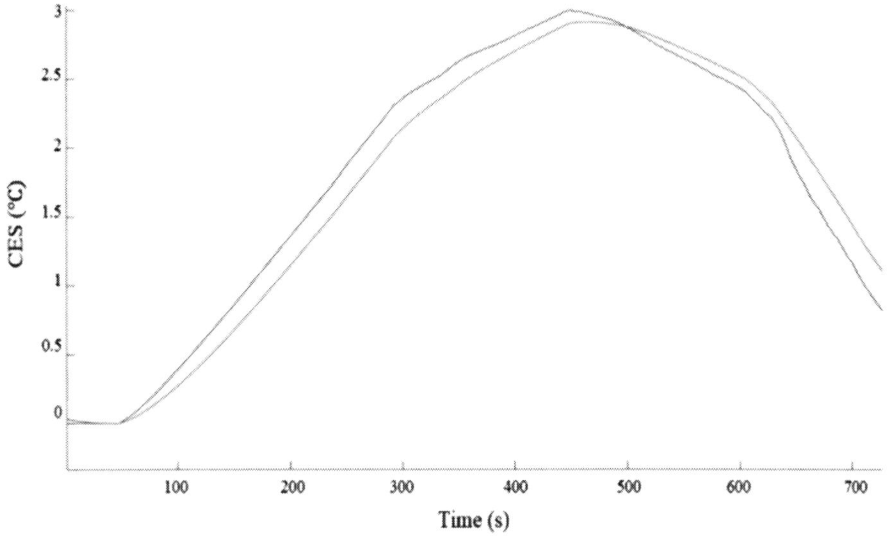


Figure 15-6: Ramp response at 75% power level

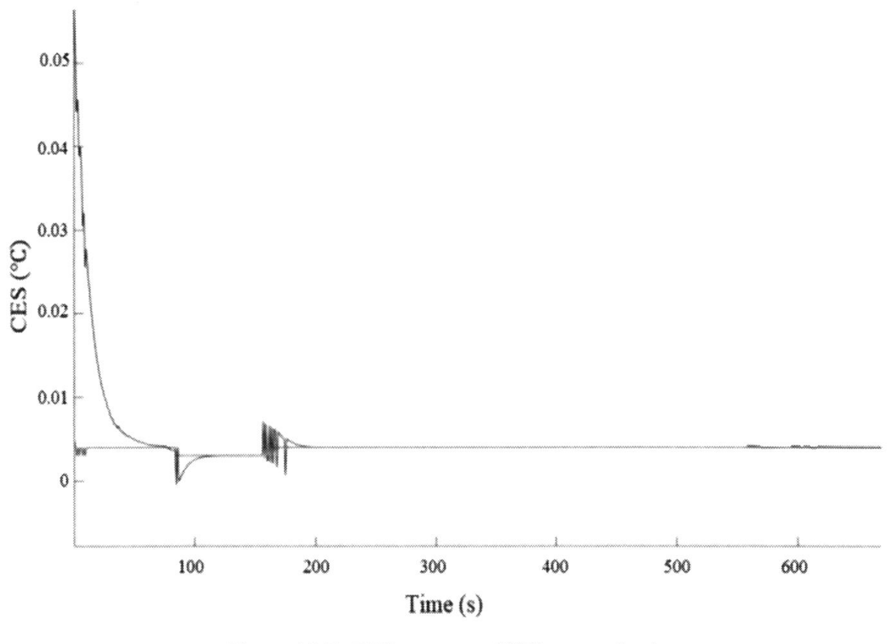


Figure 15-7: RRS output at 100% power level

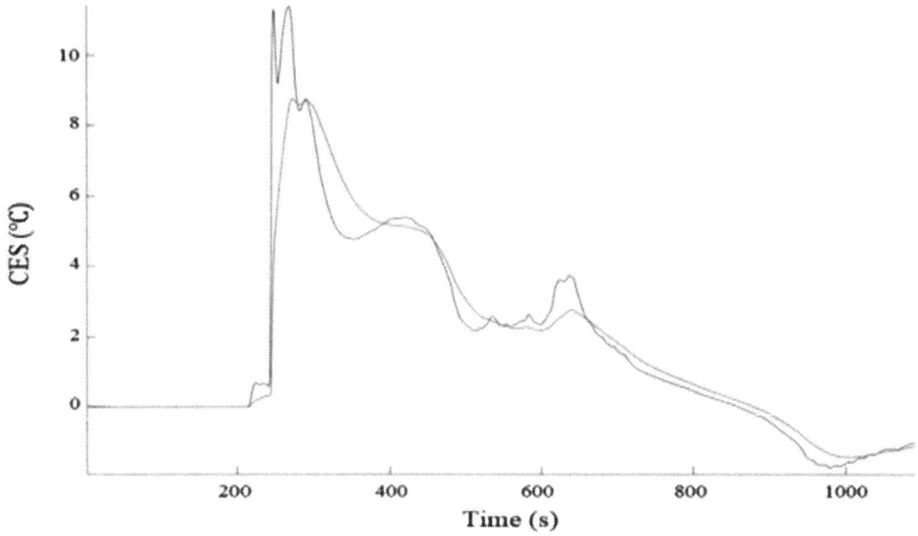


Figure 15-8: RRS output at 100% power level with a transient

The current system meets the requirements from NPP licensers. The system works as expected but due to further analysis and suantifying simulations via a simulator, the following research areas were identified. These areas are listed in Table 15-1.

The deadband section of RRS, within which CEA motions are disabled to prevent noise-induced and non-essential operations, is set at ±0.1°C. It seems that the margin between the deadband and the initial CES motions at 25 power level is relatively small, factoring into the probability of the fore-mentioned uncontrolled CEA and the fact that the total energy at low power is greater than during the subcritical case.

Table 15-1: RRS Research areas

| No | Research areas (Issues) | Measured values | Remarks |
|---|---|---|---|
| 1 | Initial CES | -0.664℃ | 25% power level [Figure 15-5] |
| 2 | Average input_1-CES deviation | 0.17923℃ | 75% power level [Figure 15-6] |
| 3 | Average input_1-CES deviation | 0.01756℃ | 100% power level with a transient by two FWP trips [Figure 15-8] |
| 4 | Maximum overshoot | 47.104℃ | 100% power level [Figure 15-7] |

The lower deviation advances the time for RRS to reduce the Tavg-Tref deviation, contributing to the system's accuracy and stability. In a similar manner, the smaller overshoot revamps the stability. The lower deviation and overshoot result in releasing AWP (Automatic Withdrawal Prevention) earlier and enable RRS to manipulate CEA motions. AWP prevents the automatic control of RRS when CES exceeds the range from -1.1°C to 1.96°C. When a large deviation between Tavg and Tref occurs, the combination of lower discrepancy, smaller overshoot and the fine manipulation accelerates the process to reverting reactor to the normal operation.

This section pays a keen attention to the RRS inputs as well as output in order to quantify how finely RRS operates and to explore ways to enhance the system further in terms of stability and accuracy. The findings of the approach shows that the withdrawal of CEA from subcritical or low-power conditions adds reactivity to the reactor core, giving rise to both the core power level and the core heat flux that raises the reactivity and coolant temperatures of reactor [1].

Among the subcritical and low-power conditions, the low-power condition is emphasized in the accident analysis as the total energy generated during the power excursion at low power is greater than during the subcritical case. It is also stated that automatic systems inclusive of RRS should provide the capability to accommodate load rejections of any magnitude. The low-power condition requires that CEA motions at low power should be operated accurately with sufficient conservatism to prevent malfunctions such as uncontrolled CEA motion. The uncontrolled motions triggered by a single failure in CEA and RRS add reactivity to the core, causing both the core power level and the core heat flux to increase.

The accommodation of load rejections requires that the RRS output minimizing the error signal in a stable and accurate manner should be generated. Control system stability is defined as making its output reach its set value as rapidly and closely as possible with the smallest possible overshoot. In order to achieve the objectives, the reverse-engineering and re-engineering phases which are revealed in Figure 15-9 are applied in this chapter.

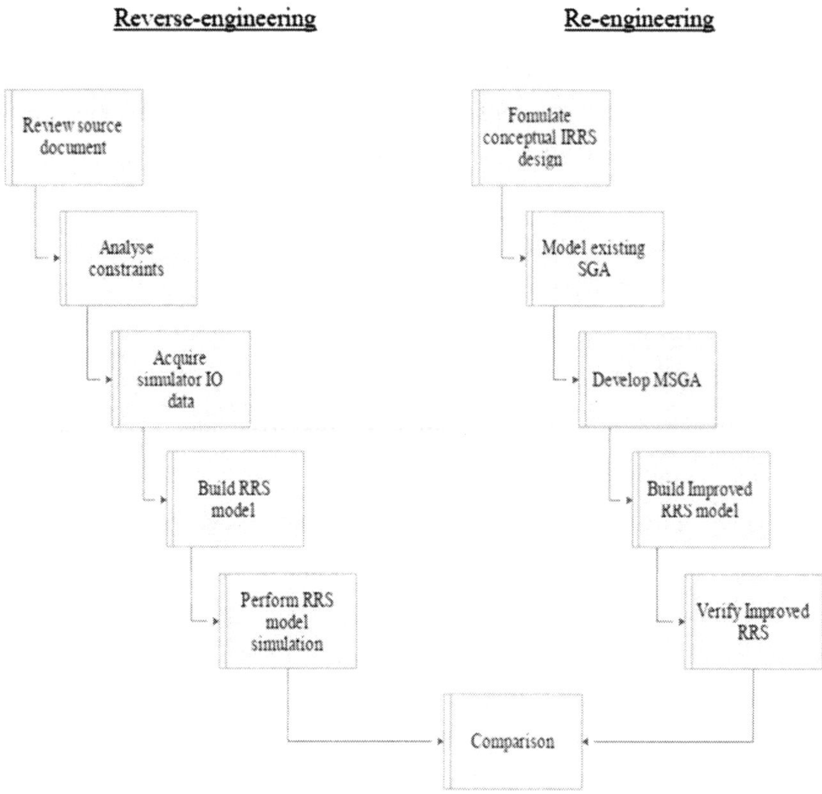


Figure 15-9: Overall scheme for GA-based RRS logic architecture improvement

The current RRS improvement methods featured reverse-engineering and reengineering as proposed in Figure 15-9. The former one was exploited to identify the due constraints and controllers of this research by reviewing stakeholders requirements and later to construct the Simulink-based RRS logic architecture (SRRS) by retrieving the inputs and CES data from the simulator. A comparison between the simulator CES and the SRRS CES was conducted and the result demonstrated a discrepancy. To cope with the discrepancy, the parameters of SRRS were manually fine-tuned. Following the modeling of the current system, the re-engineering was performed 1) to devise the conceptual design of the improved RRS (IRRS) and 2) to build IRRS using an existing GA module which were in turn tailor-made for this research. At the end, the empirical verifications for IRRS were executed in order to substantiate the stability of IRRS.

## 15.3 Genetic Algorithms

Genetic algorithms (GAs) are global, parallel, search and optimization methods, founded on Darwinian principals. They work with a population of potential solutions to a problem. Each individual within the population represents a particular solution to the problem, generally expressed in some form of genetic code. The population is evolved over generations, to produce better solutions to the problem. A schematic of the algorithm is shown Figure 15-10 [28].

Each individual within the population is assigned a fitness value, which expresses how good the solution is at solving the problem. The fitness value probabilistically determines how successfully the individual will be at propagating its genes (its code) to subsequent generations Better solutions are assigned higher values of fitness than worse performing solutions.

Evolution is performed using a set of stochastic genetic operators, which manipulate the genetic code. Most genetic algorithms include operator that select individuals for reproduction, produce new individuals based on those selected, and

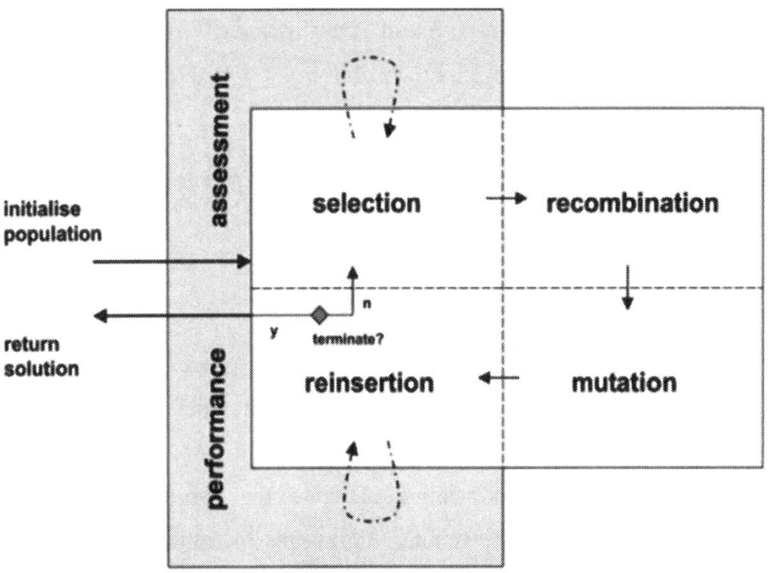


Figure 15-10: Flow of Genetic Algorithms

determine the composition of the population at the subsequent generation. Crossover and mutation are two well-known operators.

The crossover operator involves the exchange of genetic material between chromosomes (parents), in order to create new chromosomes (offspring). Various forms of this operator have been developed. The simplest form, single-point crossover, is illustrated in Figure 15-11. This operator selects two parents, chooses a random position in the genetic coding, and exchanges genetic information to the right of this point, thus creating two new offsprings.

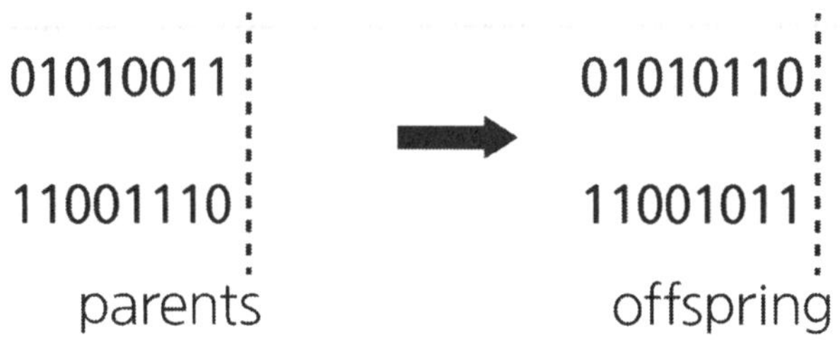


Figure 15-11: Crossover

The mutation operator in its simplest form, makes small, random, changes to a chromosome. For a binary encoding, this involves swapping genes 1 for gene 0 with small probability (typically around one percent) for each bit in the chromosome, as shown in Figure 15-12.

Figure 15-12: Mutation

Once the new generation has been constructed, the processes that result in the subsequent generation of the population are begun once more. The genetic algorithm explores and exploits the search space to find good solutions to the problem, due to its use of a population. Despite the simple concepts involved, the development and analysis of genetic algorithms is quite complicated. Many variations have been proposed since the first GA was introduced. Rigorous mathematical analysis of the GA is difficult and is still incomplete.

Genetic algorithms are robust tools, able to cope with discontinuities and noise in the problem landscape. Inclusion of domain-specific heuristics is not a pre-requisite, although it may improve the performance of a GA. They have proved useful at tackling problems that cannot be solved using conventional means.

## 15.4 Review Result of DCD Chapter 7 (Instrumentation and Control)

As illustrated in Figure 15-13, the diagram (derived from DCD Tier 2: chapter 7: Instrumentation and Control) shows how the RRS is implemented as one of the non-safety control systems.

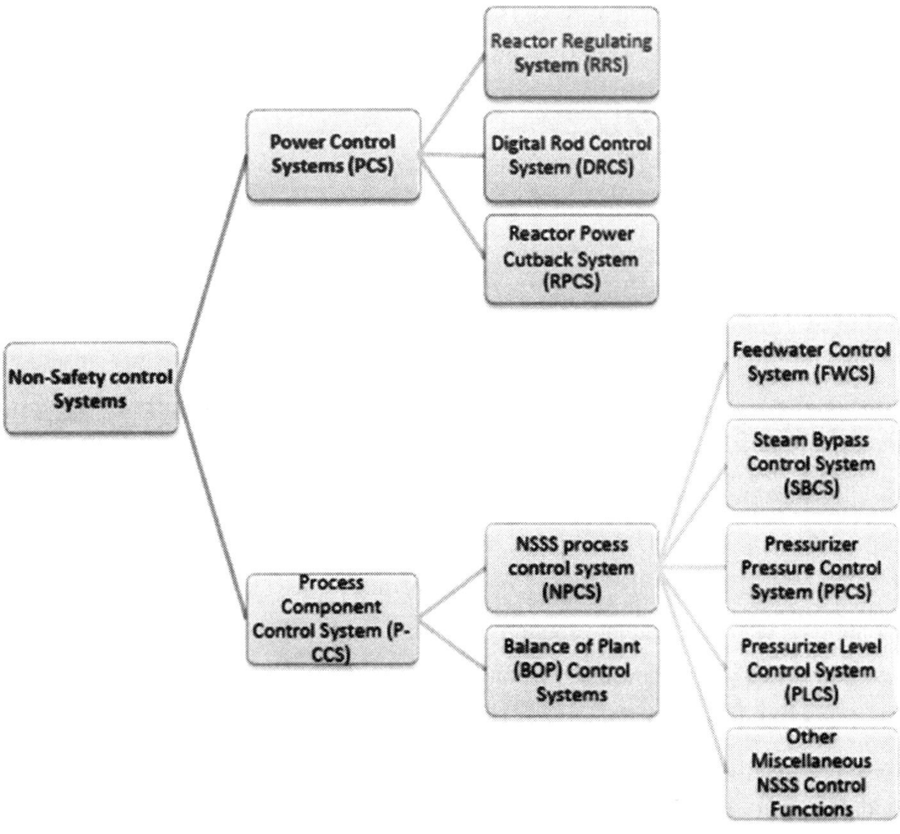


Figure 15-13: Structures of Non-safety control systems

It is important that NSSS follows the turbine load changes automatically or by operator action within normal operating limits. Except where xenon conditions may influence the capability of this automatic follow up process, these automatic systems are also able to accommodate load rejection of any magnitude. Below is a summary of PCS operation for power level control as an outlined in the DCD T2 document. Generally, reactivity is controlled either by: adjustments of the CEA for rapid reactivity changes or adjustment of boric acid concentration for long-term effects of fuel burn-up and changes in fission product concentration; since changes in acid concentrations happen slowly, operator action may be used for boric acid concentration.

CEAs can be controlled manually by operator or automatically to maintain the programmed reactor coolant temperature and power level. PCS comprises RRS, DRCS, and RPCS which work together to control the reactor power level. Control limits and interlocks are provided to prevent abnormal power and temperature conditions that could result from excessive control rod withdrawal initiated by a control system malfunction or by an operator. The CEA motion procedures with parameters are classified in Table 15-2.

Modes of DRCS CEA controls include 1) sequential group movement in automatic control, 2) sequential group movement in manual control, 3) manual group movement 4) manual individual CEA movement and 5) Standby.

Sequential group movement functions such that, when the moving group reaches a programmed low (or high) position, the next group begins inserting (or withdrawing), thus providing for overlapping motion of the regulating groups. The initial group stops upon reaching its lower (or upper) limit. Applied successively to all regulating groups, the procedure allows a smooth continuous rate of change of reactivity. A sequencing logic exists so that the preceding group reaches a specified limit before the next group is permitted to move. The DRCS and IPS monitor proper sequential motion and provide an alarm for out-of-sequence conditions. Other parameters for movement set by the DRCS include 1) normal CEA control limits (for both CEA and PSCEA) to automatically terminate CEA motion upon reaching CEA limits of travel (UEL or LEL), 2) upper group stop (UGS) and 3) lower group stop (LGS).

CEA interlocks for all CEAs and part-strength CEAs (PSCEAs) – when the CEA reaches its UEL or LEL, an interlock is provided to prohibits its further withdrawal or the insertion and terminate its motion.

Table 15-2: CEA motion procedures with parameters

| No | Procedure | Parameter |
|---|---|---|
| 1. | The desired average temperature is determined by a reference temperature (TREF) program that is inputted with the turbine load index (TLI). | TREF |
| 2. | The hot leg and cold leg temperature signals are averaged (TAVG) in the RRS. | TAVG |
| 3. | The TREF signal is then subtracted from the TAVG signal to provide a temperature error signal. | TAVG - TREF |
| 4. | The resulting signal is fed to a CEA rate program to determine whether the CEAs are to be adjusted to a high or low rate as well as to the CEA motion demand program that determines if the CEAs are to be withdrawn, inserted, or held. The outputs of the rate and motion demand programs are used by the DRCS. | TERR |
| 5. | If the temperature error signal is high (TAVG > TREF), RRS provides an automatic withdrawal prohibit (AWP) signal to the DRCS. A withdrawal of the CEAs can cause TAVG hence prohibiting it prevents an increase in the error signal. | AWP |
| 6. | DRCS uses either automatic CEA motion demand signal from RRS or manual signals from the DRCS soft control display on the IFPD convert these signals to direct current pulses that are transmitted to the control element drive mechanism (CEDM) coils to cause CEA motion. | |
| 7. | An AWP is also generated by the SBCS whenever SBCS demands turbine bypass valve opening. The AWP blocks response to RRSS demands hence preventing increase of reactor power when there is already excess energy in the NSSS. | |
| 8. | If a reactor trip is initiated, (either by RPS or DPS), input motive power will be removed from DRCS ad all CEAs will consequently be inserted by gravity | |

The shutdown CEAs are moved in the manual control mode only, with either individual or group movement. The DRCS soft control permits withdrawal of no more than one shutdown group at any time. The PSCEAs are normally moved manually, with either individual or group movement. During plant startup and shutdown, and all cases where power is below a preset value, manual control is used. Automatic control of the regulating CEAs by the RRS can be selected by the operator only when power exceeds the preset value. Manual control can be used to override automatic control at any time.

RPCS: If a large plant power imbalance occurs such as a large turbine load rejection turbine trip, or loss of two of the three operating main feed water pumps, RPCS is used to maintain NSSS within the control band range by rapid reduction of NSSS power at a rate that is greater than that provided by high speed CEA insertion.

After the RPCS has initiated the dropping of the pre-selected pattern of CEA to induce the power cut back action, subsequent insertion of other groups can be done either automatically by the RRS or manually by the operator Ex-core neutron flux monitoring system (non-safety channel).

The ENFMS includes two channels for startup and control each of which consists of one ENFMS detector assembly, preamplifier assembly, and one start-up and control signal processing drawer. Two control signal processing drawers consist of linear amplifier and test circuitry. Each drawer provides the neutron flux information in the power operating range of 0 percent to 125 percent to the RRS for use during automatic turbine load-following operation.

Postulated Event Due to Single Failure of the RERS Rod Control is Uncontrolled CEA Withdrawal. It is to be noted however that the postulated events due to a single failure of the non-safety control system does not cause plant conditions more severe than those described in the analysis of AOO in Chapter 15, DCD. An interlock signal is provided for DRCS single uncontrolled CEA withdrawal single failure from a separate control group or safety systems, to limit the failure effect of the RRS failure to provide AWP.

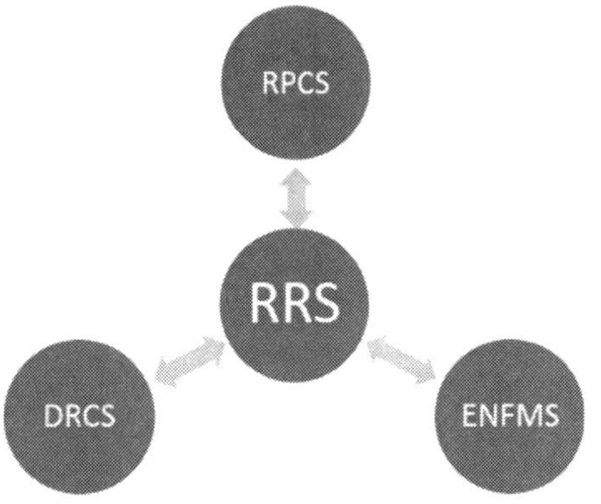


Figure 15-14: RRS interfaces

## 15.5 Review Result of DCD Chapter 15 (Transient and Accident Analyses)

This section presents the analysis of the response of APR1400 NSSS to postulated transients in process variables and to postulated malfunctions or failures of equipment. Such incidents (or events) are postulated, and their consequences are analyzed despite the many precautions that are taken in the design, construction, quality assurance, and plant operation to prevent their occurrence. The effects of the incidents are examined to determine their consequences and to evaluate the capability of the plant design to control or accommodate such failures and conditions.

Table 15-3: Review results of APR1400 DCD Chapter 15

| Section No | | | Description |
|---|---|---|---|
| 15.4 | 15.4.1 | 15.4.1.1 | Identification of Causes and Frequency Classification<br>An uncontrolled withdrawal of CEAs is assumed to occur as a result of a single failure in CEDM, CEDMCS, RRS, or as a result of operator error. In conformance with GDC 17, the loss of offsite power (LOOP) is assumed to occur concurrent with a reactor trip. |
| | | 15.4.1.2 | Sequence of Events and Systems Operation<br>The withdrawal of CEAs from subcritical or low-power conditions adds reactivity to the reactor core, causing both the core power level and the core heat flux to increase together with corresponding increases in reactor coolant temperatures and reactor coolant system (RCS) pressure. The withdrawal motion of CEAs also produces a time- dependent redistribution of core power. These transient variations in core thermal parameters result in a system approach to the specified fuel design limits, requiring the protective action of the reactor protection system (RPS). The total energy generated during the power excursion at low power is greater than during the subcritical case; therefore, only the low power case is presented here. |
| | | 15.4.1.6 | Conclusions<br>The uncontrolled CEA withdrawal from subcritical or low-power conditions with a LOOP meets GDC 20 and 25. These criteria require that the specified acceptable fuel design limits are not exceeded and the protection system action is initiated automatically. The withdrawal of CEAs from low-power conditions with a LOOP meet the following fuel design limits, which serve as the acceptance criteria for this event: (1) the transient terminates with a minimum DNBR greater than or equal to 1.29, and (2) the peak linear heat generation rate during the transient is less than 656 W/cm (20 kW/ft). The reactor coolant system pressure remains within 110 percent of its system design pressure. Therefore, the integrity of the reactor coolant pressure boundary is maintained. |
| | 15.4.2 | 15.4.2.1 | Identification of Causes and Frequency Classification<br>An uncontrolled withdrawal of CEAs is assumed to occur as a result of a single failure in CEDM, CEDMCS, RRS, or as a result of operator error. In conformance with GDC 17, the loss of offsite power (LOOP) is assumed to occur concurrent with a reactor trip. |
| | | 15.4.2.2 | Sequence of Events and Systems Operation<br>The uncontrolled withdrawal of a CEA at power conditions adds reactivity to the core, causing both the core power level and the core heat flux to increase, followed by corresponding increases in reactor coolant temperatures and reactor coolant system (RCS) pressure. The withdrawal of CEAs also produces a time-dependent redistribution of core power. These transient variations in core thermal parameters may result in an approach to the specified acceptable fuel design limits (SAFDLs) on DNBR and fuel centerline melt temperatures, requiring the protective action of the reactor protection system (RPS). |

| Section No | | | Description |
|---|---|---|---|
| | | 15.4.2.6 | Conclusions<br>The uncontrolled CEA withdrawal from subcritical or low-power conditions with a LOOP meets GDC 20 and 25. These criteria require that the specified acceptable fuel design limits are not exceeded and the protection system action is initiated automatically. The withdrawal of CEAs from low-power conditions with a LOOP meet the following fuel design limits, which serve as the acceptance criteria for this event: (1) the transient terminates with a minimum DNBR greater than or equal to 1.29, and (2) the peak linear heat generation rate during the transient is less than 656 W/cm (20 kW/ft). The reactor coolant system pressure remains within 110 percent of its system design pressure. Therefore, the integrity of the reactor coolant pressure boundary is maintained. |
| 15.5 | 15.5.2 | 15.5.2.1 | Identification of Causes and Frequency Classification<br>All events and events plus single failure that cause an increase in RCS inventory are examined with respect to the RCS pressure and fuel cladding performance.<br>The failures that affect the RCS behavior during this interval are as follows:<br>a. Failure of the pressurizer pressure control system<br>b. Failure of the feedwater control system<br>c. Failure of the steam bypass control system<br>Failure of the feedwater control system can only result in an excess cooldown, resulting in a lower peak pressure. After turbine trip, a LOOP concurrent with a turbine trip and reactor trip is considered as a basic assumption. The failures of RCPs, condenser pumps, circulation pumps, PLCS, PPCS, RRS, FWCS, and SBCS are followed after a LOOP.<br>PLCS malfunction causes the minimum DNBR to be reached within 3 seconds after the reactor trip. No single failure is identified from Table 15.0-4 that would have a significant effect on DNBR prior to the reactor trip. Therefore, any single failure that would result in a lower DNBR during the transient would have to affect at least one of the above parameters during the first 3 seconds following trip. The failures that affect the RCS behavior during this interval are as follows:<br>a. Failure of the PPCS<br>b. Failure of the RRS<br>Failure of the PPCS or RRS cannot appreciably affect any of the major factors that determine DNBR during the first 3 seconds following trip |
| | | 15.5.2.6 | Conclusions<br>The peak RCS and steam generator pressures reached during the PLCS malfunction with a LOOP at turbine trip are 186.95 kg/cm2A (2,659 psia) and 91.0 kg/cm2A (1,294.34 psia), respectively. These pressures are less than 110 percent of the primary and secondary design pressures, 193.34 kg/cm2A (2,750 psia) and 92.83 kg/cm2A (1,320 psia), respectively. The minimum DNBR is calculated to be 1.51, which is above the DNBR SAFDL value of 1.29. The acceptance criterion regarding fuel performance is met. |

### 15.5.1 RRS Interfaces

RRS interfaces with other systems as shown in Figure 15-15. RRS is the primary control portion of PCS which is largely comprised of IPS (Information Process System), RPCS (Reactor Power Cutback System), DRCS and SBCS (Steam Bypass Control System).

- RPCS: It is classified as a normal operation and it does not require action of safety equipment (i.e., the Plant Protection System or lifting of the primary and secondary safety valves), following large turbine load rejections including turbine trips from full power. It also provides an operator interface to inform the reactor operator of automatic RPCS actions to allow the operator to intervene and manually initiate a power reduction. It precludes initiating spurious power reductions which might be caused by credible equipment failures.

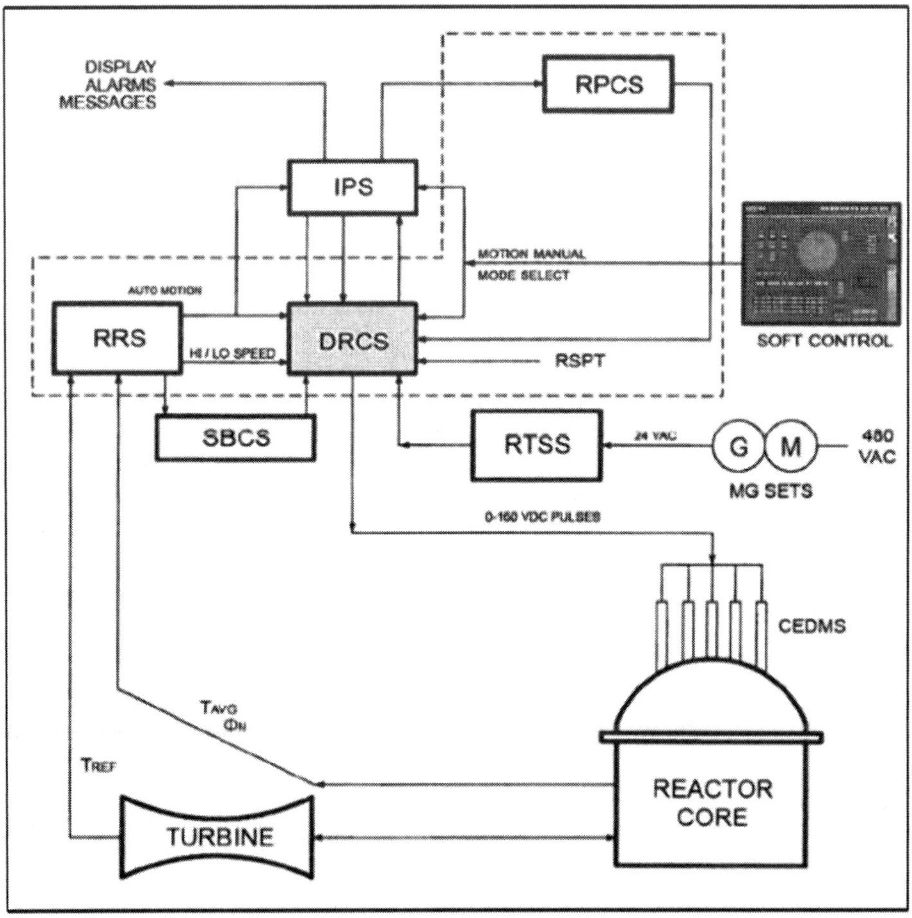


Figure 15-15: Schematic of Power Control System

- DRCS: The CEAs themselves are moved by magnetic jacks which is moved by magnetic coils to control CEA motion. When no motion is demanded, a gripper coil is energized, forcing holding latches into teeth located on the CEA extension shaft, holding the CEA in a stationary position. When CEA motion is demanded, the coils are energized in sequence to cause either insertion or withdrawal of the CEA. The coils are energized by DC power obtained by phase-firing SCRs across a three-phase A.C. source from the Motor-Generator set. CEAs are divided into one set of five and the remaining CEAs are in the sets of four. The sets are symmetrically located about the center of the reactor core. Each set of four or five CEAs is defined as a subgroup.

## 15.5.2 CEA Rate and Direction Circuits of RRS

CES is used in determining the speed of CEA movement. As shown on Figure 15-16, equivalent temperature errors of 1.1°C result in low rate CEA motion while errors of 1.96 °C result in high rate CEA motion. CES is also supplied to the CEA direction circuit which is used to determine the need for CEA insertion or withdrawal. When the temperature error is less than ±0.1°C, no CEA motion will be demanded in either direction. This 0.2°C deadband prevents unnecessary CEA motion. When the error exceeds +0.2°C (Tavg > Tref) inward CEA motion will be demanded at the rate determined by the CEA rate circuit. When the temperature error exceeds –0.2°C (Tavg < Tref), CEA withdrawal will be demanded at the rate determined by the CEA rate circuit. These values and directions are shown on Figure 15-16.

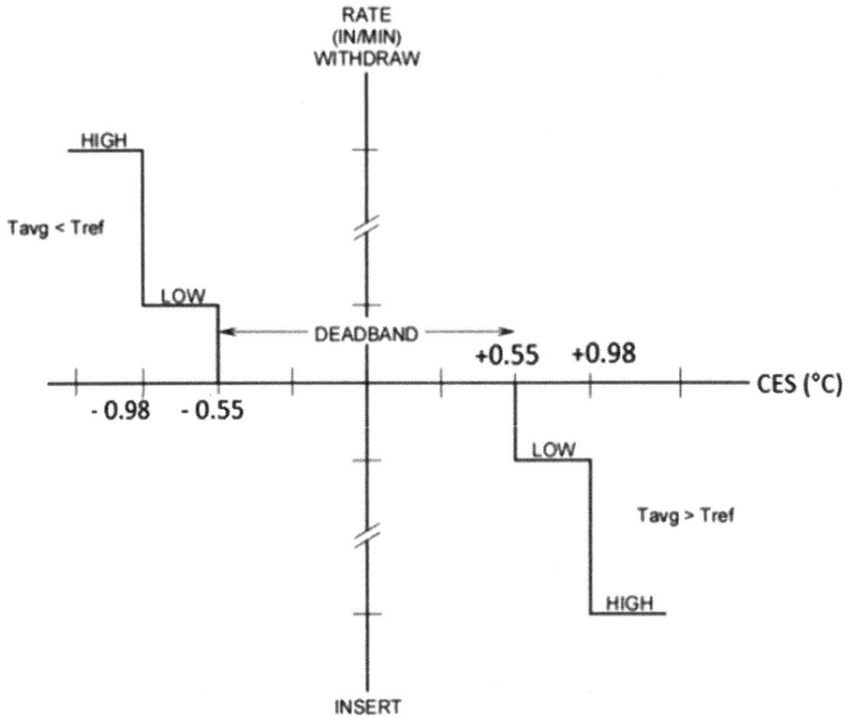


Figure 15-16: CEA rates and direction upon CES change

## 15.6 Theories and Solution Methods

The objective of this section was to methodologically address the research areas in Table 15-1 and the improvement methods for the current system featured reverse engineering and re-engineering as proposed in Figure 15-9. The former one was exploited to identify the due constraints and controllers of this research by reviewing stakeholders requirements and later to construct the Simulink-based executable RRS logic architecture (SRRS) by retrieving the inputs and CES data from the simulator.

The review results were reflected in Section 15.4 and 15.5 which then were adopted to generate Requirement Validation Traceability Diagram (RVTD) [29] illustrated in Figure 15-17. A comparison between the simulator CES and the SRRS CES was conducted and the result demonstrated a discrepancy. To cope with the discrepancy, the parameters of SRRS were manually fine-tuned.

Following the modeling of the current system, the re-engineering was performed 1) to devise the conceptual design of the improved RRS (IRRS) and 2) to build IRRS using an existing GA module which were in turn tailor-made for this research. At the end, the empirical verifications for IRRS were executed in order to substantiate the stability of IRRS.

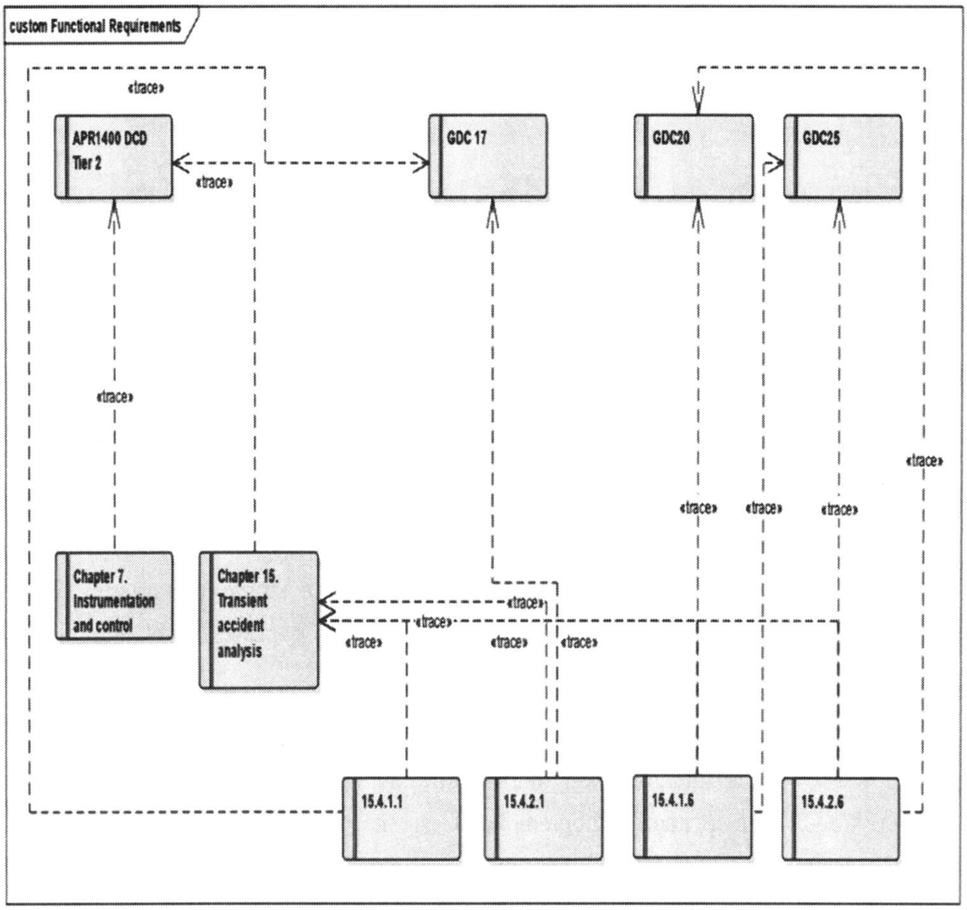


Figure 15-17: Requirement Validation Traceability Diagram

## 15.6.1 Reverse-Engineering

This approach requires reviewing the systems requirements stated in the RRS related source document [1] and identifying constraints under which the enhancement should be performed. The identification was followed by retrieving the inputs and CES data in Figure 1 from the simulator. On the basis of the collected data, a Simulink-based RRS model (SRRS) was generated. The discrepancy between the simulator and SRRS were present and it was attributed to the disparate platforms of the two.

The simulator is established with 3KEYMASTER which is the first simulation environment developed ground-up for the Microsoft Windows® operating system. Its open architecture, fully object-oriented approach, support for flexible human interface design, and leveraging of the Windows environment, offers distinct advantages in speed and usability [9]. SRRS was modelled with Simulink. The discrepancy was fixed by fine-tuning the mathematical expressions of the compensators and filter in Figure 15-2. The fixed SRRS was renamed as MRRS (Modified RRS).

### (1) Review Source Documents

The source document [1], which incorporates the requirements for APR 1400 and its accident analysis results, was reviewed and the RRS-related descriptions were searched. The major findings encompassed that automatic systems inclusive of RRS should provide the capability to accommodate load rejections of any magnitude and that the uncontrolled CEA caused by a single failure in CEA and RRS adds reactivity to the core, causing both the core power level and the core heat flux to increase.

In addition, the fact that the total energy generated during the power excursion at low power is greater than during the subcritical case was stated. These descriptions were primarily contemplated as the fulfilment of the requirements which secures the systems stability and accuracy.

### (2) Acquire Simulator Input and CEA Data

The date acquisition process was performed using MATLB Simulink which was used to develop a model of the RRS. This model was used to simulate and obtain data. The data include Tavg [°C], Tref [°C], Rx_PWR (Reactor Power, %) and TLI (%) as illustrated in Table 15-4. All but TLI from the data required are available at the KINGS full scope simulator. In other to obtain this parameter, two approaches were adopted as described below:

- **First approach**

TLI (Turbine Load Index) is a value produced based on the total steam flow and its unit is percent. In the initial stage of this work, it was discovered that the simulator does not calculate the TLI value. TLI is the value of the blade pressure of the high pressure turbine and it is of paramount importance to the simulation of the RRS logic architecture. The TLI unit is in percentage (%) and the unit of the blade pressure is in psi. Two experimental approaches were conducted to determine this parameter.

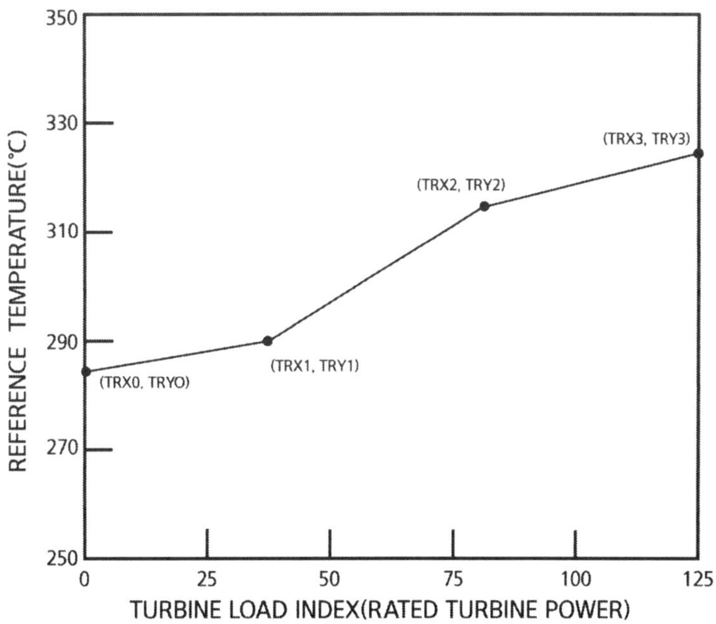


Figure 15-18: TLI-Tref conversion chart

In the first approach, a pre-determined chart converting TLI to Tref was employed. Since this research needed the automated conversion from Tref to TLI in order to continuously monitor each input and CES for a certain duration. With the acquisitions of the simulator-provided Tref and the chart, in this section a single polynomial equation applicable to 3 separate proportional slopes as shown in Figure 15-19 was developed.

The equation was formulated with a numerical analysis and its resultant equation was represented in equation x. The blue line of the chart illustrates the value of the given chart and the dotted line portrays values computed by the polynomial equation. This outcome was remarkable as TLI can be automatically computed when Tref is given.

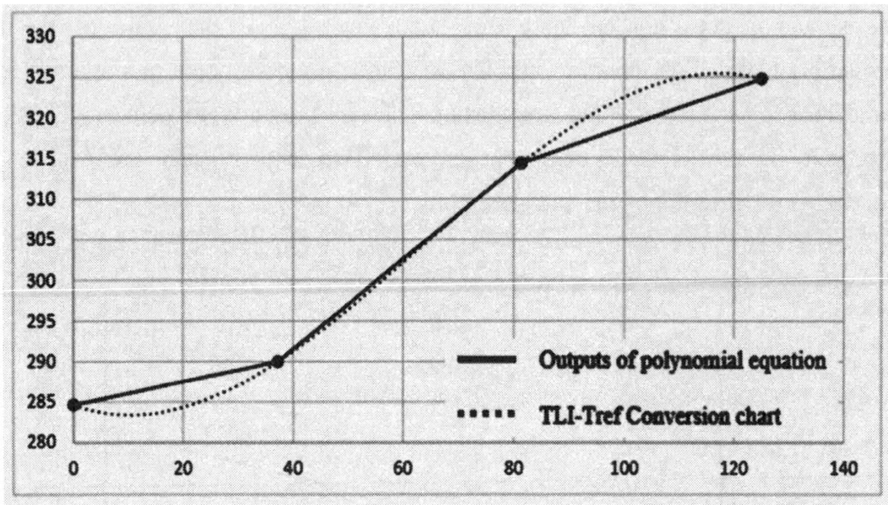


Figure 15-19: Comparison results between conversion chart and outputs of polynomial equation

- **Second approach**

The second approach observed the three distinct proportional slopes. Based on the proportional characteristic, it appeared reasonable that three slopes could be divided by the degree of the proportions and that the selection of the division would be facilitated by conditional statements. Firstly, the rough scale of Figure 15-18 was densely sectionalized as shown Figure 15-20 using Excel of Microsoft Office. Grid lines were placed on both X and Y coordinates. Judging from the alignment, it was ascertained that each of the three slopes of Figure 15-20 were of proportional range. The first slope ranged from 0 to 37.2146 and the second slope from the end of the first to 81.3364 and the third slope from 81.3364 to 124. As all of the slopes were proportional, a certain Y axis value was singled out of each slope and then the corresponding X axis value was determined.

The proportion was computed by dividing Y with X. The second and third proportions were calculated in the same manner. Each proportion was respectively named as X1 (Section A, 0~37.2146), X2 (Section B, 37.2146~81.3364) and X3 (Section C, 81.3364~124). When an inputted Tref value belongs to Section A, the value is converts to a TLI value with the proportion of X1. This conversion method was translated into the Simulink model using the IF-THEN ELSE function of Simulink.

The second approached was proven with the Simulink model in Figure 15-22. The model was run numerous times to demonstrate whether the approaches would be valid over the comprehensive ranges. More precision can be obtained using a more detailed chart converter compared to the one utilized in this second approach. This approach had some deviations due to the scaling implemented in this second approach.

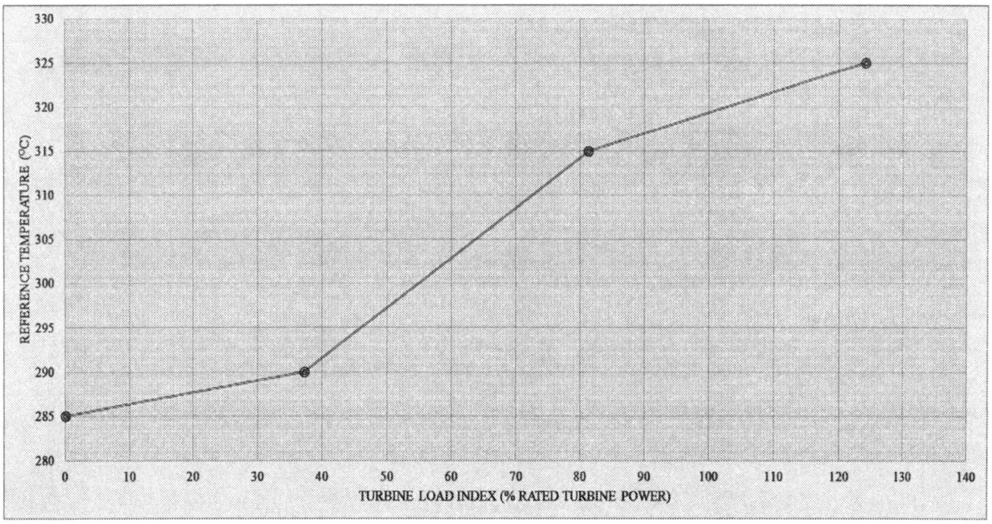


Figure 15-20: Densely sectionalized TLI-Tref conversion chart

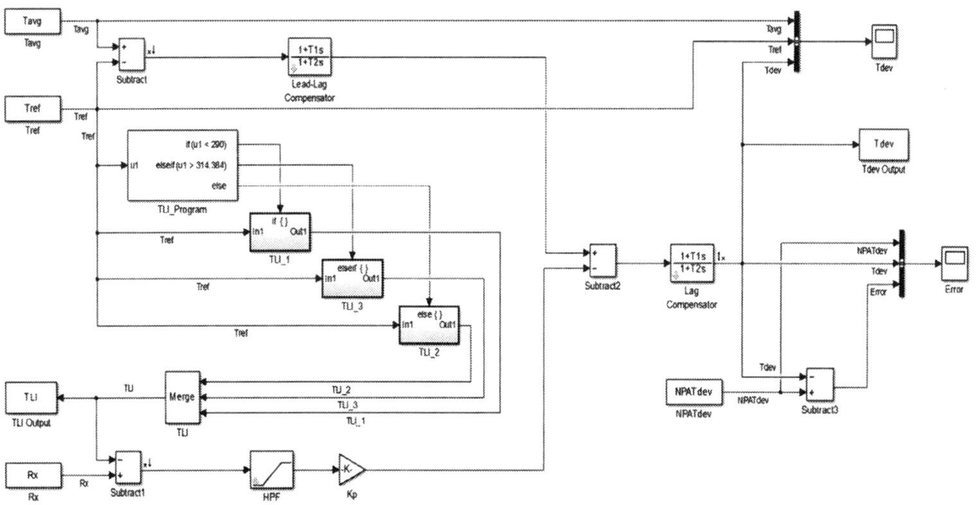


Figure 15-21: Simulation of TLI program

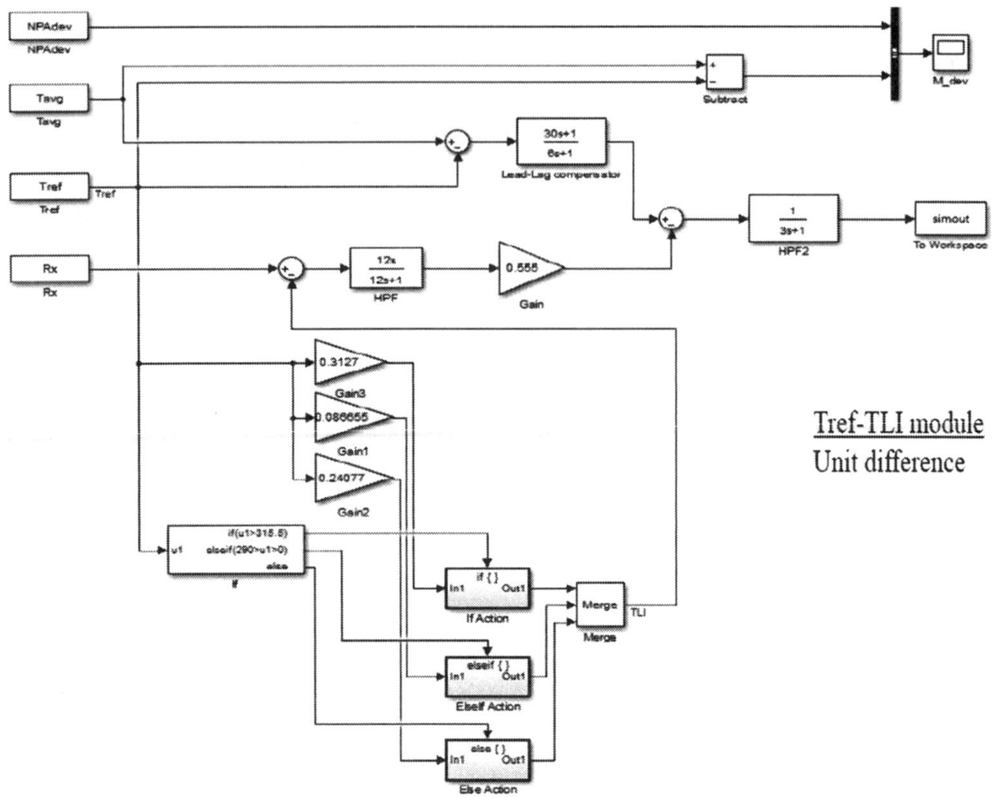


Figure 15-22: Simulink-based RRS logic with ELSE-IF condition for automatic TLITref conversion

The simulator offers a real-world NPP MCR (Main Control Room) environment, complete with LDP (Large Display Panel) on the front and numerous operator stations and one instructor station. The scenarios introduced earlier can be performed from the instructor station. These scenarios include operating the simulator in 25%, 50%, 75% and 100%. The station offers a trend function where input and output variables are called to the trend screen and monitored real-time. A 75% power opeation scenario was performed and monitored. The trend was set to have a step of 1 second and was made to run for a total of 870 seconds. In addition, at 750 seconds, the RCP (Reactor Coolant Pump) was rendered inoperable to trigger a transient. The trend results were saved in a txt format and imported to Simulink environment. The partial trend results are shown in Table 15-4.

Table 15-4: Acquired input and CES data from simulator

| Time | TLI | Tavg | Rx | Tavg | Tref | Tdev | in1 | in2 |
|---|---|---|---|---|---|---|---|---|
| 1 | 75.163 | 303.253 | 75.065 | 303.253 | 303.251 | 0.003 | 0.002 | -0.098 |
| 2 | 75.163 | 303.253 | 75.065 | 303.253 | 303.251 | 0.003 | 0.002 | -0.098 |
| 3 | 75.163 | 303.253 | 75.065 | 303.253 | 303.251 | 0.003 | 0.002 | -0.098 |
| 4 | 75.163 | 303.253 | 75.065 | 303.253 | 303.251 | 0.003 | 0.002 | -0.098 |
| 5 | 75.163 | 303.253 | 75.065 | 303.253 | 303.251 | 0.003 | 0.002 | -0.098 |
| 6 | 75.163 | 303.253 | 75.065 | 303.253 | 303.251 | 0.003 | 0.002 | -0.098 |
| 7 | 75.163 | 303.253 | 75.065 | 303.253 | 303.251 | 0.003 | 0.002 | -0.098 |
| 8 | 75.163 | 303.253 | 75.065 | 303.253 | 303.251 | 0.003 | 0.002 | -0.098 |
| 9 | 75.163 | 303.253 | 75.065 | 303.253 | 303.251 | 0.003 | 0.002 | -0.098 |
| 10 | 75.163 | 303.253 | 75.065 | 303.253 | 303.251 | 0.003 | 0.002 | -0.098 |
| 11 | 75.163 | 303.253 | 75.065 | 303.253 | 303.251 | 0.003 | 0.002 | -0.098 |
| 12 | 75.163 | 303.253 | 75.065 | 303.253 | 303.251 | 0.003 | 0.002 | -0.098 |
| 13 | 75.163 | 303.253 | 75.065 | 303.253 | 303.251 | 0.003 | 0.002 | -0.098 |
| . . . | . . . | . . . | . . . | . . . | . . . | . . . | . . . | . . . |
| 847 | 18.097 | 292.468 | 0.163 | 292.468 | 294.177 | -7.485 | -1.709 | -17.934 |
| 848 | 18.07 | 292.277 | 0.157 | 292.277 | 294.173 | -7.472 | -1.896 | -17.913 |
| 849 | 18.042 | 292.093 | 0.152 | 292.093 | 294.169 | -7.455 | -2.076 | -17.89 |
| 850 | 18.015 | 291.917 | 0.147 | 291.917 | 294.164 | -7.43 | -2.247 | -17.868 |
| 851 | 17.987 | 291.75 | 0.143 | 291.75 | 294.16 | -7.395 | -2.41 | -17.844 |
| 852 | 17.96 | 291.592 | 0.138 | 291.592 | 294.156 | -7.349 | -2.564 | -17.822 |
| 853 | 17.932 | 291.442 | 0.134 | 291.442 | 294.151 | -7.291 | -2.709 | -17.798 |
| 854 | 17.905 | 291.302 | 0.13 | 291.302 | 294.147 | -7.222 | -2.845 | -17.775 |
| 855 | 17.877 | 291.171 | 0.126 | 291.171 | 294.142 | -7.143 | -2.971 | -17.751 |
| 856 | 17.85 | 291.047 | 0.122 | 291.047 | 294.138 | -7.059 | -3.091 | -17.728 |
| 857 | 17.822 | 290.931 | 0.119 | 290.931 | 294.134 | -6.972 | -3.203 | -17.703 |
| 858 | 17.794 | 290.821 | 0.115 | 290.821 | 294.129 | -6.884 | -3.308 | -17.679 |
| 859 | 17.767 | 290.719 | 0.112 | 290.719 | 294.125 | -6.792 | -3.406 | -17.655 |
| 860 | 17.739 | 290.625 | 0.109 | 290.625 | 294.12 | -6.691 | -3.495 | -17.63 |
| 861 | 17.711 | 290.538 | 0.106 | 290.538 | 294.116 | -6.584 | -3.578 | -17.605 |
| 862 | 17.683 | 290.459 | 0.103 | 290.459 | 294.112 | -6.473 | -3.653 | -17.58 |
| 863 | 17.656 | 290.385 | 0.1 | 290.385 | 294.107 | -6.364 | -3.722 | -17.556 |
| 864 | 17.628 | 290.316 | 0.097 | 290.316 | 294.103 | -6.26 | -3.787 | -17.531 |
| 865 | 17.6 | 290.25 | 0.095 | 290.25 | 294.098 | -6.165 | -3.848 | -17.505 |
| 866 | 17.573 | 290.188 | 0.092 | 290.188 | 294.094 | -6.078 | -3.906 | -17.481 |
| 867 | 17.545 | 290.129 | 0.09 | 290.129 | 294.09 | -5.998 | -3.961 | -17.455 |
| 868 | 17.517 | 290.074 | 0.088 | 290.074 | 294.085 | -5.922 | -4.011 | -17.429 |
| 869 | 17.489 | 290.024 | 0.085 | 290.024 | 294.081 | -5.843 | -4.057 | -17.404 |
| 870 | 17.461 | 289.978 | 0.083 | 289.978 | 294.076 | -5.76 | -4.098 | -17.378 |

The 75% power operation was considered desirable to build SRRS as the level has 25% margin to the maximum normal operating limit of 100% and 25% margin to the middle power level of 50%. The data obtained from the 75% scenario which includes the transient provides enough information to: 1) build the RRS model, 2) verify the RRS model, and 3) measure the overshoot and ramp respone of the system.

### (3) Analyze Lead Lag Compensator (LLC), High Pass Filter (HPF) and Lag Compensator Designs

Simulink models of lead lag compensator(LLC), high pass filter (HPF) and lag compensator re-shown in Figure 15-23. These blocks form part of the RRS being developed. The analysis was targeted at examining the mathematical compositions and at identifying the level of a gain, time constant and β in the compositions. The performance of lead or lag compensators is governed by β. LLC is the combination of lead and lag compensators, each of which produces different results.

Lag compensator enhances systems stability and accuracy, compromising its response time and lead compensator improves the response time, aggravating the stability. LLC is present to capitalize on the best of both worlds. LLC is placed at the beginning to refine the systems stability and response time. HPF is integrated to counteract external noise interference and the function of the final lag compensator is to supplement the stability.

### (4) Modeling and Simulation

This section was committed to developing the Simulink-based RRS model working in the same way as the simulator. Following the acquisition of the simulator data vital to building the SRRS in Simulink, the initial version of SRRS was configured as depicted Figure 15-24. In the version, '$T_{avg}$ - $T_{ref}$' was replaced with input_1 and 'TLI - Rx_PWR' was substituted for input_2. The replacement and simplification were ascribed to the user-friendly functions of the simulator which separately displays the TLI and Tref variables. The second version reflecting the replacement included input_1 and input_2 as indicated in Figure 15-24.

In addition, the simulator CES was linked to one of the three scope inputs to be compared with SRRS for verifications. The acquired txt data was imported into the simulink environment which was configured with similar settings as the data acquisition stage. Figure 15-25 shows the comparison between the simulator and the SRRS.

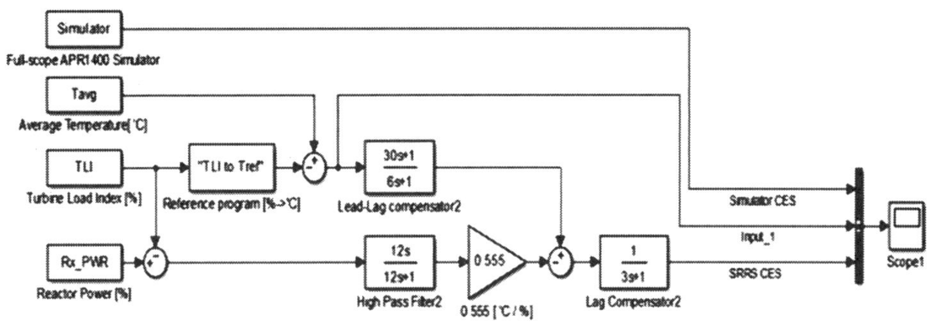


Figure 15-23: Simplified simulator model

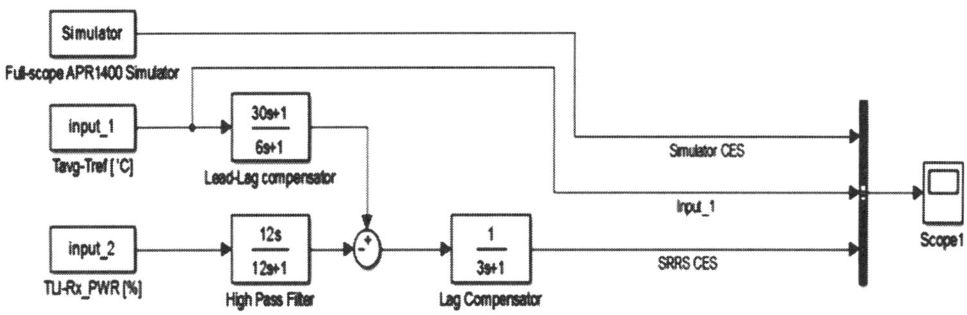


Figure 15-24: Simplified SRRS model

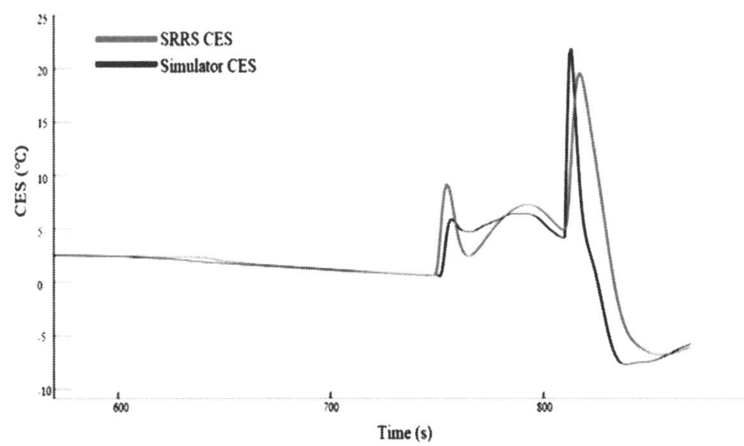


Figure 15-25: CES discrepancy between simulator and SRRS

### (5) Synchronize CEA of Simulator and SRRS

The discrepancy in Figure 15-25 was ascribed to the different platform of the simulator and SRRS. The simulator was built on 3KEYMASTER, simulation software and SRRS was modelled in Simulink. To resolve this issue, the parameters of LLC, HPF and lag compensator were adjusted and at each time, the scope output was observed. The process of reducing the discrepancies was an iterative process which involved adjusting and observing of the outputs from the scope.

The adjusted parameters were described in Figure 15-26 and the final process generated a scope output with the significantly diminished discrepancy as reflected in Figure 15-27. Following this achievement, the SRRS was renamed MRRS. This research utilised the Simulink model for its planned simulations and experiments. Therefore, the simulator had to be precisely modelled in Simulink to function as criteria for verifying IRRS.

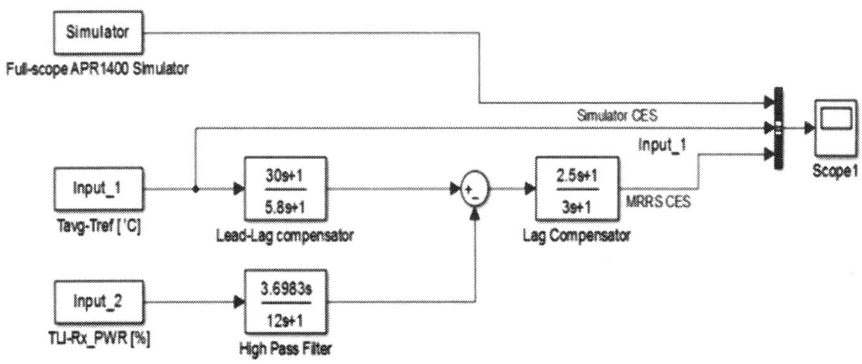


Figure 15-26: MRRS model

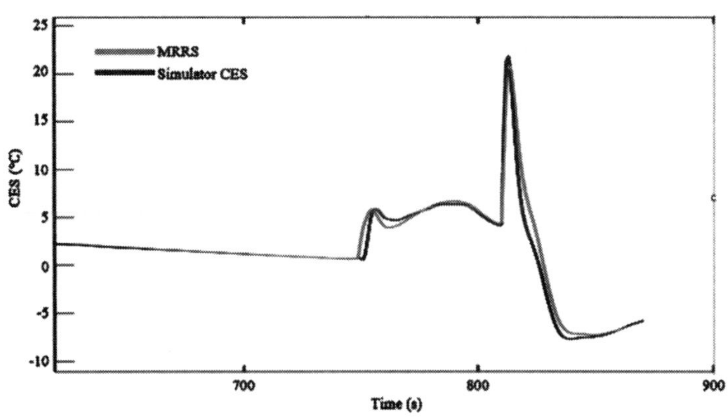


Figure 15-27: Scope output of MRRS

## 15.6.2 Re-Engineering

Having IRRS reasonably synchronized with the simulator, the conceptual design of IRRS was formulated. The formulation featured an additional lead-lag compensator (ALLC) at the end of the lag compensator to comprehensively enhance the current RRS using GA. An existing Labview-oriented GA module dubbed SGA (Simplified GA) was exhaustively reviewed to build up the insight of GA and to tailor-make it for this research. This module is an integration of various sub-modules.

Following the success of the test, the simplified genetic algorithm was modified and then named simplified genetic algorithm (MSGA) which was integrated with the additional lead-lag compensator. The modified simplified genetic algorithm was iterated numerous times to find the optimized parameters for the additional lead-lag compensator. The additional lead-lag compensator was integrated with the improved RRS and the results were compared to the modified RRS.

### (1) Formulate Improved RRS Logic Architecture

The primary objective of this research was to design an IRRS whose output was meant to finely follow input_1. IRRS was proposed to enhance 1) the initial CES at 25% power level, 2) input_1-CES deviation at 50-100% power level and 3) overshoot in the transient. ALLC was integrated with MRRS and it was positioned at the end of the lag compensator. The application and position of ALLC were attributed to the fact that IRRS was required to revamp the overall system behaviours in terms of accuracy and overshoot, resolving the three issues. Factoring into these deliberations, the conceptual configuration of IRRS was proposed in Figure 15-28.

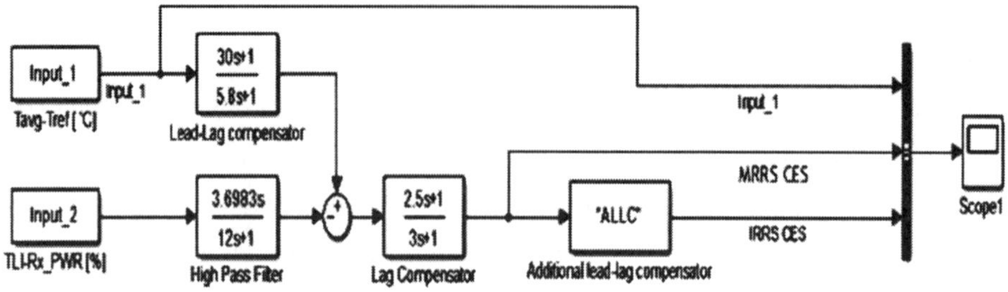


Figure 15-28: Conceptual configuration of IRRS model

### (2) Construct SGA for Development of Additional LLC

An existing SGA module introduced in [30] was reviewed to devise ALLCs mathematical expression using SGA. The module was carefully studied and modelled to comprehend the principles of SGA and ALLC development scheme as shown in Figure 15-29.

Figure 15-29: ALLC development scheme

- **Model Existing SGA module**

The technical book named “GA learning with Labview [30]” was thoroughly studied and the book precisely describes the basic functions of Labview and fundamental knowledge of GA. The GA in the book is comprised of multiple sub-modules and later the modules were integrated at the end to build the Simplified GA and to compute the PIDs primary parameters of gain, integral time and derivative time. SGA ends up with establishing successfully the PIDs parameters.

The steps to construct the sub-modules and the integrations are illustrated in a user-friendly way. The simplified genetic algorithm was implemented according to the instructions in [30]. Labview was chosen as it is a graphical programming syntax that makes it simple to visualize, create, and code engineering systems [11]. The GA model encompasses numerous sub-modules listed below. Their internal structures are described as below:

1) Initial population [bInt Pop]: as shown in Figure 15-30, the number of initial population should be inputted before running the GA and the GA produces the initial population inclusive of a preset number of chromosomes.

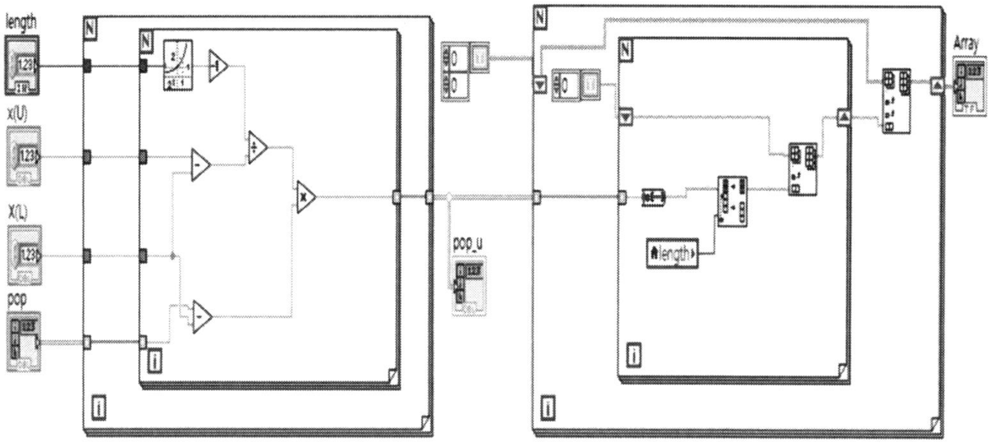


Figure 15-30: Internal structure of initial population producing sub-module

2) Fitted population: Once upper and lower limits are defined, only chromosomes within the ranges are singled out for the next steps
3) Reproduction [Rw Sel]: Objective functions and its resultant fitness functions are predefined before the module is run. The fitness function prioritizes the chromosomes and the reproduction sub-module determines which of the first chromosomes will be reproduced by 14% in the second population as shown in Figure 15-31.

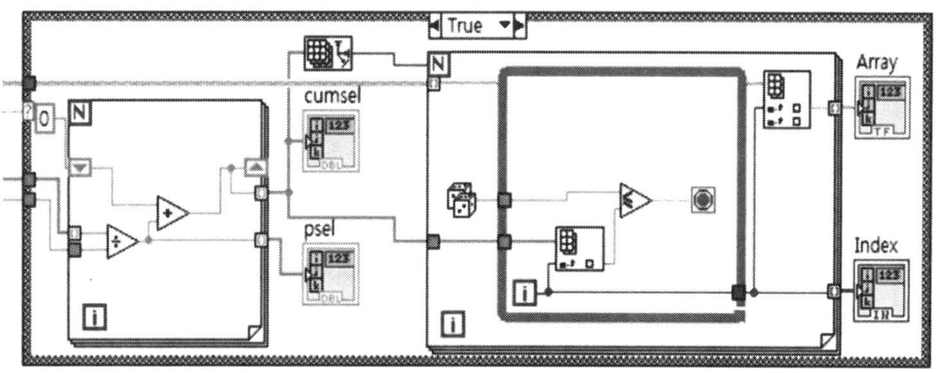


Figure 15-31: Internal structure of reproduction sub-module

4) Crossover [(3) bOp Xover]: Reproduction does not fill the initial population, some of whose chromosomes are discarded in response to the reproduction rule of eliminating low-performing populations. The crossover probability (Pcross) is generally set to 85% as shown in Figure 15-32.

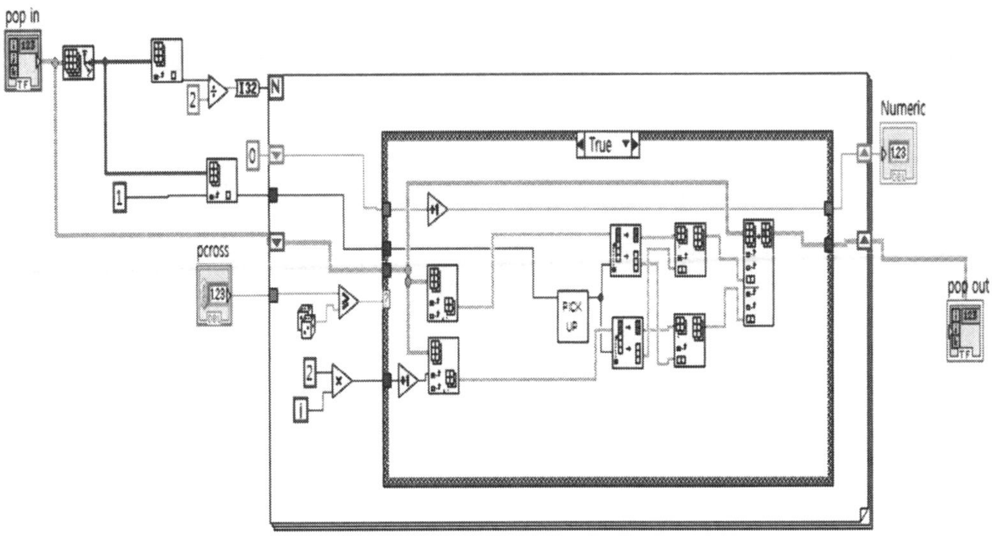


Figure 15-32: Internal structure of crossover sub-module

5) Mutation [bSim Mut]: as shown in Figure 15-33, mutation is the last leg among the three rules of GA and its suggested occurrence probability is 1%.

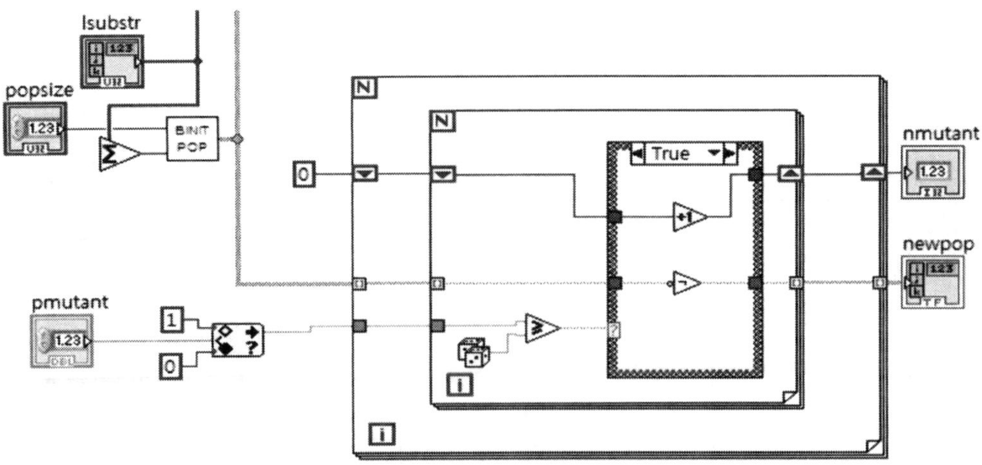


Figure 15-33: Internal structure of mutation sub-module

6) Decoding [Binary Decode]: It converts binary expression to real type numbers. Figure 15-34 shows the internal structure of decoding sub-module.

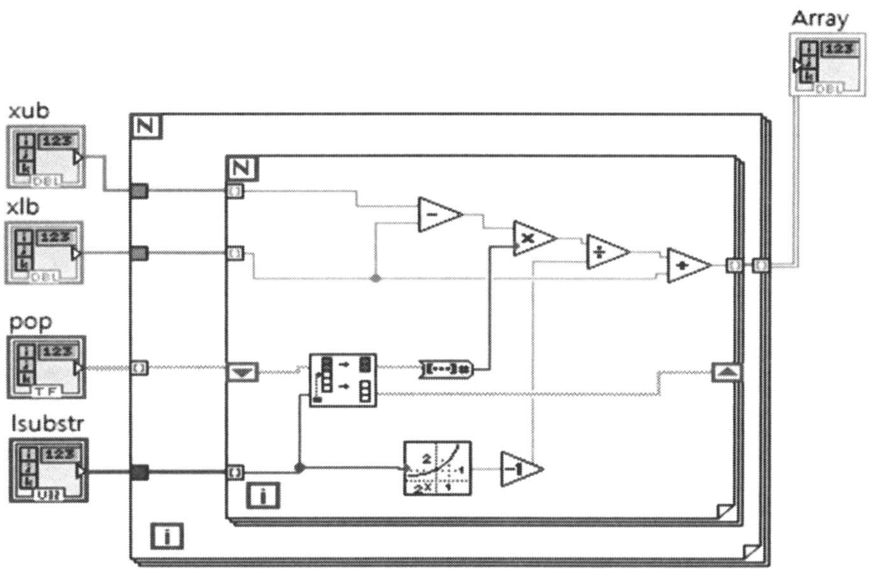


Figure 15-34: Internal structure of decoding sub-module

7) Performance indexing [per]: It judges the performance level of the outputted chromosomes from the reproduction, crossover, and mutation and yield the best chromosome among the population.

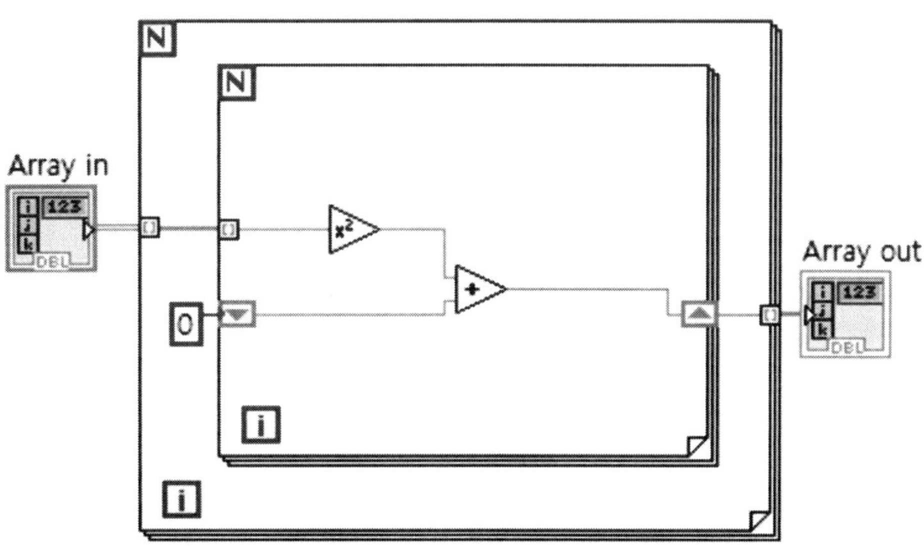


Figure 15-35: Internal structure of fitness evaluation sub-module

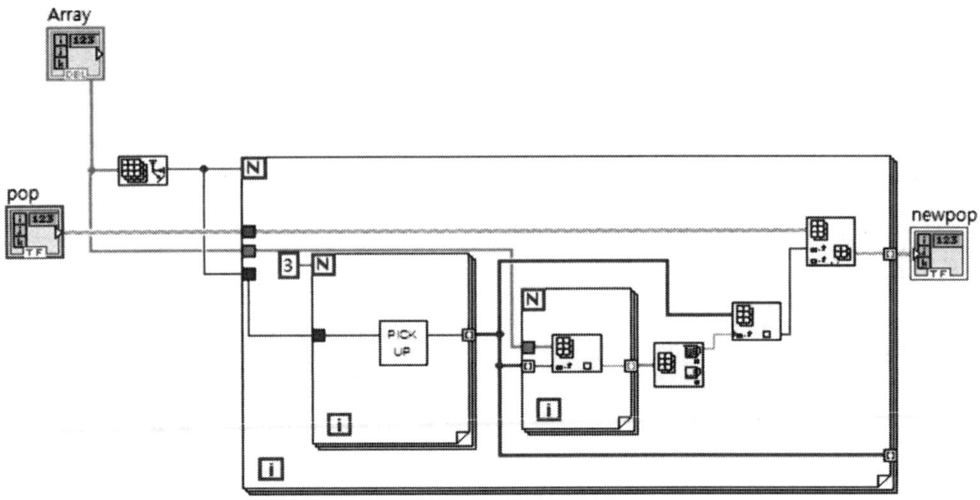


Figure 15-36: Internal structure of tournament (TOURN SEL) sub-module

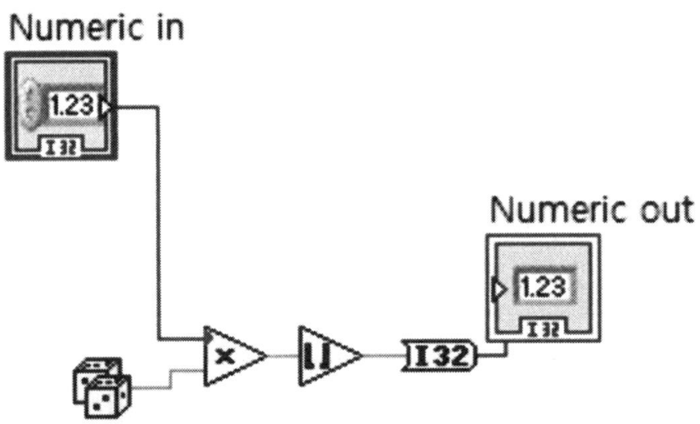


Figure 15-37: Internal structure of pick-up sub-module

From Figure 15-35 to Figure 15-39 illustrate three submodules; fitness evatuation, tournament, pick-up, objective function evatuation, and performance indexing.

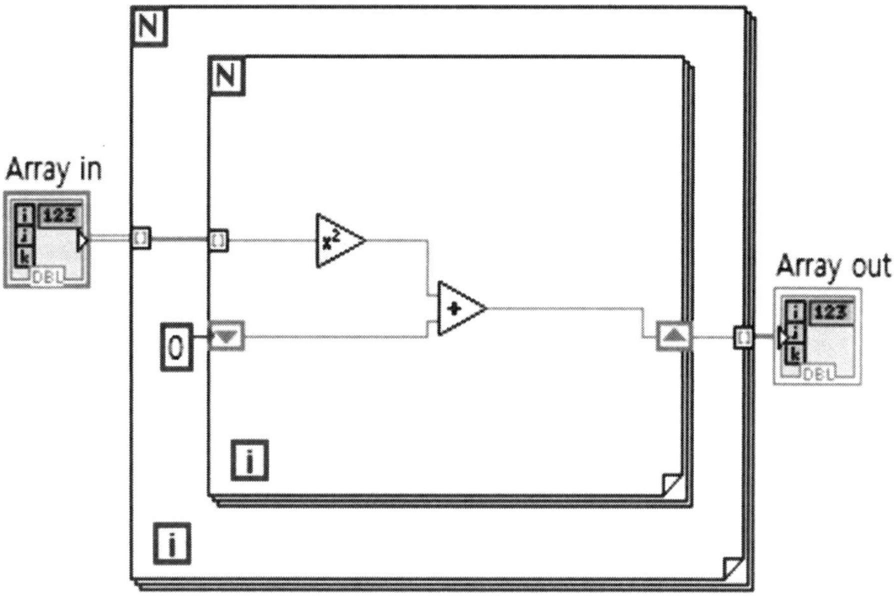


Figure 15-38: Internal structure of objective function evaluating sub-module

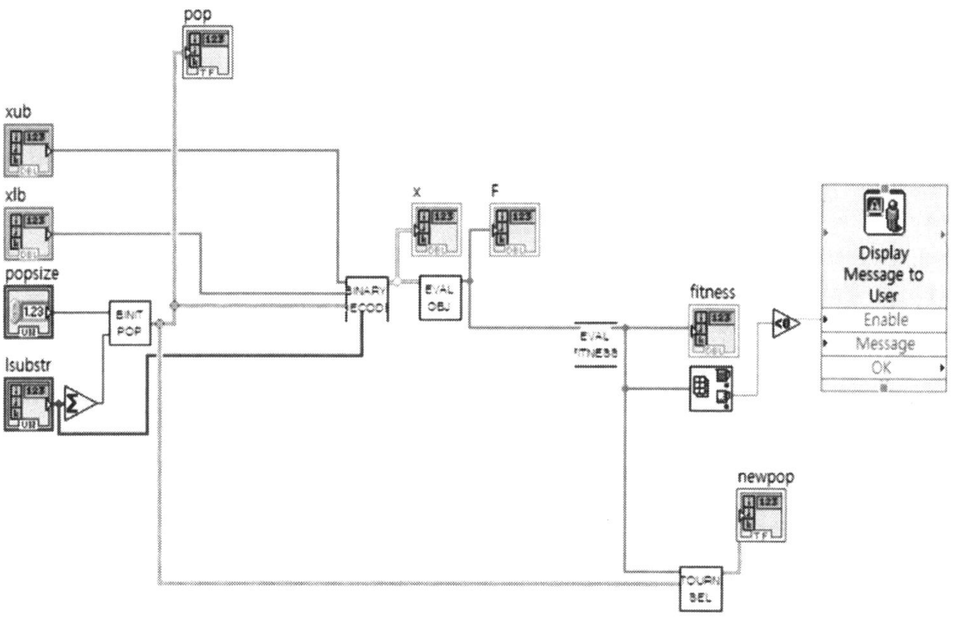


Figure 15-39: Internal structure of performance indexing sub-module

8) Elitism [binary Elitism]: It belongs to the recovery process of deceased superior chromosome in the previous population are discarded following the crossover or mutation. The chromosome substitutes for a low-performing chromosome in the current population.(shown in Figure 15-40)

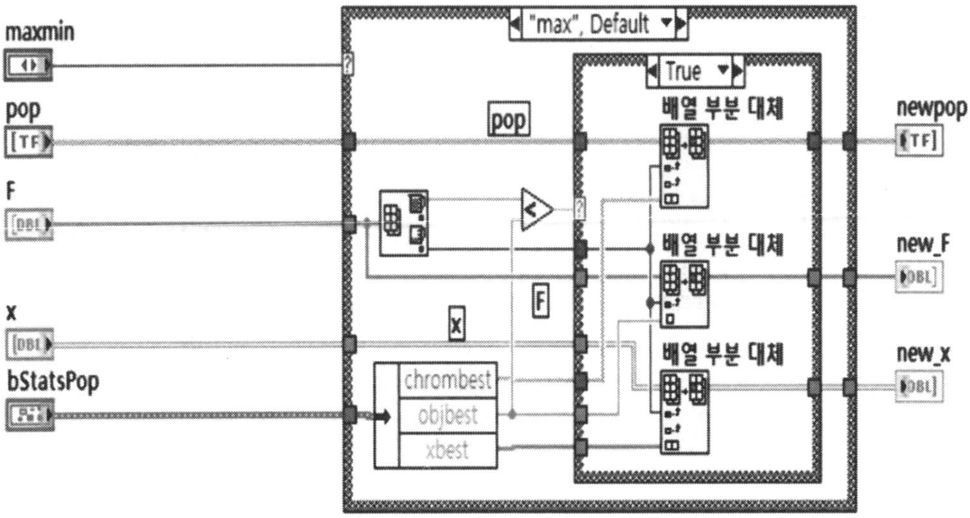


Figure 15-40: Internal structure of elitism sub-module

9) Scale [Scale Fit]: At the almost end of the optimization, most of the chromosomes have similar finesses, implying fewer choices are available and thereby hampering further optimization.(Figure 15-41)

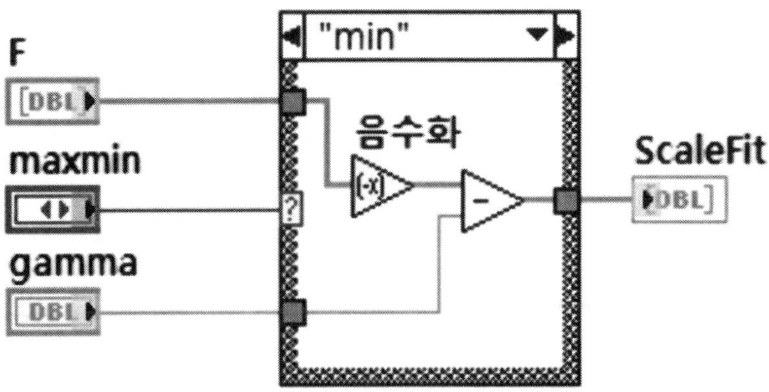


Figure 15-41: Internal structure of scale-fit sub-module

Figure 15-42 shows Labview version of GA algorithm developed in this work. Figure 15-43 illustrated internal structure of PID submodule.

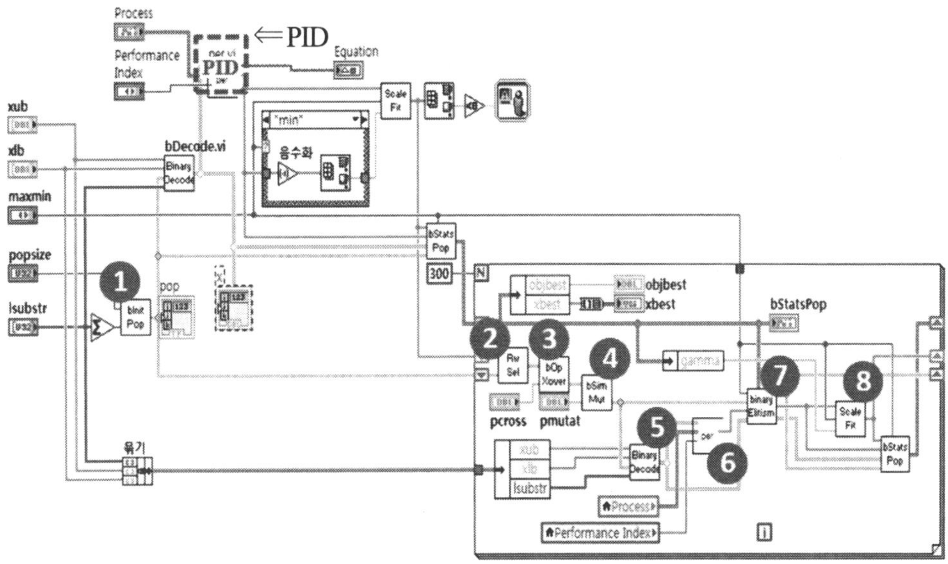


Figure 15-42: Labview version of GA (Block Diagram)

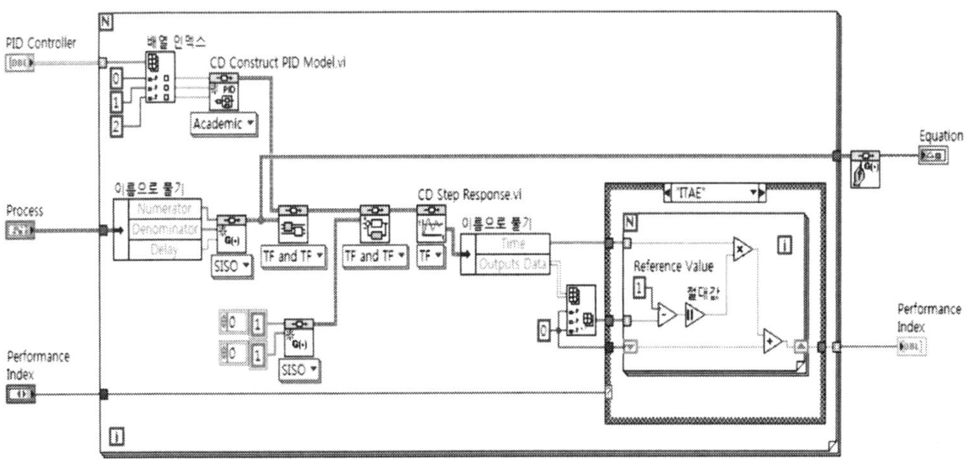


Figure 15-43: Internal structure of PID sub-module (Block Diagram)

The inputs of the diverse sub-modules were made in the control panel of the Labview as shown Figure 15-44 which featured animated parts marked green and grey in order to display the changing bit arrays of the best chromosome and populations. At the bottom, a chart was also placed to reflect the continuous temporal motion changes of the parameters.

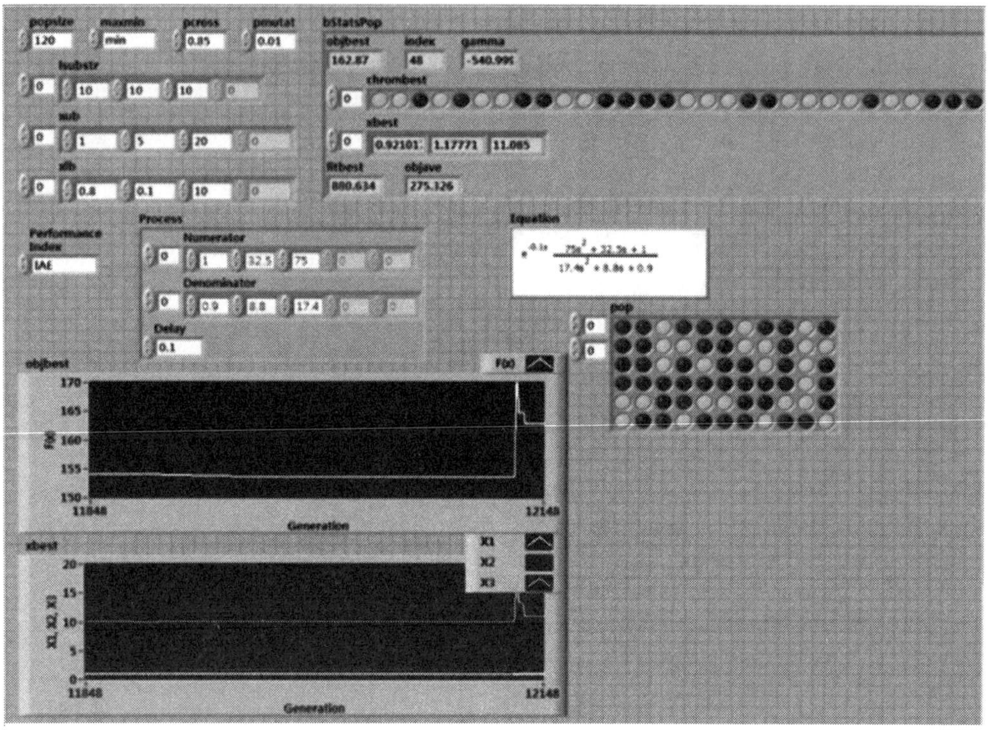


Figure 15-44: Labview version of GA (Control Panel)

### (3) Replace PID with developed ALLC

Labview library only provides blocks for lead compensator and lag compensator. As such, the need to develop a lead-lag compensator for the ALLC arose. This was done by combining both available compensators. The two compensators were connected in series as shown in Figure 15-45 and was executed for its verification. The red dotted box in Figure 15-42 represents the PID module in the SGA and was replaced with the ALLC module in Figure 15-45. ALLC incorporates three parameters of a gain, time constant and $\beta$ (the ratio of pole to zero).

In response to the replacement, the modified SGA was renamed MSGA which was projected to generate the optimized three parameters. The optimized parameters from the modified simplified genetic algorithm were converted to mathematical expressions using the Labview equations calculator as shown in Figure 15-46.

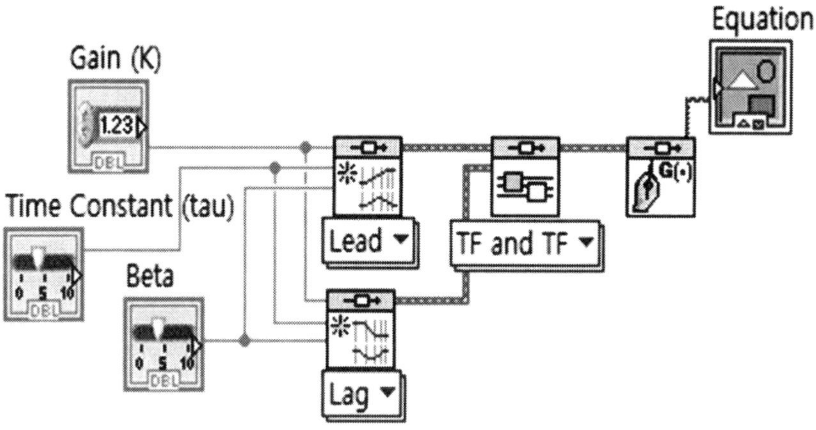


Figure 15-45: ALLC block diagram

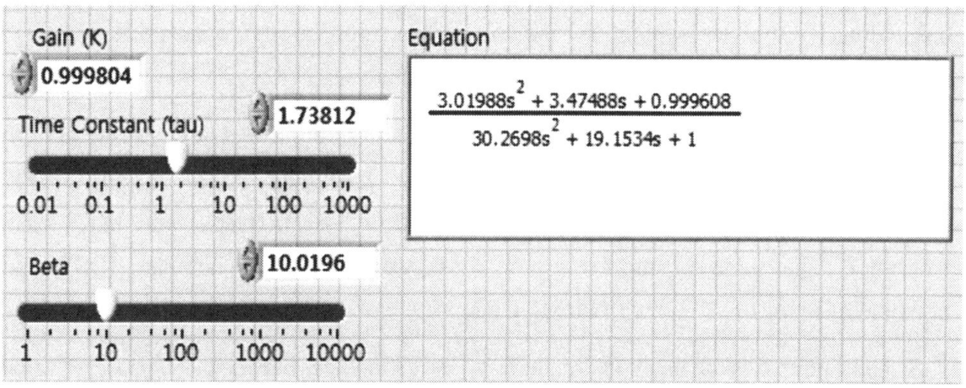


Figure 15-46: ALLC equation calculator (Control Panel)

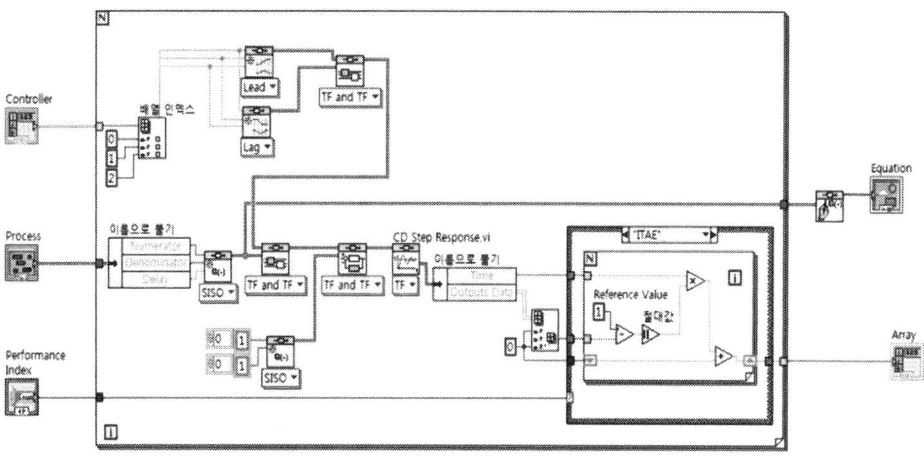


Figure 15-47: Internal structure of ALLC sub-module (Block Diagram)

### (4) Execute MSGA and search Effective Upper and Lower Ranges for Parameters

MSGA was executed numerous times to search for the most effective ranges of inputs to generate the optimized parameters of ALLC. The seven inputs enumerated in Table 15-5 were repeatedly scaled up and down until the CES discrepancy between MRRS and IRRS remained as marginal as possible. The values within the the input ranges were plugged into the Labview front panel of MSGA and the MSGA execution was iterated to derive the final setting.

The ranges are generally established by experience or the existing experiment results of analogous control systems. It is paramount to have oprimized parameters for the RRS. Therefore, the repeated iterations of the GA application was necessary. The process of one execution involved 1) the insertion of the seven inputs, 2) the insertion of the MSGA-calculated parameters into the ALLC calculator, 3) the deployment of the ALLC equation to the conceptual IRRS in Figure 15-27 and 4) the final Simulink scope analysis.

The ALLC parameters obtained after the iteration processes are represented in equation (2) and it was deployed to IRRS in Figure 15-27. Figure 15-49 illustrated three signals: input_1 (in pink), MRRS CES (in black) and IRRS CES (in blue). The same inputs for Figure 15-6 were reused to simulate IRRS and compare the MRRS CES with IRRS CES.

System tuning is defined as rendering the difference between its process value and reference value as marginal as possible by varying the parameters of a system. The tuning requires a qualitative index to judge the performance of the system and the index is called performance index [30]. The following performance indexes were implemented in this section: ISE (Integral Squared Error), IAE (Integral Absolute Error), and ITAE (Integral Time-weighted Absolute Error). This section implements all the indexes and IAE was observed to be the most suitable of the indexes for the additional lead-lag compensator. IAE is given in Eq. 15-1.

PART 3

$$IAE: J(K_p, \tau_i, \tau_d) = \int_0^t e(t)^2 dt \quad \textbf{Eq. 15-1}$$

Where, $K_p$ : Proportional gain

$\tau_i$ : Integral time

$\tau_d$ : Derivative time

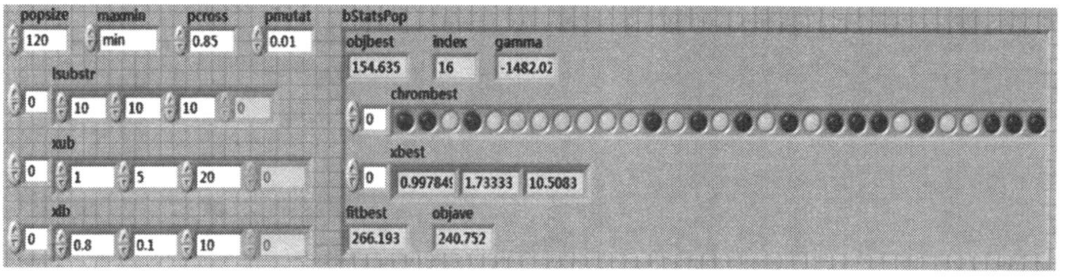


Figure 15-48: Labview front panel of MSGA

Table 15-5: Final settings of MSGA

| No | Inputs | Description | Final setting |
|---|---|---|---|
| 1 | popsize | The number of population created in each generation | 60 |
| 2 | maxmin | The type of optimization between maximized and minimized optimization | Min |
| 3 | pcross | The crossover probability | 0.85% |
| 4 | pmutat | The mutation probability | 0.01% |
| 5 | lsubstr | The number of bit string for each parameter | x1=10, x2=10, x3=10 |
| 6 | xub | The upper range of each parameters | x1=1, x2=5, x3=20 |
| 7 | xlb | The lower range of each parameters | x1=0.8, x2=0.1, x3=10 |

The ALLC equation is obtained as:

$$\frac{3.01988s^2 + 3.47488s + 0.99608}{30.2698s^2 + 19.1534s + 1} \quad \textbf{Eq. 15-2}$$

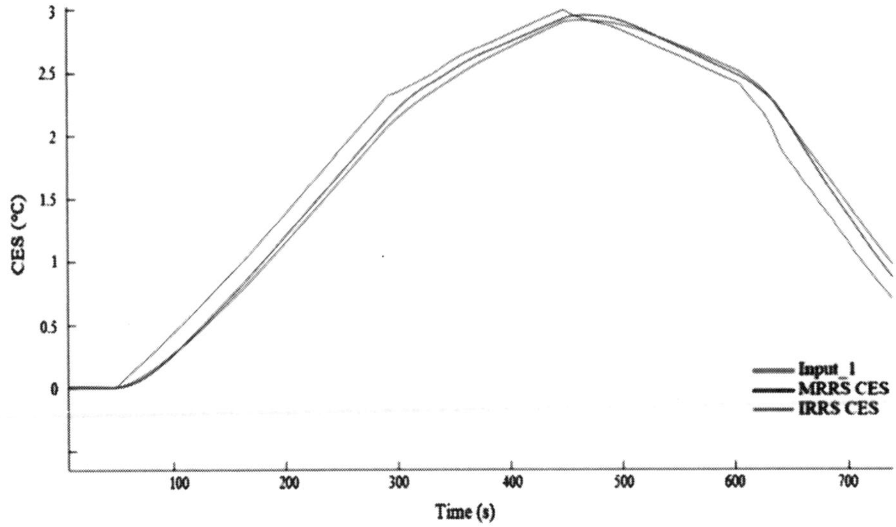


Figure 15-49: Scope output of completed IRRS at 75% power level

## (5) Derive Total Transfer Function of IRRS

At this point, the accuracy of IRRS was empirically substantiated with one scenario of 75% power level. This approach should entail theoretical and mathematical stability analysis as the stability is of paramount importance in control systems [31]. In addition, the input and output analysis of a system is only valid on the prerequisite of the systems stability [31].

There are well established stabality analysis methods, some of which are: Routh Hurwitz, Root locus and Bode diagram. All of them require the total transfer function (TTF) of a system. The IRRS logic was depicted in Figure 15-50 and its TTF was computed with two distinct approaches: Matlab linear analysis and equivalent block diagrams for transfer functions.

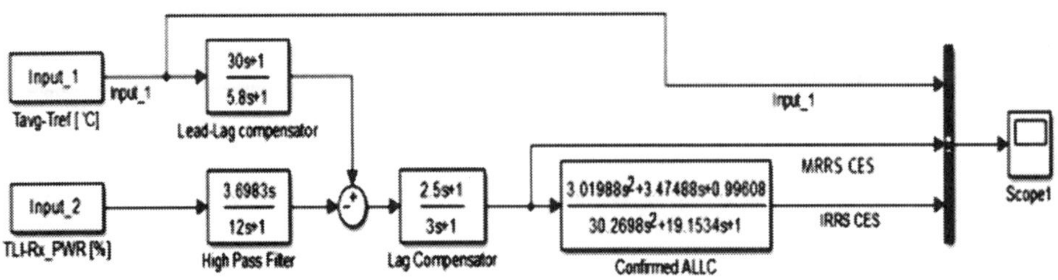


Figure 15-50: IRRS model with confirmed ALLC

### (6) MATLAB linear analysis

Two linear analysis points (Open-loop input) were applied at the end of input_1 and input_2 and at the end of ALLC, one linear analysis point (Open-loop output) was set. In response, the linear analysis tool was executed and the step plot function of the tool was chosen. The tool created a file and the file was transferred to the Matlab workspace. The file was converted to two transfer function as shown in Eq. 15-3, and Eq. 15-4 and the summation of the two gave the TTF of IRRS.

$$\frac{-0.02562s^4-0.03973s^3-0.02024s^2-0.003381s-2.794E-20}{s^4+1.049s^3+0.3245s^2+0.03134s+0.0009177} \quad \textbf{Eq. 15-3}$$

$$\frac{0.43s^4-0.6812s^3-0.362s^2-0.06806s-0.001891}{s^4+1.139s^3+0.4105s^2+0.05307s+0.001899} \quad \textbf{Eq. 15-4}$$

### (7) Equivalent block diagrams

Figure 15-50 was reshaped using equivalent block diagrams and the outcome was represented in Figure 15-51.

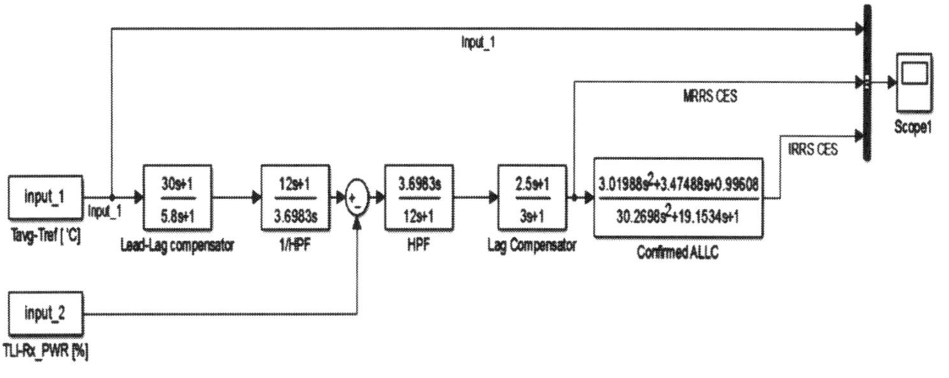


Figure 15-51: Reshaped IRRS model using equivalent block diagram

The Input_1, LLC, 1/HPF, HPF, lag compensator and ALLC were connected in series. Each of the mathematical expressions was written in the Matlab editor and their multiplication was run to derive the TTF of IRRS depicted in Eq. 15-5.

$$\frac{1.005E04s^6+1.675E04s^5+9783s^4+2295s^3+176.7s^2+3.682s}{2.336E04s^6+2.854E04s^5+1.181E04s^4+2039s^3+147.7s^2+3.696s} \qquad \textbf{Eq. 15-5}$$

## 15.7 Simulation Results of the Improved RRS

The primary objective of this section was to optimally enhance the accuracy and overshoot of the RRS output (CES) under the systems requirements specificaations in the source document by means of implementing GA which has been adopted in the control system-related research fields as discussed in the literature review. Moreover, GA has contributed to the success of the researches cited in the introduction section of this paper. Through the source document review and analysis, the significance of the accuracy and overshoot of RRS were highlighted. Then the current RRS was modelled in Simulink with the input and output data acquired from the full-scope APR1400 simulator.

### 15.7.1 Theoretical Verification

The TTFs of IRRS were obtained using the Matlab linear analysis tool as well the equivalent block diagram [32] for transfer function. The bottom line of the TTF was intended to determine the stability of IRRS as the empirical verification is only valid after the stability is validated [31]. The TTF computed by the linear analysis was named TTF-1 and the one developed by the equivalent block diagram was dubbed TTF-2. TTF-1 and TTF-2 were compared to eachother for the theoretical verification and it was observed that the TTF-2 shows a more accurate fit to the IRRS CES output as shown in Figure 15-23. Therefore, TTF-2 was selected for the stability verification.

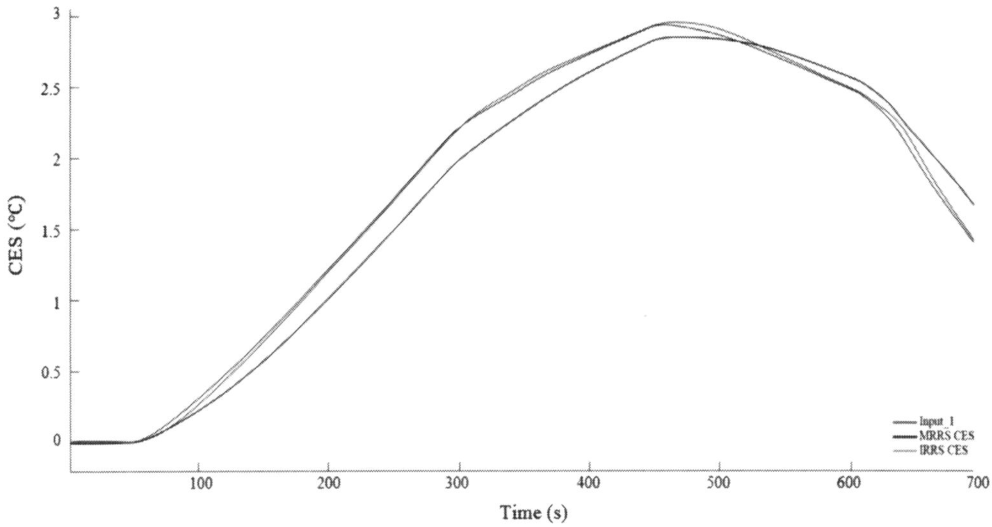


Figure 15-52: TTF fitness comparison with IRRS

### (1) Routh Hurwitz Stability Criterion

To determine stability, this criterion requires the characteristic equation of a systems total transfer function or its denominator whose coefficient signs are determined by the calculation rule of the criterion. When the resulting signs of the coefficients are all positive, the system is validated as stable. The denominator of Eq. 15-5 was set as the characteristic equation and each coefficient was put in an array as shown in Figure 15-53 and their signs calculated by the criterion turned out all positive, inferring that IRRS is stable.

| | | | | |
|---|---|---|---|---|
| + | $S^6$ | 2.34E+04 | 1.18E+04 | 1.48E+02 |
| + | $S^5$ | 2.85E+04 | 2.04E+03 | 3.70E+00 |
| + | $S^4$ | 1.01E+04 | 1.45E+02 | |
| + | $S^3$ | 1.63E+03 | 3.70E+00 | |
| + | $S^2$ | 1.22E+02 | | |
| + | $S^1$ | 3.70E+00 | | |
| + | $S^0$ | 0.00E+00 | | |

Figure 15-53: Result of Routh Hurwitz stability criterion

### (2) BODE Diagram

Matlab provides the bode diagram function which illustrates the gain and phase margins of a system in separate sections and helps ascertain the systems stability using the systems TTF. Once the function is executed with a TTF and the result reveals that the phase margin is a positive value at the gain of 0dB, the system is considered stable [33].

The function was executed to compute the margins of IRRS with Eq. 15-5. The equation was coded in the Matlab editor and the editor was run with a command of bode. The resultant diagram was illustrated in Figure 15-54 where both of the gain and phase margins had positive values. The fact that the phase margin stood at 147 and the gain margin was infinite accounted for the stability of IRRS.

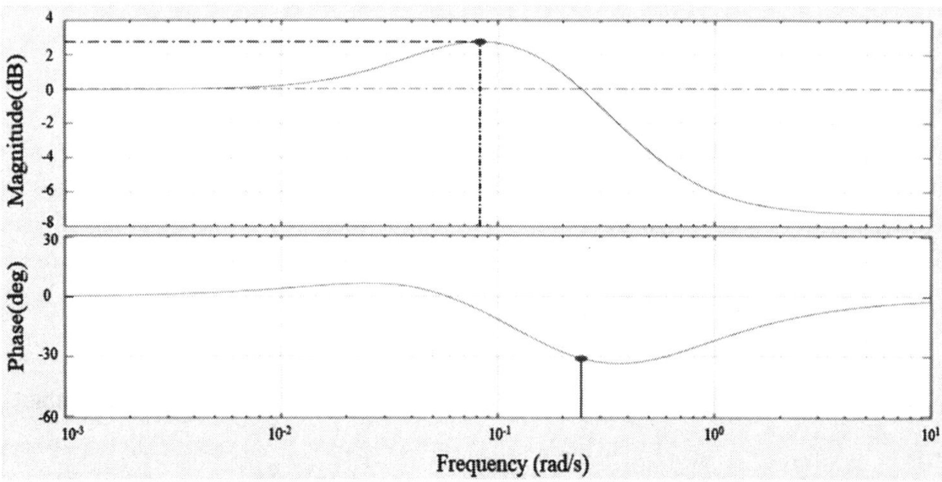


Figure 15-54: Gain and phase margins of IRRS

### (3) Root Locus Analysis

This approach is a graphical method for examining how the roots of a system change with variation of a certain system parameter, commonly a gain within a feedback system. This is a technique used as a stability criterion in the field of classical control theory developed by Walter R. Evans which can determine stability of the system.

The root locus plots the poles of the closed loop transfer function in the complex s-plane [3]. When all of the poles are positioned on the left side on the basis of the imaginary-axis, the system is proven stable. Eq. 15-5 was typed into the Matlab editor and the Matlab bode diagram was executed. The result in Figure 15-55 demonstrated that IRRS was stable with all of the poles mapped on the left side.

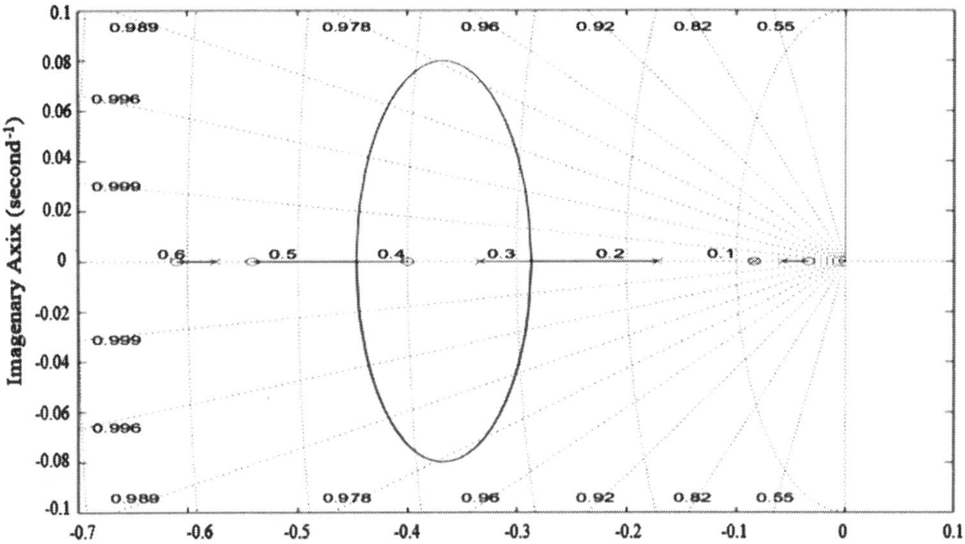


Figure 15-55: Mapped poles and zeros of IRRS

## 15.7.2 Empirical Verification

The highest IRRS CES during the initial stage of the 25% power level was recorded at -0.249°C, up 0.415°C as compared to the MRRS CES and the result was reflected in Figure 15-56 and the enhancement resulted in consolidating the conservatism of RRS and in further deterring the potential uncontrolled motion of CEA at low power levels.

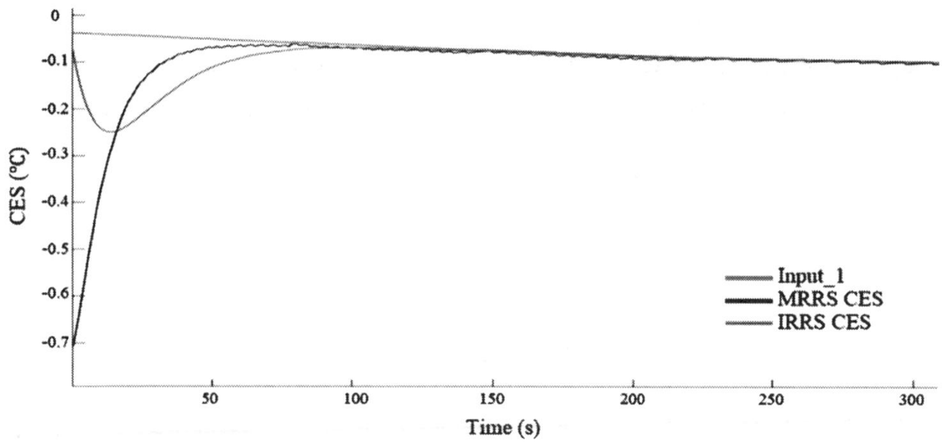


Figure 15-56: 25% power level

Figure 15-57 represents the result of the IRRS outputs using MSGA whose output and IRRS attained the minimized steady state error by 0.19°C (=0.259-0.069). Subsequently, it was observed that optimizing the error close to the temperature difference between Tavg and Tref resulted in further increase in the settling time of RRS.

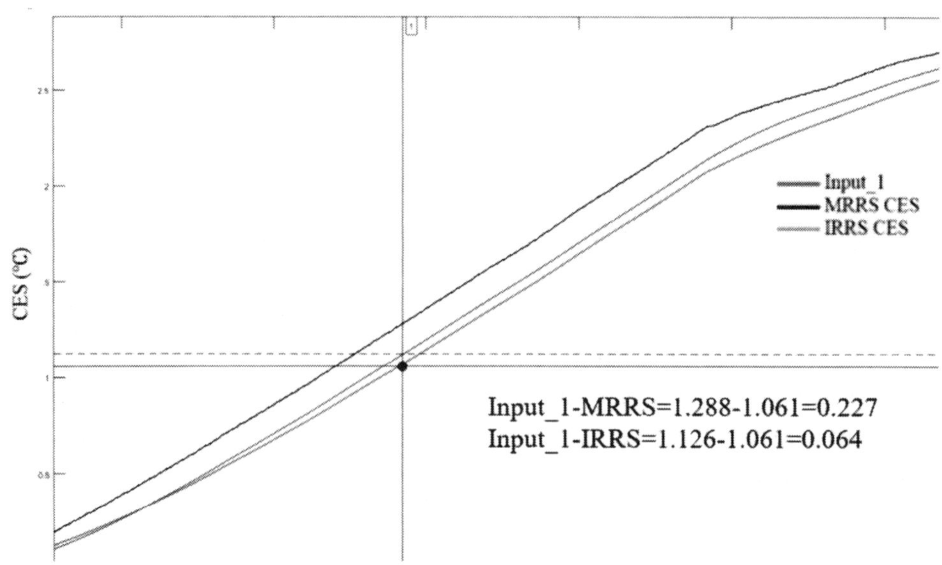


Figure 15-57: Steady state comparison

In addition, Figure 15-58 demonstrates a great deal of decline of the overshoot level in IRRS. The MRRS overshoot level was gauged at 179 with the improved RRS at 103%. The attained lower overshoot indicated that uncontrolled and unnecessary motion of CEDM would be significantly reduced.

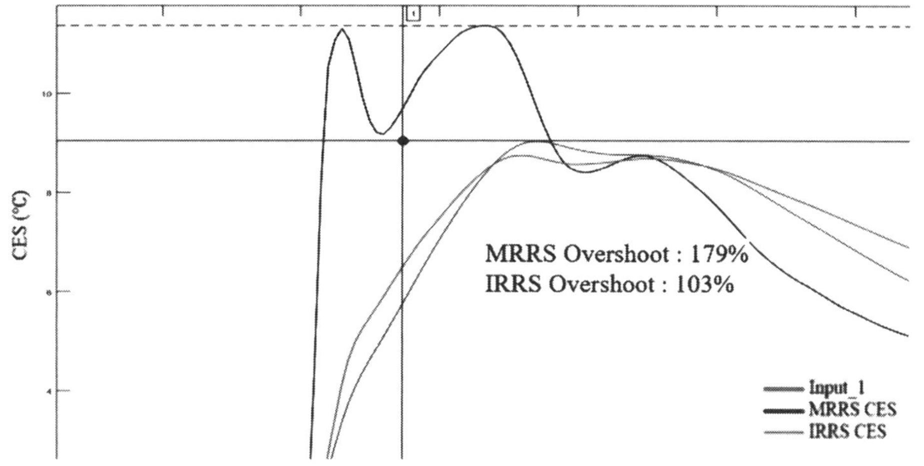


Figure 15-58: Overshoot level comparison

Figure 15-59, Figure 15-60, and Figure 15-61 show the signal output at 50%, 75% and 100% respectively. The plots show no discrepancies between MRRS and IRRS. Unlike 50 and 70%, the 100% power level generated conspicuous oscillations at 85 and 154 seconds which were observed in MRRS while IRRS smoothed out the oscillations. As regards each overshoot of the power levels, they were tabulated in Table 15-6 and IRRS diminished the overshoots of MRRS by over 3 times.

Table 15-6: Overshoot comparison between MRRS and IRRS

| No | Items | Power levels | | | | | |
|---|---|---|---|---|---|---|---|
| | | 50% | | 75% | | 100% | |
| | | MRRS | IRRS | MRRS | IRRS | MRRS | IRRS |
| 1 | Elapsed time (sec) | 1 | 13 | 1 | 12 | 1 | 14 |
| 2 | Input_1 (℃) | 12 | 12 | 3 | 3 | 3.945 | 4 |
| 3 | CES (℃) | 86.860 | 29.860 | 35.720 | 12.630 | 51.049 | 18.067 |
| 4 | Maximum Overshoot (℃) | 74.86 | 17.86 | 32.72 | 9.63 | 47.104 | 14.067 |

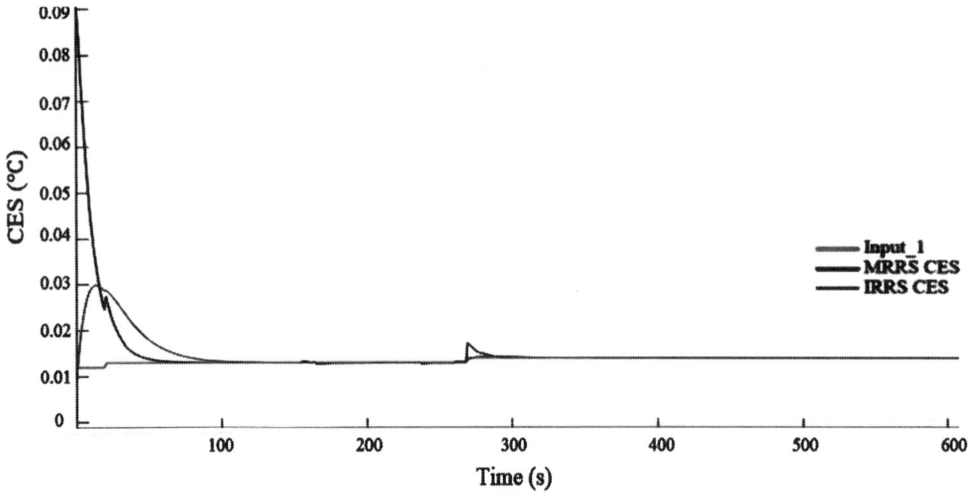


Figure 15-59: 50% power level

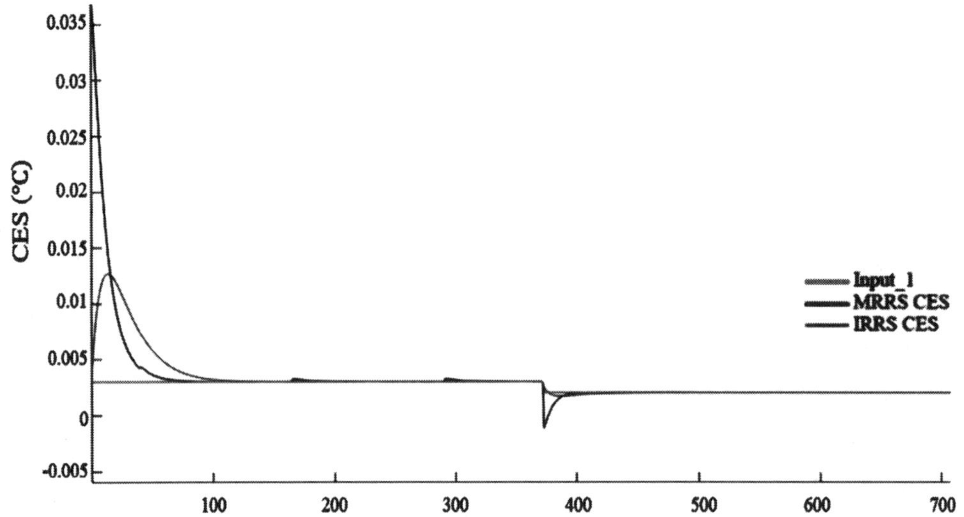


Figure 15-60: 75% power level

FWP system is comprised of 3 FWPs and the rated power of each pump stands at 155%. The power of the two FWP power forms 310% and the power can sufficiently stand in for one FWP loss. The two FWP trips were deliberately initiated to shape a load rejection like transient. The first FWP was tripped at 219 seconds and 25 seconds later, the second FWP was made inoperable to actuate RPCS which reduces the reactor power level to 50% by automatically and sequentially inserting CEAs.

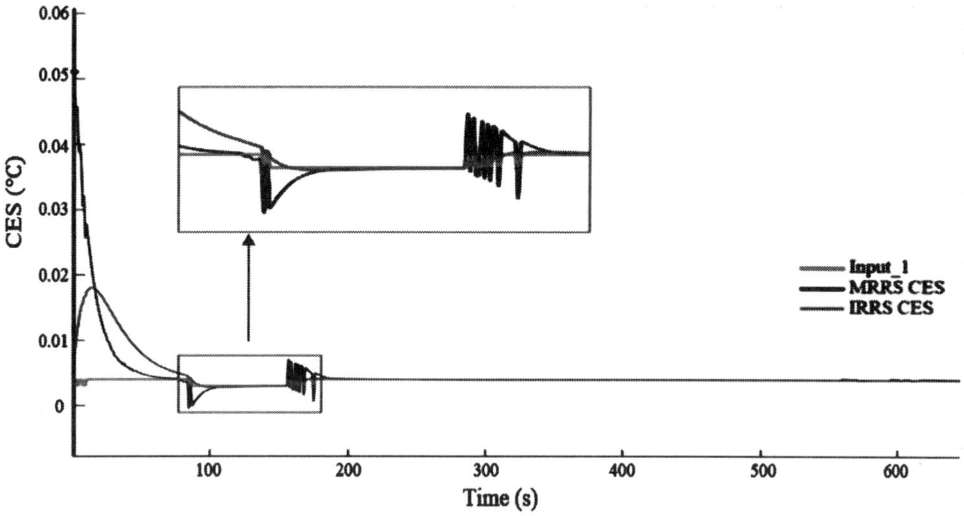


Figure 15-61: 100% power level

As soon as the two FWP all stopped, the PCS actuation commenced as illustrated in Figure 15-62. In the face of the transient IRRS was able to generate refined CES. This simulation continued for 1106 seconds and the average deviation between input_1 and the IRRS CES was 0.009726°C, down 0.00784°C from 0.01756°C in Table 15-1.

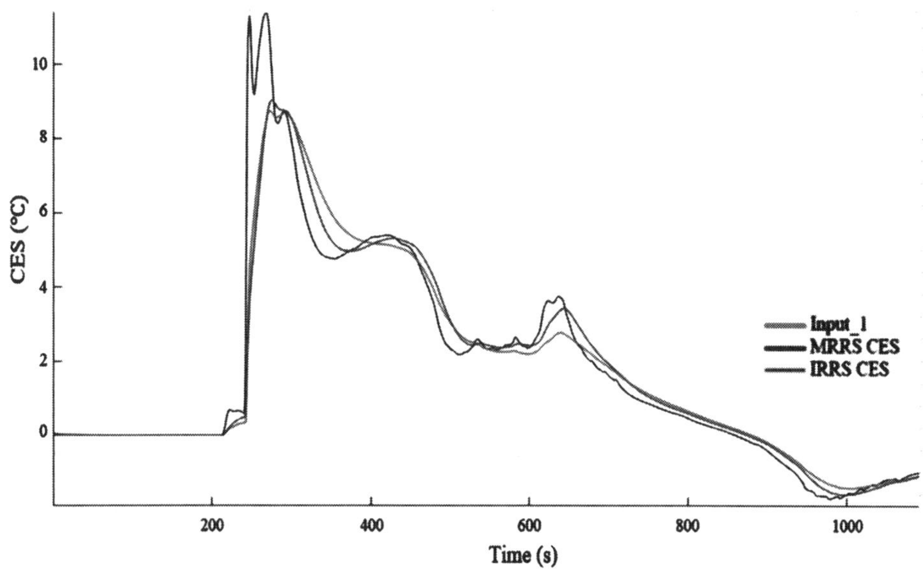


Figure 15-62: 100% power level with a transient by two FWP trips

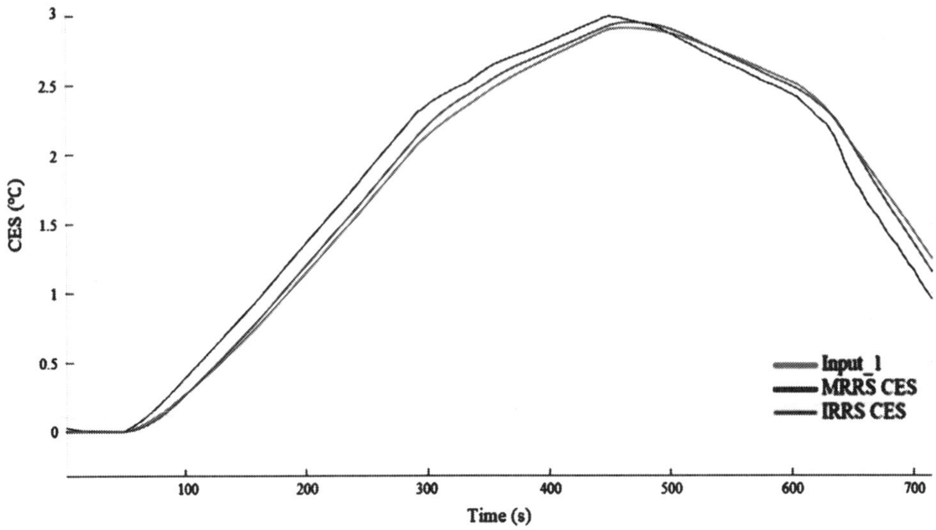


Figure 15-63: Ramp response at 75% power level

In the introduction section, the ramp response measurement at 75% power level was conducted by scaling up and down the turbine load. Figure 15-63 represented that input_1 signal was on the increase and decease as the result of the scaling. The average MRRS input_1-CES deviation was indicated at 0.17923°C. The measurement was executed once again with IRRS. The average IRRS input_1-CES deviation was gauged at 0.033352°C, down 0.145881°C from 0.17923°C in Table 15-7.

Table 15-7: Accomplished results with IRRS

| No | Research areas (Issues) | Measured values | | Remarks |
|---|---|---|---|---|
| | | MRRS | IRRS | |
| 1 | Initial CES | -0.664℃ | -0.249℃ | 25% power level [Figure 15-2] |
| 2 | Average input_1- CES deviation | 0.17923℃ | 0.033352℃ | 75% power level [Figure 15-3] |
| 3 | Average input_1- CES deviation | 0.01756℃ | 0.009726℃ | 100% power level with a transient by two FWP trips [Figure 15-5] |
| 4 | Maximum overshoot | 47.104℃ | 14.067℃ | 100% power level [Figure 15-4] |

PART 3

### 15.7.3 RRS-IRRS Comparison Results on Single Test-bed

Figure 15-64 shows the overall scheme of RRS-IRRS comparison and verification.

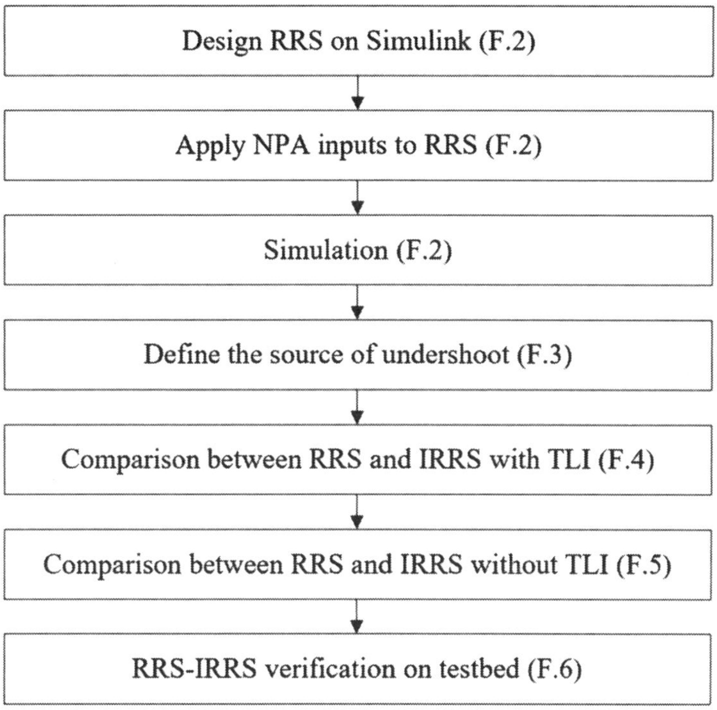


Figure 15-64 Overall scheme of RRS-IRRS comparison and verification

#### (1) 25% power output simulation results of RRS with TLI

Figure 15-65 depicts the pre-synchronization logic structure of RRS. The synchronization is explained in sub-section 15.7.1.5 and is carried out to fix the CES discrepancy between RRS and NPA with the discrepancy resolved by manually tuning each of the logics. This section is prepared to reflect that IRRS outperforms even on a single testbed requiring no artificial tuning involved.

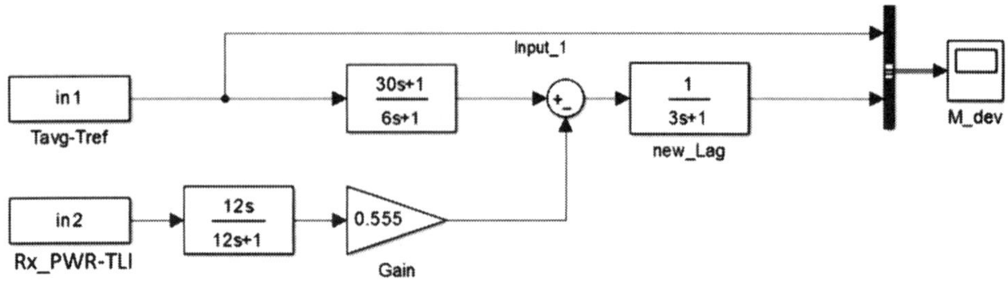


Figure 15-65: RRS logical architecture with TLI

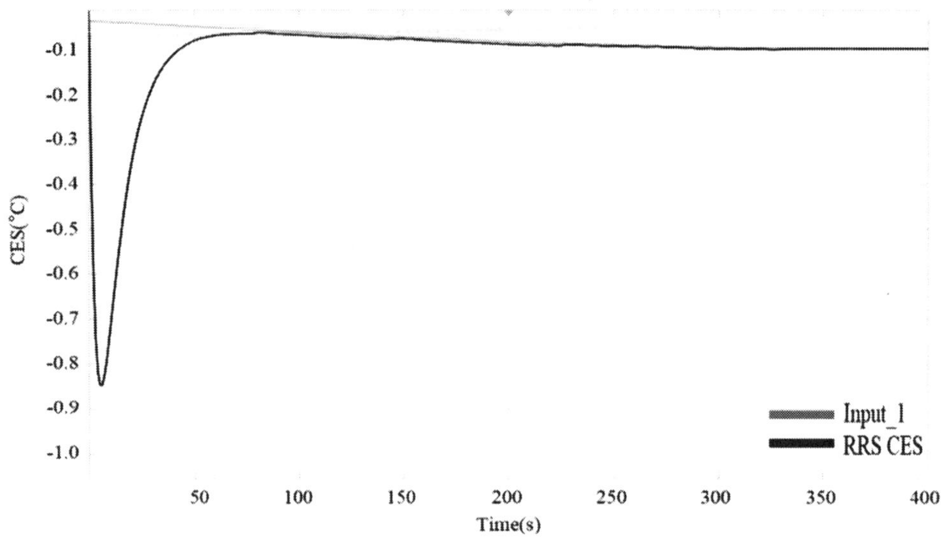


Figure 15-66: RRS scope output of 25% power simulation results with TLI

Figure 15-66 represents the output of Figure 15-65 and the undershoot level of around -0.85°C. The level is larger than the one of the synchronized RRS whose undershoot stands at –0.664°C.

### (2) 25% power output simulation results of RRS without TLI

This experiment is designed to discover the reason for the problematic undershoot level of RRS. For this, Rx_PWR-TLI(in2) is disconnected as illustrated in Figure 15-67 and its substantially reduced output of -0.2°C in Figure 15-68 signifies that Rx_PWR-TLI(in2) is the main contributing factor in triggering the undershoot.

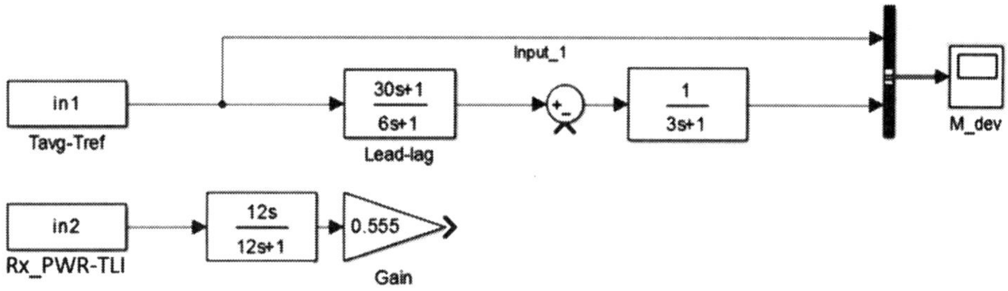


Figure 15-67: RRS logical architecture without TLI

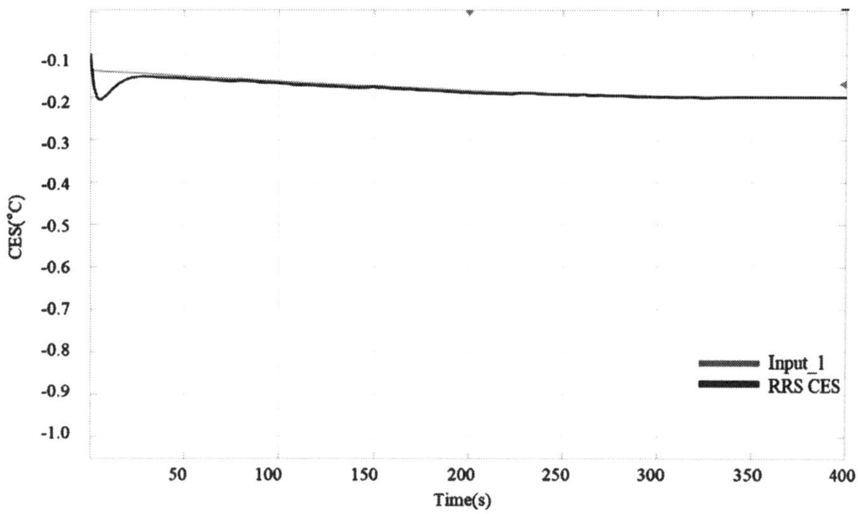


Figure 15-68: RRS scope output of 25% power simulation results without TLI

The changes of Rx_PWR-TLI(in2) values over time are monitored by adding the display function of Simulink to multiple points in Figure 15-69. The displays in Figure 15-68 depict variable changes at 6 seconds. Then, Tavg-Tref(in1) of -0.03879 is subtracted with Rx_PWR-TLI(in2) of 2.124 and the CES output comes to -0.8053. This reveals again Rx_PWR-TLI(in2) is the root cause of the undershoot level and Figure 15-69 explains how the level is computed. Rx_PWR-TLI(in2) is devised to promptly trigger CEA motions and to strike the power balance between the primary and secondary side especially in a situation where Tavg-Tref(in1) deviation is marginal but Rx_PWR-TLI(in2) deviation is high. Consequently, CES outputs in that situation should be enhanced to prevent uncontrolled CEA motions.

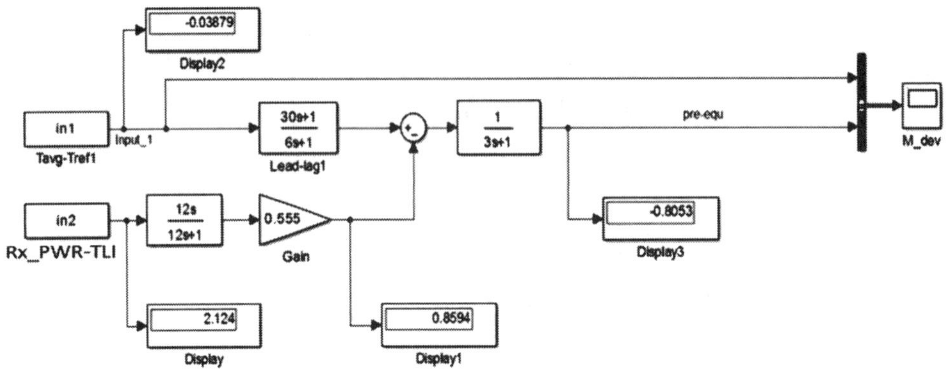


Figure 15-69: RRS logical architecture with TLI and value displays

### (3) 25% power output simulation results of IRRS with TLI

Figure 15-71 also reflects that IRRS outperforms RRS in terms of the undershoot level and satisfies the requirement of uncontrolled CES motion as the measured level is below -0.5°C while Tavg-Tref(in1) hovers around -1.5°C. In Figure 15-70 portrays IRRS complete with GA-enabled ALLC.

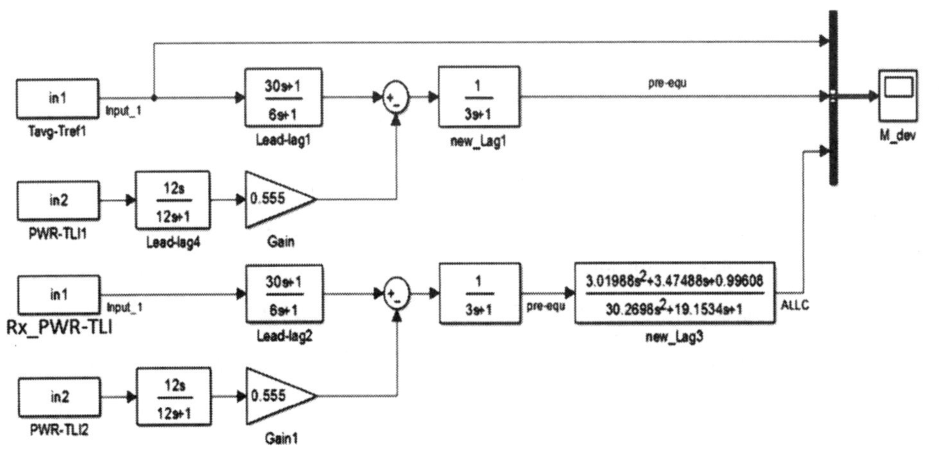


Figure 15-70: IRRS logical architecture with TLI

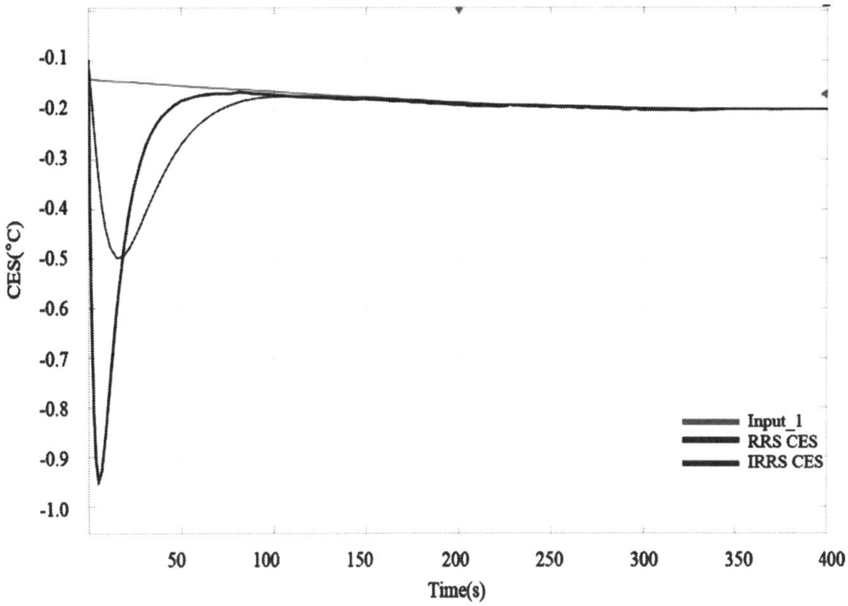


Figure 15-71: RRS scope output of 25% power simulation results without TLI

In Figure 15-71, RRS scope output of 25% power simulation results without TLI is shown.

### (4) 25% power output simulation results of IRRS without TLI

When no TLI is concerned, the CES outputs reveals that the transient conditions (undershoot or overshoot) as well as the overall accuracy are enhanced as shown in Figure 15-73. In addition, IRRS demonstrates smooth CES outputs while RRS represents uneven outputs.

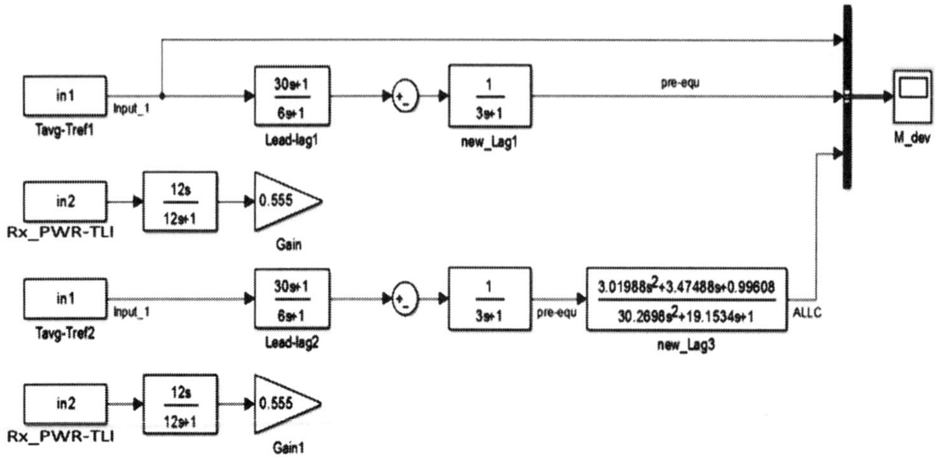


Figure 15-72 IRRS scope output of 25% power simulation results without TLI

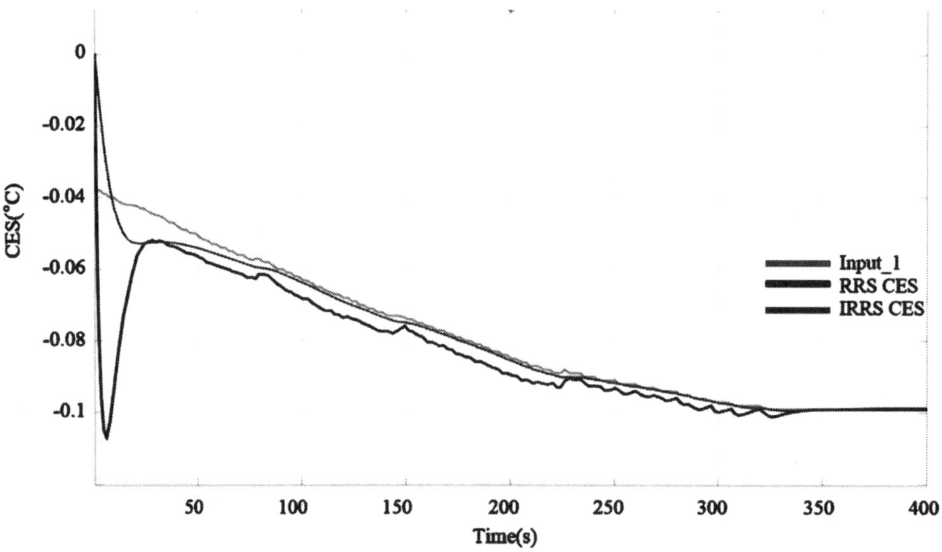


Figure 15-73: RRS scope output of 25% power simulation results without TLI

P A R T 3

### (5) RRS-IRRS verification on single testbed

This section is produced to verify that the performance of IRRS remains better than RRS and include several experiments on TLI involvement. The transient of both RRS and IRRS appears inevitable due to the mathematical composition of each compensator. The results are tabulated in Table 15-8 and qualitatively proves IRRS outperforms RRS.

Table 15-8: RRS-IRRS comparison results on single testbed

| No | Subject | Measured values | Difference (IRRS-RRS) | |
|---|---|---|---|---|
| | RRS | IRRS | | |
| 1 | Transient level with TLI | 0.84775℃ (At 6 seconds) | 0.39989℃ (At 17 seconds) | 0.44786℃ |
| 2 | Average deviation with TLI | 0.116098℃ | 0.111162℃ | 0.004936℃ |
| 3 | Transient level without TLI | 0.068464005℃ (At 6 seconds) | 0.03161948℃ (At 1 second) | 0.36845℃ |
| 4 | Average deviation without TLI | 0.004224℃ | 0.001134℃ | 0.00309℃ |

## PROBLEMS

1. Explain the operation of the RRS using the schematic diagram.
2. Describe the overall scheme for GA-based RRS logic architecture improvement
3. What are the CEA rate and direction circuits of RRS?
4. Explain how the TLI is computed.
5. How many steps are there in GA approach? Describe the each step.
6. Describe the Lead-Lag Compensator, High Pass Filter, and Lag Compensator Designs. In which situation would you use each of these compensators?
7. What is the Routh Hurwitz Stability Criterion
8. Describe the MATLAB linear analysis approach.
9. What is a BODE Diagram? What is the significance of bode diagrams in control system design and analysis?
10. Describe the modeling and simulation of the RRS performed.
11. What improvements do you think could be made to the existing RRS?
12. In this study, Genetic Algorithms were chosen. What alternative AI model do you think could be used? Justify your answer.

## FURTHER READING


- USNRC, "APR1400 Design Control Document : Tier 2," 2018. Available from: http://www.nrc.gov.
- K. Krishnakumar, and DAVID E. GOLDBERG, "Control System Optimization using Genetic Algorithms," Journal of Guidance, Control, and Dynamics 15 (1992) 735-740.
- Tzuu-Hseng S. Li, Ming-Yuan Shieh, "Design of a GA-based Fuzzy PID Controller for Non-Minimum Phase Systems," Fuzzy Sets and Systems, 111 183-197. 2000
- National Instruments, Labview. Available from: http//www.ni.com [Cited 2016 June 12]
- Shim, Hyo-sub, Jung, Jae-cheon, "Improved RRS logical architecture using genetic algorithm," Nuclear Engineering and Technology 49, 2017.

# Chapter 16 CYBER SECURITY

## SUMMARY

Computer security, also known as cybersecurity, is "the protection of computer systems from the theft or damage to the hardware, software or the information on them, as well as from disruption or misdirection of the services they provide. It includes controlling physical access to the hardware, as well as protecting against harm that may come via network access, data and code injection and due to malpractice by operators, whether intentional, accidental, or due to them being tricked into deviating from secure procedures" [34]. USNRC defines cybersecurity as "the protection of digital assets including digital instrumentation and control systems or equipment. Cybersecurity at nuclear power plants promotes performance-based programmatic approaches supported by defense-in-depth strategies, including efforts to detect, prevent, delay, mitigate, and recover from cyberattacks" [35].

Generally, protecting data systems against unauthorized intrusion is necessary to maintain the data availability and integrity. However, no system can be made invulnerable. A vulnerability is a system susceptibility or flaw. Many vulnerabilities are documented in the CVE database [36] and NVD database [37]. An exploitable vulnerability is one for which a cyberattack, intrusion or exploit exists. Cyberattack is any type of adversary action employed by individuals or whole organizations that targets computer data systems, infrastructures, and data networks by various means of malicious acts that either steals, alters, or destroys a specified target by hacking into a vulnerable system. Cyberattacks have become increasingly sophisticated and dangerous. Therefore, to protect a data system and maintain its availability against potential cyberattacks, it is necessary to understand the cyberattacks that can be initiated against them. Mind-map in Figure 16-1 depicts the overview of this chapter.

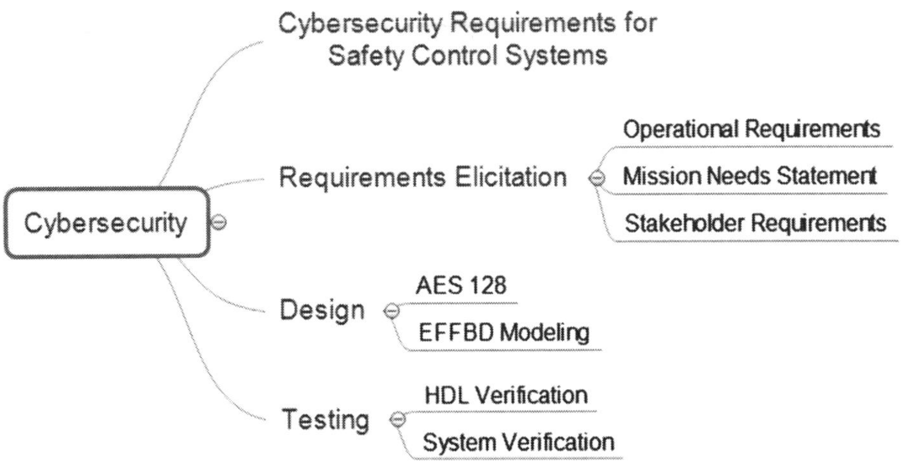


Figure 16-1: Mind Map for Chapter 16

## OBJECTIVES

- Introduce the cybersecurity requirements for NPPS
- Present the design and modeling of a hardware based encryption and decryption system

PART 3

# 16.1 Cybersecurity Requirements for Safety Control Systems

Data communication systems in APR1400 mainly encompass safety systems data network (SDN) and non-safety systems data communication and information network (DCN-I). Unidirectional gateways and data diodes (e.g., fiber-optic modems) are serving as network security zones allowing data transfer from the safety to the non-safety networks and block the reverse communication path.

A unidirectional gateway server is a non-safety related system. It protects the safety critical digital systems from unauthorized access. The gateway server is computer-based, its availability could be compromised if a potential cyberattack initiated from the non-safety side. The unavailability (i.e., system failure or malfunction) of gateway server does not prevent the safety systems from performing its required functions. The failed gateway server could not supply the real-time status of plant performance which causing a Loss of View (LoV) event. Cyberattacks maliciously affect the availability, integrity, and confidentiality of data and systems.

A gateway server is used to guarantee the unidirectional data transmission from safety side to non-safety side of MMIS system. The Gateway Server interfaces with other systems and data networks as shown in Figure 16-2.

In order to develop an FPGA-based Gateway server, aimed at improving reliability and performance, the requirements must first be established from existing laws, regulatory guides, and standards. Thus reverse and re-engineering methodology is applied to Know-why and develop the requirements for cybersecurity plan and develop an FPGA-based Gateway server.

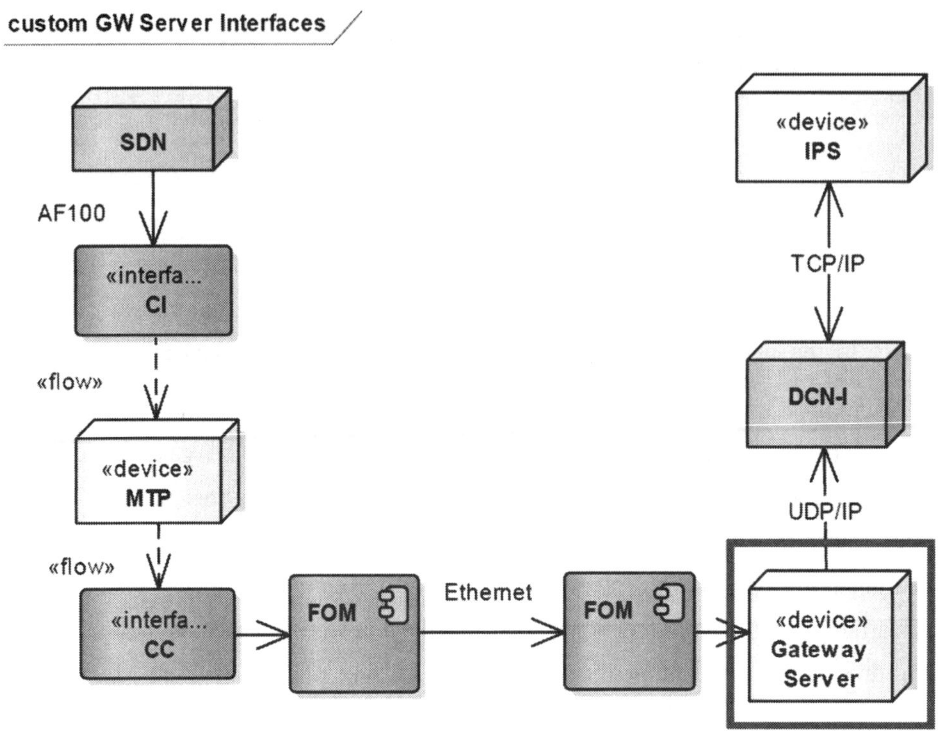


Figure 16-2: Gateway server interface with MMIS

## 16.2 Requirements Elicitation

Figure 16-3 shows the requirements traceability diagrams of design criteria for using computers in safety control systems of nuclear power plants based on the RG 1.152 and RG-5.71.

According to NRC RG 73.54, the design should provide high assurance that digital networks and communication systems' equipment are adequately protected against cyber-attacks, up to and including the design basis threat [38]. All systems and networks should be protected against any probable cyber-attacks that have adversely impact the integrity or confidentiality of data and/or software, such as MITM(Man In the Middle) attacks, by employing cryptographic mechanisms.

PART 3

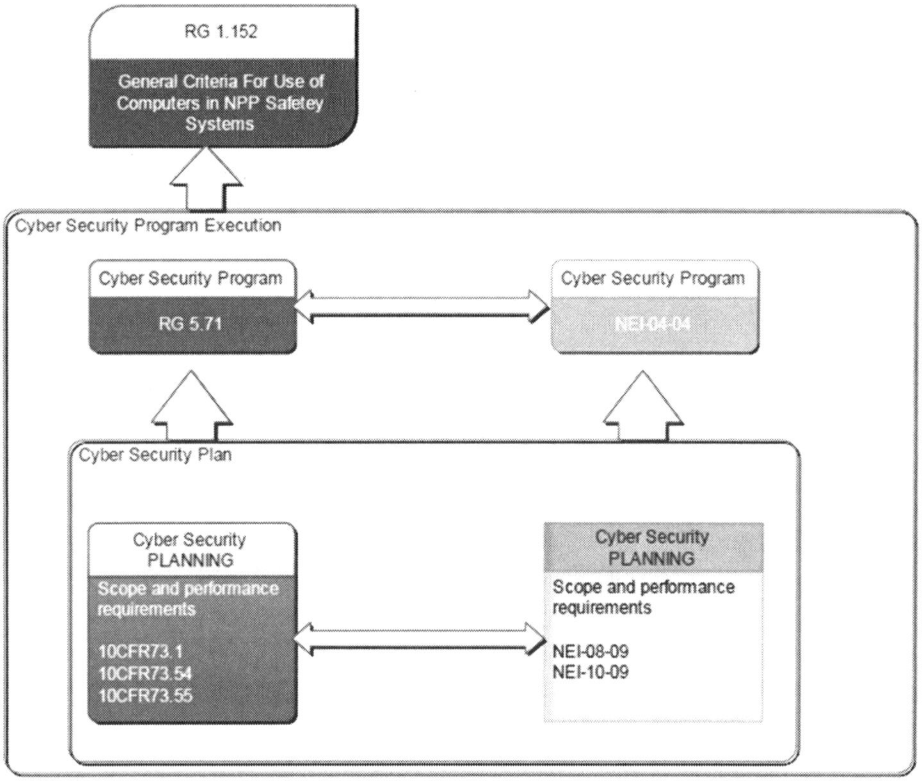


Figure 16-3: Cybersecurity regulatory and utility requirements identification

These mechanisms should be able to recognize alteration of the information at both the transmitting and receiving ends of the communication link. All systems and networks should deny an unauthorized access to systems, services, and/or data. The cyber security plan must establish, implement, and maintain to satisfy the cyber security program requirements of regulation guide. According to the NRC [39], this plan should describe how to controls, protect the assets and maintain defense-in-depth protective strategies to ensure the capability to detect, respond, and recover from cyber-attacks, mitigate the adverse effects of cyber-attacks, and ensure that the functions of protected assets identified are not adversely impacted due to cyber-attacks. In addition, the system should be capable of timely detection and response to cyber-attacks.

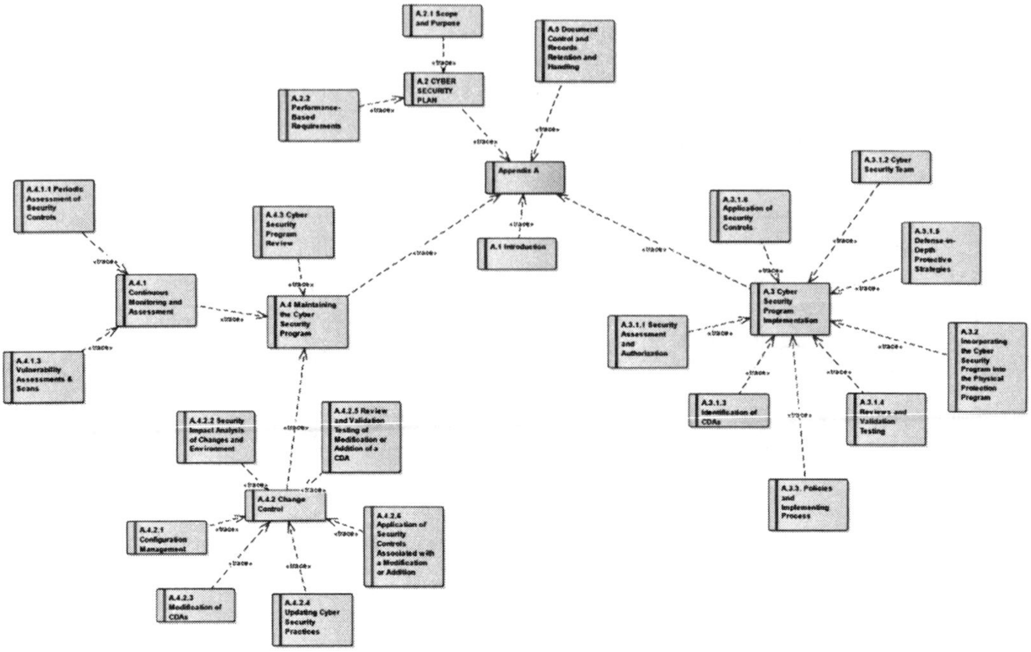


Figure 16-4: Regulatory requirements to cybersecurity (RG-5.71 to SRP 13.6.6)

Referring to RG 5.71; Figure 16-4, cyber security must comply with the licensing requirements of 10 CFR 73.54; Figure 16-5 and Figure 16-6 gives an acceptable set of security controls, which are developed by using the NIST cyber security standards and security controls. These controls depend on well defined and well- understood vulnerabilities, threats, and attacks, in addition to well-understood and vetted countermeasures and protective techniques [40], [41].

Acceptable security controls are transmission integrity and transmission confidentiality and should be achieved by employing cryptographic mechanisms. When security requirements are considered as an integral subset of other information system requirements, the resulting system has fewer weaknesses and deficiencies, and therefore, fewer vulnerabilities that can be exploited in the future [42] [43].

The Cryptographic Key is one of the critical security parameters. NIST Special Publication, 800-57 Part 1 is a key management recommendations guide for handling cryptographic keys. These keys are managed by using automated mechanisms, and the key entry into the cryptographic module should use key loading device and secure transport of cryptographic keys. Key storage shall not be accessible from outside of the cryptographic module to unauthorized operators.

For security strength, cryptoperiod should be less than 2 years. The cryptographic key should be changed before the end of the cryptoperiod [12].

In addition, HDL coding rules are detailed in NUREG 7006[13]. These rules are aimed at ensuring that the hardware design practices and HDL code result in a system that is reliable, robust, traceable, and maintainable. Further requirements, specific to the design process are detailed in IEC 62566 [14]. This is currently the only standard that deals with FPGA-based safety I&C systems. However, this standard is not endorsed by the U.S NRC, and therefore serves only as a reference.

Referring to RG 5.71 Figure 16-5 and Figure 16-6, the developed FPGA based encryption module guarantees that only one-way data flow is allowed from safety network to the non-safety network and block the data flow from the non-safety network to the safety network, so it enforces the security level and works as a unidirectional diode. System and Information Integrity and confidentiality control are achieved by the encryption ability of FPGA based encryption module to perform a high secure encryption algorithm such as AES, in addition, key management requirements are strongly guaranteed by using FPGA, It is a safe and flexible mechanism to distribute the key securely and to manage cryptoperiod.

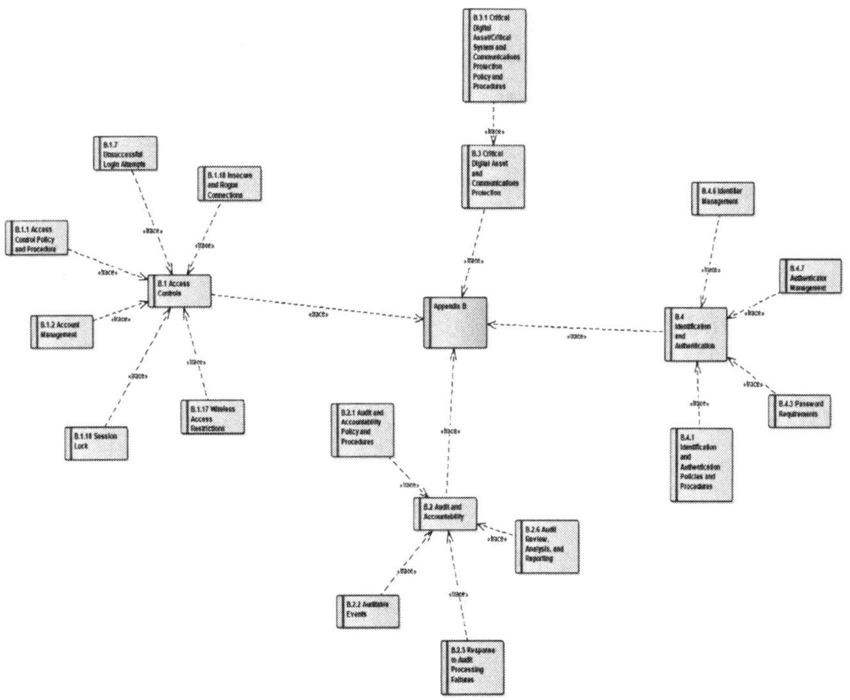


Figure 16-5: Regulatory requirements to cybersecurity (RG-5.71 to SRP 13.6.6)

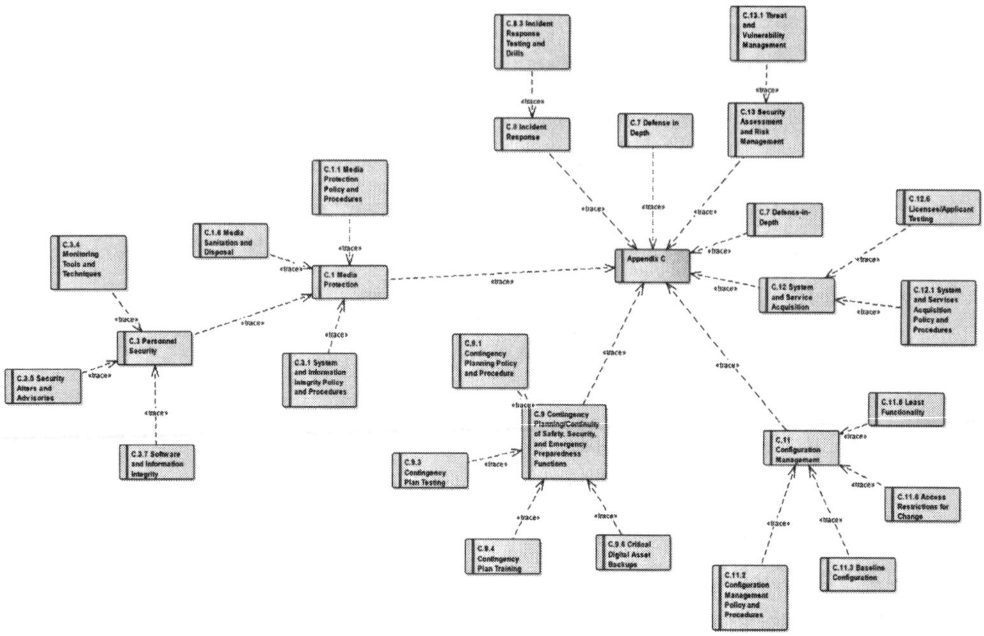

Figure 16-6: Regulatory requirements to cybersecurity (RG-5.71 to SRP 13.6.6)

## (1) Operational Requirements

Operational requirements identify measures of effectiveness for the stated mission. They are summarized as:

- Defining cybersecurity requirements for safety control systems.
- Developing a cybersecurity plan
- Analyze Risk/ Hazard management
- Execute Re-engineering for FPGA-based Gateway server

The activities carried out in order to accomplish this are:

- Identifying general design criteria for using computers in safety systems
- Developing cybersecurity plan through reverse engineering process
- Analyzing DCS Gateway server function
- Developing alternative scenarios for critical data security
- Designing FPGA-based Gateway ensuring unidirectional data transmission and encryption

### (2) Mission Needs Statement

The operational requirements and activities are condensed to arrive at a generalized needs statement for the DCS Gateway server as:

- The Gateway should provide invulnerable unidirectional data transmission.
- High reliability, availability and maintainability should be demonstrated
- Man-in-the-middle (MITM) avoidance should be demonstrated

### (3) Stakeholder Requirements

The stakeholder requirements are defined in Figure 16-7

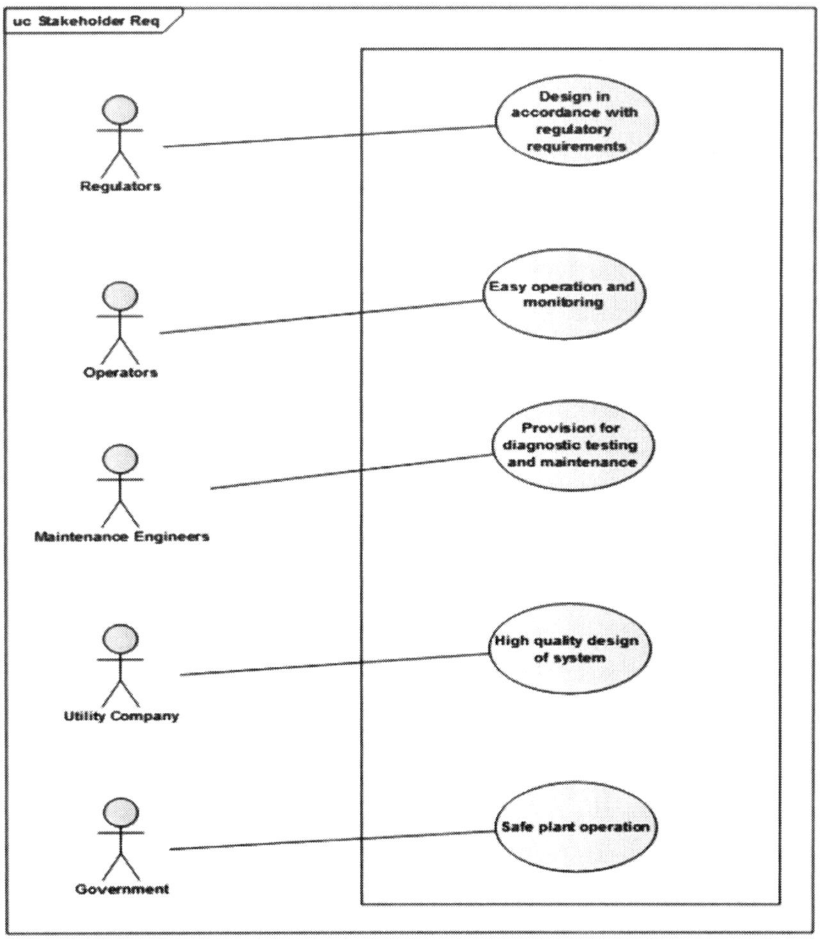


Figure 16-7: Stakeholder Requirements

## 16.3 Design

NUREG-7117 provides "technical criteria concerning the features contributing to secure network designs at nuclear power plants. Regulatory guidance concerning the design and review of digital systems is provided in a variety of other resources" [44], including RG- 5.71, RG-1.152, DI&C-ISG-04, and Chapter 7 of the SRP (NUREG-0800). Relevant standard and guidelines have been studied and reviewed to address the performance requirements of cybersecurity plan [See Appendix C].

Additionally, the current DCS gateway server function was analyzed and the functional requirements identifies through the reverse engineering process. Figure 16-8 shows the data transmission from safety to non-safety networks

The design is aimed at developing an FPGA-based Gateway to guarantee secured unidirectional data transmission from safety control system to non-safety data network that is secured against cyber-attacks. The gateway server operates as a network device to control the flow of data into or out of networks. It can serve multiple purposes such as the conversion of protocols and the routing of data between different networks. A gateway can provide protocol conversions between ports on different networks and can perform a routing function that directs packet flows from one port to another. It can also perform a proxy service, which facilitates the creation of independent connections between devices on separate "isolated" networks (e.g., safety and non-safety networks) [44].

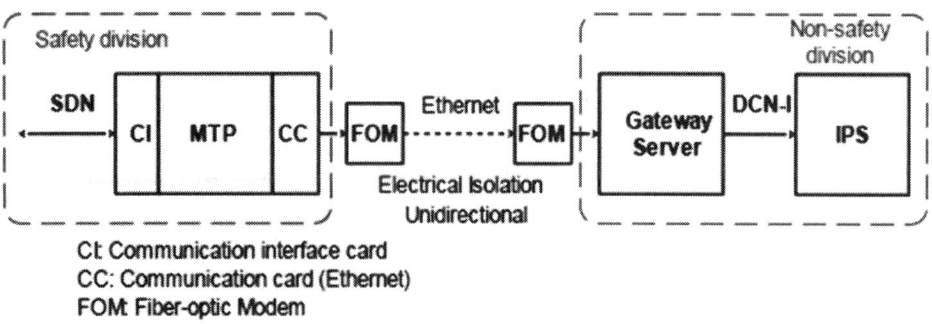


Figure 16-8: Data transmission from SDN to DCN-I

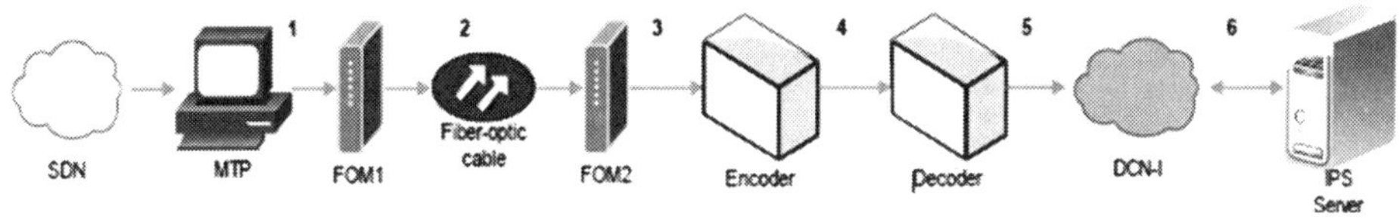


Figure 16-9: FPGA-based Gateway Design

Figure 16-9 shows the control of a unidirectional data transmission from safety to non-safety networks via an FPGA-based Gateway as follows:

1. Safety control systems (i.e., PPS, ESF-CCS, and CPCS) send control signals to the shared MTP display via the AF100 Fieldbus SDN safety network. The MTP sends the control parameters data to the FOM via Ethernet communication card. UDP data transmission may be utilized because it is connectionless, no handshaking or acknowledgement is required like that at TCP, so that UDP protocol is practical to transmit unidirectional data from safety network to non-safety network.
2. FOM is to physically separate and electrically isolate the safety and non-safety networks for cybersecurity protection. FOM is to convert UTP to Fiber-optic (FO) and vice versa. FOM1 is a UTP/FO converter while FOM2 is an FO/UTP converter. FOM is based on data security diode to ensure the
3. The FPGA-based Gateway server operates as network encryption device to protect the unidirectional data from Man-in-the-middle (MITM) attack. In cryptography and cybersecurity, an MITM attack is an attack where the attacker secretly relays and possibly alters the communication between two parties who believe they are directly communicating with each other. Stuxnet malware is considered as the most hazardous MITM attack in the nuclear industry.
4. FPGA-based Gateway utilizes two FPGA boards, one developed as an Encoder (for data encryption), the second as Decoder (for data decryption). Encoder is to guarantee that it only receives data from the safety side, encrypt the received data, and forward the encrypted data to the Decoder, which in turn decrypt the data and forward it to the DCN-I network. The Decoder is to guarantee that it transmits only and does not allow communication in the reverse direction (i.e., from non-safety back to safety). The purpose of using two FPGA boards as data encryption device is to avoid MITM attack.
5. DCN-I is comprised of Ethernet LAN switches that support Layer 2, data link, data frame communications via fiber-optic or copper (UTP) cables, DCN-I supports TCP/IP protocol data transfer where bi-directional data transmission is existing. DCN-I receives the UDP packets from safety side and forward them to the IPS.

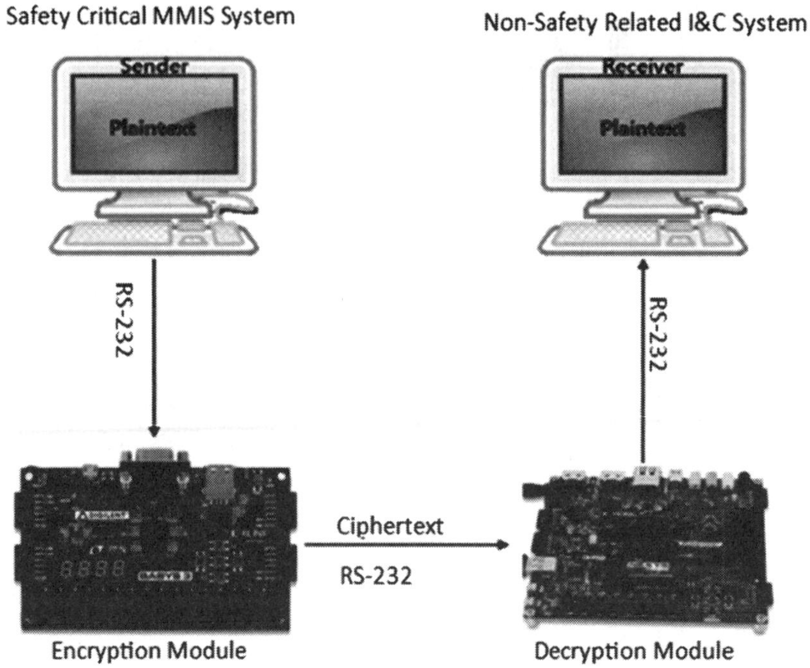


Figure 16-10: FPGA based Encryption and Decryption modules

6. The IPS is to process the received data from safety side, forward it to the monitoring, alarm, and display systems at the MCR.

To accomplish this re-engineering process, a BASYS 3 FPGA board was selected to implement the Encryption module and a Nexys Video board for the decryption module, as illustrated in Figure 16-10.

### 16.3.1 Advanced Encryption Standard 128

The algorithm chosen for the data encryption is the Advanced Encryption Standard. The AES algorithm is the most important, secure, and widely used algorithm in cryptography, the algorithm usage period is up to 2030 [45]. It is a block cipher process where all 128 bits of the data block (plaintext) is encrypted in parallel using an encryption key. There are three encryption key lengths used for AES algorithm and the key length determines the number of rounds needed for the encryption process, for key lengths (bits) =128, 192, 256 the required rounds =10, 12, 14 respectively. The key length that was chosen for this design is 128 bits.

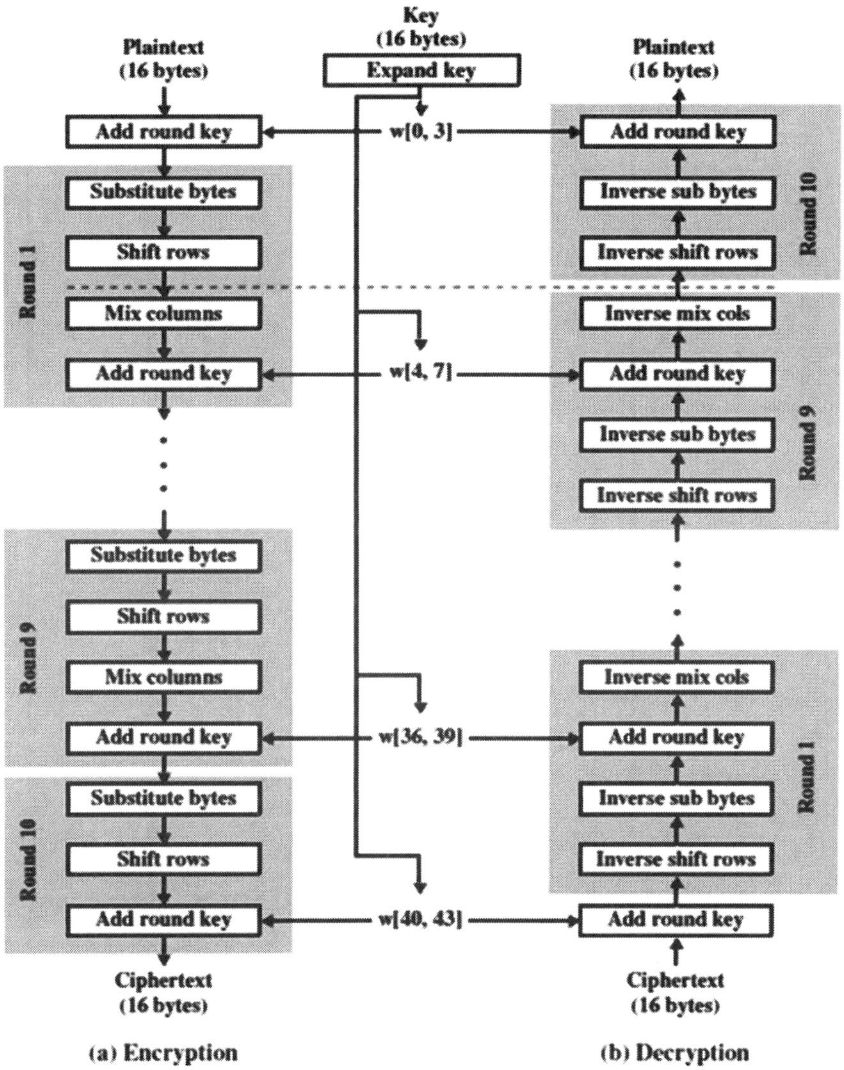


Figure 16-11: AES 128 Encryption and Decryption Process

The encryption and decryption processed are shown in Figure 16-11.

Due to the capability of parallel processing in the FPGA, the hardware based encryption and decryption modules are expected to perform faster than software-based modules. In order to compare the speed of the FPGA based system to the software based system. EFFBDs for both systems were developed and simulated as shown in the following figures.

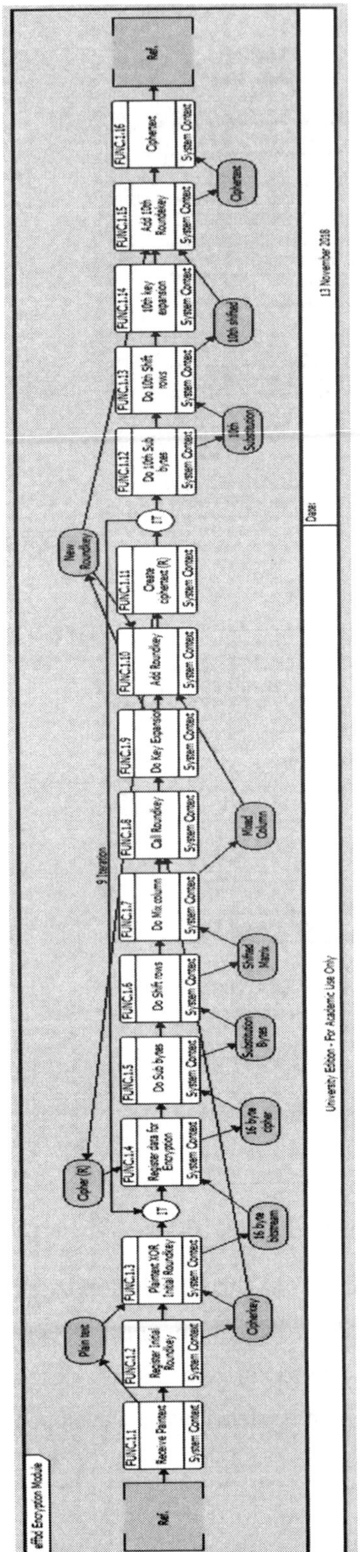

Figure 16-12: EFFBD of Software based Encryption

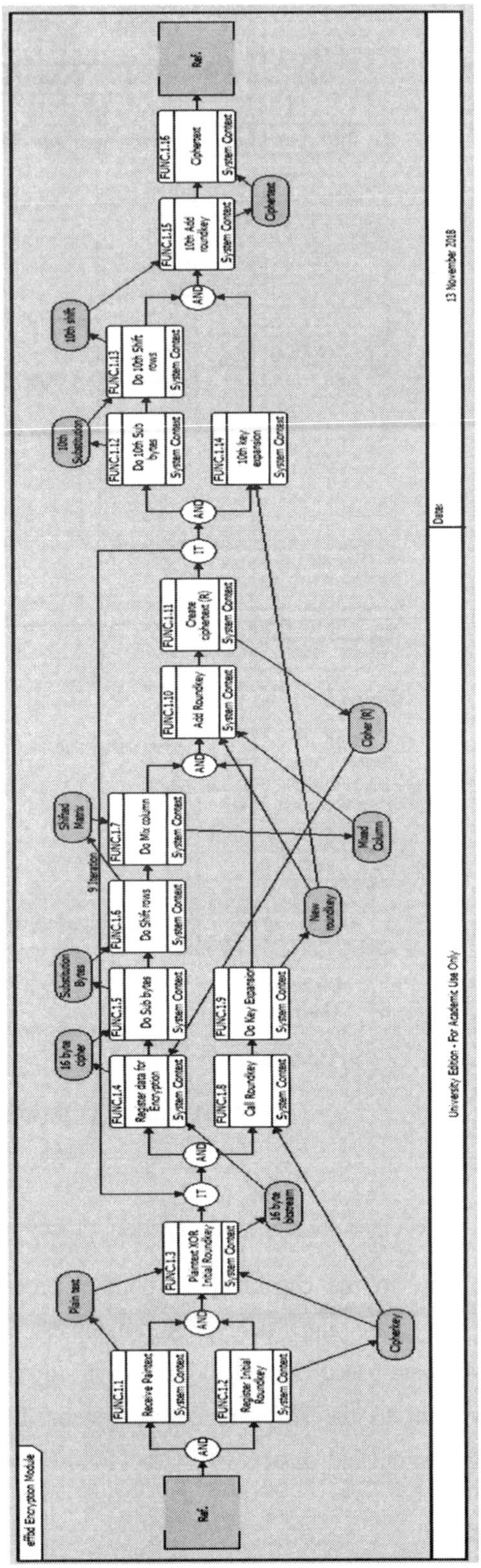

Figure 16-13: EFFBD of FPGA based Encryption

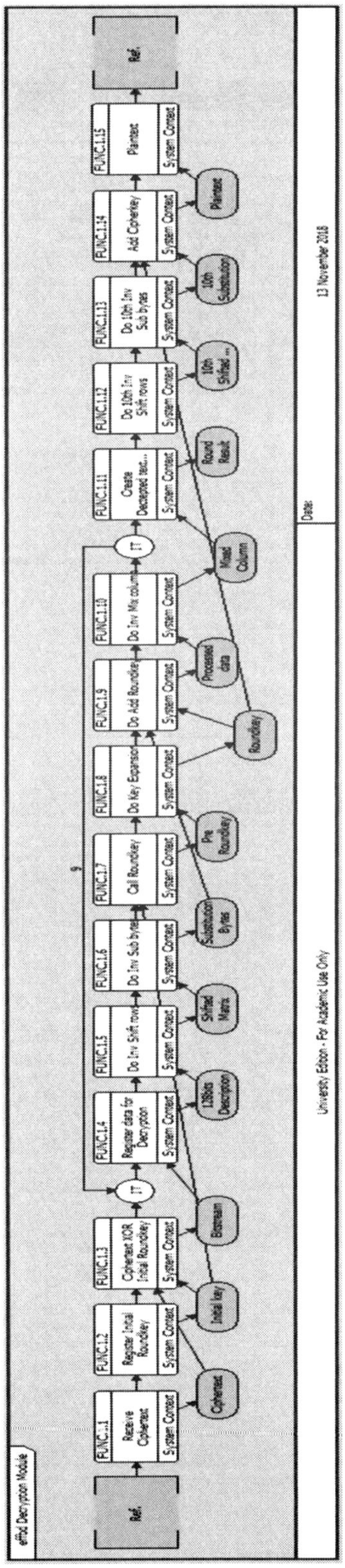

Figure 16-14: Simulation of EFFBD of Software based Decryption

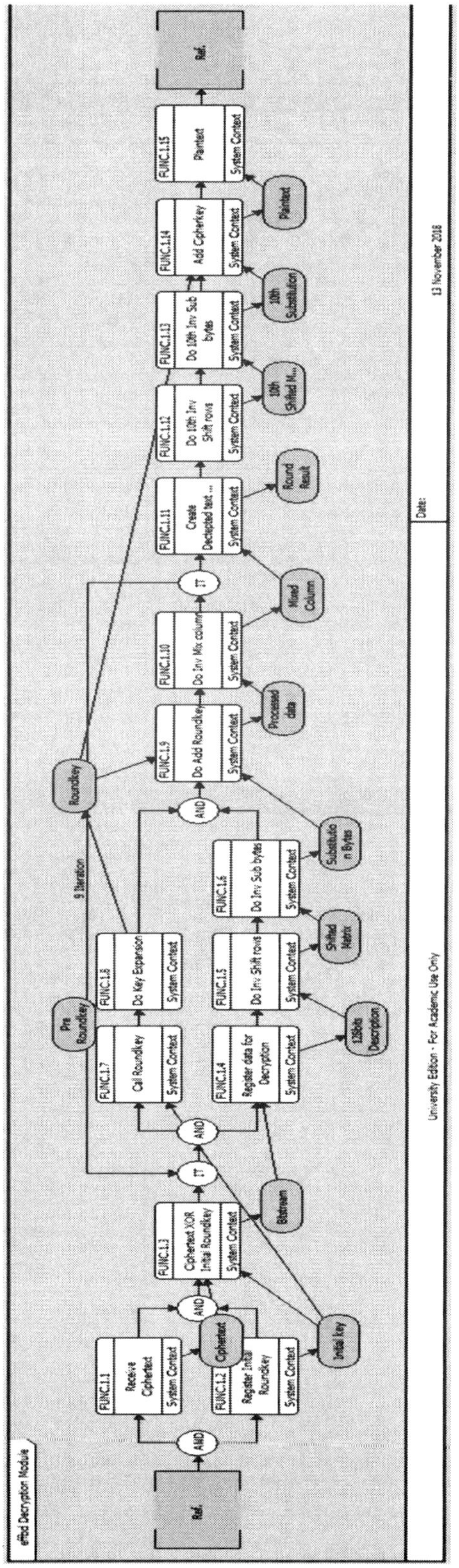

Figure 16-15: Simulation of EFFBD of FPGA based Decryption

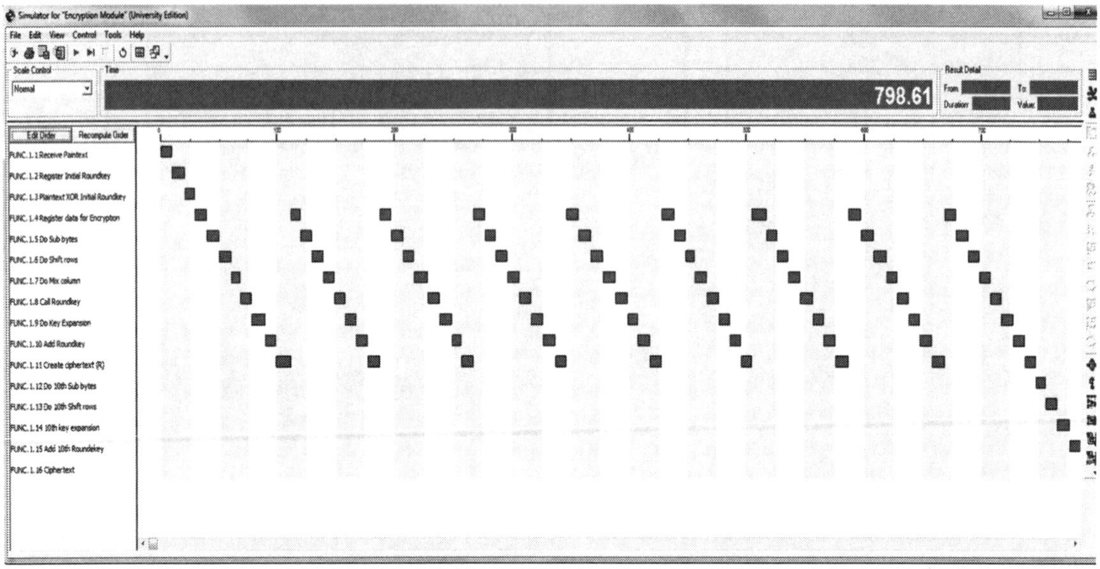


Figure 16-16: Simulation of software based Encryption

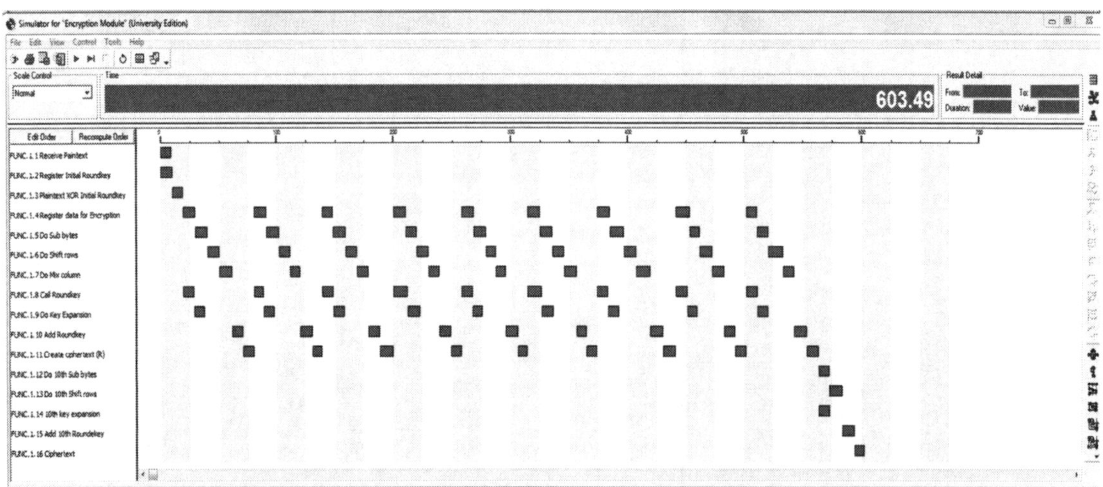


Figure 16-17 Simulation of FPGA based Encryption

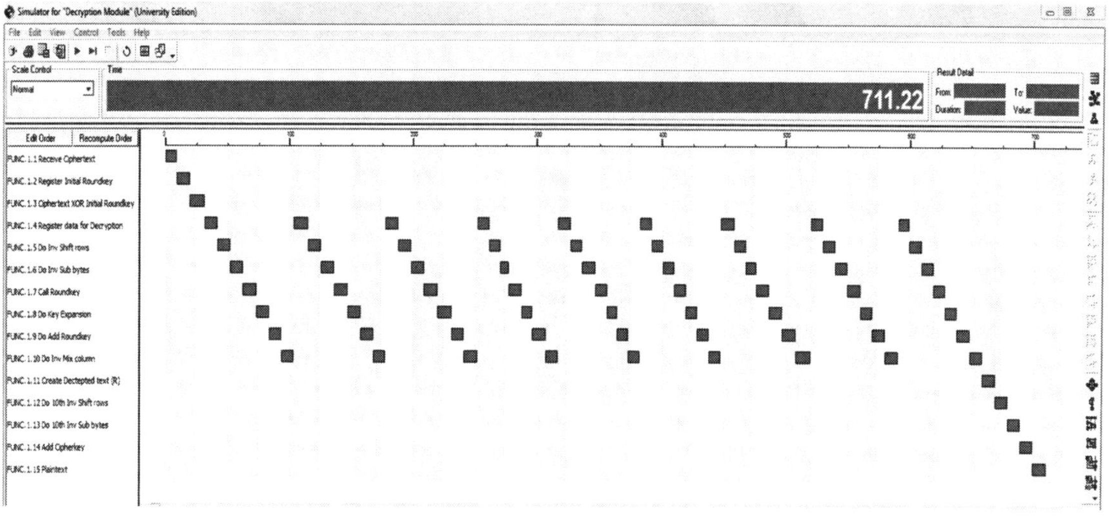


Figure 16-18: Simulation of software based Encryption

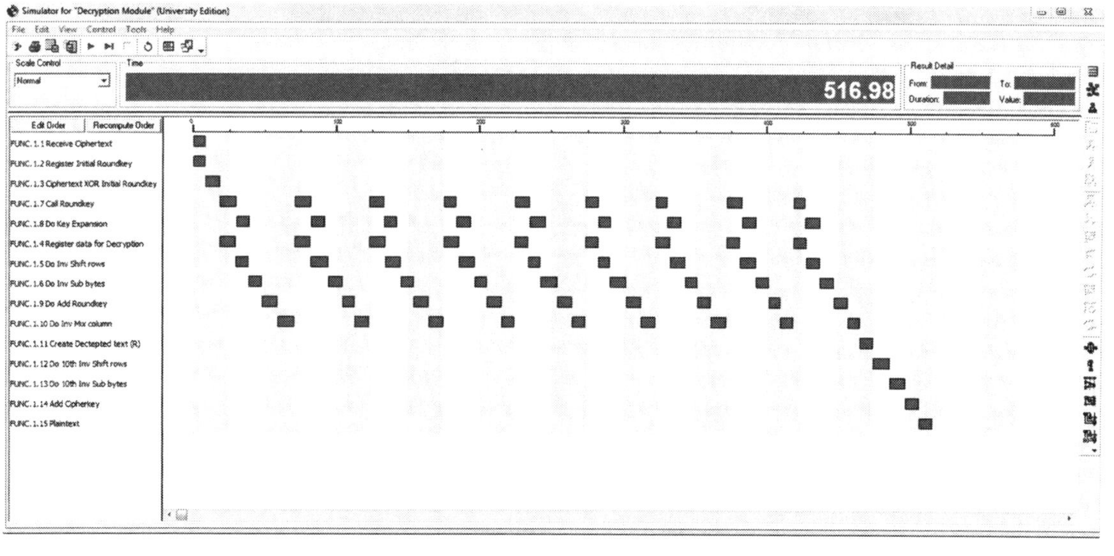


Figure 16-19: Simulation of FPGA based Decryption

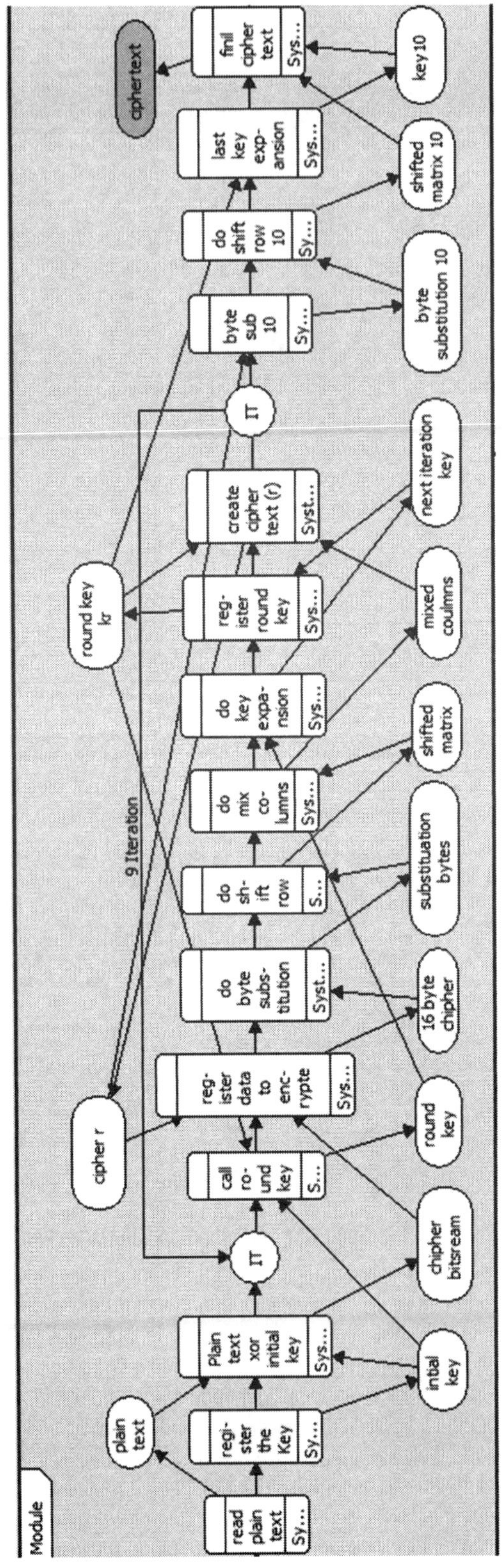


Figure 16-20: EFFBD for the Software based Encryption Module

PART 3

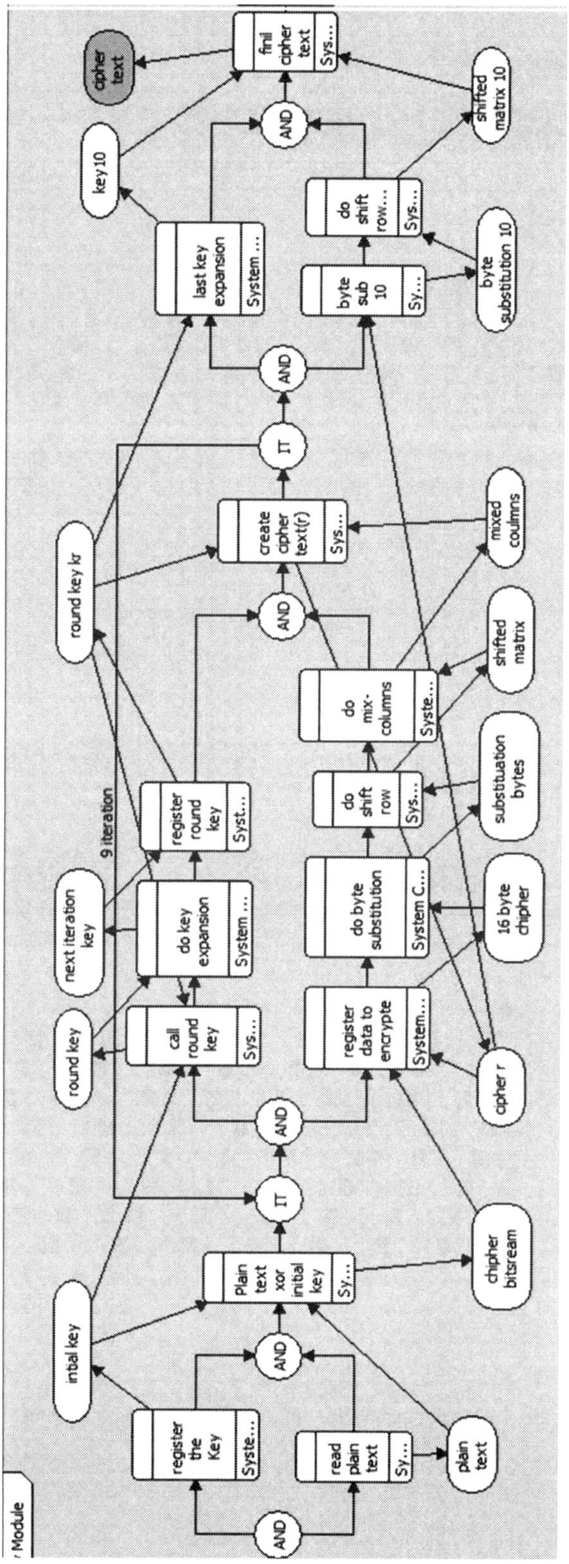


Figure 16-21: EFFBD for the FPGA based Encryption Module

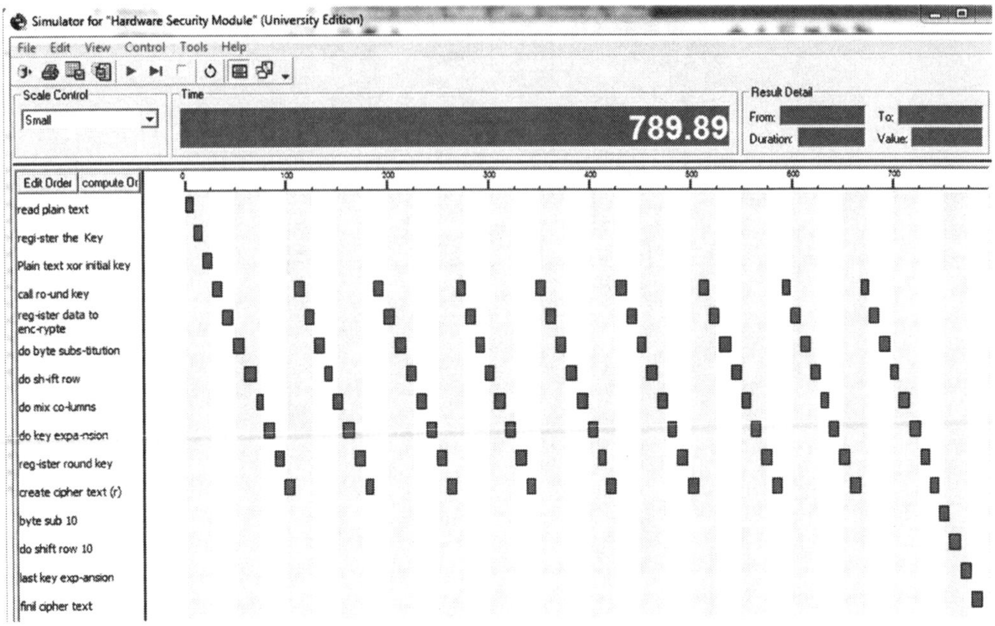


Figure 16-22: Simulation result for the Software Based System

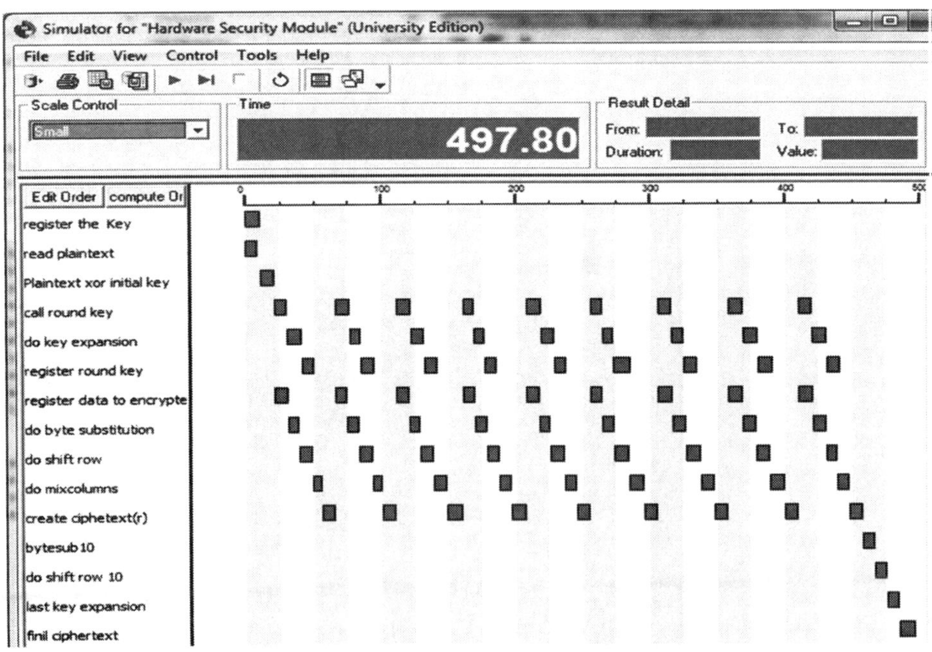


Figure 16-23: Simulation Result for the FPGA-based System

As shown in Figure 16-16 to Figure 16-19, Figure 16-21 and Figure 16-23, a dimensionless and non-real-time simulation is run by using CORE 9, which yields a dimensionless time unit, allowing for comparison between the FPGA based AES-128 system and the software-based AES-128 system. These figures show that the FPGA-based system is faster than the software-based system. Comparing the two simulation results shows that FPGA can reduce the encryption execution time by 37%. Similarly, for the decryption module, the execution time is 24.43% faster in the FPGA based system.

## 16.4 Testing

### 16.4.1 HDL Verification

In order to verify the developed code, test benches are used for both the encryption and decryption modules. The HDL was developed in Vivado, which has in-built capability to simulate the HDL and associated netlists using test benches. In the test benches, the input data and expected results are defined. On running the test benches, the result from the HDL code is compared to the expected result. Figure 16-24 below shows the log generated while running the test bench. Any discrepancy between the actual result and the expected result would be flagged as unsuccessful.

```
Tcl Console | Messages | Log ×
Vivado Simulator 2018.2
Time resolution is 1 ps
Note: ==== Second test decrypting data only (Key expansion included) ====
Time: 1500 ns  Iteration: 0  Process: /aes_cipher_block_128_TB/testbench_clrl  File: C:/Users/kings/Desktop/ENC_DEC/DECF
Note: Successful decoding from the vector 0 of the ciphertext table
Time: 4230 ns  Iteration: 1  Process: /aes_cipher_block_128_TB/result_validation  File: C:/Users/kings/Desktop/ENC_DEC/I
Note: Successful decoding from the vector 1 of the ciphertext table
Time: 4250 ns  Iteration: 1  Process: /aes_cipher_block_128_TB/result_validation  File: C:/Users/kings/Desktop/ENC_DEC/I
Note: Successful decoding from the vector 2 of the ciphertext table
Time: 4270 ns  Iteration: 1  Process: /aes_cipher_block_128_TB/result_validation  File: C:/Users/kings/Desktop/ENC_DEC/I
Note: Successful decoding from the vector 3 of the ciphertext table
Time: 4290 ns  Iteration: 1  Process: /aes_cipher_block_128_TB/result_validation  File: C:/Users/kings/Desktop/ENC_DEC/I
Note: Successful decoding from the vector 4 of the ciphertext table
Time: 4310 ns  Iteration: 1  Process: /aes_cipher_block_128_TB/result_validation  File: C:/Users/kings/Desktop/ENC_DEC/I
Note: Successful decoding from the vector 5 of the ciphertext table
Time: 4330 ns  Iteration: 1  Process: /aes_cipher_block_128_TB/result_validation  File: C:/Users/kings/Desktop/ENC_DEC/I
Note: Successful decoding from the vector 6 of the ciphertext table
Time: 4350 ns  Iteration: 1  Process: /aes_cipher_block_128_TB/result_validation  File: C:/Users/kings/Desktop/ENC_DEC/I
Note: Successful decoding from the vector 7 of the ciphertext table
Time: 4370 ns  Iteration: 1  Process: /aes_cipher_block_128_TB/result_validation  File: C:/Users/kings/Desktop/ENC_DEC/I
Note: Successful decoding from the vector 8 of the ciphertext table
```

Figure 16-24: Test bench Log

### 16.4.2 System Verification

Once the function is verified using the test benches, the HDL is synthesized and implemented on the FPGA boards. In order to enable the hardware to be tested an interface was developed such that the user can enter different sets of data and the encryption and decryption results viewed. This interface was developed to enable the testing and validation process. A simple graphical interface was developed using Java as illustrated in Figure 16-25 and Figure 16-26.

Figure 16-25: User Interface for Encryption Module

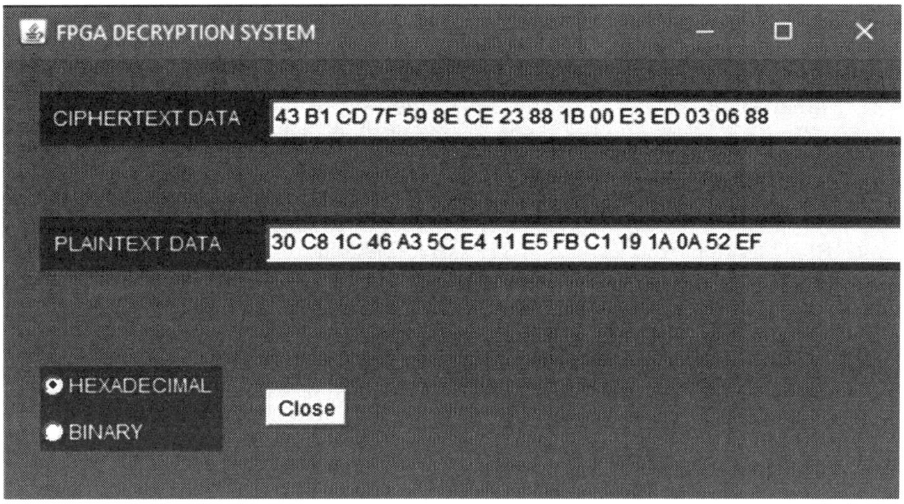


Figure 16-26: User Interface for decryption Module

The data is validated and transmitted to the FPGA board using the RS232, Serial Communication protocol. The settings used were:

- Baud Rate: 19200
- Data Bits: 8
- Stop Bit: 1

The same settings were used on both the Encryption and Decryption FPGA boards. This method of communication was chosen due to its relative simple implementation. Table 16-1 illustrates the plain text input into the encryption module, the ciphertext output from the encryption module and input to the decryption module, and the plaintext that is output by the decryption module. These data sets are recommended by NIST for the verification of encryption and decryption systems) were input to the system and the results observed and is listed in Table 16-1. From Vivado simulation, the performance of the developed module gives the same results.

Table 16-1: System Test Results

| Plain Text: Initial Input | Encryption Module Output | Decryption Module Output |
|---|---|---|
| 0000000000000000<br>0000000000000000 | 7DF76B0C1AB899B<br>33E42F047B91B546F | 0000000000000000<br>0000000000000000 |
| 6BC1BEE22E409F96<br>E93D7E117393172A | 3AD77BB40D7A366<br>0A89ECAF32466EF97 | 6BC1BEE22E409F96<br>E93D7E117393172A |
| AE2D8A571E03AC9C<br>9EB76FAC45AF8E51 | F5D3D58503B9699D<br>E785895A96FDBAAF | AE2D8A571E03AC99<br>EB76FAC45AF8E51 |
| 30C81C46A35CE411<br>E5FBC1191A0A52EF | 43B1CD7F598ECE23<br>881B00E3ED030688 | 30C81C46A35CE411<br>E5FBC1191A0A52EF |
| F69F2445DF4F9B17<br>AD2B417BE66C3710 | 7B0C785E27E8AD3F<br>8223207104725DD4 | F69F2445DF4F9B17<br>AD2B417BE66C3710 |

## PROBLEMS

1. Why is Cybersecurity in the NPP important?
2. Discuss the famous Stuxnet virus. What cyber-security measures could have been put in place to avoid it?
3. What are the advantages of a hardware based cybersecurity system?
4. Which algorithms other than AES 128 are in use?
5. -Figure 164 gives a simple Use-case model of the stakeholder requirements. Can you think of any additional requirements?
6. In this chapter, Enhanced Functional Flow Block Diagrams were used for system modeling and analysis. What other modeling method do you think is applicable?

## FURTHER READING


- USNRC, "Cyber Security in Digital Instrumentation," October 2016. [Online]. Available: http://www.nrc.gov/about-nrc/regulatory/research/digital/key-issues/cyber-security.html.
- MITRE, "Common Vulnerabilities and Exposures (CVE) database," [Online]. Available: https://cve.mitre.org/.
- NIST, "National Vulnerability Database (NVD)," [Online]. Available: https://nvd.nist.gov/.
- NIST, "Recommendation for Key Management", NIST Special Publication 800-57 Part 1, Revision 3, 2012

Chapter 17

# HUMAN SYSTEM INTERFACE (HSI)


SUMMARY

Human Safety Interfaces (HSI) is comprised of compact workstation–based operator consoles that use soft control and a distributed digital control system [46]. HSI provides an environment that displays plant status information as well as allows the operator to carry out control operations while reducing the probability of human error. In this chapter the main concern with respect to human factors in the Nuclear Power Plant are introduced. In addition, an analysis of the requirements of the HSI in NPPs is performed. The key documents are reviewed, decomposed, and traced to the originating requirements. Further, the concept of situational awareness is introduced and a simple design for performing Situational Analysis of the Emergency Operating Procedures presented.

A mind map of this chapter is shown in Figure 17-1.

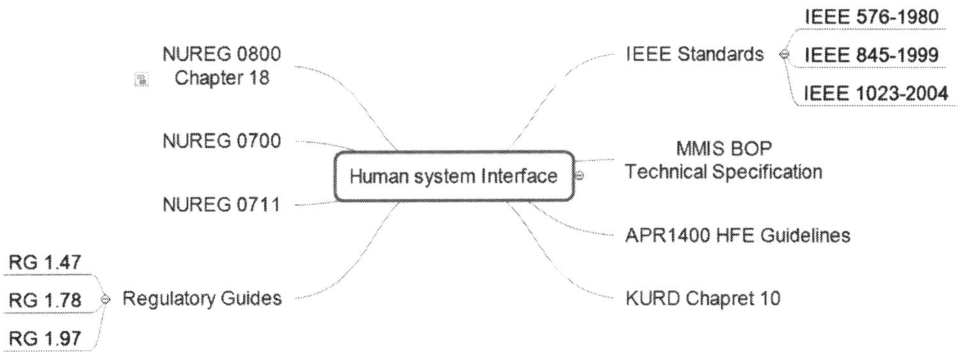


Figure 17-1: Mind Map of Chapter 17

**OBJECTIVES**

- Discuss the main human factors concerns in NPP systems
- Define situational awareness and introduce several methods for its computation

## 17.1 Overview of the HSI

The HSI is made up of three key facilities that offer support to the operating staff for the efficient and safe operation of the plant:

- Compact workstation-based operator consoles
- An LDP for displaying overall plant operation and safety assessment data
- Soft Controls for providing control to maintain the plant in safe condition and which are independent of the operator consoles.

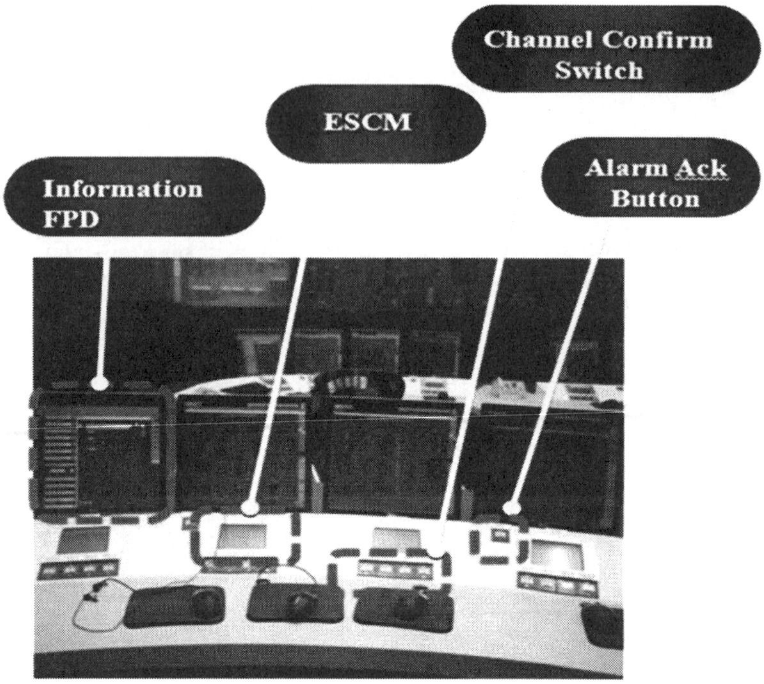


Figure 17-2: Composition of the Operator Workstation in the MCR

The operator console is designed as redundant workstations which means that either one of the consoles can access all of the information and controls necessary for the operator to monitor and control all processes associated with safety system operation. Each compact operator console as seen in Figure 17-2 comprises of four Information FPDs (IFPDs), four ESCM (4 channel confirm switches for each), an alarm acknowledgement button and pointing devices. The IFPDs display plant status information via graphical diagrams, flow diagrams, alarm displays, plant summary displays, computerized procedures, and non-safety soft controls. The LDP continuously displays plant information as status of the plant critical safety functions, the operating mode, the operating parameters trends, and indication of the high priority alarms [46].

Figure 17-3: Arrangement of Operators and HSI in a typical digital MCR

The monitoring and control resources used in the HSI and some of their major characteristics are given in the table below:

| Monitoring and Control Resources | Characteristics |
|---|---|
| Large Display Panel | a) The LDP is legible not only from the operator consoles but also from the expected locations of observers or support personnel in the MCR.<br>b) Selected parameters and component status that represent the critical safety functions and critical power production functions, and their preferred normal and emergency success paths, are provided through Spatially Dedicated Continuously Visible (SDCV) indications<br>c) System-level alarms and component-level alarms of high priority are provided<br>d) Operators can display any format that is available at information displays on the variable display area. |
| Console Information Display Hierarchy | a) The console information display is an integrated presentation of the plant process information. The operator console information display provides access to displays incorporating system and component status, process parameters, and alarm status and acknowledgement<br>b) The information display permits selectable access to any display page on the same VDU.<br>c) The console information display permits selection of component controllers or process controllers at the associated soft control display. |

Table 17-1: Overview of HSI Design and Key HSI Resources

| Monitoring and Control Resources | Characteristics |
|---|---|
| Soft Control | a) All non-safety components can be controlled from non-safety console information displays. Safety components can be selected for control from those same displays; however, actual control actions are activated from separate flat panel display units for each safety division<br>b) Soft control is used to control the system and components of the component control system, power control system, and turbine-generator control system.<br>c) Soft control provides both continuous process control and discrete component control |
| Alarms | a) An alarm list grouped by priority is provided in the operator console information display<br>b) Alarms are prioritized and presented so that operator responses can be made based on importance or urgency<br>c) The alarm system is designed to minimize the number of alarms using alarm reduction methods |
| Computer-Based Procedure (CBP) Display | a) The CBP provides an overview pane where the current operation step as well as past and future steps of the procedure are presented<br>b) The CBP provides detailed instructions of the current step<br>c) The CBP provides an integrated presentation of process information and the instructions |
| Safety Aspects of the HIS | a) Safety function monitoring<br>b) Periodic testing of protection system actuation functions<br>c) Manual initiation for protective actions<br>d) HSIs for the emergency response facilities<br>e) instrumentation required to assess the plant and environmental conditions during and following an accident |
| Minimum Inventory Control | The minimum inventory controls are conventional SDCV safety and non-safety component controls required during emergency operating procedure execution. Minimum inventory controls provide defense against operator console failure. The minimum inventory controls are selected from the results of task analysis (TA) to identify all controls necessary to perform the tasks required for EOP execution. |

| Monitoring and Control Resources | Characteristics |
|---|---|
| System-level Controls | Reactor trip and engineered safety features system-level actuation switches are provided to execute RT and ESF system-level actuation. Four channels of switches are provided at the safety console, with two-out-of four initiation for each function. These switches are provided primarily to give operators the ability to take preemptive manual actions for degrading plant conditions. The switches also provide backup initiation to accommodate failures in some portions of the automated RT and ESF initiation functions. |
| Diverse HIS | The diverse HSI is fulfilled by:<br>(1) The diverse indication system that includes soft displays with key parameters to monitor the critical safety functions, and<br>(2) Diverse manual actuation switches, which allow operators to take the necessary manual actions to mitigate accidents and maintain hot shutdown, and to actuate plant systems to control the critical safety functions. |

# 17.2 Human Factor Concerns in NPP Systems

## 17.2.1 Understanding sources of Human Errors

According to the IAEA data on outage statistics in the cumulative period between 2011 and 2015 [47], human factor related causes were the third most prevalent reason for energy generation and time loss after equipment failure and refurbishment. In 2015 alone, human factor related reasons were responsible for the approximately 10% of the total time lost due to outages. This was only second to the equipment failure at 89%. From this data, we can surmise that human factors play a significant role in plant incident and failure related shutdowns. Human factor optimization of safety systems is therefore relevant and pertinent in addressing plant safety and performance concerns.

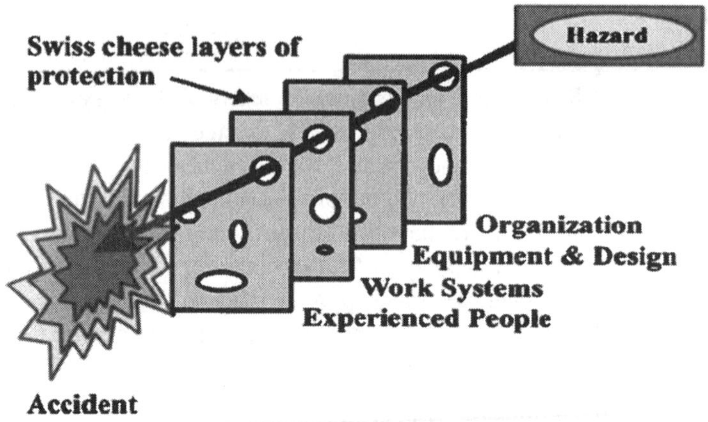


Figure 17-4: Swiss cheese analogical model for human factor related accidents

As shown in Figure 17-4, several human factors collectively provide protection from hazards, as multiple multi barriers of protection. Human errors committed in Nuclear Power Plant safety control system that can compromise these protection barriers are directly attributed to equipment design and work systems failure [48]

As seen in Figure 17-4, errors can cause unwanted conditions that interfere with expected task results. Some error precursors that have been identified by research are:

- Task demands (Equipment and design): High mental, physical or team requirements that may exceed the capabilities or challenge the limitations of the human assigned to the task. Some task demand include excessive task difficulty, hurrying and concurrent actions.
- Work Environment: This describes the errors that can arise as a result of distractions, awkward equipment, complex procedures and incomprehensible human machine interfaces.

Human Information processing, memory and attention (collectively known as cognition) define the mental process of knowing. Information that human operators interact with in the environment needs to be perceived, processed, stored, and retrieved from memory. Errors resulting from these processes lead to breakdown in awareness.

Figure 17-5 below describes information processing and storage in human memory.

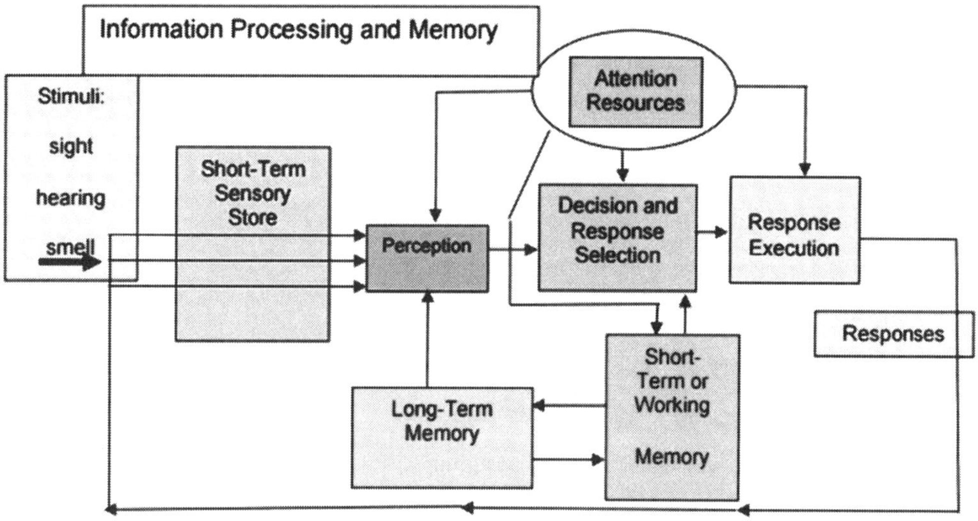


Figure 17-5: Information processing in the human mind

The Attention Resources depicted in the model describes how the mind distributes resources (attention) while performing one or more tasks. How much attention is required to satisfactorily perform a given task is defined as the individual mental workload because some tasks require more attention than others do. Knowledge, skill, and experience can decrease the demand for attention. Insufficient allocation of attentional demand is responsible for human errors.

Table 17-2 shows the most common error precursors (in order of impact) compiled from a study of INPO's (Institute of Nuclear Power Operations) event database and human performance, ergonomics and human factor data sources [49].

Table 17-2: Human Error Causes

| Task Demands (Placed by the Equipment and the Design) | Work Environment |
|---|---|
| 1. Time pressure (In a hurry)<br>2. High workload<br>3. Simultaneous, multiple actions<br>4. Repetition/ Monotony<br>5. Irreversible actions<br>6. Interpretation requirements | 1. Distractions/ Interruptions<br>2. Change/ departure from Routines<br>3. Confusing displays and controls<br>4. Work-around in instrumentation<br>5. Hidden system responses<br>6. Unexpected equipment conditions<br>7. Lack of alternative indication |

Human interaction with control system hardware and software involves varying levels of monitoring, detection, control, diagnosis, and data entry tasks, which are all potentially susceptible to human failure. Some ways in which errors may occur in these systems include:

- Incorrect reading of information
- Incorrect response to system prompts and alarms
- Incorrect diagnosis of the situation
- Wrong data entry or omission of such data when required
- Incorrect issuance of commands or instructions
- Failure to issue commands when required

In order to mitigate these errors, it is pertinent that display HMIs for control systems:

- Have a sufficient number and quality of displays
- Ensure rapidity and ease of access to information
- Make changes to system parameters easy to notice
- Enable confirmation of normal process operation
- Display information at a pace that is adapted to the operator
- Make it easy for operator to find information requested and identify faults to avoid alarm flooding.

### 17.2.2 Emergency Operating Procedures in the MCR

USNRC RG 1.33 [50] Quality Assurance program requirements (Operations) identifies EOPs as follows: "EOPs are plant procedures that direct operators' actions necessary to mitigate the consequences of transients and accidents that have caused plant parameters to exceed reactor protection system set points or engineered safety feature set point, or other established limits."

Plant Procedures describe the operating and emergency procedures that the plant staff uses to provide reasonable assurance that routine operating, off-normal, and emergency activities are conducted in a safe manner. The Emergency Operating Procedures must be classified and the group responsible for performing and maintaining the procedures, identified. The requirement from R.G 1.33 is only to provide a reasonable assurance that the procedures help the staff to accomplish the stated functions.

The IAEA provides the functional requirements for EOP [51] as;

- The objective of EOPs is to return NPPs to a condition covered by normal procedures or a safe and stable shutdown condition.
- Expected emergency conditions should be identified and EOPs for dealing them should be prepared for use when required.
- Since emergencies may not follow anticipated patterns, EOPs should provide for sufficient flexibility of actions to accommodate variations, including multiple and sequential failures."

Emergency Operating Procedures are supposed to provide measures to control the power plant by considering events (event-based approach) or considering the symptoms (Symptom oriented approach). The event-based approach relies on the analytical method of categorizing emergency events based on interpretation of models, simulation of thermohydraulic codes and/or investigation of historical data. Design Basis Accident are an example of an event that belongs to this category because an unwanted effect of such an accident can be prevented or minimized by the application of a set of well-known event-based recovery actions. This method is useful to cover diagnosable events [52].

As seen in Figure 17-6, the symptom based approach applies where event based procedures cannot be applied meaningfully because of complications that may be associated and which may make it impossible to pinpoint the exact nature of an accident. Examples of this include multiple events occurring simultaneously or instrumentation failures that could distort or hide the nature of the emergency event. Development of a symptoms based procedure systems is challenging because of the difficulty in developing the theoretical basis for identifying a set of symptoms.

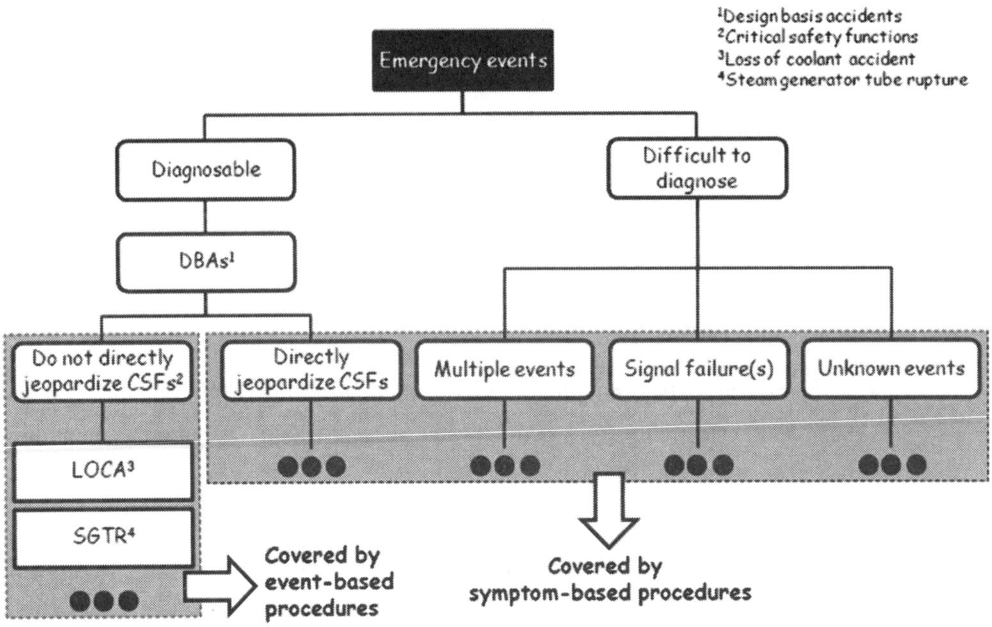


Figure 17-6: Structure of EOPs [52]

The (Critical Safety Functions) CSFs that are described and controlled by the event-based EOPs, are prioritized according to the defense in depth strategy for the provision of the physical barriers needed to ultimately prevent release of radioactive materials into the atmosphere. An optimal set of emergency is developed from these CSF. The prioritization of Reactivity Control functions is considered as seen in Figure 17-7. Contingency actions are provided if the pre-planned activities are not working. EOPs are usually presented in two-column format with operators expected to carry out procedures on the left column if the expected responses from the system are observed. If the expected responses are not satisfied, the operators are expected to move to the right column and perform the contingency action [53].

When the emergency action occurs the operating team in the MCR of NPP is responsible for carrying out the EOP. The team consists of four qualified operators: (1) a senior reactor operator (SRO), (2) a reactor operator (RO), (3) a turbine operator (TO), and (4) an electrical operator (EO).SRO has the oversight responsibility of the performance of emergency tasks. If we consider an example where the SRO has to perform a verification task. In order to perform this action, the SRO needs to know the current operating component process value. At this moment, the SRO tells the either TO or RO to read the current required process value. When the TO/ RO gives the SRO the desired information after reading the appropriate indicator, the SRO ultimately decides whether the process value corresponds to the specified task control pre-requisite requirement [54].

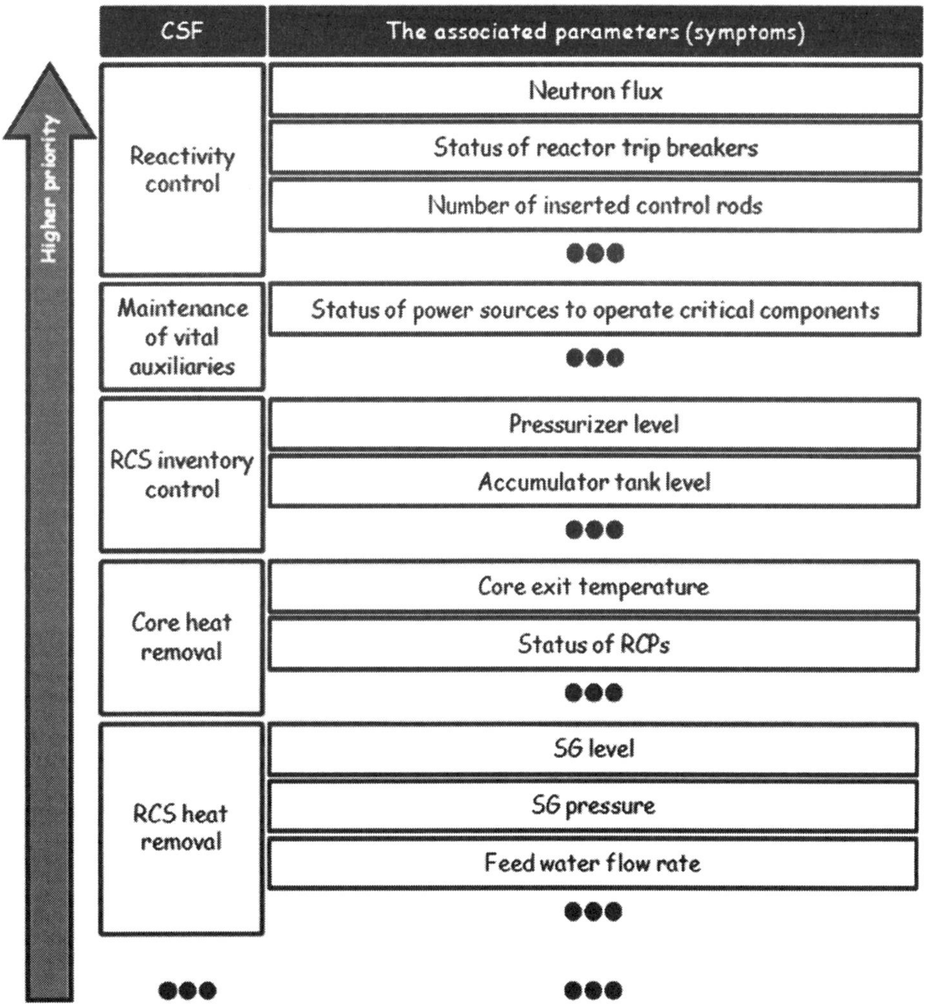


Figure 17-7: Prioritization of safety critical functions [53]

### Computer Based Procedures (CBP)

CBP system is a type of system that uses dynamic process data to evaluate conditions, check procedure logic, and display the plant condition in order to monitor the plant during procedurally defined intervals [55].

The CPS (Computerized Procedure System) is an online type of CBP implemented in KHNP APR1400 as a computerized operator support system, which covers all operating procedures and provides information through workstations using flowcharts and tree structures with the capability of checking steps logically. Passive means that the operator cannot override the computerized monitoring. Online means that it differs from the EBP and PBS by accessing plant data in real time [56].

Some limitations of the CPS system identified by review have been highlighted are:

- The CPS does not have the capability to automate decision-making and manipulate the plant control according to procedures rather it offers up plant information thereby leaving decision making to the plant operator. This can exasperate the performance of the decision maker because the system is incapable of supporting him should he perform a wrong action.
- Although the current system displays not only the procedure overviews but also step by step automated processing of the procedure task as they are performed, the utilization of the small display space remains an issue. It forces the HMI to be displayed on multiple display panels in front of the operator. This coupled with the large number of continuously applied steps can lead to cognitive degradation.
- Software quality: due to the non-safety grading of CPS software system, the V&V plans applied to software quality grading are similar to those applied for important to safety category of equipment. It is recognized however that CPS is necessary in performing protection system control actions to monitor critical plant safety functions. Since the proposed optimized system would invariably not only be active but also be accessing safety plant information to perform its functions, it would follow that it would be better categorized as safety grade information system.

## 17.3 Situational Awareness

The enhancement of operator situation awareness (SA) has become a major design goal for those developing operator interfaces, automation concepts, and training programs in a wide variety of fields, including aircraft, air traffic control (ATC), power plants, and advanced manufacturing systems. SA has always been needed in order for people to perform tasks effectively. This is because with rise of the computer determined and controlled tasks in use today, emphasis has shifted away from improving performance of physical tasks to creating a new class of tools to help people perform information based perceptive and cognitive tasks better.

SA put simply is knowing what is going on around you. Tied to this knowledge is an understanding of what is important to know and what is not. In the scope of HFE design efforts, SA is concerned with the requirement of awareness for very specific reasons. For a given operator, therefore, SA is defined in terms of the goals and decision tasks for that job. The A plant operator for example does not need to know everything (e.g., the Shift supervisor's shoe size and spouse's name), but does need to know a great deal of information related to the goal of safely controlling the NPP. SA is generally described as "the perception of the elements in the environment within a volume of time and space, the comprehension of their meaning and the projection of their status in the near future" [57]

Today the level of SA that teams and operators possess is now recognized as a critical factor in system design and assessment [58]. It is therefore critical that system practitioners accurately describe and measure individual (or team) SA in these environments. The need for SA measurement is twofold: to find reliable and valid methods for testing SAs and assuring that SA is not degraded by new artifacts, systems, interfaces, or procedures. Many SA measurement approaches have been described and used for describing varying constructs related to SA.

### 17.3.1 Nasa-Task Load Index (NASA-TLX)

NASA-TLX is a multi-dimensional scale designed to obtain workload estimates from one or more operators while they are performing a task or immediately afterwards. The definitions below are as defined by NASA [59] [60]

### (1) Workload

Workload is a term that represents the cost of accomplishing mission requirements for the human operator. If people could accomplish everything they are expected to do quickly, accurately, and reliably using available resources, the concept would have little practical importance. Since they often cannot, or the human cost (e.g., fatigue, stress, illness, and accidents) of maintaining performance is unacceptably high, designers, manufacturers, managers, and operators, who are ultimately interested in system performance, need answers about operator workload at all stages of system design and operation.

### (2) Dimensions

The NASA Task Load Index (NASATLX) consists of six subscales that represent somewhat independent clusters of variables: Mental, Physical, and Temporal Demands, Frustration, Effort, and Performance. The assumption is that some combination of these dimensions are likely to represent the "workload" experienced by most people performing most tasks.

These dimensions were selected after an extensive analysis of the primary factors that do (and do not) define the subjective experience of workload for different people performing a variety of activities ranging from simple laboratory tasks to flying an aircraft. Coincidentally, these dimensions also correspond to various theories that equate workload with the magnitude of the demands imposed on the operator, physical, mental, and emotional responses to those demands or the operator's ability to meet those demands.

### (3) Weighting

A weighting scheme is introduced to consider such individual differences when computing an overall workload score (Figure 1). Essentially, overall workload represents the total areas of the six bars. The weights are derived for each participant at the beginning of a study by requiring simple decisions about which member of each paired combination of the 6 dimensions are more related to their personal definition of workload. Each subscale rating provided by that person during the study is then multiplied by the appropriate weight, developing a composite tailored to individual workload definitions. The benefit of this weighting scheme was an increase in sensitivity (to relevant variables) and a decrease in between- rater variability. The development and theoretical rationale for the scale were described in a chapter published in 1988 by Hart & Staveland.

Table 17-3: Rating Scale Definitions

| TITLE | ENDPOINTS | DESCRIPTION |
|---|---|---|
| Mental Demand | Low/High | How much mental perceptual activity was required (e.g. thinking, deciding, calculating, remembering, looking, searching, etc.)? Was the task easy or demanding, simple or complex, exacting or forgiving? |
| Physical Demand | Low/High | How much physical activity was required (e.g. pushing, pulling, turning, controlling, activating, etc.)? Was the task easy or demanding, slow or brisk, slack or strenuous, restful or laborious? |
| Temporal Demand | Low/High | How much time pressure did you feel due to the rate or pace at which the task or task elements occurred? Was the pace slow and leisurely or rapid and frantic? |
| Effort | Low/High | How hard did you have to work (mentally and physically) to accomplish your level of performance? |
| Performance | Good/Poor | How successful do you think you were in accomplishing the goals of the task set by the experimenter (or yourself)? How satisfied were you with your performance in accomplishing these goals? |
| Frustration Level | Low/High | How insecure, discouraged, irritated, stressed and annoyed versus secure, gratified, content, relaxed and complacent did you feel during the task? |

### (4) Description

The degree to which each of the six factors contribute to the workload of the specific task to be evaluated from the raters' perspectives is determined by their responses to pair-wise comparisons among the six factors. Magnitude ratings on each subscale are obtained after each performance of a task or task segment. Ratings of factors deemed most important in creating the workload of the task are given more weight in computing the overall workload score thereby enhancing the sensitivity of the scale.

The Task Load Index has been tested in a variety of experimental tasks that range from simulated flight to supervisory control simulations and laboratory tasks (e.g. the Sternberg memory task, choice reaction time, critical instability tracking, compensatory tracking, mental arithmetic, mental rotation, target acquisition, grammatical reasoning, etc.). The results of the first validation study are summarized in Hart and Steveland (1988). The derived workload scores have been found to have substantially less between-rater variability than unidimensional workload ratings and the subscales provide diagnostic information about the sources of load.

### (5) Overview of the program

The NASA Task Load Index workload evaluation procedure is a two-part procedure requiring the collection of both weights and ratings from the subject and the manipulation of the collected data to provide weighted subscale ratings and an Overall Workload score.

**Advantages:**

- NASA TLX is widely accepted in the research community.
- Quick and easy method of estimating workload
- Flexible
- Well-established
- Easy to apply, software available
- Multi-dimensional approach

**Disadvantages:**

- Can be intrusive and disruptive to primary task performance if applied online
- Participants may have forgotten relevant workload aspects, as it is filled out after the task
- Response bias (halo effect, primacy-recency effect)
- Workload ratings may be correlated with task performance e.g. subjects who performed poorly on the primary task may rate their workload as very high and vice versa
- Sub scale weighting which is laborious and time consuming
- Sub-scale ratings could be repetitive, i.e. the same or similar rating applied to each sub-scale

### 17.3.2 Situation Awareness Rating Technique (SART)

SART is a post-trial subjective rating technique that was originally developed for the assessment of pilot SA. It uses ten dimensions to measure operator SA and participants are rating each dimension on a seven point rating scale.

The situation awareness rating technique (SART) (Taylor 1990) is a simplistic post-trial subjective rating technique that was originally developed for the assessment of pilot SA. SART uses the following ten dimensions to measure operator SA: Familiarity of the situation, focusing of attention, information quantity, information quality, and instability of the situation, concentration of attention, complexity of the situation, variability of the situation, arousal, and spare mental capacity. SART is administered post trial and involves participants subjectively rating each dimension on a seven point rating scale (1 = Low, 7 = High) based on their performance of the task under analysis. The ratings are then combined in order to calculate a measure of participant SA.

A quicker version of the SART approach also exists, known as the 3D SART. The 3D SART uses the 10 dimensions described above grouped into the following three dimensions:

1. Demands on attentional resources,
2. Supply of attentional resources, and
3. Understanding of the situation

#### (1) Purpose

SART is a self-rating technique, which elicits the subjective opinion on how aware a person was during task performance.

The reason that SART measures different components is that the SART developers feel that, similar to workload, SA is a complex construct; therefore, to measure SA in all its aspects, separate measurement dimensions are required. Because information processing and decision making are inextricably bound with SA (because SA involves primarily cognitive rather than physical workload),

### (2) Description of the method

SART technique that consists of a series of questions that have bipolar responses. The number of dimensions varies between different SART forms. In its original form, there are ten dimensions. It is administered post trial and involves participants subjectively rating each dimension on a seven point rating scale (1 = Low, 7 = High) based on their performance of the task under analysis. The ratings are then combined in order to calculate a measure of participant SA. The original SART uses the following ten dimensions (construct) to measure operator SA, as defined in Table 17-4.

The ten SART dimensions can also be condensed into the quicker version of the SART, known as the 3D SART. The 3D SART uses the 10 dimensions described above grouped into the following three dimensions:

(1) Demands on attentional resources (complexity, variability, and instability of the situation),
(2) Supply of attentional resources (division of attention, arousal, concentration, and spare mental capacity), and
(3) Understanding of the situation (information quantity, and information quality).

Table 17-4: SART Dimensions

| Domains | Construct | Definition |
|---|---|---|
| Attentional demand | Instability of situation | Likeiness of situation to change suddenly |
| | Variability of situation | Number of variables that require attention |
| | Complexity of situation | Degree of complication of situation |
| Attentional supply | Arousal | Degree that one is ready for activity |
| | Spare mental capacity | Amount of mental ability available for new variables |
| | Concentration | Degree that one's thoughts are brought to bear on the situation |
| | Division of attention | Amount of division of attention in the situation |
| Understanding | Information quantity | Amount of knowledge received and understood |
| | Information quality | Degree of goodness of value of knowledge communicated |
| | Familiarity | Degree of acquaintance with situation experience |

### (3) Technical requirements for using the method

None, SART is applied using pen and paper only. However, it can also be administered using Microsoft Excel, which can also be used to automate the SART score calculation process.

### (4) Measure/Response Type

SART is administered post-trial and involves the participant rating of each of the 10 dimensions on a seven point rating scale (one = Low, 7 = High, see figure below). The ratings are then combined in order to calculate a measure of participant SA.

A composite SART score is calculated using the following formula:

SA = U – (D – S),

Where:

U = summed understanding

D = summed demand

S = summed supply

The quicker 3-D SART uses a 100-point scale from zero (low) to 100 (high), for each measure (demand on attentional resources, supply of attentional resources and understanding of the situation). The overall SART score is then calculated similar to above:

SA = Understanding – (Demand – Supply)

**Advantages:**

- Quick and easy to administer, requires very little training. Also low cost.
- Generic – can be used in other domains.
- Applicability to multiple domains.
- High ecological validity.
- A certain level of diagnosticity.
- Accounts for attentional supply and demand.
- Self-ratings of SA can be obtained from different team members and so offer a potential approach into the assessment of team SA.

- SART is non-intrusive to task performance.
- SART is a widely used method and has been applied in a range of different domains.
- It provides a quantitative assessment of SA.

SITUATION AWARENESS RATING TECHNIQUE (SART; Taylor, 1990)

**Instability of Situation**
How changeable is the situation? Is the situation highly unstable and likely to change suddenly (High) or is it very stable and straightforward (Low)?

1 2 3 4 5 6 7

**Complexity of Situation**
How complicated is the situation? Is it complex with many interrelated components (High) or is it simple and straightforward (Low)?

1 2 3 4 5 6 7

**Variability of Situation**
How many variables are changing within the situation? Are there a large number of factors varying (High) or are there very few variables changing (Low)?

1 2 3 4 5 6 7

**Arousal**
How aroused are you in the situation? Are you alert and ready for activity (High) or do you have a low degree of alertness (Low)?

1 2 3 4 5 6 7

**Concentration of Attention**
How much are you concentrating on the situation? Are you concentrating on many aspects of the situation (High) or focussed on only one (Low)?

1 2 3 4 5 6 7

**Division of Attention**
How much is your attention divided in the situation? Are you concentrating on many aspects of the situation (High) or focussed on only one (Low)?

1 2 3 4 5 6 7

**Spare Mental Capacity**
How much mental capacity do you have to spare in the situation? Do you have sufficient to attend to many variables (High) or nothing to spare at all (Low)?

1 2 3 4 5 6 7

**Information Quantity**
How much information have you gained about the situation? Have you received and understood a great deal of knowledge (High) or very little (Low)?

1 2 3 4 5 6 7

**Familiarity with Situation**
How familiar are you with the situation? Do you have a great deal of relevant experience (High) or is it a new situation (Low)?

1 2 3 4 5 6 7

Figure 17-8: Situational Analysis Rating Technique [61]

**Disadvantages:**

- Recall - Problems of gathering SA data post-trial e.g. correlation with performance, i.e. participants may be prone to 'forgetting' periods of the task when they possessed a poor level of SA, and may more readily remember the periods when they possessed a superior level of SA.
- Issues regarding sensitivity of the technique - Participants' ability to rate their own SA is questionable, regarding whether or not participants can accurately rate their own awareness (e.g., how can one be aware that they are not aware?).
- Potential influence of workload - Data collection post task has some problems associated, including memory degradation and poor recall.

## 17.3.3 Augmented Virtual Reality

Virtual reality has been defined as a system or an environment created to provide a synthetic experience for its user (s) also known as illusionary or virtual because sensory stimulation to the user is simulated or generated by the system. Conventional VR systems comprise of several types of displays that will deliver the simulation, sensors to detect user action and computers that process the user action. In order to create a virtual experience, a developer will create a virtual environment to enable an interaction that can allow the user to change the virtual world, as shown in Figure 17-9.

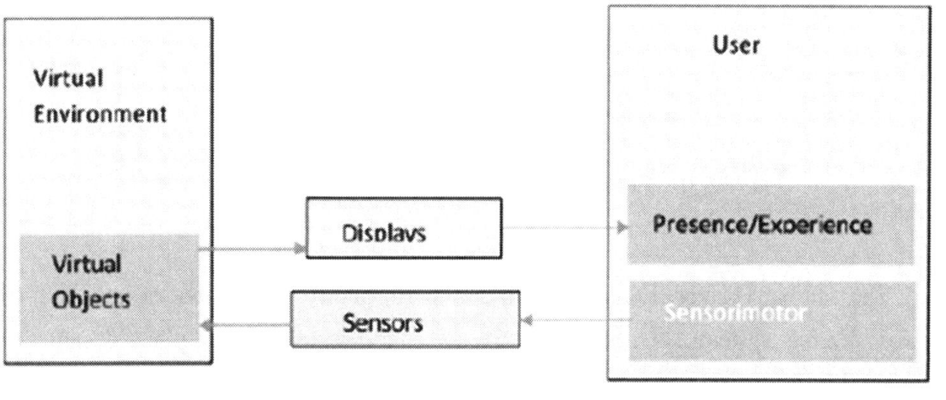


Figure 17-9: Virtual Environment interacting with user

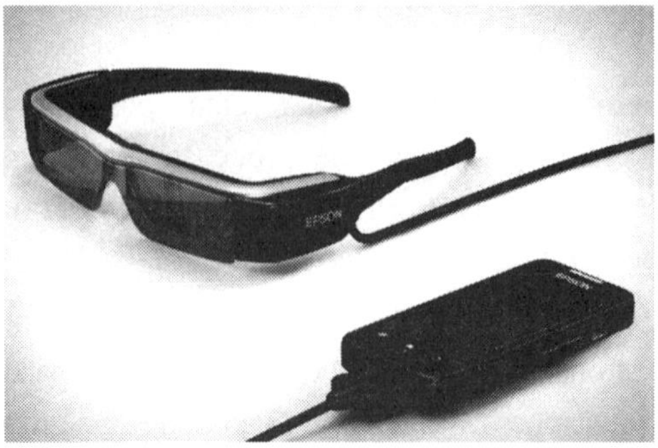

Figure 17-10: AVR interface gear

Augmented Virtual Reality (AVR) interface technology is achieving ease of availability and widespread use in many applications today as illustrated by the launch of several AR and VR devices aimed at media consumption. As such, environments that incorporate such AVR hardware(Figure 17-10) would be invaluable tools in designing a human interface system because of the high fidelity and intuitive response to natural human interaction that can be achieved [62]. It is the intention of this study to explore how such systems can be used in NPP control systems to determine whether Situational Awareness can

1. Enhance Operator(s) performance by introducing an improved cognitive method of monitoring plant information during an Emergency Operating Procedure (EOP) and
2. In correlation, inform the performance of the diverse safety systems based on human factors.

## 17.4 Design Practice

The Human Factor Engineering based performance evaluation of the ESF-CCS and diverse systems is not addressed by either regulatory or plant designers. In the APR1400 DCD documents, ESF-CCS diversity which is considered as a measure of the availability of digital I&C systems, is a performance requirement that is evaluated from an I&C point of view. On the other hand, the operating procedures in a power plant, which also provide a measure of the availability of those systems, are evaluated from a HFE point of view. However, the availability of the ESF-CCS, being a system that is critical during emergency operations and its subsystems is largely determined by human operations, behavior and awareness during abnormal plant conditions as described in NUREG 0711. It is worthwhile therefore to consider how both HFE and I&C can be used together to evaluate the performance of safety critical systems to further assure their availability.

The scope of the design presented here will be concerned with performing a human factors evaluation on a digital I&C system. The performance of the system shall be evaluated by considering the human factor engineering contribution towards its availability during an accident scenario. The objective is to propose an improvement to the system that can enhance its availability. In addition, the design seeks to improve SA through incorporation of Augmented Reality. The measures of success are defined as:

1. Measurement of the awareness of the system and confirmation of its validity
2. Improved design incorporating HFE factors that will improve the system performance.

The scope of the design will be limited to evaluating one system (the ESF-CCS) and only the first ten EOPs performed during a LOCA in order to keep the study focused. Further development of the AVR tool will be performed after the result from the SA measurement is validated.

Parameters of the study will be determined by availability of background information to develop an improved system, usability of the system that will be modelled as well as the improved method to be developed, scalability of the system, accuracy of the SA measurements, time constraints. The only measurable variable of interest is the SA data, which will vary as the performance evaluation is carried out.

### 17.4.1 Situational Awareness as a Means of Performance Evaluation

Two SA measurement methods were considered [63] :

1. The NASA Task Load Index (NASA TLX): a subjective multidimensional assessment tool that rates perceived workload on six different subscales; Mental Demand, Physical demand, Temporal demand, Performance, Effort and Frustration.
2. Situational Awareness assessment using the (3D) SART [SART; Taylor, 1989]; an experimentally validated, retrospective measure which requires participants to rate themselves on three dimensions that include attentional demands (D), attentional supply (S), and understanding (U) immediately following task performance. The ratings on each of the three dimensions are combined into a single SART value according to a formula [61] :

   Situation Awareness = U − (D − S).

### 17.4.2 Choice of EOP LOCA as Reference Point of Measurement

The EOP was chosen as the scenario for modeling because it is one of the critical plant conditions that requires human intervention. As previous studies have demonstrated, situational awareness is directly linked with the system response to an emergency. The system is expected to realize an improvement in the level of Situational awareness and mental workload in the MCR. These two factors will be used to validate the measure of performance.

The primary goal of the designed system is to leverage the existing infrastructure and system data sources to implement a method that can reliably measure the efficacy and performance level of the emergency response system using SA measurement analysis.

The human related component of performance of ESF-CCS input is considered partly because the measure must be sufficiently demonstrable in the worst-case situation and partly because human interactions with emergency carry the risk of being unpredictable and inconsistent.

### 17.4.3 Modeling the Existing System to Incorporate the SA Measurement Method Chosen

Scenario 1 was conducted as detailed by Figure 17-9 below, which is run in the APR1400 simulator training environment. The objective of which is to obtain the baseline data. During the experiment the STA will execute the EOP LOCA as detailed by the plant operations procedure documentation and the CPS. Only the first ten (10) procedures of EOP-LOCA are to be carried out because they represent the diverse operation of ESF-CCS during the LOCA scenario.

Scenario 2 would be conducted as detailed in the first scenario but incorporating additional AVR data interface.

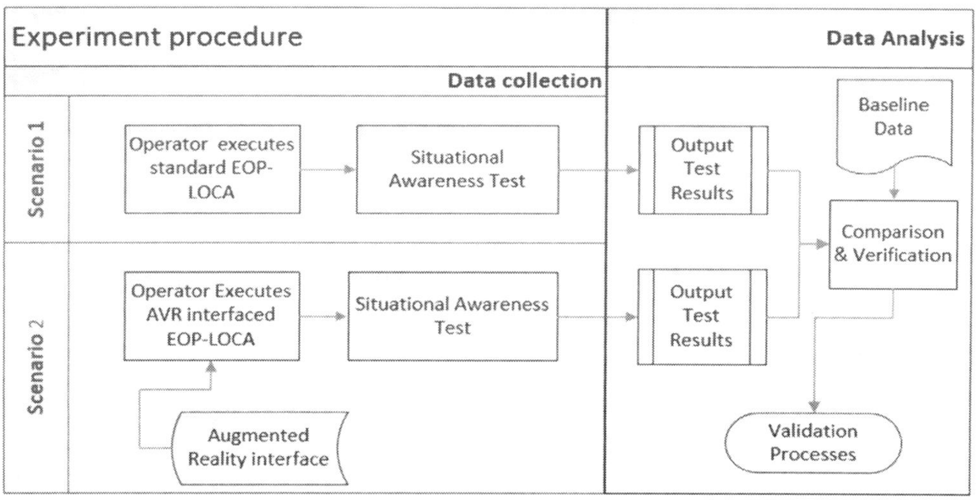


Figure 17-11: Design Modeling

### 17.4.4 Model for SA Data Collection

An interface was designed for the purpose of the Situational Awareness data collection. This system was designed using the free MS Visual Basic windows based graphical development tool. The interface would allow the user to carry out a specific set of instructions closely modelled after the EOP and immediately following it to conduct two situational awareness tests as illustrated in Figure 17-12 and 17-13.

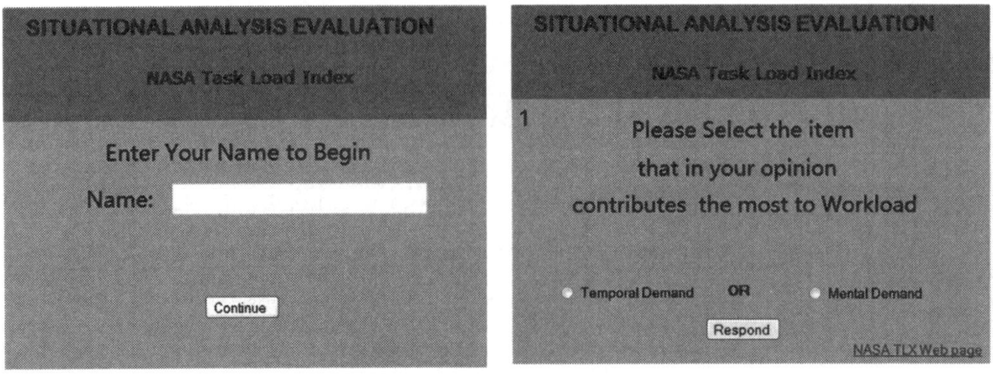


Figure 17-12: NASA TLX for SA

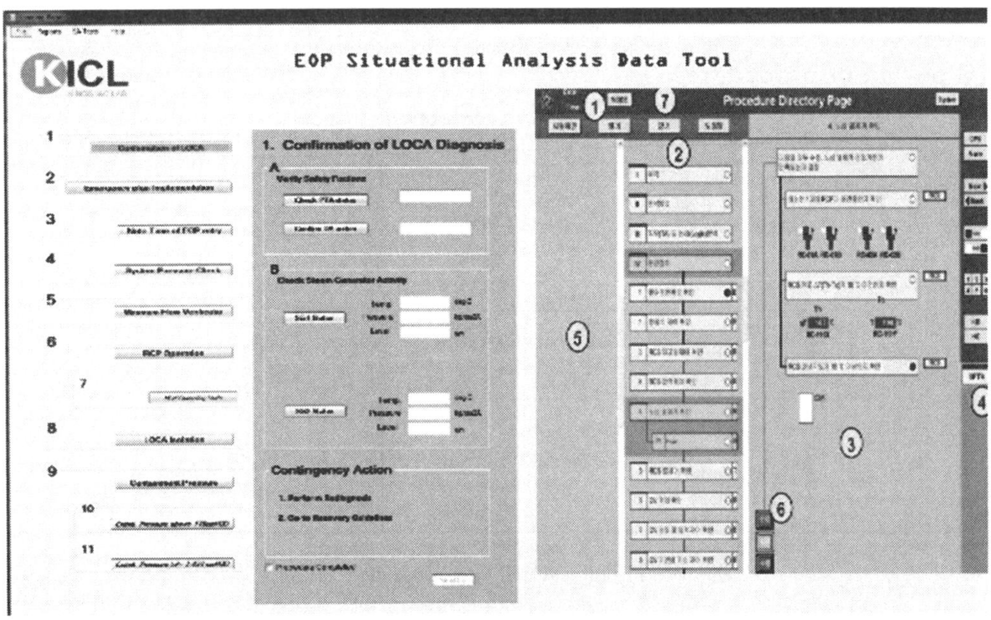


Figure 17-13: EOP SA Data tool developed by KICRI Lab in KINGS

Data from the evaluation result was collected in an Excel sheet and used to calculate both Situational Awareness and Mental Workload calculations. A level of fault tolerance in the design was incorporated in the code so that the data entered was verified to avoid errors during SA data collection.

In the course of conducting this design, the critical information/data needed by a Shift Technical Advisor (STA) during an Emergency Operation Procedure (EOP) for his monitoring/ checking/advisory tasks was identified. After analyzing the current system by examining EOP LOCA, it was discovered that the performance requirements of ESF- CCS could be evaluated from a HFE pint of view. By conducting a SA analysis, it was determined that the work burden could be reduced/ released from the operators during the selected EOP. Figure 17-14 and 17-15 indicate the SA Rating and SART result respectively.

Further work is needed to develop an improved system that will make operations conducted in the MCR by operators more intuitive through the introduction of improved situational awareness using AR-VR design. Other design aims of improved system would be reduced navigation by operators, assurance of validation of entry conditions and correct transition of the plant procedures.

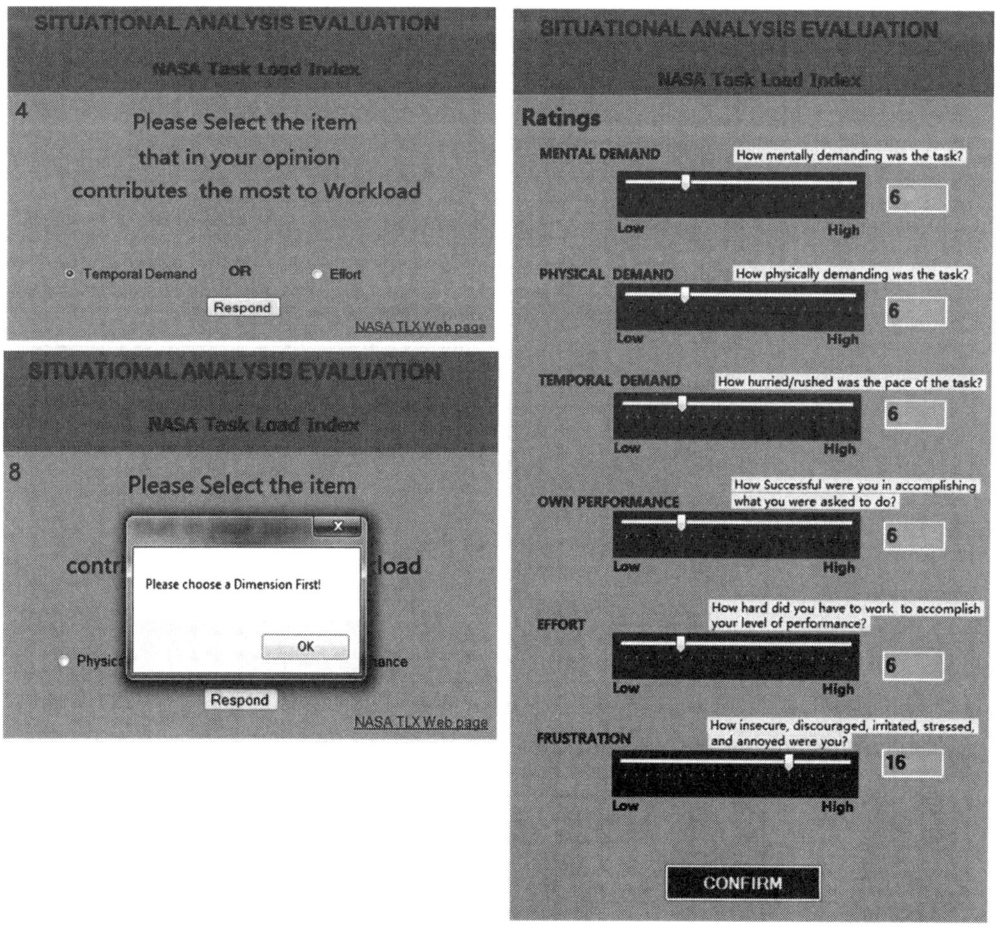


Figure 17-14: SA Rating

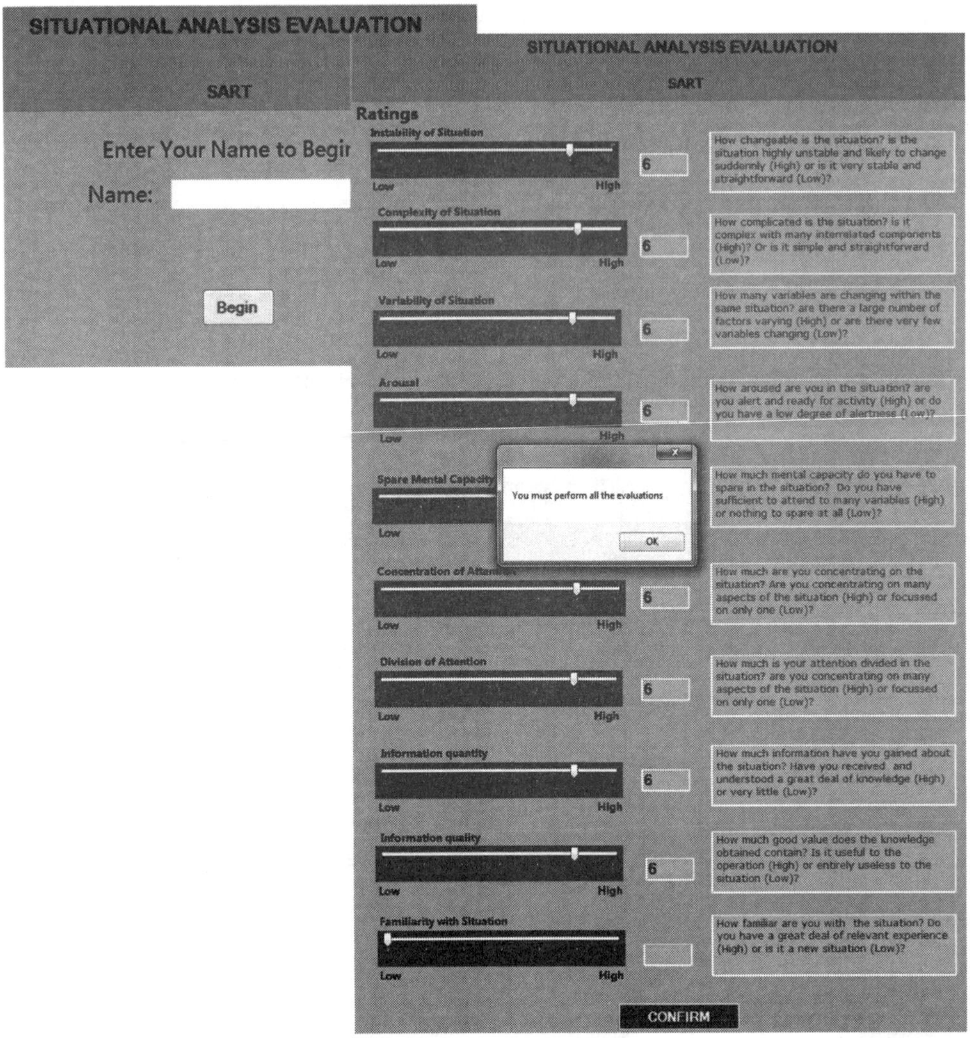


Figure 17-15: Result of SART

It can be concluded that improved SA would not only provide comprehensive plant status information as graphical interface but reduce human factor related operator performance degradation during safety critical operations (EOP) especially those stemming from human errors, and reduced readability

## PROBLEMS

1. What is the importance of HIS in the nuclear power plant?
2. Discuss the main sources of Human Error in the NPP.
3. Explain how multiple barriers (Swiss cheese model) can prevent human error.
4. What is situational awareness?
5. Describe Nasa TLX
6. Describe SART
7. In this chapter, AVR is proposed for use in assessing the EOP. What are the advantages of this technology? What might be its shortcomings?

## FURTHER READING


- Attwood D, et al., "Human Factors Methods for Improving Performance in the Process Industries," Wiley, 2007.
- USDOE, "Human Performance Improvement handbook: Volume 1: Concepts and Principles," 2009.
- IAEA, "Development and Review of Plant Specific Emergency Operating Procedures," 2006.
- Park, J. G. and Wondea Jung, "A Study on the Development of a Task Complexity Measure for Emergency Operating Procedures of NPPs," Reliability Engineering & System Safety, Volume 92, Issue 8, Pages 1102-1116, 2007,
- Mica R. Endsley, "Design and Evaluation for Situation Awareness Enhancement,"
- Proceedings of the Human Factors and Ergonomics Society Annual Meeting, 1988
- Cecília Sik Lányi, "Applications of Virtual Reality," https://www.intechopen.com/books/2277
- S. J. Selcon, R. M. Taylor and Koritsas, "Workload or Situational Awareness?: TLX vs. SART for Aerospace Systems Design Evaluation," *Proceedings of the Human Factors Society Annual Meeting*, vol. 35(2), p. 62–66, 1991.
- Sandra G. Hart, "NASA-TASK LOAD INDEX (NASA-TLX); 20 YEARS LATER". *Proceedings of the Human Factors Society Annual Meeting*, 2006.
- S. G. Hart and L. E. Staveland, "Development of NASA-TLX (Task Load Index): Results of empirical and theoretical research.," in *Human Mental Workload*, North Holland Press, 1988.

# REFERENCES


[1] USNRC, KHNP, "APR1400 Design Control Document ; Tier 2," 2018

[2] S. Baik, "Design Realizes the Nuclear Reactor," KAIST, 2014. 6.10.

[3] K. Ogata, "Modern Control Engineering," Pearson, Fifth Edition, 2009.

[4] "Tutorialspoint," [Online]. Available: https://www.tutorialspoint.com. [Accessed 20 5 2019].

[5] "Electrical 4U," [Online]. Available: www.electrical4u.com. [Accessed 20 5 2019].

[6] KHNP, "APR 1400 Control Systems," Training Textbook.

[7] KEPCO and KHNP, "Control System CCF Analysis," Rev. 0, November, 2014.

[8] KEPCO and KHNP, "CCF Coping Analysis," November, 2014.

[9] M. S. Jung, J. C. Park, J. H. Choi and G. C. Lee, "Evaluation of Control System Software Common Cause Failure in APR1400," in *Transactions of the American Nuclear Society*, Washington, D.C., 2017.

[10] T. C. Tran and J. C. Jung, "Development of Anti-windup PI Control and Bumpless Control Transfer Methodology for Feedwater Control System," *Annals of Nuclear Energy,* vol. 131, pp. 233-241, 2019.

[11] J. G. Ziegler and N. B. Nichols, "Optimum Settings for Automatic Controllers," *trans. ASME,* vol. 64, pp. 759-768, 1942.

[12] T. Hägglund and K. J. Åström, "Revisiting the Ziegler-Nichols Tuning Rules for PI Control," *Asian Journal of Control,* vol. 4, no. 4, pp. 364-380, 2002.

[13] K. J. Åström, H. Panagopoulos and T. Hägglund, "Design of PI Controllers based on Non-Convex Optimization," *Automatica,* vol. 34, no. 5, pp. 585-601, 1998.

[14] I. M. Horowitz, Synthesis of Feedback Systems, Academic Press, 1963.

[15] H. Taguch and M. Araki, "Two-Degree-of-Freedom PID controllers-their functions and optimal tuning," *IFAC Digital Control: Past, Present and Future of PID Control,* pp. 91-96, 2000.

[16] H. Panagopoulos, T. Hägglund and K. J. Åström, "Design of PID controllers based on constrained optimisation," *IEE Proc.-Control Theory Appl,* vol. 149, no. 1, pp. 32-40, 2002.

[17] F. G. Shinskey, Process-Control Systems: Application, Design, and Tuning, fourth edition, New York: McGraw-Hill, 1996.

[18] F. G. Shinskey, "How good are our controllers in absolute performance and robustness?," *Measurement and Control,* vol. 23, pp. 114-121, 1990.

[19] K. J. Åström and T. Hägglund, Advanced PID controller, ISA - Instrumentation, Systems, and Automation Society, 2006.

[20] S. J. Ha, C. E. Park, K. D. Kim and C. H. Ban, "Development of The Space Code for Nuclear Power Plants," *Nuclear Engineering and Technology,* vol. 43, no. 1, pp. 45-62, 2011.

[21] J. J. Sohn and P. H. Seong, "A steam generator model identification and robust H∞ controller design with v-gap metric for a feedwater control system," *Annals of Nuclear Energy,* vol. 37, pp. 180-195, 2010.

[22] L. Ljung, System Identification: Theory for the User, second edition, New Jersey: Prentice Hall, 1999.

[23] L. Ljung, "Identification for Control: Simple Process Models," *Proceedings of the 41st IEEE Conference on Decision and Control,* pp. 4652-4657, 2002.

[24] The Mathworks Inc., 2018a. [Online]. [Accessed 22 10 2018].

[25] Mathworks.com, "Getting Started with the Control System Designer," 2018a. [Online]. [Accessed 5 8 2018].

[26] T. C. Tran and J. C. Jung, "Controller tuning constants for the Advanced Power Reactor 1400 Feedwater Control System based on Ms-Constrained Integral Gain Optimization Method," *Annals of Nuclear Energy,* vol. 127, pp. 39-52, 2019.

[27] H. S. Shim and J. C. Jung, "Improved reactor regulating system logical architecture using genetic algorithm," *Nuclear Engineering and Technology,* vol. 49, no. 8, pp. 1696-1710, 2017.

[28] D. E. Goldberg and K. Krishnakumar, "Control System Optimization using Genetic Algorithms," *Journal of Guidance, Control, and Dynamics,* vol. 15, pp. 735-740, 1992.

[29] "https://en.wikipedia.org/wiki/Requirements_traceability," [Online]. [Accessed 20 5 2019].

[30] H. I. Kim, G. G. Jin, S. H. Jeon, “Genetic algorithm learning on Labview, 1st ed. ISBN 978-89-92649-80-3.

[31] K. Seul, “Control System Analysis and Simulink applications,” 2nd ed. ISBN 978-89-6364-063-1.

[32] "Wikibooks," [Online]. Available: https://en.wikibooks.org . [Accessed 20 5 2019].

[33] C. Yeonuk, “Simulink comprehension and actual controller design,” 1st ed. ISBN 979-11-5600-386-1.

[34] Wikipedia, Available: https://en.wikipedia.org/wiki/Computer_security.

[35] USNRC, NUREG/CR-7141, “Cyber Security Regulatory Framework for Nuclear Power Reactors”

[36] [Online]. Available: https://cve.mitre.org/.

[37] [Online]. Available: https://nvd.nist.gov/.

[38] J. LaChance, G. Baum, F. Durán, K. O. Farnum, S. Jordan, B. Middleton and T. Wheeler, "Evaluation of the Applicability of Existing Nuclear Power Plant Regulatory Requirements in the U.S. to Advanced Small Modular Reactors," Sandia National Laboratories, California, 2013.

[39] USNRC, "Protection of digital computer and communication systems and networks.," U.S NRC, 29 August 2017. [Online]. [Accessed 10 June 2019].

[40] USNRC, 10 CFR 73.54, Protection of digital computer and communication systems and networks.

[41] USNRC, "Regulatory Guide 5.71, Cyber Security Programs for Nuclear Facilities," January, 2010.

[42] NIST, "Cybersecurity Framework," February 2014. [Online].

[43] NIST, "Guide for Applying the Risk Management Framework to Federal Information Systems: A Security Life Cycle Approach," NIST, 2016.

[44] J. T. Michalski and F. J. Wyant, "Secure Network Design," Sandia National Laboratories, 2012.

[45] NIST, "Recommendation for Key Management, NIST Special Publication 800-57 Part 1, Revision 3, July".

[46] KEPCO and KHNP, "Safety I&C System," [Online]. [Accessed 5 June 2019].

[47] IAEA, "Nuclear Power Reactors in the world 2016 Edition Report," 2016.

[48] Attwood D, et al., "Human Factors Methods for Improving Performance in the Process Industries," Wiley, 2007.

[49] DOE, "Human Performance Improvement handbook: Volume 1: Concepts and Principles," 2009.

[50] USNRC, Regulatory Guide 1.33, "Quality Assurance Program Requirements (Operation)," 2013.

[51] IAEA, "Development and Review of Plant Specific Emergency Operating Procedures," 2006.

[52] Park et al, "Study on the development of a task complexity measure for emergency operating procedures of nuclear power plants," *Elsevier,* 2006.

[53] J. Park, The Complexity of Proceduralized Tasks, Seoul: Springer, 2009.

[54] J. Park, The Complexity of Proceduralized Tasks, Springer.

[55] NUREG/CR-6634, "Computer-based Procedure Systems: Technical Basis and Human Factors Review Guidance," U.S NRC, Washington D.C, 2000.

[56] Yang et al., "Preliminary Evaluation of Computerized Procedure from Safety Viewpoints," Korea Institute of Nuclear Safety, Daejeon, 2002.

[57] Endsley, "Design and Evaluation for Situation Awareness Enhancement," in

*Proceedings of the Human Factors and Ergonomics Society Annual Meeting*, 1988.

[58] Mica R. Endsley, Designing for Situation Awareness: An Approach to User-Centered Design, ISBN 9781420063554, CRC Press, 2012.

[59] S. G. Hart, "NASA-TASK LOAD INDEX (NASA-TLX); 20 YEARS LATER".

[60] S. G. Hart and L. E. Staveland, "Development of NASA-TLX (Task Load Index): Results of empirical and theoretical research.," in *Human Mental Workload*, Amsterdam, North Holland Press., 1988.

[61] S. J. Selcon, R. M. Taylor and Koritsas, "Workload or Situational Awareness?: TLX vs. SART for Aerospace Systems Design Evaluation," *Proceedings of the Human Factors Society Annual Meeting,* vol. 35(2), p. 62–66, 1991.

[62] C. S. Lanyi, Applications of Virtual Reality, Croatia, 2012.

[63] M. Murungi and J. C. Jung, "Situational Awareness based Performance Evaluation of APR 1400 Engineered Safety Features – Component Control Actuation Systems using Augmented Reality," *INCOSE International Symposium,* no. 26, 2016.

# Jung Jae-cheon Profile

Professor Jung Jae-Cheon is a Professor at the KEPCO International Nuclear Graduate School (KINGS), where he teaches graduate courses in Nuclear Power Plant (NPP) Instrumentation and Control (I&C) systems. He is the Manager of KINGS Digital I&C Research Institute (KICRI), which operates under the KINGS Research and Development Foundation, where pertinent I&C-related research activities are conducted in the nuclear domain. In addition, he is the current President of the Korea Council on Systems Engineering (KCOSE).

Professor Jung Jae-Cheon received his Ph.D. from the Department of Nuclear and Quantum Engineering at the renowned Korea Advanced Institute of Science and Technology (KAIST). Since then, he has been involved in many national and international projects. During his early career, he worked at the Korea Atomic Energy Research Institute (KAERI) as a Project Development Team Leader. He also worked at the KEPCO Engineering & Construction Company (KEPCO E&C) as the Engineering Group Supervisor in the NSSS Division. Currently, as a Professor at KINGS, he is teaching master's degree students from 28 countries around the world in NPP design based on the systems engineering management technical processes.

As for his research and publications, Professor Jung Jae-Cheon has published several SCI-level papers in famous journals and has authored quite a few books related to nuclear technology. The research works at KINGS I&C laboratory and KICRI have contributed enormously to the body of knowledge in the fields of FPGA programming, artificial intelligence, reliability engineering, and NPP human-machine interface systems. The research works support capabilities in design analysis, software & system design, system testing, verification & validation, and commissioning of both domestic and international projects. In addition, several registered patents delineate his successful academic and research career.

# Handbook for I&C System Development Process using Systems Engineering Approach


2022년 6월 20일 초판 발행

저 자 정재천

발 행 인 송기수

발 행 처 도서출판 GS인터비전

편 집 처 도서출판 GS인터비전

편 집 인 김명신, 공예서

표지디자인 디자인붐

인 쇄 처 대명프린팅

등록번호 제 25100-2016-000050호

I S B N 979-11-5576-323-0 (93550)



주 소 서울 은평구 증산로 15길 69 2층

전 화 02-976-7898, 02-3272-7898.

팩 스 02-6468-7898

홈페이지 gsintervision.co.kr

E-Mail gsinter7@gmail.com



정 가 40,000원